D0211028

The Facts On File

DICTIONARY of EARTH SCIENCE

The Facts on File

DICTIONARY of EARTH SCIENCE

Edited by
John O. E. Clark
Stella Stiegeler

Facts On File, Inc.

The Facts On File Dictionary of Earth Science

Facts On File, Inc.
132 West 31st Street
New York NY 10001

Library of Congress Cataloging-in-Publication Data
The Facts on File dictionary of earth science. / [edited by]
John O. E. Clark and Stella Stiegeler.
p. cm.
ISBN 0-8160-4287-X (acid-free paper)
1. Earth sciences—Dictionaries. I. Clark, John Owen Edward.
II. Stiegeler, Stella. III. Facts on File, Inc. IV. Title: Dictionary of earth science.
OH302.5.F38 2000
570'.3—dc21 99-17788

Facts On File books are available at special discounts when purchased in bulk quantities for businesses, associations, institutions, or sales promotions. Please call our Special Sales Department in New York at (212) 967-8800 or (800) 322-8755.

You can find Facts On File on the World Wide Web at
http://www.factsonfile.com

Compiled and typeset by Market House Books Ltd, Aylesbury, UK

Printed in the United States of America

MP 10 9 8 7 6 5 4 3 2

This book is printed on acid-free paper

FOREWORD

This dictionary is one of a series designed for use in schools. It is intended for students of earth sciences, but we hope that it will also be helpful to other science students and to anyone interested in science. The other books in the series are *The Facts On File Dictionary of Biology*, *The Facts On File Dictionary of Chemistry*, *The Facts On File Dictionary of Physics*, *The Facts On File Dictionary of Mathematics*, *The Facts On File Dictionary of Computer Science*, and *The Facts On File Dictionary of Astronomy*.

This book is based on an edition first published by the Macmillan Press in 1976. This edition has been extensively revised and extended. The dictionary now contains over 3400 headwords covering the terminology of modern earth science. A short Appendix of chemical elements is included.

We would like to thank all the people who have cooperated in producing this book. A list of contributors is given on the acknowledgments page. We are also grateful to the many people who have given additional help and advice.

ACKNOWLEDGMENTS

Editor (First Edition)

Stella Stiegeler B.Sc.

Contributors

B. M. Abbott B.Sc., Ph.D., F.G.S.
Anna Clyde B.Sc.
Andrew Hill B.Sc., Ph.D., F.G.S.
I. P. Joliffe B.Sc., M.Sc, Ph.D., C.I.C.E., F.C.S., F.R.G.S.
R. B. Lanwarne B.Sc.
P. A. Smithson B.Sc., Ph.D., F.R.Met.S.
T. J. Speechley B.Sc.
F. A. Sultan B.Sc., M.Sc., D.I.C.
S. D. Weaver B.Sc., Ph.D., F.G.S.
A. C. Wornell B.A.

A

aa A LAVA with an extremely rough spinose surface. *Compare* block lava; pahoehoe.

Aalenian The earliest part of the Middle JURASSIC period.

abandoned cliff *See* cliff.

abiotic Describing a nonliving factor in an ECOSYSTEM. Light, rainfall, soil type, and temperature are all abiotic factors. Chemical residues from artificial fertilizers and pollution constitute harmful abiotic factors. *See also* biotic.

ablation The disappearance of snow and ice by melting and evaporation. This can refer to ice crystals or snow flakes in the atmosphere but is most commonly used for glacier ice and surface snow cover, when it can also include avalanching. The rate of ablation is primarily controlled by air temperature but sunshine, rainfall, humidity, and wind speed also affect the process.

ablation cone A cone of FIRN, ice, or snow covered in rock debris caused by differential ABLATION.

ablation moraine Debris of rock fragments left at the side of a glacier following ABLATION.

ablation till TILL material formerly present on the surface of glaciers and subsequently deposited as a result of the melting of the ice beneath. These deposits tend to contain relatively little fine material because this may be removed by meltwater prior to deposition. *Compare* lodgment till.

Abney level A simple surveying instrument for measuring slope angles. The level is held by hand, objects being viewed through a sighting tube. The observer may stand at the top or bottom of a slope, but must sight onto a point at a height equivalent to his or her eye-level. Once an object is sighted, adjustment is made to a tilting spirit level, an image of the bubble being visible in the sighting tube. When the bubble image coincides with the object sight the angle of slope can be read off an attached scale, which is provided with a vernier to permit reading down to ten minutes of arc. The instrument is not very accurate, but is small and quick to use.

abnormal erosion (accelerated erosion) Erosion acting faster than normal as a result of the removal of the vegetation cover by human agencies. It is abnormal in the sense that it is superimposed upon natural processes. *See also* blowout; human influence on geomorphology.

abrasion The wearing away of rocks by an agent of transportation charged with a load of already eroded material, which acts as a tool for cutting, grinding, scratching, and polishing. All the major transportation agents (running water, wind, moving ice, and sea waves and currents) can abrade so long as they carry debris. Abrasion by water and ice characteristically produces rounded forms, and abrasion by ice also produces striations, while the sand-blasting effect of wind abrasion is most effective in a narrow zone just above ground level, resulting in undercut features. As abrasion takes place, the corresponding reduction in size of the initial load debris is known as ATTRITION. A distinction is sometimes made between abrasion and CORRASION, which refers to the erosional processes that result in abraded rock surfaces.

abrasion platform A gently sloping ledge of rock at the base of a steep cliff extending from the high-tide level to the low-tide level. It is caused by ABRASION by waves at the base of the cliff, where undercutting results in rock falls from above. It is also called a wave-cut platform.

absolute age The age of a fossil, rock formation, or individual rock, usually stated in years. Such ages are usually determined by DENDROCHRONOLOGY or RADIOMETRIC DATING. *See also* relative dating.

absolute drought A prolonged period (usually at least 15 days) during which no more than 0.25 mm of rain falls on any day.

absolute humidity The mass of water vapor in a unit volume of air (usually stated in units of $g\ m^{-3}$ or $kg\ m^{-3}$). It depends on the temperature and pressure of the air. *See* humidity.

absolute temperature A temperature scale based on the coldest temperature that is physically possible. This absolute zero of temperature is –273.15°C, but for meteorological purposes the absolute temperature is taken as the Celsius temperature plus 273°, so that the freezing point of water is 273° and boiling point 373°. Formerly measured in degrees absolute (°A), it is now measured in KELVINS.

absorption (in meterology) The conversion of short- or long-wave radiation to a different form of energy by gases in the atmosphere. Absorption is highly selective in terms of wavelength; some wavelengths are entirely absorbed and others are totally unaffected. The main atmospheric gases, oxygen and nitrogen, are not very important as absorbers of radiation, but minor gases such as carbon dioxide, water vapor, ozone, and nitrous oxide have a significant effect. *See also* atmospheric window.

abstraction 1. The fraction of PRECIPITATION that does not immediately run off. It is absorbed, evaporated, stored, or transpired.
2. The union of two streams resulting from erosion of the land (DIVIDE) between them. It generally occurs in gullies and ravines.

abyss *See* deep.

abyssal hill A large dome-shaped submarine hill on the ABYSSAL PLAIN. Such hills rise to heights of 1000 m in up to 6000 m of water and may be several kilometers wide. There are many along the MID-ATLANTIC RIDGE and in the Pacific Ocean.

abyssal plain An extremely smooth portion of the deep-sea floor. The gradients across abyssal plains fall within the range of 1:1000 to 1:10 000, which means that variations in depth amount to only a few meters. This remarkable degree of flatness has come to light because of deep-sea photographs and high-precision sounding techniques. In the Atlantic Ocean, abyssal plains range between 200 and 400 kilometers in width, but they can be several hundred kilometers wide. They tail upwards into the continental rise and frequently, in a seaward direction, merge into abyssal hills. Their sediments are varied; while most are thinly veneered with pelagic sediment, perhaps globigerina ooze and red clay, they also display sediments and plant and animal remains that normally characterize shallow-water environments. The reason for this may be the operation of TURBIDITY CURRENTS.

abyssal rock A type of intrusive igneous rock formed deep within the Earth's crust.

abyssal zone A zone of greatest ocean depth, lying seaward of and deeper than the BATHYAL ZONE of the continental slope, that is below a depth of some 1000 m, and including the deeper parts of the oceans and the deep-sea trenches. Lying between the abyssal and bathyal zones is the CONTINENTAL RISE, which is often bordered on its seaward flank by abyssal plains and abyssal hills. The pelagic-abyssal environment (*see* pelagic (def. 1)), which is not reached for at least several hundred kilometers from the coast, globally represents an area of 250 × 10^6 sq km, i.e. roughly half the area of the Earth. Few organisms live in these depths, where pressure is high and light does not reach, and deposition of sediment is very slow.

accelerated erosion *See* abnormal erosion.

acceleration of free fall (acceleration due to gravity) The acceleration (g) of a body falling freely in a vacuum in the Earth's gravitational field. The standard value is 9.806 65 m s^{-2}, although actual values depend on the distance from the Earth's center of mass.

accessory mineral A mineral that is present in small quantities in a rock and does not affect the overall character of that rock for classification purposes.

acclimatization The process by which humans become adapted to living in a markedly unfamiliar climatic regime. This normally refers to a change to hot and humid or especially cold climatic conditions. If acclimatization has not taken place, severe physiological stress may result. The full process may take up to ten years.

accordant (conformable) Describing a drainage pattern that is controlled by the structures over which it flows. Actual patterns vary greatly, depending on the nature of the structures or lithologies. *Compare* discordant (def. 2).

accordian folding A type of folding in rocks, in which the beds of the hinge area are markedly thickened and sharply folded, while on the limbs the beds are straight and of uniform thickness.

accretion **1.** The process of ice crystal growth by collision with water droplets in clouds. This is one of the mechanisms by which minute cloud droplets achieve sufficient size to give rainfall.
2. *See* nucleation.

accumulated temperature A method of indicating the excess or deficit of temperature with reference to a specified value for a specified period. Two temperature bases that have been frequently used are 6°C for the commencement of plant growth and 16°C for heating requirements. The accumulated temperature is calculated by taking the number of hours in a specific period the temperature was above or below the set value and multiplying by the mean temperature during this time to give the number of degree hours. More commonly, daily or monthly values of accumulated temperature are obtained from daily or mean values rather than hourly ones.

accumulation **1.** The total amount of PRECIPITATION that gathers on a snowfield or glacier.
2. The overall result of all processes that add mass to a snowfield, glacier, or ice floe, including snow from avalanches, snowfall, and windblown snow.

achnelichs Fragments of glassy smooth PYROCLASTIC ROCK derived from basaltic LAVA that has sprayed in the air while molten and solidified. Pele's tears are achnelichs.

achondrite A stony meteorite that does not contain CHONDRULES. *Compare* chondrite.

acicular Describing a crystal having a needle-like habit. *See* crystal habit.

acid brown soil A type of soil found in the BROWN EARTH zone on base-deficient parent materials. Such soils are strongly acid, with a moder humus, and although the B horizon is more clearly differentiated by color than in the true brown earth there is no appreciable eluviation of clay or iron oxides. With time it seems likely that these soils would become podzolic brown earths.

acid lava Slow-moving viscous LAVA containing a high proportion of silica. Produced by so-called acidic volcanoes, it generally solidifies very quickly. *See also* basic lava.

acid rain Any PRECIPITATION (including fog, rain, sleet, and snow) that is acidic due to the presence of sulfur dioxide and nitrogen oxides in the atmosphere. Most of these pollutants derive from the burning of FOSSIL FUELS. Acid rain kills trees, poisons rivers and lakes, and corrodes buildings. Runoff of acid rain affects the mineral composition of the soil.

acid rock The dominant chemical constituent of igneous rocks is silica, SiO_2, which typically ranges from 35–75% (by weight). Arbitrary divisions are made as follows: acid – $SiO_2 > 66\%$, intermediate – SiO_2 55–66%, basic – SiO_2 45–55%, and ultrabasic – $SiO_2 < 45\%$. These strict divisions have been largely abandoned but the general descriptive terms acid, basic, etc., remain. In current usage, an acid rock contains in excess of about 10% free QUARTZ, e.g.: granite, granodiorite, rhyolite.

acid soil Soil acidity is measured on the pH scale, which is related to the concentration of hydrogen ions in the soil. A neutral soil is given a pH value of 7.0; if the value is less than this it is termed an acid soil and if greater an alkaline soil. Thus there is an inverse relationship between pH value and the concentration of hydrogen ions. Acid soils develop where, for some reason, there is a lack of the exchangeable bases in the soil, such as calcium and sodium. The bases have been largely replaced by the two cations hydrogen and aluminum, which control soil acidity. Possible acid-forming factors are leaching, organic matter containing few bases, and an acid parent material. Acid soils are therefore particularly common in the humid tropics and the humid temperate lands. Examples of typical acid soils are podzols, brown earths, and latosols. From the agricultural viewpoint most cultivated crops will thrive on mildly acid soils.

acmite (aegirine) *See* pyroxene.

acre A unit of area equal to 4840 yd^2 (equivalent to 4046.86 m^2 or 0.4047 hectares).

acre foot A unit of volume equal to an area of 1 acre to a depth of 1 foot (equivalent to 43 560 ft^3 or 1233.5 m^3). It is used for expressing the volumes of lakes and reservoirs.

actinolite A monoclinic AMPHIBOLE.

actinomycetes Aerobic bacteria that have a filamentous habit as do fungi. Optimum conditions are a moist well-aerated soil with a pH between 6.0 and 7.5. They are more prolific than other bacteria and fungi in drier areas and are important also in that they can decompose the more resistant soil organic matter.

Actinopterygii Ray-finned bony fish in which the paired fins are not fleshy (*compare* Crossopterygii) but have narrow bases and are supported by thin fin rays. The Actinopterygii also have a single dorsal fin. The group includes the modern TELEOSTEI. *See also* Osteichthyes.

active layer *See* permafrost.

actualism *See* uniformitarianism.

actuopaleontology The branch of paleontology in which investigations into modern organisms, including their effects and remains in modern environments, are directed toward the understanding of fossil analogs. Based upon uniformitarian principles (*see* uniformitarianism), actuopaleontology provides a means of relating various TRACE FOSSILS. to the particular animals from which they are derived and of understanding the changes that an organism undergoes between death and fossilization (*see* taphonomy; thanatocoenosis). BIOSTRATONOMY is one aspect of actuopaleontology.

adamantine Describing a mineral that has a brilliant diamondlike luster.

adamellite A variety of granite consisting of about equal proportions of potassium feldspar and sodic plagioclase feldspar together with one or more ferromagnesian minerals.

adiabatic Denoting an atmospheric process in which there is no exchange of heat between the system and its environment. In the more rapid exchanges, such as thermals rising from the ground surface, this is a reasonable approximation. In these circumstances, the change in temperature of rising air is determined by the physical properties of the air and the external pressure. As air pressure decreases with height above the ground surface, rising air expands, and exerts mechanical work on its environment. This necessitates a loss of heat energy from the rising air and its temperature falls. The rate of fall of temperature in an adiabatic process is constant for our atmospheric composition, being 0.98°C/100 meters, and is known as the DRY ADIABATIC LAPSE RATE.

adit A horizontal or slightly upward-sloping closed-ended passage into a mine, usually constructed to intersect a seam of coal or vein of mineral. The slope is to ensure that any water drains out of the adit.

adjacent seas Marginal seas and inland seas. They lie adjacent to the oceans but generally they are much smaller. Because many adjacent seas are largely encircled by land, the characteristics of the water and sediments they contain, and to a certain extent their topography, are strongly influenced by the landmass, for example in the amounts and nature of terrigenous sediments transported down rivers to the coast and the climatic effects associated with the land-sea margin.

adobe A mixture of silt and clay, common in Mexico and the southwestern USA. It has long been used for making bricks because it dries to a hard weatherproof mass. *See also* loess.

adret The sunny or, in the N hemisphere, south-facing slope of a valley. It is the side favored for farming, as in the European Alps.

adularia A variety of alkali FELDSPAR.

advection The horizontal component in the transfer of air properties. For example, the heat and water vapor content of the air at the Earth's surface varies appreciably and by the wind systems these properties are transferred to other areas. With winds from tropical latitudes there is advection of warm air, and from polar latitudes, advection of cold air.

advection fog Fog formed by the horizontal transfer of moist air over a cold surface, which sufficiently cools the lower layers of the atmosphere to give saturation and condensation. In summer it occurs over cool seas, such as the North Sea, the Labrador area, and off the coast of California, frequently affecting the adjacent coasts. In winter the advection of warm moist air over a cold snow-covered ground can also produce this type of fog.

aegirine (acmite) A sodic clinopyroxene. *See* pyroxene.

aegirine-augite *See* pyroxene.

aeolian erosion The direct erosive action of wind. This is the least important aspect of wind action, generally of little consequence in landscape formation when compared with the role of wind transport and deposition. Sand blast, from the impact of saltating grains, is limited to below one meter or so from the ground surface; it can undercut rocks leaving pedestals, although in some instances increased weathering at the foot of the rock weakens it beforehand. More significant than aeolian erosion by abrasion is the production of DEFLATION hollows. These can reach 100 km across, and although partly due in some cases to faulting or rock solution, they are mainly due to wind removal of preweathered material down to the water table, which halts further removal and produces OASES.

aeolian form A landform produced by material transported by wind. Large-scale features include dunes, sand shadows, and sand sheets, while small-scale features include sand ripples and ridges. Sand shadows, unlike true dunes, are deposited in the shelter of an obstacle, while sand sheets or seas are amorphous sheets with gentle swellings reaching 3–6 meters. Ripples are the products of irregularities in the surface over which the sand is passing: these initiate local concentrations of sand grain impacts on the slopes facing the wind, which become built up as ripples. Further, each ripple acts as a take-off point for grains in SALTATION, and since the average length of travel per jump is related to wind speed, an even repetition of areas of concentrated grain impacts occurs downwind, leading to regular ripple patterns.

aeolian transport The movement of sediment by wind. Below a threshold of 16 km per hour wind is incapable of moving sand, but thereafter sand movement rises as a cubic function of wind speed. Short periods of high-velocity winds are therefore very much more effective than longer periods of gentle winds.

Aeolian transport takes place concurrently in three forms: suspension, SALTATION, and surface creep. A small proportion of grains of less than 0.2 mm in diameter can be carried wholly in suspension; the particle is totally buoyed up by the rising eddies in the air and carried along parallel to the air stream. Surface creep accounts for about 25% of actual sand movement, and involves the movement of grains of coarse sand along the surface by the impact of the saltating grains. Saltation accounts for about 75% of sand flow and involves the bouncing of grains along the surface at heights of less than 1 meter, and mostly within 200 mm of the ground surface.

aerodynamic roughness An index of the nature of airflow near the ground surface. A surface is aerodynamically smooth if there is a layer of air immediately above it that has laminar flow. However, in meteorological terms, nearly all surfaces are aerodynamically rough, producing turbulent flow down to the ground surface, even for the lightest winds.

aerolite A stony METEORITE made up of silicate minerals.

aeronautical chart A form of map produced essentially for air navigation or pilotage. In addition to showing the relevant topographical features, such as contours, vegetation, roads, and cultural detail, the map shows supplementary information for specialized use, such as detailed vertical obstruction information, flight areas, air corridors, etc.

aeronomy The science of the upper atmosphere, where dissociation and ionization of gas molecules takes place. The lower limit for these processes to occur is about 30 km. As the relationships between the upper and lower atmosphere are as yet little understood, there has been considerable research recently into the possible effects of these upper levels in tropospheric weather.

aerosol A particle of matter that is larger than a molecule but small enough to remain suspended in the atmosphere. Aerosols may be solid or liquid and play an important part in many atmospheric processes, such as precipitation formation, atmospheric electrification, radiation balances, and visibility. The origins of aerosols are diverse. Over sea areas, sea spray provides large salt nuclei and over land, weathering dusts of clay particles are probably the major source.

afforestation The large-scale planting or replanting of trees. They may be planted to prevent soil EROSION, to act as WINDBREAKS, or as a source of timber (usually for wood pulp).

aftershocks A series of minor shocks or vibrations that follow the main series of earthquake shocks. In general they originate at or near the focus of the main earthquake as a result of the readjustment of rocks that have over-reacted during the main period of movement. Depending upon the size of the earthquake they can continue for a few days or months.

agate An extremely fine-grained type of QUARTZ consisting of alternate bands of colored CHALCEDONY. It may be white, gray, red, brown, or black, and can be dyed other colors. It generally occurs in rock cavities. It is used for making ornaments and as a semiprecious gemstone. *See also* moss agate; silica minerals.

age An interval of geologic time in the Chronomeric Standard scale of chronostratigraphic classification (*see* chronostratigraphy). The equivalent Stratomeric Standard term, indicating the body of rock formed during this time, is the STAGE. Ages may be grouped together to form EPOCHS and may themselves be comprised of several CHRONS.

ageostrophic wind The vector difference between the actual wind and the geostrophic wind. Without an ageostrophic component there would be no changes in pressure systems and atmospheric circulation would be a perfect balance between the pressure gradient and the CORIOLIS EFFECT.

agglomerate A mixture of coarse angular fragments of rock and finer-grained material formed during a volcanic explosion. *See* pyroclastic rock.

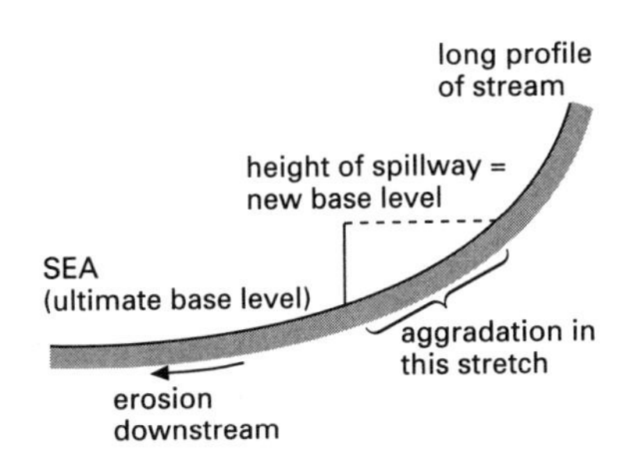

Aggradation in a dammed stream

aggradation The raising of the level of the land surface, especially by rivers. The cause of aggradation is incompetence (*see* competence) or incapacity (*see* capacity) of the river to transport its load, leading to deposition. This may arise through an increase in the volume of load supplied to the river, a loss of speed or volume of flow, or most commonly a rise in base level. Another common cause of aggradation is the damming of a stream, artificially creating a new base level for the upper reaches of a river. Originally graded to sea level, the upper reaches will aggrade to the new base level, the height of the dam's spillway. *See also* degradation.

aggregate Mineral fragments and/or rock particles that combine to form a hard mass.

agmatite A MIGMATITE in which melanocratic material occurs as angular inclusions in a leucocratic granitic host, giving an overall appearance resembling a breccia.

Agnatha Primitive jawless fish, from which all more advanced vertebrate types have presumably evolved. Modern species are few and represent groups unknown as fossils. The diverse extinct forms can be grouped together as the OSTRACODERMI, which were abundant in the Silurian and Devonian and are used in stratigraphic CORRELATION. *Compare* Gnathostomata.

agonic line A line joining all points on the surface of the Earth where the ANGLE OF DECLINATION is zero. Along such a line, magnetic north and true north coincide. *See also* isogon.

A horizon The uppermost layer of soil, or topsoil. It consists of fine soil particles and HUMUS. Some soluble material will have been dissolved out and passed downward to the B HORIZON beneath. *See also* horizon.

air The mixture of gases in the atmosphere. Its composition is almost uniform throughout the troposphere and is shown in the table overleaf. Only water vapor and carbon dioxide vary appreciably, the former in relation to evaporation and precipitation and the latter through plant photosynthesis.

COMPOSITION OF AIR
(% by volume)

nitrogen	(N_2)	78.08
oxygen	(O_2)	20.95
carbon dioxide	(CO_2)	0.03
argon	(Ar)	0.93
neon	(Ne)	1.82×10^{-3}
helium	(He)	5.24×10^{-4}
methane	(CH_4)	1.5×10^{-4}
krypton	(Kr)	1.14×10^{-4}
xenon	(Xe)	8.7×10^{-5}
ozone	(O_3)	1×10^{-5}
nitrous oxide	(N_2O)	3×10^{-5}
water	(H_2O)	variable, up to 1.00
hydrogen	(H_2)	5×10^{-5}

air-fall deposit A sediment composed of fallout from a cloud of airborne material from an erupting volcano, as opposed to LAVA (which generally flows from a volcano).

air mass An area of the lower atmosphere with similar properties of temperature and moisture in the horizontal field. At the margins of the air mass, temperature gradients become steep at a transition zone known as a FRONT. The uniformity of properties is achieved by prolonged contact with the underlying surface and little horizontal or vertical mixing. These requirements are experienced in areas of high pressure or anticyclones, which are the main source areas for air masses. Away from their source areas, air masses undergo modification by coming in contact with different surfaces with the result that in a short period of time they can become indistinguishable. As most parts of the world represent modification zones rather than source areas, air mass terminology is less frequently used than formerly. There have been many attempts to classify air masses, but the most frequently quoted is that by Bergeron. Two basic air masses are identified on thermal properties – polar (P) and tropical (T), and two by moisture categories – maritime (m) and continental (c). The temperature of the air mass relative to the surface over which it is passing is included as warm (w) or cold (k) to give a wide range of combinations – mPw, cTk, etc. Willett modified this system slightly to add stable (s) or unstable (u) to indicate the likelihood of precipitation in the system. Other classifications distinguish equatorial, monsoon, and Arctic (or Antarctic) but difficulties can arise in identification.

air stream A flow of air that is not necessarily homogeneous but has a distinctive origin. Air streams are therefore distinguished by their direction of approach rather than assuming any specific thermal or stability properties. The mid-latitude westerlies can be regarded as being a mixture of slightly baroclinic air streams bounded by sharp frontal zones.

Airy's hypothesis of isostasy George Bedell Airy proposed that in order for isostatic equilibrium to exist, mountain ranges must have roots proportional to their height, i.e. the highest mountains have the deepest roots. These roots are composed of sialic material and displace an equivalent volume of sima, thereby causing the gravity anomalies present near mountain chains.

Aitken nucleus *See* nucleus.

Aitoff's equal-area projection A cylindrical map projection of a hemisphere. The major axis, the Equator, is twice the length of the minor, central meridian, axis. The projection is bounded by an ellipse. The main characteristics of this projection are that it is an equal-area projection and landmasses near the center of the area covered are of quite good shape, although the distortion increases towards the east and west limits of the projection.

alabaster A fine-grained white, sometimes translucent, variety of GYPSUM, used for making ornaments.

albedo An index of reflection comprising the ratio of reflected radiation to the total incident radiation. Usually this value is expressed in a percentage form for visible wavelengths. Typical values for surface albedo are: forest 5–10%, wet soil 10%, sand 20–30%, grass 25%, old snow 55%, concrete 17–27%, fresh snow 80%. Water surfaces vary from about 5% with high sun and calm seas to 70% at low elevation and rough seas. The planetary albedo of the Earth measured from artificial satellites is approximately 34%, which means that over one third of the Sun's radiation is returned to space without a change of wavelength.

albite A sodic plagioclase FELDSPAR.

Aleutian Low The mean low pressure center in the N Pacific Ocean. It represents a statistical average of pressure value and location, which in turn are determined by the tracks of the depressions and the point at which they reach their lowest pressure. It is most marked in the winter.

alexandrite A transparent emerald-green type of CHRYSOBERYL, used as a semiprecious gemstone. It has the unusual property of turning red in artificial light.

alfisol One of the ten soil orders of the SEVENTH APPROXIMATION, covering pedalfers that are equivalent to the gray-brown podzolic, gray-wooded, gray-forest, sol lessivé, degraded chernozem, and planosol soils of the old American classification. They are found in the humid regions of the world under deciduous woodland or grassland vegetation. The dominant soil-forming process is leaching, which is more intense in these soils than in the inceptisols but less than in the spodosols. They are productive soils and favor the more common agricultural crops.

algae A group of largely aquatic organisms formerly classified as plants. It includes both microscopic forms, such as the DIATOMS, as well as the multicellular seaweeds, which may grow to a large size. Algae are subdivided into different phyla that are now usually placed in the kingdom Protista (or Protoctista). The only geologically significant algae are those having hard parts, which may form bioherms, either by trapping sediment or by secreting massive laminated structures of calcium carbonate. Such structures commonly constitute lower Paleozoic REEFS. Compared with bacteria and fungi, algae are relatively unimportant in soils, but they are often pioneers in colonizing new ground and may number as many as 100 000 per gram of dry surface soil.

algal bloom A sudden increase in the numbers of ALGAE in a lake or river, caused by an increase in the amount of nitrates, phosphates, and other nutrients. *See* eutrophication.

alidade A surveying instrument used for sighting onto objects of detail and for defining the rays to be drawn to them in PLANE TABLING. The alidade is basically a ruler of metal or wood with a vertical slit sight at the observer's end and a vertical stretched wire sight at the other. (A telescope is fixed parallel to the ruler in more sophisticated types of alidade.) The ruler edge is placed against the point marked on the table over which the apparatus is standing, and detail to be fixed is sighted onto. A ray is then drawn on the plane table sheet along the ruler edge. Although either side of the ruler may be used initially, once one ray has been drawn, only that side may be used until that sheet is completed.

alkali (alkaline) (in petrology and mineralogy) Denoting igneous rocks and minerals that have high contents of the alkali metal oxides, Na_2O and K_2O. For a given silica content such rocks are relatively richer in sodium and potassium and poorer in calcium compared with calc-alkaline rocks. *See* alkali basalt; alkali gabbro; granite; ijolite; nephelinite.

alkali basalt A basic undersaturated volcanic rock that is the fine-grained equivalent of ALKALI GABBRO. The essential minerals of all basalts are plagioclase feldspar of labradorite-bytownite composition and pyroxene. In alkali basalts the pyroxene is augite or titanaugite and olivine is present in abundance. Olivine is frequently rimmed or pseudomorphed by the alteration products iddingsite and/or serpentine. Small amounts of alkali feldspar and/or feldspathoid (nepheline or analcite) may be present. Alkali basalts are typically porphyritic. Basalts containing large plagioclase phenocrysts are referred to as *feldsparphyric* or *big feldspar* basalts. Those rich in olivine and augite are termed picrite basalts. Oceanite and ankaramite are varieties in which olivine and augite respectively have become concentrated.

Alkali basalts are usually holocrystalline and have ophitic or intergranular textures. Nodules of gabbro and peridotite are often found in alkali basalt lavas. With an increase in the amount of nepheline to greater than 10%, alkali basalts pass into basanites (olivine-bearing) and tephrites (olivine-free). In some rocks, the place of nepheline is taken by analcite or leucite and such terms as analcite-basanite and leucite-tephrite are appropriate. *See also* basalt.

alkali feldspar A member of a series of minerals with composition varying between the two end-members albite ($NaAlSi_3O_8$) and orthoclase ($KAlSi_3O_8$). *See* feldspar.

alkali gabbro Alkali gabbros and syenogabbros are basic plutonic rocks containing, in addition to the normal gabbro mineralogy, alkali feldspar and/or feldspathoids. Syenogabbros contain approximately equal amounts of alkali and plagioclase feldspar with TITANAUGITE, ANALCIME, and/or NEPHELINE plus or minus olivine. Having more sodium and potassium than gabbros, syenogabbros are, as the name implies, related to SYENITES.

Essexite consists of labradorite, titanaugite, and olivine together with small amounts of nepheline and/or analcime. Alkali feldspar, apatite, and ilmenite may also be present. *Teschenite* and *crinanite* are analcime-bearing varieties from which nepheline is excluded. *Theralite* is a nepheline-bearing gabbro containing no analcime. *Kentallenite* is a saturated rock containing augite, olivine, biotite, labradorite, and orthoclase and is equivalent to olivine-monzonite.

The names of the plutonic rocks are applied to the medium-grained varieties of doleritic aspect but with the prefix 'micro-'. The volcanic equivalents include basanites, tephrites, and trachybasalts. Teschenite and theralite are found in differentiated sills and dikes.

alkaline soil *See* acid soil.

allanite (orthite) One of the EPIDOTE group of minerals.

Alleröd A phase of warming of about 1000 years during the period of deglaciation after the Würm/Weichsel/Wisconsin ice age. In many parts of NW Europe this was followed by a sudden cooling from 8800 to 8300 BC. The type-site from which the period takes its name is in Denmark.

allivalite A gabbro consisting of olivine and plagioclase feldspar of bytownite-anorthite composition.

allochem A discrete calcareous particle that has usually been transported at some stage, including FOSSILS, ooliths (*see* oolite), INTRACLASTS, and PELLETS, found in LIMESTONES; the terminology used in the petrographic description and classification of limestones is based upon these constituents and the matrix in which they are set (*see* micrite; sparite). Thus a limestone composed of fossil fragments set in a micrite matrix is a biomicrite; a pellet limestone with a sparite matrix is a pelsparite.

allochthonous Denoting an isolated mass of rock displaced over a considerable distance from its original source by low-angle thrusting. *Compare* autochthonous.

allogenic (allothigenous) Describing rock constituents that were formed at some distance from their present location, particularly minerals or rock fragments derived from existing rocks. *Compare* authigenic.

allotriomorphic Describing a rock in which the majority of crystals are euhedral. *Compare* hypidiomorphic; idiomorphic.

allotrope A form of a chemical element that differs (usually in its crystal structure) from another form of the element. Graphite, diamond, and buckminsterfullerene, for example, are allotropic forms of carbon.

alluvial cone A steep alluvial deposit that forms where a fast-flowing stream debouches onto a lowland plain. Apart from its steepness, it in other respects resembles an ALLUVIAL FAN.

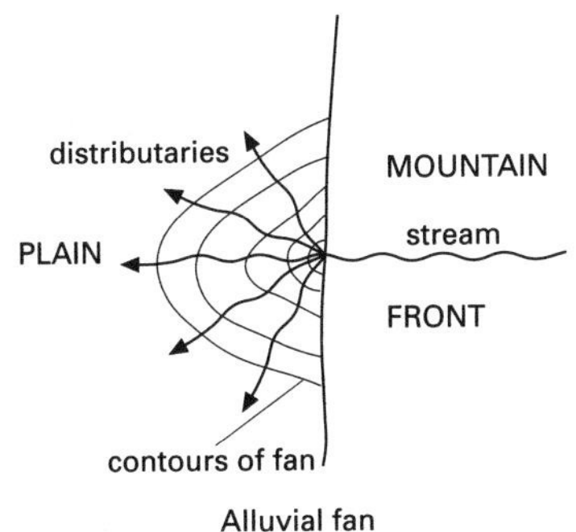

Alluvial fan

alluvial fan A fan of material deposited by a stream where it debouches from a mountain front onto a plain, with the apex of the fan at the point of emergence from the mountains. In its mountain tract, the stream will have been confined into a single flow in a rock-cut gorge, but on reaching the plain this control is lost, and it breaks up into a number of distributaries. This increases the wetted perimeter, or area of contact between water and land, and hence friction increases; with more of its energy used in overcoming friction, less is available for sediment transport, so deposition occurs, in the form of a fan. This is the direct landborne analogy to DELTA formation, and alluvial fans are characteristic landforms of arid and semiarid environments.

alluvium The products of SEDIMENTATION by rivers, sometimes including deposits in estuaries, lakes, and other bodies of fresh water. Alluvium includes material of a wide range of PARTICLE SIZES, usually restricted to the silt size fraction of 0.006–0.02 mm

There is a marked decrease in the size of alluvial material down-valley, with finer material in the lower reaches. This may be due to SORTING, which leaves coarse material behind and carries finer material down-valley, or it may be due to progressive wear of the material as it travels downstream. Another cause may be that smaller valley-side slopes downstream supply a finer caliber of material to the stream.

Within the FLOODPLAIN, alluvium (in the looser sense) varies in size from the finest clay and silt-sized material through sands to coarse angular gravels. The gravels form the basal portions of alluvial valleys, and may have originated at the end of the PLEISTOCENE period under PERIGLACIAL conditions; in some rivers they are attributed to deposition within the channel in conditions in which severe scour removed the fine material. Silts, sands, and gravels are more imortant, constituting 75% of the Mississippi alluvium. Sands constitute the point bars formed on the inside of meander bends, and are an important constituent of levées. Silts and clays are deposited in the lee of point bars or as overbank deposits in times of flood, covering the far reaches of the floodplains with backswamp deposits.

almandine A light red to red-brown member of the GARNET group of minerals, $Fe_3Al_2(SiO_4)_3$. Deep red crystals are valued as semiprecious gemstones.

alnoite A basic or ultrabasic dike rock composed largely of melilite and biotite with subordinate pyroxene, calcite, and olivine. Alnoite is found in ijolite-carbonatite complexes.

alp A region of grassland on a high mountainside. Above the timberline and covered with snow in the winter, alpine grassland may be used as pasture for grazing animals in summer. It is named after the Alps in west-central Europe.

alpine A type of climate found in mountainous areas above the timberline but below any PERMAFROST level. Grass and other low-growing plants make up the typical vegetation. *See also* alp.

Alpine folding A recent type of large-scale FOLDING that created such mountain ranges as the Alps, Andes, Himalayas, and Rockies.

Alpine-Himalayan chain A predominantly east-west trending orogenic belt, extending from Spain through Europe to S Asia. It was formed mainly in the Tertiary by the closure of the former Tethys Sea, mainly as a result of the northerly drift of the African continental plate. It consists of a series of igneous, metamorphic, and deformed sedimentary rocks, which now form the Alps and Himalayan Mountains.

altiplanation A process associated with a PERIGLACIAL environment in which terraces or benches are cut in solid rock in hillside or summit locations. The weathering and erosion of the rocks is achieved by a combination of FROST shattering, SOLIFLUCTION, and CONGELITURBATION. Terrace formation is probably initiated by the production of a NIVATION HOLLOW in hard rock, the backwall of which then retreats, causing an enlargement of the flat or gently sloping terrace in front. Altiplanation terraces may have a cover of debris, which can exhibit PATTERNED GROUND.

altiplano A high plateau that is surrounded by mountains. A typical example is the South American Altiplano between two extensions of the Andes, mainly in Bolivia and southwestern Peru. It contains volcanic material and eroded sediments, with evidence of ancient lakes in the form of salt basins.

altitude A measure of height, usually taken to be the height above mean sea level. Sometimes the altitude of a hill or mountain is given in terms of its vertical height from base to summit.

altocumulus cloud A type of cloud that indicates some form of vertical motion at medium levels in the atmosphere. It includes a wide range of cloud origins from genuine convection to billow and orographic clouds where the atmosphere is essentially stable, but the uplift is forced by mountains. Altocumulus clouds generally occur as globular masses in bands across the sky.

altostratus cloud A grayish cloud sheet normally composed of a mixture of ice crystals and water droplets. It is distinguished from cirrostratus by lying at lower levels, being somewhat thicker, and not exhibiting halo phenomena. The Sun may be seen through the thinner parts of the cloud. It is frequently followed by rain because the approach of a warm front is heralded by this cloud type.

alum **1.** In general terms, any double salt consisting of the sulfates of a trivalent metal and a monovalent metal, with 24 molecules of water of crystallization; its general formula is $A_2(SO_4)_3.B_2SO_4.24H_2O$, where A is the trivalent metal and B is the monovalent metal.
2. Specifically, aluminum potassium sulfate, $Al_2(SO_4)_3.K_2SO_4.24H_2O$, also known as potash alum. It occurs naturally as the mineral kalinite.

alumina The mineral form of aluminum oxide, Al_2O_3.

aluminum silicates There are three aluminum silicates with the composition Al_2SiO_5: the polymorphs ANDALUSITE, KYANITE, and SILLIMANITE. They are found in metamorphic rocks, mostly of argillaceous composition. Each has a different yet closely related crystal structure and is stable over a different range of pressures and temperatures.

Andalusite is orthorhombic and usually pink or white in color. The variety *chiastolite* shows a regular arrangement of impurities in the form of a cross. Andalusite is characteristic of the low-pressure and high-temperature conditions associated with contact metamorphism around igneous intrusions. Kyanite is triclinic and often blue-green to white in color. It is stable at higher pressures than andalusite and is found in intermediate- to high-grade regionally metamorphosed rocks. Sillimanite is orthorhombic, usually white in color, and commonly occurs as acicular crystals, hence the alternative name *fibrolite*. It is stable at higher temperatures than kyanite or andalusite and is found in the highest grades of thermally and regionally metamorphosed rocks. The orthorhombic aluminum silicate *mullite* has a composition $Al_6Si_2O_{13}$. It is found in argillaceous xenoliths (*buchites*) in basic igneous rocks.

alum shale A claylike rock containing potash ALUM, formerly much used as a source of alum for industry. It forms from shales when sulfides they contain decompose to produce sulfuric acid, which reacts with mica to form aluminum sulfate.

amazonite (amazonstone) A type of microcline FELDSPAR, blue-green or green in color, sometimes used as a semiprecious gemstone.

amber Yellow translucent fossilized resin once exuded by trees and often enclosing insects that have been trapped prior to hardening.

amblygonite A greenish or white lithium-containing mineral, $(Li,Na)Al(PO_4)(F,OH)$. It crystallizes in the triclinic system, occurs in PEGMATITES, and is used as a source of lithium.

amethyst A transparent purple variety of QUARTZ, used as a semiprecious gemstone. The color is due to impurities of iron oxide. *See* silica minerals.

ammonite One of the more advanced mollusks of the subclass AMMONOIDEA, whose shells had extremely convoluted crinkled suture lines. They are known from the Permian, are valuable as Mesozoic zone fossils, and became extinct at the end of the Cretaceous Period.

Ammonoidea An extinct subclass of marine mollusks of the class Cephalopoda. Ammonoids had external shells, usually coiled in a plane spiral and divided by septa into chambers. The shells ranged from about 25 mm in diameter to over 2 m in some species. The animal inhabited the terminal, and most recently formed, of these. The chambers were connected by a sometimes discontinuous calcareous tube known as a *siphuncle*. The septa met the inner wall of the shell at intersections called *suture lines*, and the character of these in fossil shells is used in taxonomic classification. The Ammonoidea had relatively complex suture lines compared with those of the NAUTILOIDEA; early ammonoids had angular folded suture lines (*see* goniatite) but in the later AMMONITES the suture lines were highly convoluted and crinkled. Ammonoids probably evolved from the nautiloids and the earliest are known from rocks of Silurian age. They reached their peak of development in the Mesozoic, for which they are important ZONE FOSSILS, and became extinct at the end of this era.

amorphous Having no regular atomic structure; noncrystalline.

Amphibia The first class of vertebrates to colonize land, evolving from crossopterygian fishes in the late Devonian. These fish had a bony skeleton that could provide support out of water; they also had lungs and their fleshy paired fins evolved into the more substantial limbs of the Amphibia. Amphibian eggs remain unprotected and must be laid in water, where the animals, too, spend their early life as tadpole larvae before undergoing metamorphosis into the terrestrial adults. The Amphibia were abundant in Carboniferous swamps, some reaching the size of modern crocodiles, but the class was in decline by the end of the Paleozoic. Modern representatives include newts, salamanders, toads, and frogs. Amphibians were the ancestors of the reptiles.

amphibole Any member of a group of rock-forming minerals that have $(Si,Al)O_4$ tetrahedra linked to form a double chain. PYROXENES are a similar group but with a single chain structure. The general amphibole formula is $X_{23}Y_5Z_8O_{22}(OH)_2$, where X = Ca, Na, K, Mg, or Fe^{2+}, Y = Mg, Fe^{2+}, Fe^{3+}, Al, Ti, or Mn, and Z = Si or Al. The hydroxyl ions may be replaced by F, Cl, or O. Most amphiboles are monoclinic but anthophyllite and gedrite are orthorhombic. Subgroups are based on the dominant cation occupying the X position.

1. *anthophyllite-cummingtonite subgroup*:
   anthophyllite $(Mg,Fe^{2+})_7(Si_8O_{22})(OH,F)_2$
   gedrite $(Mg,Fe^{2+})_6Al((Si,Al)_8O_{22})$-$(OH,F)_2$
   cummingtonite $(Mg,Fe^{2+})_7(Si_8O_{22})(OH)_2$
   grunerite $(Fe^{2+})_4(Fe^{2+},Mg)_3(Si_8O_{22})$-$(OH)_2$
2. *hornblende subgroup*:
   tremolite $Ca_2Mg_5(Si_8O_{22})(OH,F)_2$
   actinolite $Ca_2(Mg,Fe^{2+})_5(Si_8O_{22})(OH,F)_2$
   hornblende $NaCa_2(Mg,Fe^{2+},Fe^{3+},Al)_5$-$((Si,Al)_8O_{22})(OH,F)_2$
   edenite $NaCa_2(Mg,Fe^{2+})_5(Si_7AlO_{22})$-$(OH,F)_2$
   hastingsite $NaCa_2(Fe^{2+},Mg,Al,Fe^{3+})_5$-$(Si_6Al_2O_{22})(OH,F)_2$
   kaersutite $Ca_2(Na,K)(Mg,Fe^{2+},Fe^{3+})_4Ti$-$(Si_6Al_2O_{22})(O,OH,F)_2$
3. *alkali amphibole subgroup*:
   glaucophane $Na_2,Mg_3,Al_2(Si_8O_{22})(OH,F)_2$
   riebeckite $Na_2,Fe_3{}^{2+}Fe_2{}^{3+}(Si_8O_{22})(OH,F)_2$
   richterite $Na_2Ca(Mg,Fe^{2+},Mn,Fe^{3+},Al)_5(Si_8O_{22})$-$(OH,F)_2$
   katophorite $Na_2CaFe_4{}^{2+}(Fe^{3+},Al)(Si_7AlO_{22})(OH,F)_2$

The angle between the prismatic cleavages of amphiboles is 124°, the corresponding angle for pyroxenes being 87°. Some amphiboles occur in fibrous forms; asbestos is a fibrous form of actinolite. The members of the anthophyllite–cummingtonite subgroup together with tremolite, actinolite, and hornblende occur in metamorphic rocks. Hornblende is also common in igneous rocks. The alkali amphiboles occur in alkali igneous rocks with the exception of glaucophane, which is an index mineral of the glaucophane schist facies.

amphibolite A metamorphic rock consisting predominantly of amphibole. *See also* metamorphic facies.

amphidromic system A type of tide or tidal system in which the high water rotates around a central point (the *amphidromic point*). It is one consequence of the modifying influence that the Earth's rotation has on a standing oscillation. The range of the tide is nil, or very small, at the amphidromic point itself but increases outward from the point. The times of low and high water progress in a counterclockwise or clockwise direction around the amphidromic point. In the N hemisphere, high water rotates counterclockwise round the central point.

amygdale A spheroidal or ellipsoidal VESICLE within a lava, filled with deuteric or secondary minerals often in a zonal arrangement. Typical amygdaloidal minerals include calcite, zeolites, and quartz.

anabatic wind An upslope breeze often developing when mountain slopes are heated by the Sun during calm conditions. As turbulence is greater during the day it is more often suppressed than the night-time equivalent, the KATABATIC WIND.

anafront Any frontal surface at which the warm air is rising. As air cools on rising, condensation and precipitation are more extensive with this type of front than on a KATAFRONT where air is descending.

analcime (analcite) *See* feldspathoid.

analogs Similar patterns of the surface atmospheric pressure field that occur at different times or different places. The basic assumption in analog forecasting is that if two pressure situations are identical then the weather sequences that followed the first occasion will also follow the second. However, since analogs are never identical in all important features, only general trends can be deduced.

anatase A brown to black tetragonal polymorph of titanium dioxide, TiO_2, found in vein deposits and pegmatites. *See* brookite; rutile.

anatexis The partial melting of rock, which can then change composition by being contaminated, by mixing with other rocks, or by MAGMATIC DIFFERENTIATION. *See also* assimilation.

anauxite A claylike aluminosilicate mineral, $Al_2(SiO)_7(OH)_4$. It resembles kaolinite in composition (*see* clay minerals).

andalusite A hard pink, gray, or brown mineral form of aluminum silicate, Al_2SiO_5. It crystallizes in the orthorhombic system and occurs in metamorphic rocks such as gneiss and schist. It is used as a refractory and as a semiprecious gemstone. *See also* aluminum silicates.

andesine A variety of plagioclase FELDSPAR.

andesite The fine-grained volcanic equivalent of DIORITE, oversaturated or saturated in composition (*see* silica saturation). These rocks are characterized by the presence of plagioclase in the range oligoclase-andesine, often occurring as strongly zoned phenocrysts. The mafic minerals, augite, hypersthene, hornblende, biotite, and olivine, occur both as phenocrysts and in the groundmass. Small amounts of quartz and alkali feldspar may also be present. With an increase in the amount of quartz, andesites pass into DACITES and with an increase in the proportions of alkali feldspar, into LATITES (trachyandesites). Most andesites are porphyritic and have pilotaxitic or trachytic textures. Andesites are the intermediate members of the calc-alkaline volcanic suite and are associated with basalts and rhyolites in island arcs and orogenic regions.

Members of the alkaline basalt-trachyte suite containing dominant oligoclase and andesine are called *mugearite* and *hawaiite* respectively. These rocks may be distinguished from andesites by their basic generally undersaturated compositions, higher olivine contents, and differing field associations. Classification of volcanic rocks is often very difficult in petrographic terms because of the fine grain size, and divisions are made on a chemical basis.

Andesite line A line that delimits those parts of the Earth's surface that are of true oceanic structure as compared with true continental structure. The line can be drawn throughout the Pacific Ocean: on the ocean side of it are volcanic rocks that are entirely basic in character, but no in-situ volcanic rocks of the continental type. In the case of the W coast of North and South America, the Andesite line runs parallel to and comparatively close to the coast.

andosol A soil that has developed recently on base-rich volcanic materials. The upper A horizon of these soils is dark owing to the presence of organic matter that is well mixed by earthworm activity. This is a low-density friable layer and it overlies a more compact B horizon in which there has been little appreciable change in the content of clay. With increased duration of soil formation these soils become deeper and profile differentiation becomes more apparent. In humid temperate regions they may alter to form podzols and in the humid tropics they may form grumusols. *See also* inceptisol.

andradite A yellow, green, or brown type of GARNET, $Ca_3Fe_2(SiO_4)_3$. It occurs in metamorphosed limestones and is used as a semiprecious gemstone.

anemometer An instrument for measuring wind speed and often wind direction. The most common type has three cups mounted symmetrically about a vertical rotating axis, so that the rate of rotation is proportional to the wind speed. Wind can also be measured by the pressure it exerts on a plate or in a tube, or by the degree of cooling of a hot wire.

aneroid barometer The most commonly used instrument for measuring air pressure. In its simplest form it consists of a corrugated metal box from which the air is evacuated. The top and bottom of the box are separated by a spring and changes of atmospheric pressure produce movements in the distance apart of the faces. The changes are then amplified by a series of levers and can be recorded on a chart.

angiosperms The flowering plants. They differ from the GYMNOSPERMS in that their seeds are protected by an outer casing known as a *carpel*. Angiosperms are thought to have arisen in the Triassic; they became common during the Cretaceous and by the end of this period had replaced the gymnosperms as the dominant land plants, a position they still hold today. They occupy a great variety of terrestrial and freshwater environments and range in size from small herbaceous plants to giant trees. Angiosperms form an important source of food and habitats for both birds and mammals, reflected in the fact that these groups, too, began their main evolutionary radiation during the Cretaceous. The evolution and spread of the grasses, probably during the Miocene, had a particularly important effect on the environment, and therefore on animal life.

There are two main divisions of angiosperms, the classes Monocotyledoneae and Dicotyledoneae.

angle of declination (magnetic declination) The angle between geographical north and the direction of the magnetic meridian.

angle of incidence The angle between a light ray arriving at a surface and the normal to that surface at that point.

angle of inclination (angle of dip, magnetic inclination) The angle between the horizontal and the direction of the Earth's magnetic field at a given place. It is 90°; at the magnetic poles and zero along the magnetic equator.

angle of reflection The angle between a light ray reflected from a surface and the normal to that surface at the point of reflection. Other types of waves may also be reflected.

angle of refraction The angle between a refracted light ray and the normal from the surface at which the ray is refracted at the point of refraction. Other types of waves may also be refracted.

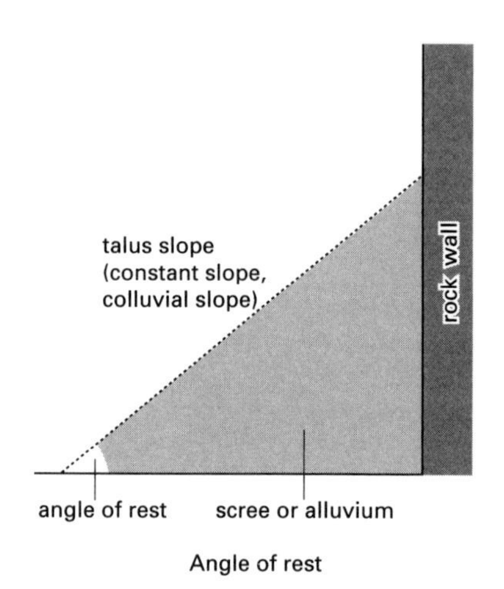

Angle of rest

angle of rest (angle of repose) The characteristic maximum angle of slope at which a pile of unconsolidated material is stable, the angle depending on the particle size of the material. Above that angle, the slope will be unstable and subject to slides and flow, which will reduce the angle of slope until it is again stable.

At rest, the particles interdigitate with each other, so that the weight of each one is balanced against the one below. As the angle becomes too steep, the particles upslope tend to lack support at the base and slide over the particles in front, moving downslope until the debris pile is again balanced. Larger material has a higher angle of rest than fine material; in most rocks and sands the angle is 30–35°. Where the material forming a slope is unconsolidated, its angle of rest determines the angle of slope; such slopes are called *talus slopes*, *constant slopes*, or *colluvial slopes* (*see* colluvium).

anglesite A white mineral form of lead sulfate, $PbSO_4$. Generally formed by the oxidation of GALENA, it crystallizes in the orthorhombic system. It is a common secondary mineral, used as a source of lead.

angular momentum The product of the linear velocity of a point on the Earth's surface, as it moves from west to east owing to the Earth's rotation, and the perpendicular distance of that point from the Earth's axis of rotation, a line through the north and south poles. Its value is greatest at the Equator, where the perpendicular distance is at a maximum, and zero at the poles.

angular unconformity A plane of erosion, generally related to earth movements, marking the boundary between two rock units of different age. The beds above and below the plane of erosion have different angles of dip, the lower beds generally dipping more steeply.

angular velocity The rate of turning of a rotating body, expressed in degrees, radians, or revolutions per unit of time. In meteorology, it can be used to indicate the horizontal rate of rotation around an area of low pressure or, on a large scale, the rotation of the Earth. As the Earth makes one complete rotation in a day, its angular velocity is 15° longitude per hour, or 7.29×10^{-5} radians per second.

angulate drainage A drainage pattern in which streams are developed equally in two directions, like the RECTANGULAR DRAINAGE pattern, but in this case the streams meet at an acute angle rather than right angles. The pattern results from structural guidance by two sets of joints or faults in an area with homogeneous strata.

anhedral Describing crystals having irregular boundaries, and having no crystal faces. *Compare* euhedral; subhedral.

anhydrite An orthorhombic mineral form of calcium carbonate, $CaSO_4$. It is usually white or colorless, occurring in massive or fibrous forms in evaporite deposits. *See also* gypsum.

anisotropic Having different physical properties when measured in different directions. Crystals belonging to systems other than cubic are optically anisotropic and light is transmitted at differing velocities along different directions within each crystal. When a ray of light enters an anisotropic crystal it is split into two rays, the ordinary and extraordinary rays, which travel in different directions and are polarized in perpendicular planes. Such crystals are said to exhibit double refraction or BIREFRINGENCE.

ankaramite A basaltic rock containing a high proportion of augite phenocrysts. *Compare* oceanite.

ankerite A gray to brown variety of the mineral DOLOMITE that contains iron, $Ca(Fe,Mg,Mn)(CO_3)_2$. *See also* carbonate minerals.

Annelida The phylum of animals containing the true worms, such as the common earthworm, whose bodies are divided into a number of segments. Annelids of one group secrete calcareous tubes, and fossil specimens of these exist, but most annelids are soft bodied and not often preserved after death. Certain small siliceous or chitinous remains known as *scolecodonts*, occurring in rocks of Ordovician age onwards, are thought to be annelid jaw parts. The effects of worms are often seen in rocks as TRACE FOSSILS, trails, burrows, or other characteristic disturbances of the sediment, features present even in some Precambrian rocks.

anomaly 1. (in climatology) The departure of any climatic element from the long-period average for that location. The spatial distribution of an anomaly is an interesting feature because it helps provide a greater understanding of the interaction between the atmosphere and the surface. A good example of a temperature anomaly is the area of NW Europe where mean annual temperatures are much higher than the lat-

itudinal average by as much as 12°C in parts of coastal Norway.
2. (in geophysics) The difference between observed and computed geophysical values, especially with regard to the thermal, gravitational, and magnetic properties of the Earth. *See* gravity anomaly.

anorthite A calcic plagioclase FELDSPAR.

anorthoclase A variety of alkali FELDSPAR.

anorthosite A type of coarse-grained leucocratic igneous rock that occurs as vast intrusive bodies or as members of layered basic complexes. Plagioclase feldspar constitutes over 90% of anorthosite and is generally of labradorite composition but anorthosites composed of plagioclase ranging from andesine to anorthite are known. Minor mafic minerals include augite and hypersthene and with an increase in the amount of mafic minerals, anorthosites pass into gabbros.

Antarctic Circle The line of latitude 66° 30′ S. In December there is continuous daylight for 24 hours and in June continuous darkness for 24 hours along this line.

antecedence A type of DISCORDANT drainage, originating when a preexisting drainage pattern cuts down into a rising fold, or a series of rising folds, in its path. The cause of the folding will be tectonic, and in the best developed examples, the rate of uplift of the folds is equaled by the rate of downcutting of the streams, eventually leaving them in deep CANYONS. As an explanation of discordance, it is less commonly used than SUPERIMPOSITION, but is often used to explain the origin of streams cutting through mountain ranges in deep gorges, e.g. the streams that cut through the Himalayas.

anthophyllite An orthorhombic mineral of the AMPHIBOLE group.

Anthozoa A class of the phylum CNIDARIA that includes the sea anemones and CORALS. The corals are geologically the most important anthozoans because many of them possess hard parts that may be fossilized. Their evolution may be used in BIOSTRATIGRAPHY and the colonial forms are important reef-builders. Typically, corals secrete a cup (*corallum*), from whose wall (*theca*) supporting platelike septa project radially toward the center. The Tabulata (tabulate corals) are known from the Ordovician Period to the Jurassic, although the group had greatly declined by the end of the Paleozoic. They were colonial forms having only rudimentary septa or none at all. The walls were reinforced by *tabulae*, calcareous plates extending across each coralite. The Rugosa, or Tetracoralla, (rugose corals) are first known from the Ordovician and are confined to the Paleozoic. Their septa occur in multiples of four and are strengthened by many small plates known as *dissepiments*. The group included both solitary and colonial species. The Scleractinia, or Hexacoralla, includes the modern reef-building corals, whose septa are in multiples of six. They range from the Triassic Period onwards.

anthracite Coal that has a very high fixed carbon content and a low amount of volatiles.

anthraxolite A hard type of BITUMEN that occurs in sedimentary rocks. It is often found associated with OIL SHALES.

anthropomorphic soil An intrazonal soil that has been formed as a direct result of human activity. Farming practices through the ages have, in some cases, produced distinctive soils or distinctive surface horizons. An important anthropomorphic soil is the paddy soil. This is similar to a gley and its features have been formed by the process of alternate wetting and drying, which has been controlled by farmers. Many soils have anthropic EPIPEDONS (surface horizons) or plaggen epipedons. The former have a high phosphate content and the latter a high organic matter content.

anticline An arch-shaped fold into which rock strata have been compressed, the oldest rocks occurring in the core. *See* fold.

anticlinorium A large-scale regional feature, many kilometers in diameter, consisting of an anticlinal structure with several minor folds on its limbs.

anticyclone (high) An area in the air of higher pressure than the surrounding air, with a closed isobar of approximately circular form at its center. Winds circulate in a clockwise manner around the high pressure center in the N hemisphere, and counterclockwise in the S hemisphere, but are generally light. Once established, the anticyclone moves only slowly and normally is a significant feature of surface pressure charts for a much longer period than cyclones. Anticyclones can be subdivided into two categories, cold and warm. Cold anticyclones are shallow features produced by strong radiational cooling at the surface or in the cold air behind a depression. The tropopause tends to be low above a cold anticyclone. Far more important are the warm anticyclones, which are characterized by warmer temperatures throughout a deep troposphere, with a cold stratosphere above. Their most frequent locations are in the subtropical high-pressure belts, but they do move polewards and block the normal westerly circulation producing temperature and precipitation anomalies in these areas. Descending motion is characteristic of anticyclones. They represent the parts of the atmosphere in which air slowly subsides from higher levels, warming and drying as it does so. This has the effect of stabilizing the atmosphere within an anticyclone so that rain is infrequent. In the surface layers, divergence of airflow takes place to maintain the continuity of subsidence. *Compare* cyclone.

anticyclonic gloom Under anticyclonic conditions, the layer of air near the ground undergoes little descent and if moist can easily become saturated to give a uniform layer of cloud. This is a stable situation with the cloud slowly gaining in thickness through radiational cooling of the layer aloft. Above this level the air is much warmer due to descent; a temperature INVERSION and stability are maintained. At the surface, little light penetrates the cloud and pollution can be trapped to give poor illumination. If prolonged, these circumstances can give high pollution concentrations, which may reach dangerous levels.

antiform An upward-closing structure, the precise stratigraphic relationships of whose strata are not known, i.e. the core need not be the oldest rocks. It results from complex folding in orogenic areas. *See* fold.

antigorite A mineral form of hydrated iron-containing magnesium silicate, $(Mg,Fe)_3Si_2O_5(OH)_4$, a variety of SERPENTINE. It occurs as fibers or undulating plates.

antiperthite An intergrowth of orthoclase and albite feldspar in which the orthoclase occurs as patches in the albite host. *See also* exsolution; perthite.

antithetic fault A minor normal fault associated with a major fault but whose planes dip into the main fault plane.

antitrades Originally considered to be the returning flow in the upper atmosphere for the surface TRADE WINDS, antitrades now more generally indicate the airflow above the NE or SE trade winds, blowing in a general westerly direction at heights of about 1800 m.

antitriptic wind A wind in which the forces controlling air movement are the pressure gradient force and friction. The CORIOLIS EFFECT and centripetal acceleration are neglected either because of the small scale of the system, such as a sea breeze, or because the Coriolis effect is negligible, as it is near the Equator.

anvil If the top of a cumulonimbus cloud reaches the tropopause, a temperature inversion, or strong wind shear, the main cloud updrafts are deflected horizontally causing the ice crystals at that level to spread out. From the ground, the whole cloud has the appearance of a blacksmith's anvil.

apatite A hexagonal phosphate mineral of composition $Ca_5(PO_4)_3(OH,F,Cl)$, the commonest phosphorus mineral. It may be formed as an accessory mineral in igneous rocks of most compositions. Apatite also occurs as a detrital mineral in sedimentary rocks and in metamorphic rocks.

aphanitic Denoting an igneous rock that is so fine-grained that individual crystals cannot be resolved with the naked eye. *Compare* phanerocrystalline.

aphelion The point on the orbit of a planet that is farthest from the central axis of rotation. *Compare* perihelion.

aphotic zone Ocean depths below the maximum depth at which photosynthesis takes place because of lack of light, stretching from 200 m to the sea floor. *See also* disphotic zone; euphotic zone.

aphyric Denoting an igneous rock that is not PORPHYRITIC.

aplites Leucocratic acid igneous rocks of medium to fine grain size occurring as thin veins and dikes around granite intrusions. Aplites consist largely of quartz and alkali feldspar and have micrographic or SACCHAROIDAL textures. They are thought to have crystallized from residual liquids lower in volatiles than pegmatites.

apophyllite A rare mineral of composition $KFCa_4(Si_8O_{20}).8H_2O$ found in AMYGDALES in basalts.

apparent dip The angle between a structural surface, e.g. a bedding plane, and the horizontal, measured vertically in any direction except that which is at 90° to the strike of the surface. This results in a value of dip that is less than that of true DIP.

appinite A melanocratic variety of DIORITE.

applanation The processes that reduce the heights of land features and cause an area to become more like a plain. It results from the erosion of high areas and the deposition of sediments in low areas.

apron A deposit of unconsolidated fragments that forms a broad extension in front of a glacier or at the base of a mountain.

aquamarine A transparent blue-green variety of BERYL that occurs, sometimes as very large crystals, in pegmatite. It is valued as a semiprecious gemstone.

aquatic Describing any organism that is found in water. In an aquatic environment temperatures generally vary little, dehydration is virtually impossible, and the water provides physical support for plants and invertebrate animals.

aquifer Any water-saturated rock horizon that has sufficient porosity and permeability to yield economic supplies of groundwater, either as springs or in wells.

aragonite A white or gray mineral form of calcium carbonate, $CaCO_3$. It crystallizes in the orthorhombic system and occurs near the surface in sedimentary rocks and in deposits in caves and from hot mineral springs. It also occurs in PEARLS and in the shells of some invertebrate marine animals. *See also* carbonate minerals.

Archaeopteryx The earliest known bird, whose fossils come from late Jurassic strata. Good fossil preservation reveals that *Archaeopteryx* had feathers but in its skeletal structure it resembled the archosaur reptiles from which it had evolved. The wings were primitive and the three separate fingers had claws. Other reptilian features included teeth, a long tail, and solid bones (*compare* Aves). Its flying ability was probably poor.

Archean *See* Precambrian.

archipelagic apron The moatlike rock filling that may surround a group of volcanic islands or seamounts. The moat filling is usually gently sloping and smooth, the upper surface being called an *archipelagic plain*. These groups of islands or seamounts probably depress the Earth's crust to the extent that a moat or depression is developed, but the depression may be concealed by a sedimentary overlay. Part of the Hawaiian Islands display a moatlike form surrounding them.

archipelago A group of islands in fairly close proximity. The term is sometimes used to describe sea areas that contain numerous scattered islands.

Archosauria A subclass of the Reptilia consisting principally of the extinct DINOSAURS and PTEROSAURIA and the modern crocodiles. Primarily carnivorous, with simple pointed teeth, they show a tendency to bipedalism involving development of the hind limbs, a corresponding reduction of the fore limbs, and a well-developed tail for balancing the body. Archosaurs became the dominant form of terrestrial life during the Jurassic and Cretaceous periods; some became secondarily quadrupedal and vegetarian.

Arctic Circle The line of latitude 66° 30′ N. In June there is continuous daylight for 24 hours and in December continuous darkness for 24 hours along this line.

Arctic sea smoke A type of fog formed by evaporation from a relatively warm sea surface into cold air aloft. Condensation takes place in the cold air, but unless an inversion develops, further mixing prevents the extensive development of fog. In the Aleutian Islands, fog depths up to 1500 m may occur under suitable conditions.

Arctic warming The marked warming of the Arctic area that took place between the 1920s and 1950s. The ice-free period increased and mean annual temperatures rose by about 4–10°C. It was primarily the result of the more northerly tracks of the Atlantic and Pacific depressions carrying moist warm air toward the poles. Although the amount of warming has decreased considerably, Arctic temperatures are still higher than pre-1920 and may be rising again.

arc-trench gap At a destructive PLATE BOUNDARY, a region 50–400 km wide between an ocean trench and a volcanic arc, often associated with an ISLAND ARC. Generally the wider the gap, the faster is the rate of convergence between the two tectonic plates involved.

arcuate delta A DELTA with a rounded convex margin.

arenaceous Describing a CLASTIC sedimentary deposit or rock in which the constituent fragments are of sand grade in size. This includes particles from 0.06 mm to 2 mm in diameter. Arenaceous rocks are also referred to as SANDSTONES. The grains are often fragments of crystals, although lithic sandstones are known (*see* graywacke).

arenite **1.** A type of SANDSTONE that has little or no matrix material binding the grains together.
2. A sedimentary rock that consists of sand-sized grains (up to 2 mm across). The grains need not be silica or silicates, and may include fragments of carbonate rocks.

arête A sharp ridge bounded by steep slopes and found in glaciated or formerly glaciated areas. These forms seldom develop in isolation and headwall erosion of two adjacent cirques causes the formation of an arête. Continued headwall recession will result in the destruction of the arête and the cirques will merge.

argentite A mineral form of silver sulfide, Ag_2S. It crystallizes in the cubic system but is stable only at high temperatures; below 180°C it changes to the monoclinic mineral acanthite. It is the most important ore of silver.

argillaceous Describing a CLASTIC sedimentary deposit or rock in which the constituent fragments are of SILT or CLAY grade in size. This includes particles smaller than 0.06 mm in diameter. These consist of finely ground rock as well as the various clay minerals that have been produced in the course of weathering of the parent rock. Siltstones and MUDSTONES are rocks formed of sediment in this size range.

argillite A slatelike sedimentary rock that is formed from shale or mudstone by the effects of pressure and cementation. Unlike slate, however, it has no distinct cleavage planes.

argon An inert gas comprising about 1% of the atmosphere by volume. It has no meteorological significance.

arid climate A climate that experiences a moisture deficiency sufficient to inhibit but not prevent natural vegetative growth and a mean annual rainfall usually taken as below 250 mm. Attempts to produce a precise definition of aridity involve an assessment of the efficiency of precipitation. Arid climates are found in areas of semipermanent anticyclones, where cold ocean currents stabilize the lower atmosphere, within extensive mountain ranges, and in locations at enormous distances from the sea. *See also* desert.

aridisol One of the ten soil orders of the SEVENTH APPROXIMATION classification, including saline and alkaline mineral soils of desert areas. They are characterized by low organic matter contents and a horizon of calcium or sodium accumulation within 1 meter of the surface. They are infertile because of lack of moisture, coarse particle size, and their susceptibility to erosion due to lack of vegetation. In some profiles, groundwater may concentrate sodium to toxic proportions, forming a salic horizon in which sodium salts exceed 2% of the mineral matter. The aridisols include true desert soils, sierozem, solonchak, and solonetz.

aridity index An assessment of the degree of dryness of a climate. There are a variety of such indices, the best known being devised by Thornthwaite in 1948. Most involve the relationships between temperature, total rainfall, and humidity.

arkose An ARENACEOUS sedimentary rock that includes more than 25% FELDSPAR in its composition.

Armorican orogeny A period of mountain-building during the Upper Paleozoic affecting W Europe, named after Armorica (Brittany). It is part of the more complex VARISCAN orogeny but characterized by a roughly northwest-southeast trend.

arrival time The first recording of a seismic disturbance. Close to an earthquake, there is little difference in the arrival times of primary (P) waves and secondary (S) waves. Farther away, the faster P waves arrive first. The distance to the earthquake can be calculated from this difference in arrival times. *See* primary wave; secondary wave.

arroyo An ephemeral stream of the semiarid USA and Latin America. Arroyos originate as discontinuous gullies on a hillside where the vegetation has been locally weakened by trampling, grazing, or fire. Headward erosion during periods when the gullies contain water leads to coalescence, and the formation of a continuous gully, or arroyo. *See also* abnormal erosion; gully erosion.

arsenopyrite An orthorhombic arsenic mineral of composition FeAsS, found in hydrothermal veins.

artesian Describing water that has moved underground from its original source. This may occur by percolation upward along a sloping AQUIFER with the result that the artesian water is above the level of the water table.

artesian basin A SYNCLINE that has a layer of permeable rock between two impermeable layers. Water can be obtained by drilling boreholes into the permeable layer. If this layer is below the water table in nearby hills, the water may flow up the borehole under hydrostatic pressure, like a fountain, forming an artesian well.

Arthropoda The phylum of animals having jointed limbs and a segmented body protected by a chitinous exoskeleton. The Arthropoda is the largest phylum in the animal kingdom and its six most important classes are the CRUSTACEA, Arachnida (spiders and scorpions), Myriapoda (centipedes and millipedes), Merostomata (king crabs and the extinct eurypterids), INSECTA, and the extinct TRILOBITA. Fossils of arthropods represent mainly aquatic benthonic forms, which lived in the most favorable environments for preservation; such arthropods are thought to have been recognized in the Cambrian, perhaps even the Precambrian.

Artiodactyla The order of herbivorous hoofed mammals comprising those with an even number of toes, such as antelopes, cattle, deer, and pigs. Originating in the Eocene and at first outnumbered by early perissodactyls, they are now the more successful group. The most primitive members are the pigs, which still possess four toes. In the ruminants the limbs are developed for fast running with the loss of all but two of the digits, and the teeth are modified to deal with a coarse vegetable diet. *Compare* Perissodactyla.

asbestos A fibrous variety of AMPHIBOLE, usually tremolite or actinolite. The SERPENTINE mineral chrysotile is used in the manufacture of commercial asbestos.

aseismic plate Any of the large areas of the Earth's crust above the asthenosphere, within which there are relatively few earthquakes.

aseismic ridge A ridge on the seabed that, unlike a MID-OCEAN RIDGE, has no volcanic activity along it.

ash *See* pyroclastic rock.

ash flow deposit A volcanic deposit resulting from a NUÉE ARDENTE, in which an avalanche of glowing ash flows rapidly down the side of a volcano. The resulting sediment may be partly blocky, but also contains tuffs where shards of glassy lava, carried by the flow, have become welded together.

asphalt A type of brown or black BITUMEN consisting mainly of carbon disulfide (CS_2) and hydrogen. It varies from a thick viscous liquid to a tarry solid. It occurs in oil-bearing rocks and probably represents an early stage in the formation of PETROLEUM from buried marine organisms. A similar substance remains after the distillation of crude oil.

assimilation During intrusion, magma forces its way along joints and cracks in the country or wall rock so that large blocks (XENOLITHS) may become detached and sink into the magma. This process, known as *stoping*, may occur on a large scale when large volumes of magma are emplaced. The complex interactions taking place between the magma and solid rock are known by the term *syntexis*. Melting of the rock and chemical reaction with the magma may result in the complete digestion or assimilation of the incorporated material. Xenoliths survive only when assimilation is incomplete. Contaminated magma that has assimilated large quantities of country rock is said to be hybrid. Hybridization also occurs when two different magmas are mixed. That such a process has taken place may be indicated by the occurrence of corroded XENOCRYSTS within an igneous rock.

asterism An optical phenomenon displayed by some crystals that produce starlike flashes of light. Caused by tiny needle-shaped inclusions, it may occur in transmitted light (as with some forms of mica) or in reflected light (as with star sapphire).

asthenosphere A zone within the Earth's upper MANTLE in which the velocity of seismic waves is considerably reduced. Movement between the Earth's outer lithosphere and inner mesosphere is thought to take place along this zone, which is capable of prolonged plastic deformation. It is thought to be composed of partly molten peridotite, with a liquid fraction having the composition of basalt. It is developed between 50 and 240 km beneath the Earth's surface.

astrobleme An ancient crater on the Earth's surface resulting from the impact of an extraterrestrial body.

astrogeology The application of terrestrial geology to studying the origin and history of extraterrestrial objects in the Solar System.

asymmetric fold A fold in which the axial plane is not vertical, with the result that the two limbs have different angles of dip. See diagram at FOLD.

Atlantic period The period from 5000 to 3000 BC, within the present interglacial, when the climate of much of NW Europe was warm and moist, achieving the highest mean annual temperatures since the last glaciation. It is also known as the *climatic optimum*.

atmophile An element that occurs naturally as a gas, such as helium in some uranium ores and the various gases in the ATMOSPHERE. *See also* chalcophile; lithophile; siderophile.

atmosphere The mixture of gases surrounding the Earth, which we breathe and which provides our weather. The atmosphere can be divided into a number of layers on thermal and lapse-rate properties. The troposphere is the lowest, extending up to approximately 10 to 15 km. Throughout this layer, there is a decrease of temperature with height, and it is the source of all precipitation and most of our weather phenomena. Above this layer, temperatures gradually increase through the stratosphere to about 50 km, decrease through the mesosphere reaching values of about –90°C between 80 and 90 km, and finally there is the thermosphere or ionosphere where absorption of ultraviolet radiation of shorter wavelengths than that absorbed by ozone causes a rise of temperature to values of 1500 to 2000°C at between 300 and 400 km. Gases become ionized at these high levels, which is of vital importance in radio communication.

The distribution of water vapor in the atmosphere is very variable. It depends upon the amount of evaporation from the surface and the amount of moisture advected from elsewhere. In continental interiors the value is very low. It is important as an absorber of long-wave radiation, so that moist air reduces the rate at which the ground surface cools at night by increasing counter-radiation. Moisture is also required in the atmosphere for precipitation, although the absolute amount is less important than other factors favoring precipitation formation.

The atmosphere acts as a protection to the ground surface by burning up most cosmic particles, which can be seen as meteors in the night sky. *See also* air; general circulation of the atmosphere.

atmospheric pressure The pressure at the Earth's surface that results from the weight of the ATMOSPHERE. At sea level, standard atmospheric pressure is 760 mm of mercury, 101.3 kilopascals, or 1013.25 millibars. It decreases with increasing altitude, and is measured using a BAROMETER.

atmospheric window The region in the radiation spectrum containing wavelengths between 8.5 and 11 micrometers, which are not absorbed to any great extent by atmospheric gases. In the absence of cloud, terrestrial radiation of this wavelength is lost to space, enabling the cooling of the Earth to take place. Other narrower wavebands also do not absorb long-wave radiation, but these are of less importance in this context. *See also* absorption.

atoll A ring-shaped reef, island, or islands that surround or nearly surround a lagoonal area of water, in which detrital material may collect. The surrounding rim may itself lie in shallow water, in which case the central area of water is rather deeper, sometimes very deep (the average depth according to one authority is 45 m). Not all oceanic atolls are formed entirely of coral, and certain calcareous algae may constitute the bulk of the reef material.

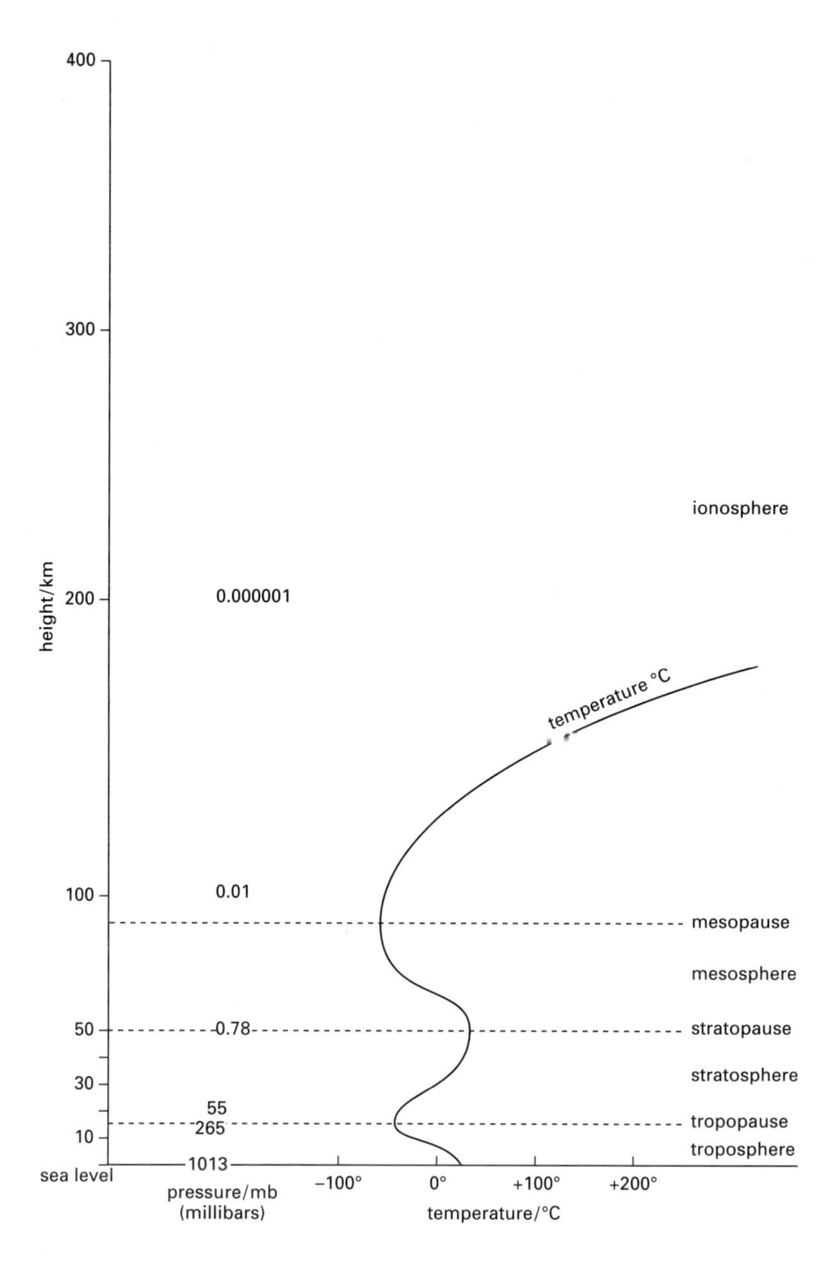
400
300
200
100
50
30
10
height/km
sea level
0.000001
0.01
0.78
55
265
1013
pressure/mb
(millibars)
temperature °C
-100°
0°
+100°
+200°
temperature/°C
ionosphere
mesopause
mesosphere
stratopause
stratosphere
tropopause
troposphere

Atmosphere

Atolls are based on some kind of platform, often an eroded platform that subsided at some stage and on which coral growth or other organic growth later occurred. An *atollon* is a small atoll lying on the flank of a larger atoll. *See also* reef.

attenuation The loss of energy of electromagnetic waves as they pass through the atmosphere. It is caused by ABSORPTION and SCATTERING by the molecules and particles of the atmosphere. In the ionosphere, free electrons absorb kinetic energy from the radiation and lose it in subsequent collisions.

attrition The reduction in particle sizes of sediment by rubbing and grinding action (*see* abrasion) during transport.

augen Large eye-shaped crystals, commonly of feldspar, that have survived the intense shearing of schist and gneiss taking place during the formation of a MYLONITE.

augite A monoclinic PYROXENE.

aulacogen A long depression on the edge of a continent bounded by FAULTS, which did not develop into an active rift as two continental plates split apart. *See also* rift valley.

aurora A transient optical phenomenon usually seen in the polar skies at the time of solar flare activity. The displays of the aurora can vary from shimmering rays to colored corona effects occupying large parts of the sky. In the N hemisphere it is called the *aurora borealis* (or *northern lights*) and in the S hemisphere the *aurora australis* (or *southern lights*). The displays are most frequent around the geomagnetic poles, with a daily occurrence maximum about midnight. There is some evidence of a seasonal maximum about the time of the equinoxes, but the major maximum follows the 11.2 year solar activity cycle.

The aurora is caused by the interaction of the Earth's tenuous upper atmosphere and charged particles streaming from the Sun. The solar particles are deflected by the geomagnetic field and so only occur in limited parts of the upper atmosphere, at a height of around 100 km.

autecology A branch of ECOLOGY that studies the interaction between a single species and its environment.

authigenic (authigenous) Describing rock constituents that were formed in situ, coming into existence during or after the formation of the rock in which they lie. *Compare* allogenic.

autochthonous Describing rocks that are still in their place of formation and have not been displaced by thrusting, e.g. a folded sequence of rocks whose roots are still connected. *Compare* allochthonous.

autolith An igneous INCLUSION in an igneous rock. It consists of material that has crystallized from magma and is thus genetically related to the surrounding rock. *See also* xenolith.

autotrophic Describing an organism that produces its own 'food' directly from inorganic substances. Chlorophyll-containing plants and algae are typical autotrophs; they produce organic compounds from carbon dioxide and water during photosynthesis, using the energy of sunlight. Some types of bacteria are also autotrophs.

autumnal equinox *See* equinox.

auxiliary mineral Any relatively rare light-colored mineral that occurs in an igneous rock. Such minerals include APATITE, CORUNDUM, FLUORITE, and MUSCOVITE.

avalanche A rapid movement of snow en masse down steep slopes, which must usually have an angle greater than 22°. It may consist of fresh powdery snow, sliding over the contact surface of compacted older snow; of slabs of consolidated snow, which roll down the slope; or of the whole ice and snow cover of a slope, in which case much rock material is also carried with the fall. If avalanches are a frequent occurrence in any one location, they can be important from the geomorphological point of view, because they maintain an exposed bedrock surface, which is therefore susceptible to continued weathering.

aventurine A spangled translucent form of quartz or feldspar, used as a semiprecious stone. Its appearance is caused by inclusions of particles of hematite, mica, or other mineral.

Aves The birds: a class of vertebrates distinguished from the REPTILIA by the presence of feathers. They probably arose from primitive archosaur reptiles (*see* Archaeopteryx) and most of their evolution is connected with adaptation for flight. The wings are formed from the whole forearm and three fused fingers (*compare* Pterosauria). They are warm-blooded and have short tails, a large breast bone for the attachment of flight muscles, and light hollow bones. All modern birds are toothless but teeth were present in their Mesozoic ancestors. Because of the fragile nature of the skeleton, fossils are few. There have been a number of secondarily flightless forms – the ostrich, emu, etc. – often of large size.

axial modulus (in geophysics) The ratio of stress to strain, when in the presence of laterally confining forces.

axial plane A plane that passes through the successive hinge lines of the beds in a FOLD. Different types of fold are characterized by different angles of inclination from this plane.

axial plane cleavage Cleavage planes that are parallel to the axial plane of a FOLD. Generally the cleavage is related to minor fold axes, but occasionally it may be more closely related to the region's fold trend.

axial rift zone *See* median valley.

axial trace The intersection of a FOLD AXIS with the Earth's surface, reflecting the trend of the fold. (See diagram at FOLD.)

axinite A lilac-brown triclinic mineral with the composition $Ca_2(Mn,Fe^{2+})$-$Al_2BO_3(Si_4O_{12})OH$, produced during the boron metasomatism of calcium-rich sediments and igneous rocks.

axis of the Earth A line that joins the North and South Poles, about which the Earth rotates every 24 hours.

azimuth **1.** (in surveying) A horizontal angle measured clockwise from true north (true azimuth) or magnetic north (magnetic azimuth) to another point.
2. (in astronomy) The angle between the plane of the meridian of the observer and the vertical plane passing through a heavenly body.

azimuthal equal-area projection A map projection differing from the other azimuthal projections in that the spacing of the parallels decreases with increasing distance from the center of the projection, producing the equal-area property.

azimuthal equidistant projection A map projection that is the same as the azimuthal projection in that the straight lines radiating from the center of the projection all have their true bearings, but it has the additional property that the distances along these lines are true to scale. It is not an equal-area projection, nor is it conformal.

azimuthal projection (zenithal projection) A map projection constructed as though a plane were placed at a tangent to the Earth's surface and the portion of the Earth covered by it were projected onto the plane. The result is that all points have their true compass bearings. The tangent plane is not always drawn at the pole; it can be constructed anywhere on the surface of the globe, the point where the tangent touches the Earth being the center of the map.

Azoic Designating PRECAMBRIAN rocks that were deposited before the origin of life. In practice, however, it is difficult to distinguish such rocks and the term is little used. It is never used for unfossiliferous PHANEROZOIC strata.

azonal soil Soil lacking a B horizon owing to insufficient time for complete pedogenesis. Profiles of azonal soils therefore mainly reflect the influence of parent material and show an A horizon slightly darker than the C horizon because of additions of organic matter. The three main groups are lithosols, regosols, and alluvial soils. In the SEVENTH APPROXIMATION azonal soils are classified as entisols.

Azores anticyclone The semipermanent anticyclone centered in the Atlantic near the Azores Islands. It is part of the subtropical high-pressure systems of the N hemisphere. Its mean latitudinal position oscillates, being farther south in winter and farther north in summer, with high-pressure cells occasionally drifting northeastward from the main center to form blocking anticyclones within the westerly circulation.

azurite A deep blue mineral form of basic copper carbonate, $Cu_3(CO_3)_2(OH)_2$. It crystallizes in the monoclinic system, and occurs in oxidixed copper deposits, often associated with malachite. It is a copper ore and also a semiprecious gemstone.

B

backing In meteorology, a counterclockwise change in wind direction, such as from easterly to north-easterly. *See also* veering.

backshore The part of a beach that lies above the level of normal high spring tides. Only when exceptionally high spring tides occur, or severe storms take place, does this zone come under the influence of wave action. Cliffs or sand dunes behind beaches are usually included as part of the backshore.

backsight *See* leveling.

backwash The return of water down a beach, under the influence of gravity, following the breaking of a wave and the associated SWASH. The difference in load-carrying capacity between the swash and backwash of prevalent waves determines whether a beach will aggrade or degrade. Whereas flat waves tend to produce a strong swash, steep waves, which break vertically onto the beach, result in a powerful backwash and cause a net seaward movement of beach materials.

bacteria Single-celled microorganisms that are fundamental to soil productivity. Where oxygen is lacking they account for most of the biochemical changes in the soil. Environmental controls such as moisture, aeration, temperature, acidity, and organic matter result in a constantly fluctuating bacterial population in the soil. It has been estimated there may be as many as 4000 million per gram of soil.

They are commonly classified into two broad groups in relation to energy supply: *autotrophic bacteria*, which obtain their energy from the oxidation of mineral substances and their carbon mostly from carbon dioxide; *heterotrophic bacteria*, which obtain their energy and carbon directly from the soil organic matter. Certain enzymic transformations in the soil are dominated by bacteria. Autotrophic bacteria are much less abundant than heterotrophic but because they include the bacteria responsible for nitrification and sulfur oxidation they are of paramount importance. *See* nitrification; nitrogen fixation; sulfur oxidation.

By their action of breaking down organic matter, bacteria are probably instrumental in the formation of oil. They are also involved in inorganic reactions, such as those leading to the origin of iron ore deposits. There is evidence indicating that bacteria were present in PRECAMBRIAN times. structures called STROMATOLITES, which are formed by the action of bacteria, notably blue-green bacteria (cyanobacteria; formerly called blue-green algae), are present in Precambrian rocks.

badlands An eroded furrowed landscape in a dry region, such as parts of Nebraska and South Dakota in the American West. There is little or no vegetation and so any rainwater runs off quickly along the short steep slopes, further eroding any exposed or soft rocks.

bahada *See* bajada.

Bai-u season A period in early June when the summer circulation of China and Japan commences, bringing with it a marked rainfall maximum. The rains are the result of weak nonfrontal disturbances within the southwesterly flow, which rapidly moves northward at this time as the upper westerlies move to the north of the Himalayas.

bajada (bahada) The gently sloping surface leading from a mountain range down to an INLAND BASIN in arid and semiarid areas. The term has been used both for coalesced alluvial fans occurring on the rock-cut PEDIMENT at the foot of the mountain front and for a gently sloping concave surface (composed of increasingly fine-grained particles) leading down from the pediment to the flat basin or playa. When the term is applied to the former case, this sloping surface, which still exists, is known as a *peripediment*. In either case, the composing material is derived from the mountains and is brought down by surface runoff, which follows the infrequent but heavy downpours.

bald-headed Denoting an anticlinal fold whose upper beds have been eroded away, exposing the older rocks of its core.

ball clay A very plastic fine-textured gray or buff-colored clay that often contains some organic matter. Also called pipe clay, it is used to make ceramics.

balled-up structure An isolated patch of silty sediment contorted into spherical nodules and enclosed within mud-sized sediments, resulting from slump movements while in a poorly consolidated state. The term was proposed by O. T. Jones in 1937.

ball lightning A rare form of lightning that is not, as yet, adequately understood. Its very existence has been debated because most reported cases follow brilliant lightning flashes and so could be physiological in origin, an afterimage being produced in the eye. Reports of the lightning suggest a brilliant sphere with a diameter of 1 cm to 1 m, which can drift almost randomly at a few meters per second.

banded agate A type of AGATE that has alternate bands of colors. The bands may be blended into each other or be sharply defined; they may be concentric or take the form of wavy lines. *See also* onyx.

banded iron formation SEDIMENTARY ROCKS in the form of beds or layers that consist of CHALCEDONY, CHERT, JASPER, or QUARTZ and at least 25% iron. The iron is usually in the form of hematite, magnetite, or other oxides and may be used as low-grade ore.

bank **1.** Either side of a river channel, best marked where the river has considerable powers of vertical erosion. At BANKFULL no banks are exposed.
2. A mass of sand, mud, shells, etc., usually below the water surface, and often elliptical in shape. Banks of unbedded limestone are distinguished from reefs by their lack of rigid framework formed by corals or other calcareous organisms.

bankfull The state of flow of a river immediately before flooding when the river's channel is full with water to its brim. In streams of very different sizes, and in different environments, it has a fairly constant recurrence interval of about once every 1.5 years.

banner cloud A stationary cloud that appears to be attached to an isolated mountain peak. It extends downwind from the peak for some distance and so resembles a flag or banner. The Matterhorn is probably the best known summit with a frequent occurrence of this type of cloud. The origin of the cloud is due to the lifting of air in the eddy of the leeward side of the peak, although an aerodynamic pressure reduction may play some part in its formation.

bar **1.** An elongated body of sediment, such as sand or shingle, occurring in the sea more or less parallel with the coastline and sometimes attached to it. It may be permanently submerged (*see* submarine bar) or be submerged for part of the tide, for instance at the mouth of a river or near the entrance to a harbor. Some such bars are much exposed at low tide. *See also* barrier

beach; bay bar; longshore bar; offshore bar; spit; tombolo.
2. A unit of pressure. *See* millibar.

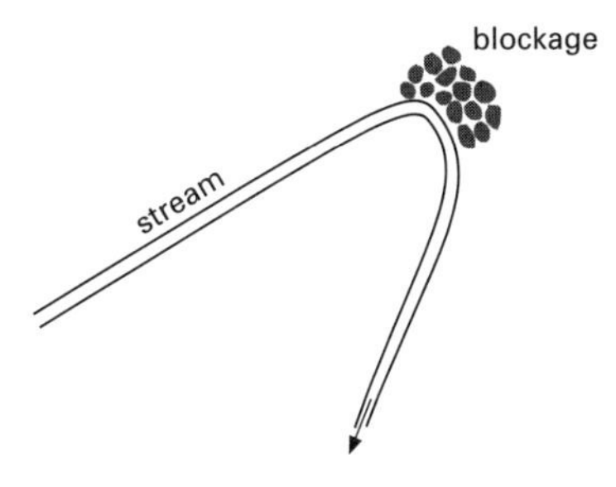

Barbed drainage

barbed drainage A drainage pattern in which a stream suddenly bends back on itself and flows in an almost reverse direction, often due to glacial blocking, river capture, or local tectonic movement in its path.

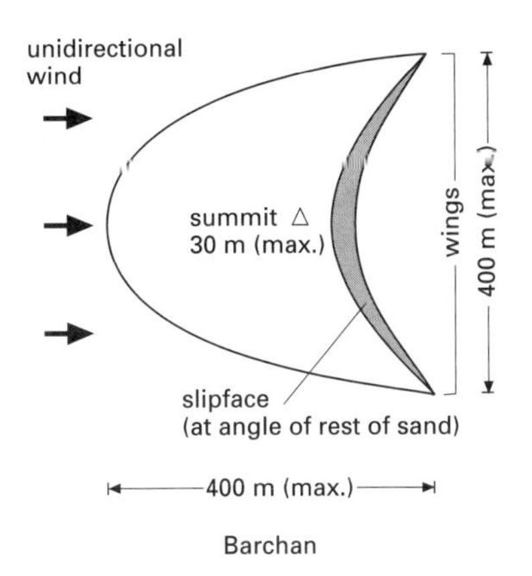

Barchan

barchan A crescentic dune characterized by an oval shape upwind, with a slip face and two wings spreading out downwind. Barchans occur in desert areas with a unidirectional wind regime. The slip face is formed by erosion of material from the upwind side, and its transport to the summit where it builds up until unstable, at which point it slumps forward to its angle of rest. The wings develop because the rate of advance of the dune is inversely proportional to its height, so the lower sides extend downwind faster than the center; at their tips they end in a dwindling of the sand pile to nil, which is reached at the point where the slip face upwind ceases to shelter the wings, and they become subject to wind erosion. Barchans typically occur in belts up to 300 km long and over 12 km wide.

barite *See* barytes.

baroclinic Describing a state of the atmosphere where the surfaces of constant pressure (isobars) intersect into surfaces of constant density (isopycnics). This state is the result of large horizontal temperature gradients and is believed to be responsible in part for the formation of midlatitude depressions.

barometer An instrument for measuring pressure, invented in 1643 by Torricelli. The standard instrument (Kew pattern or Fortin barometer) measures the height to which a column of pure mercury can be supported by the atmospheric pressure exerted upon it. Corrections have to be made to this value to allow for variations in gravity, the purity of the mercury, and the ambient temperature.

barotropic Describing the state of the atmosphere when the surfaces of constant pressure are parallel to the surfaces of constant density. Although this is a useful concept in theoretical studies, such an ideal state rarely, if ever, occurs.

barrage A large engineered structure of concrete or soil and rocks, constructed across a river. It dams the flow to create a lake, generally for use in the irrigation of adjacent farmland. A barrage across an estuary also prevents exceptionally high tides from causing flooding.

barrier beach An elongated accumulation of sand, shingle, or in-situ rock lying roughly parallel to the coast but separated from it by a channel, lagoon, or other water area. The distance offshore may vary from a few meters to several kilometers. Some lie across bay mouths or estuaries. Off low coasts, barriers are often in the form of *barrier islands*, the lagoonal areas behind being of the mangrove type in warm sea regions and of the salt-marsh type or even devoid of vegetation in the case of temperate or cold sea regions. Many barriers, such as those off the E coast of the USA and the Gulf of Mexico coast, carry dunes. Most sand and shingle barriers are destined to overwash, even breaching on occasions during severe storms. The more common sandy barriers are formed by the continued activity of CONSTRUCTIVE WAVES, which can progressively build up material above water level. Shingle barriers are formed by storm waves, the main input of material resulting from longshore movements.

barrier island *See* barrier beach.

barrier reef An elongate accumulation of coral, extending upward from the sea floor to about low-tide level, parallel with the coast but separated from it by a lagoon. A reef may lie between several meters and several kilometers offshore. They are believed to have been formed as a result of the submergence of a flat surface as the postglacial sea level rose. The growth of coral organisms kept pace with the rising water so that great thicknesses could develop. The best-known example is the GREAT BARRIER REEF.

Barrow zone *See* zone (def. 1).

barysphere All of the Earth's interior beneath the lithosphere, i.e. the asthenosphere, most of the upper mantle, and the outer and inner core.

barytes (barite) An orthorhombic mineral form of barium sulfate, $BaSO_4$. It is usually white but impure varieties are colored. Barytes has a high density, which distinguishes it from calcite. A continuous chemical series exists between barytes and celestite ($SrSO_4$). Both minerals are found in hydrothermal veins with galena, sphalerite, fluorite, and calcite as well as in cavities in limestone.

basal conglomerate A CONGLOMERATE at the base of a sequence of sedimentary rocks. It consists of a mixture of well-sorted coarse particles, usually deposited as a thin layer by an encroaching sea.

basal sapping Any of a range of processes that act to remove debris from the foot of a slope. Basal sapping is important in the creation and maintenance of FREE FACES, leading to near vertical or steep slopes that are often unstable, hence causing MASS MOVEMENTS. For this reason it constitutes the major explanation of PARALLEL RETREAT of slopes, because the debris at the foot is constantly removed and the steepness of the slope is maintained, thus preventing slope decline. Types of basal sapping include wave action at the foot of a sea cliff, lateral erosion by a river at the foot of a bluff, headward erosion of streams issuing at the foot of a slope, and mass movements produced by heterogeneity of rock strata.

basalt A basic volcanic rock, the fine-grained equivalent of GABBRO. Its essential constituents are calcic plagioclase, generally in the range labradorite-bytownite, and pyroxene. Olivine may not always be present but magnetite and apatite are invariable accessories. Basalts range from undersaturated to oversaturated compositions (*see* silica saturation) but two extremes are recognized, named ALKALI BASALT and tholeiite.

Alkali basalts are undersaturated and contain abundant olivine both as phenocrysts and in the groundmass. The pyroxenes of alkali basalt are calcium-rich augite or titanaugite. Tholeiites are saturated or slightly oversaturated in composition. They contain sufficient silica to partly or completely convert olivine to orthopyroxene and are characterized by this reaction relationship, which is indicated by the mantling of olivine by calcium-poor pyroxene (hypersthene or pigeonite). When present, olivine occurs only as phenocrysts and is absent from the groundmass. Tholeiites typically have intersertal textures and contain silica minerals in the groundmass.

A third type of basalt is now widely recognized, namely high-alumina basalt (or calc-alkali basalt) but mineralogical criteria alone are insufficient to distinguish the different basalt types and chemical criteria are applied. Even so, the three basalt types, alkali basalt, high-alumina basalt, and tholeiite, are gradational in composition. High-alumina basalt is distinguished by an Al_2O_3 content in excess of 17% in aphyric rocks. The $Na_2O + K_2O$ content lies between those of the other two types but mineralogically, high-alumina basalt has a strong affinity with tholeiite. Three volcanic associations corresponding to the three basalt types may be considered.

1. alkali olivine-basalt–mugearite–trachyte/phonolite
2. high-alumina basalt–andesite–dacite–rhyolite (calc-alkaline suite)
3. tholeiite–tholeiitic andesite–rhyolite

Members of the alkali basalt association are found on oceanic islands and continents. Calc-alkaline volcanic rocks are characteristic of island arcs and orogenic belts. Tholeiites occur on the continents, especially along continental margins. Most of the plateau or flood basalts of the world are of tholeiitic composition. The basalts of the ocean floor generated at the MID-OCEAN RIDGES are tholeiites with extremely low K_2O and TiO_2 contents.

basanite A type of olivine BASALT containing augite, plagioclase, and a FELDSPATHOID mineral.

base (in soil science) The neutral or alkaline constituent in the soil, notably calcium, magnesium, sodium, and potassium. Each soil has a theoretical maximum content of bases and the actual content is expressed as a percentage of this theoretical figure to express the degree of base saturation. A high base saturation, especially with calcium, generally promotes a neutral soil, with good structure, aeration, and fertility. A low base saturation represents an acidic soil. Different bases dominate in different environments: calcium is commonest in temperate soils, sodium in saline and alkaline soils of the arid environments.

base construction line The line drawn at right angles to the CENTRAL MERIDIAN of a map projection, from which the other meridians of the map are established.

base level An imaginary line running from SEA LEVEL under the landmass, rising slightly above the horizontal. It constitutes the controlling level down to which, but not below which, a river can cut its valley.

Local base levels may replace sea level as the controlling lower limit of erosion for certain parts of streams, or in certain regions; for example, a dam across a stream, a hard rock band, or a lake will constitute the local base levels for the upstream section of a river. In inland drainage basins, the base level will be the lake to which the streams flow; the Lake Eyre basin in Australia has a local level 14 m below sea level. No river ever actually reaches base level, except at its mouth, because it must retain some gradient in order to flow.

Sea level change, tectonic movement, or the removal of ice pressure from a formerly glaciated area can all change the position of sea level (the ultimate base level) relative to the land. Base level, along with climate and geology, is one of the independent variables that control the processes in a particular cycle of erosion. Changes in base level therefore produce fundamental changes in the geomorphological system, to the extent that a major base level change initiates a new cycle of erosion.

If the land is raised relative to the sea, base level movement is negative; if vice versa, it is positive. The former tends to increase erosional activity, the latter to initiate deposition. *See also* eustasy; rejuvenation.

baseline A line that is measured very accurately as part of a TRIANGULATION scheme. Great accuracy is required because the location of all the points within the triangulation area is based upon this one measured line, although the largest schemes may use two or three baselines.

They are usually measured using long tapes or wires, supported above the ground. Several corrections must be applied to the taped lengths to allow for such factors as slope, thermal conditions, and sag. More recent developments in surveying allow the measurements of long baselines without the need for tapes: both the GEODIMETER and TELLUROMETER provide a rapid means of measuring a baseline very accurately.

base map A map or chart used as a base to which different types of information can be added or overprinted. It usually contains basic information, e.g. major political boundaries, drainage, and coastlines. The base is used when several maps of an area are required, each covering a different topic, such as geology, vegetation, population distributions, etc.

basement The level below which sedimentary rocks do not occur. Rocks below this level are generally igneous or metamorphic.

basic lava A type of dark-colored LAVA containing basic ferromagnesian minerals and less than 50% silica. It flows freely from a volcano's crater or fissures, quickly spreading across the terrain. It becomes BASALT when it solidifies. Weathered basic lava forms a rich soil, much used for farming. *See also* acid lava.

basic rock An igneous rock with a silica content of 45–55% (by weight). Such rocks consist largely of relatively silica-poor minerals such as olivine, pyroxene, and calcic plagioclase feldspar. Typical basic rocks are basalt, dolerite, and gabbro. *See* acid rock; intermediate rock; ultrabasic rock.

basin **1.** A large sediment-filled depression often present in cratonic areas; it may be circular or elliptical.
2. A synclinal structure having a plunged depression, i.e. two directions of plunge at 180° to each other but plunging toward each other.
3. *See* drainage basin.

bastnaesite A yellow to red-brown mineral, $(Ce,La)CO_3(F,OH)$, which occurs in alkaline igneous rocks such as CARBONATITE. It is used as a source of rare-earth elements.

batholith A large body of intrusive igneous rock consisting of several plutons joined at depth and occupying many thousands of square kilometers. They are generally composed of granite material and are associated with mountain belts.

bathyal zone The CONTINENTAL SLOPE zone within the oceans. It embraces depths of some 200 m out to depths of about 1000 m and lies between the relatively shallow NERITIC zone, within which there is greater sediment deposition, and the deep ABYSSAL ZONE. Its upper limit corresponds approximately with the SHELF-EDGE zone and its lower limit is quite arbitrary, because the depth of 1000 m does not mark an abrupt change in topography or other factors. Light reaches only to the top of this zone but animal life is quite abundant and varied. Globally the bathyal environment covers approximately 40 million sq km.

bathyorographical Describing a map showing the height of the land and depth of the sea, usually by layer-coloring.

bathyscaphe A spherical or sausage-shaped device designed to enable investigators to descend into the sea and to observe, through portholes or other observation windows, the marine environment. The first of such devices to achieve useful underwater work were Beebe's bathysphere and Barton's benthoscope. By about 1953, improved bathyscaphes were reaching depths in excess of 3000 m during experimental trials. During 1960, the bathyscaphe *Trieste* landed on the floor of the Mariana Trench (10 911 m). They can be equipped with grab-sampling, coring, and other equipment.

bauxite An aluminum ore formed by the weathering of aluminum-rich, relatively iron- and silicon-poor rocks (mainly syenites) under tropical conditions. The breakdown of aluminosilicate minerals and the removal of silica by leaching leaves a residue composed mainly of boehmite (AlO(OH)), diaspore ($HAlO_2$), and gibbsite ($Al(OH)_3$).

bay An inlet or indentation in the shore of a lake or the sea. Bays are commonly formed by DIFFERENTIAL EROSION, in which softer rocks are worn away faster than the harder rocks surrounding them.

bay bar A barrier of sand or shingle that extends across the entrance to a bay, effectively straightening the coastline. The most likely causes of a bay bar are an enlargement of a single SPIT in one direction or the independent growth, and subsequent coalescence, of two spits that increase their lengths toward each other. Once closed off by a bay bar, the bay will gradually fill up with sediment.

bayou A marshy area of an estuary or where a lake outflows, as occurs typically in the southeastern states of the USA. The water flows only sluggishly or may be stagnant.

beach An accumulation of unconsolidated materials found in the zone of intersection between land and sea. For most practical purposes the beach can be considered to extend from the highest point of severe storm-wave activity (*see* storm beach) down to the point at which waves approaching the coast first start to cause extensive movement of seabed materials. Beach materials are located on a foundation of eroded solid rock (*see* wave-cut platform); they usually form a smooth profile that is gently convex and is sometimes interrupted by BEACH RIDGES. The response of a beach to the activity of waves and other agents of erosion and deposition largely depends upon the nature and thickness of the deposits, sand and shingle behaving very differently.

beach cusp *See* cusp.

beach ridge An upstanding linear accumulation found on shingle beaches (an accumulation developed in sand is known as a BERM). Whenever CONSTRUCTIVE WAVES are active on a shingle beach, a small ridge will be produced at the limit of SWASH for each high tide. As maximum tide level falls from spring to neap level, a series of ridges will be produced; these will be destroyed again on tidal rise to spring level, when just one ridge remains at high-water spring level. Much larger shingle ridges can be produced in connection with severe storms. These develop far above the swash limits of normal waves (*see* storm beach).

beach rock Consolidated and erosion-resistant sand-forms, often reeflike, that develop along some intertidal shores in tropical and subtropical regions. Beach rock has also been found in Portugal, Hawaii, Morocco, and certain temperate regions. Its origin has provoked a great deal of argument and speculation. In general, it appears to develop on account of cementation or lithification processes, the cement being of a calcareous type (often calcite or aragonite). Hence, geologists classify beach rock as a calcarenite. Its rapid consolidation may be aided by the presence of microorganisms such as bacteria and unicellular algae, and certainly by high temperatures. Beach rock, which tends to form as small outcrops, is not always found on all beaches in the same locality. Once formed, however, it tends to resist the action of waves and may act as a barrier to sediment movement.

beaded esker An ESKER comprising alternate wide and narrow segments, which reflect differential input rates of fluvioglacial sediments. The narrow sections were formed during colder periods, when there was less meltwater and hence less sediment, whereas the wider parts were deposited in warmer periods. These segments may mark successive winter and summer accumulations as an ice front (or glacier snout) retreated. Beaded eskers are comparatively rare.

bearing A horizontal angle measured clockwise from a known direction, usually north, to another point. If the initial direc-

BEAUFORT SCALE

			Speed/knots (km/hr)
0	Calm	Smoke rises vertically	<1
1	Light air	Smoke or leaves indicates movement, otherwise almost calm	1–3 (1–5)
2	Light breeze	Wind felt on face, leaves rustle, etc.	4–6 (6–11)
3	Gentle breeze	Flag extended; leaves and twigs in constant motion	7–10 (12–19)
4	Moderate breeze	Small branches moved; dust and litter raised	11–16 (20–28)
5	Fresh breeze	Small trees begin to sway	17–21 (29–38)
6	Strong breeze	Large branches in motion; whistling in telephone wires	22–27 (39–49)
7	Moderate gale	Whole trees in motion; inconvenience experienced in walking	28–33 (50–61)
8	Fresh gale (gale)	Twigs broken off; walking impeded	34–40 (62–74)
9	Strong gale	Slight structural damage experienced	41–47 (75–88)
10	Whole gale (storm)	Widespread damage to trees and buildings	48–55 (89–102)
11	Storm (violent storm)	Very rarely experienced inland, severe damage results	56–63 (103–114)
12–17	Hurricane	Very rarely experienced inland, severe damage results	>64 (>117)

tion is true north then the angles are true bearings. Angles measured with reference to magnetic north, such as those obtained with a prismatic compass, are known as magnetic or compass bearings.

Beaufort scale A scale of wind speed based on easily observable indications such as tree movement and smoke. The scale ranges from 0 to 17 (numbers 13 to 17 were added by the US Weather Bureau in 1955) with the numbers and indicators as shown in the table.

bed (stratum) The smallest division of stratified sedimentary rocks, consisting of a single distinct sheetlike layer of sedimentary material, separated from the beds above and below by relatively well-defined planar surfaces called *bedding planes*, which mark a break in sedimentation. A bed may be part of a MEMBER (*see also* marker bed). *See* lithostratigraphy; stratigraphy.

bedding The parallel layering of SEDIMENTARY ROCKS. The different layers may have different compositions or structures.

bedding plane Any of the planes that separate each of the layers or strata in a sedimentary rock formation. There are generally differences of color, composition, or structure on each side of such a bedding plane.

bedding-plane slip The displacement of one bed of rock over another, taking place along the bedding plane between adjacent beds. It is usually associated with folding, where in extreme cases it can lead to DÉCOLLEMENT.

bed load (bottom load, traction load) The material that is carried along the bed of a stream or the sea by moving water, or along the ground by the wind. It consists of particles that are too large to be transported in suspension (*see* suspended load).

beef Thin beds and veins of fibrous CALCITE. *See* carbonate minerals.

belemnite Any mollusk of the extinct subclass BELEMNOIDEA. Belemnites are important fossils from rocks of the Mesozoic Era.

Belemnoidea An extinct subclass of marine mollusks of the class CEPHALOPODA. With squids and octopuses, they are classified as Coleoidea (or Dibranchiata). They had an internal bullet-shaped shell with simple septa dividing it into chambers; this shell is usually the only part of the animal to be fossilized. Belemnites are known from Lower Carboniferous rocks and extend to the beginning of the Tertiary, but they reached their maximum development in the Jurassic and Cretaceous Periods.

bench mark A mark indicating a point of known position and height, which has been surveyed extremely accurately by the national surveying body of the country concerned.

Benguela Current A sea current that flows northward off the coast of southwest Africa. Part of the region in which it flows is important because of the existence of an UPWELLING of cold water and a zone of marked divergence. The upwelling water comes from subsurface water perhaps only two or three hundred meters deep. The most marked flow in the Benguela Current, which can be up to about 16 million cubic meters per second, occurs between the Cape of Good Hope and latitude 18°S.

Benioff zone The inclined seismic zone within the Earth's lithosphere extending down at an angle of usually around 45° from the base of an ocean trench to the asthenosphere. It is typical of destructive plate boundaries where one plate overrides another as a result of sea-floor spreading. These are zones where ocean floor is consumed. The shallowest earthquake foci along the zone occur near the base of the trench, the foci becoming progressively deeper with distance along the Benioff zone from the trench.

benitoite A silicate mineral of barium and titanium, noted for its strongly dichroic nature. It varies in color from deep sapphire blue to colorless, depending on whether it is viewed by reflected or transmitted light.

benmoreite A basic igneous rock composed of an alkali FELDSPAR, olivine, and pyroxene. *See* trachybasalt.

benthic Describing an organism that lives on the floor of the sea or a lake. *See* benthos.

benthos Plant and animal marine life, both small and large, living within the waters of the sea or on the sea bed (*see also* pelagic (def. 2)). They are a fundamental part of the complex food chains that operate within marine environments. Benthos exist in numerous forms, the greatest variety occurring within the shallow-water environments associated with continental shelves. In the photic (light-penetrating) zone, the benthos feed on both phytoplankton and zooplankton; they are especially well developed in plankton-rich areas and often poorly developed in less fertile areas. The benthos help to make available to larger marine creatures the tiny food particles that they consume or produce. *Vagrant benthos* are capable of active movement on or within the sediment; *sessile benthos* remain attached to the sea floor.

bentonite Clay formed by the alteration and weathering of tuffs and volcanic ash. *See* clay minerals.

Bergeron–Findeisen theory A theory of precipitation formation proposed by Bergeron in 1933 and subsequently modified by Findeisen. It suggested that ice crystals and supercooled water droplets could exist together at certain levels in the clouds where the temperature was between –15° and –30°C. Because the saturation vapor pressure for ice is lower than that for a water surface, there would be preferential deposition of moisture on the ice crystals, causing them to grow sufficiently for snowflakes or raindrops to result.

bergschrund A deep narrow chasm frequently found in CIRQUES in which the upper part of the CIRQUE GLACIER has become separated from the rock headwall. They were formerly believed to be the most favorable sites for FREEZE-THAW activity, which causes headwall retreat. However, recent work has shown that fluctuations of temperature within bergschrunds are infrequently of sufficient magnitude or rapidity to cause any rock breakdown. Only in the upper parts of open bergschrunds, where cold air can circulate freely, will there be any appreciable attack on the rock of the headwall.

berm An accumulation of material found a little way above the mean high-water mark on sand beaches. The berm is usually flat (although it may slope gently landward), is of variable width, and is characterized by a marked break of slope at the seaward edge. Berms are created by the action of CONSTRUCTIVE WAVES but their growth is not a very rapid process. A beach of coarse sand, having a steeper gradient than one of finer material, will develop a berm more rapidly, since wave action is restricted to a narrower width of beach. *See also* beach ridge.

beryl A hexagonal mineral of composition $Be_3Al_2Si_6O_{18}$ found in granitic rocks and PEGMATITES. Beryl is white to pale green in color. The semiprecious gem variety aquamarine is bluish green. The bright green variety is EMERALD, found mainly in metamorphic rocks.

B horizon The second highest layer of soil, or upper subsoil, immediately below the A HORIZON. It has less humus and contains less weathered material, but may contain chemicals washed down from above. *See also* horizon (def. 1).

bifurcation ratio The quantitative relationship between STREAM ORDERS, expressed as the number of streams of one order divided by the number of streams of the next highest order. Values are commonly between 2.5 and 3.5, but are higher in long thin drainage basins dominated by one master stream.

billow clouds Clouds found in a series of regular bands with clear areas between, usually of similar width to that of the cloud. They occur most frequently about 6–8 km, and appear to be due to a strong increase of wind speed with height when the airflow is stable, but the precise mechanism is still in doubt.

bimineralic Describing a rock consisting of only two kinds of minerals, e.g. websterite.

bioclastic rock A sedimentary rock that contains the fragmentary remains of once-living organisms, such as the shell fragments that make up some kinds of limestone. *See also* biogenic rock.

biocoenosis (life assemblage) An assemblage of fossil organisms associated and occurring in the same position as they occupied in life. Some bioherms and reefs approximate to this condition but such assemblages are usually rare, because most organisms suffer damage and transport after death. *Compare* thanatocoenosis. *See also* taphonomy.

biogenic rock A type of rock directly created by living organisms, their remains, or activities. Examples include coal, coral reefs, and limestones composed of the shells of mollusks. *See also* bioclastic rock.

biogeography The scientific study of the distribution of plants and animals around the world. It includes studies of soil (pedology), climate, ecology, and ecosystems.

bioherm A REEF of unstratified limestone formed by organic processes and usually having the shape of a dome rather than a linear feature. Most recent bioherms are produced by CORALS, but in the past important reef-building organisms have included sponges (*see* Porifera) and certain algae. A bioherm may include a hard skeletal structure secreted by the organism itself, the remains of animals that live in the environment it produces, and trapped sediment. They are mainly marine.

biological weathering A mechanism in which rocks are broken down by the action of plants and animals. For example, plant roots can enter cracks in rocks and break the rocks as the roots grow and expand. Bacteria, worms, mollusks, and other invertebrates can also split rocks or wear them away. *See also* weathering.

biomass The total organic matter – plant and animal – in a particular area. It is usually stated as the dried weight per unit area, such as kg m^{-2}. Most of the biomass is derived from plants. Biomass varies widely in different regions of the world, from less than 0.02 kg m^{-2} in polar regions to as much as 45 kg m^{-2} in tropical rainforests.

biome A major ecological community of plants and animals that share the same climate and vegetation. It is the largest such community recognized by ecologists and may include several HABITATS. Examples of biomes include desert, grassland, savanna, taiga, temperate rainforest, tropical rainforest, and tundra.

biosome An accumulation of sediment that has been deposited under constant biological conditions.

biosphere The region of the Earth's crust and atmosphere in which life exists; it contains a number of HABITATS. The biosphere extends from about 3 m below ground to 30 m above it. It also includes aquatic zones extending to about 200 m deep.

biostratigraphy The branch of stratigraphy that utilizes information from fossils in the CALIBRATION of stratigraphic sequences of rock in which they occur. The evolution of organisms is believed to be a continuous unidirectional process and it therefore provides an important means of correlating and comparing separate rock sections. The fundamental biostratigraphic division is the ZONE. Some geologists make no distinction between biostratigraphy and CHRONOSTRATIGRAPHY and include the biostratigraphic zone within the chronostratigraphic hierarchy of terms. Others consider biostratigraphy to be simply one method of calibrating the essentially separate chronostratigraphic scale (*see* chronozone).

biostratonomy The branch of paleontology concerned with the processes by which the remains of organisms become embedded in rock. This is important in interpreting the significance of fossil assemblages, many of which are THANATOCOENOSES. *See* actuopaleontology; taphonomy.

biostrome A structure similar to a BIOHERM but not swelling into a mound or lenslike body. It more closely resembles a broad sheet of sediment consisting of a large quantity of organic remains.

biotic Describing the living factors in an ECOSYSTEM. They result from the activities of animals and plants, and include competition and feeding. *See also* abiotic.

biotite One of the major forms of MICA which forms dark brown or black shiny slabs that split easily into transparent flakes. It is found widely in many kinds of igneous, metamorphic, and even sedimentary rocks.

bioturbation The reworking and further degradation of sediment by the action of organisms moving through it and feeding on it.

birds *See* Aves.

bird's-foot delta An elongated form of DELTA, with sediment deposited in a finger-like pattern following the courses of the various distributary streams.

birefringence The double refraction of light entering an anisotropic crystal, when it is split into two plane-polarized rays, the ordinary and extraordinary rays. The two rays vibrate in mutually perpendicular planes and travel through the crystal in different directions and with different velocities (there is a fast ray and a slow ray) and thus have different refractive indices. This may be illustrated by the well-known experiment of viewing a single spot on a piece of paper through a piece of Iceland spar. Two images of the spot are seen representing the ordinary and extraordinary rays.

When an anisotropic crystal is viewed under a petrological (polarizing) microscope, bright polarization colors are seen and the crystal is said to exhibit birefringence. The two rays produced by the mineral are recombined in the microscope and their optical interference produces the polarization colors. These colors are of great use in the identification of minerals.

bittern The brine that remains when useful more soluble minerals (such as carnallite and polyhalite) have been removed from sea water by evaporation. Bittern can itself be used as a source of bromides and iodides.

bitumen A semisolid or solid flammable substance that consists mainly of hydrocarbons, particularly those occurring in PETROLEUM. *See also* asphalt.

bituminous coal A black or dark brown COAL with a high carbon content, the type most commonly used as fuel. It burns with a yellowish smoky flame and is used for making degassed coal (smokeless fuel) and coke. *See also* anthracite; lignite.

Bivalvia (Lamellibranchia, Pelecypoda) A class of the phylum MOLLUSCA whose members are characterized by a two-valved shell protecting the body. The valves are secreted by the mantle and meet at a hinge line, where they are joined by an elastic ligament. The hinge is aligned and strengthened by a system of interlocking teeth and sockets, which are used in the taxonomic classification of the group. The shell has growth lines and often ornamentation. The animals are bilaterally symmetrical, with the plane of symmetry passing between the valves, unlike the BRACHIOPODA, which are symmetrical across the valves. Some bivalves have secondarily lost this symmetry and the valves are unequal.

Bivalves are aquatic (mostly marine), occupying a great variety of habitats, and include free-swimming, sessile, and burrowing forms. They range from the Cambrian Period to the present day and have been used biostratigraphically as ZONE FOSSILS, particularly in the Upper Carboniferous and Tertiary.

black alkali soil *See* solonetz.

blackband ironstone A SEDIMENTARY ROCK, containing iron, that occurs in association with coal deposits. It contains iron carbonate (SIDERITE) and is used as an iron ore.

black-body radiation A perfect black body is one that absorbs all radiation falling upon it and emits, for any temperature, the maximum amount of radiant energy. No real substance achieves this degree of physical perfection, but the inside of a sphere coated with carbon black approaches it. Some substances can act as black bodies for certain wavelengths only. An extreme example of this is snow, which has a high albedo for visible light, but effectively acts as a black body for wavelengths greater than 1.5 micrometers.

black earth *See* chernozem.

black ice (glazed frost) A smooth transparent thin coating of ice formed on asphalted road or other surfaces, having a black almost invisible appearance. It is formed when drizzle or light rain falls onto a surface that is at a temperature below 0°C.

black sand A type of sand found on beaches or in alluvial deposits that contains dark heavy minerals such as ILMENITE and MAGNETITE (containing iron) and RUTILE (containing titanium).

black smoker A jet of very hot water that rises from a vent in the ocean floor, colored black by dissolved sulfides of copper, iron, manganese, and zinc. Most smokers are associated with MID-OCEAN RIDGES. The minerals may be deposited from solution to form a tall sulfide-covered chimney round the vent.

blanket bog *See* organic soil.

blende Any sulfide mineral, especially one used as an ore. The best known is zinc blende (*see* sphalerite).

blizzard Heavy snow falling or drifting during strong winds.

block-faulting A series of normal faults that separate an area into a series of horsts and grabens. *See* fault.

blocking The situation arising when anomalous pressure patterns develop in the zone of the mid-latitude westerlies. It more commonly refers to anticyclone formations that act as a block to the normal depression tracks, which then move toward the northeast and southeast around the anticyclone to give spells of quiet and dry weather near the blocking high. The two favored sites for the location of blocking highs are over NW Europe and the NW Pacific. The precise reasons for these positions are not understood but they are believed to be due to the interaction between the westerly circulation and the north-south oriented mountain ranges such as the Rockies.

block lava A LAVA the surface of which is broken into large retangular blocks. *Compare* aa; pahoehoe.

block mountain A steep-sided mountain or plateau that has been uplifted between a pair of parallel FAULTS.

block stream A pile of rock debris that has accumulated at the head of a ravine, probably as a result of glacial action. It is also called a rock stream.

blowhole A vertical fissure in the roof of a sea cave or cliff through which sea water is forced as a jet at high tide. It is formed when wave action causes erosion along a joint in the rock.

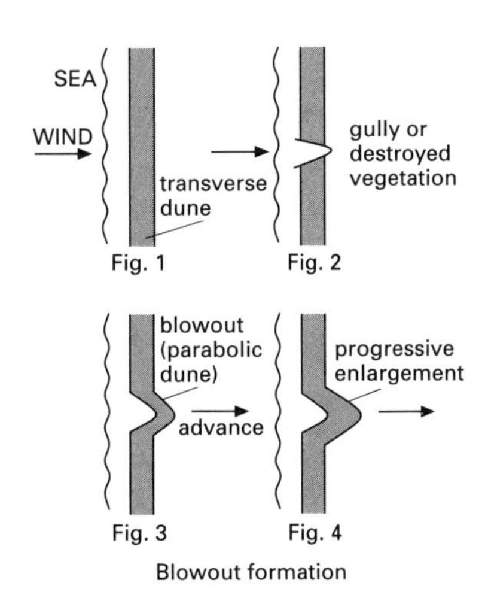

Blowout formation

blowout A break in the crest of a sand dune, caused by vegetation destruction or gullying. It leads to increased sand mobility and wind speed at the break point, with the sand being blown downwind as a parabolic dune bowing out from the transverse line. By the time stage 3 is reached, unless practical measures are taken to break wind speed and reinstate the vegetation, the blowout becomes progressively enlarged by advancing downwind and widens itself through undercutting of the vegetation on either side.

blue ground *See* kimberlite.

Blue John *See* fluorite.

bluff (river cliff) A steeply sloped river bank on the outside of a MEANDER. It is caused by erosion by the faster-flowing water on the outside of the bend. It may be as much as 100 m in height. Any other headland or inland cliff may also be called a bluff.

body wave A seismic wave, generated either by an earthquake or explosion, which reaches a recording station after having traveled through the Earth's interior. Body waves include primary (P or longitudinal) and secondary (S or shear) waves.

bog A waterlogged area of land, resulting from poor drainage, in which vegetation becomes partly decomposed. The ground is spongy and wet, and eventually the vegetation forms an acid PEAT. *See also* organic soil.

bog soil *See* organic soil.

bole Thin red weathered horizons formed between basalt lava flows. The presence of bole indicates subaerial extrusion and a tropical climate. *See* laterite.

bolson **See** inland basin.

bomb (volcanic bomb) *See* pyroclastic rock.

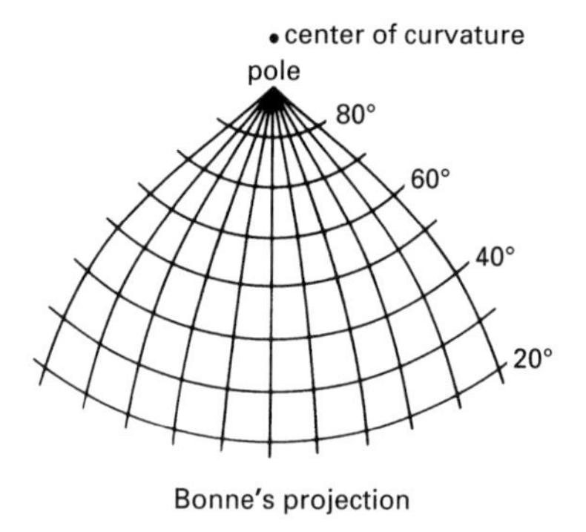

Bonne's projection

Bonne's projection A type of conical map projection, with the central meridian straight and truly divided, the parallels all concentric circles, and the other meridians composite curves joining divisions on the correctly divided parallels. The scale along the parallels is kept true. The shapes of landmasses are distorted with distance from the central meridian but the projection is equal-area. *See* conical projection.

bora A cold, dry, and gusty northeasterly wind that sweeps through the mountain gaps of the W Balkan Peninsula to the Adriatic coast. It is strongest in winter, with mean speeds of about 50 km/hr, and is the result of high pressure over central Europe and a deep depression to the southwest. The presence of the mountains helps the funneling of the wind and its rapid descent prevents adiabatic warming. *See also* mistral.

borax A mineral form of hydrated sodium borate, $Na_2B_4O_7.10H_2O$. It occurs as deposits and on the surface of the ground in dry regions, formed by the evaporation of alkaline lakes or hot springs. The transparent crystals can lose water of crystallization to become tincalconite, a chalky white mineral. It is used as a source of boron.

bore A tidal phenomenon that occurs in certain rivers or estuaries. The tidal range is usually fairly large and may lead to a tidal flood wave with a high sharply developed front. The frontal area is often characterized by subsidiary waves or furrows. With the initial flooding of the tide, a surging inflow may develop and travel rapidly some kilometers upstream. Well-known bores occur in the Amazon, the Ch'ien t'ang Kiang River in China, the Hooghly River in India, and the Bay of Fundy. The large bore that travels up the Ch'ien t'ang Kiang River has a fast-moving abrupt front up to 4 m in height, necessitating the prior removal of boats from the river.

Boreal climate A climatic type in Köppen's classification system characterized by cold snowy winters and warm summers, giving a large annual range of temperature. Precipitation totals are small. It is found over the American, European, and Asian continents between latitudes 40°N and

65°N with the S boundary being farther north on the warmer W coasts and trending WNW–ESE toward the colder continental interiors.

Boreal forest *See* taiga.

Boreal period The period from 7500 BC to 5000 BC, which was characterized by an improving climate following the deglaciation of the last Ice Age. It was named Boreal because the vegetation at this time in NW Europe, where the first investigations were conducted, was predominantly coniferous as in the Boreal areas at present.

borer A marine borer is an organism that is capable of performing marine erosion, usually by a boring process. Such erosion may be limited to fairly local areas but is significant not only because of the way in which rocks become riddled with holes but also because many boulders, smaller stones, shells, and flakes of rock may be affected in this way. Particularly susceptible are soft limestones and sandstones. Borers abound in the sea and are particularly active in the shallow littoral zone. Boring algae are effective, and certain sponges attack rocks; some worms, chiefly the annelids, together with a variety of mollusks, excavate pits and tubes in rocks. Crustaceans and fish such as the eel often enlarge the cavities they tend to occupy. Serious damage may even be caused to timber and concrete structures, for example groins.

bornhardt A dome-shaped mound of rock; a large INSELBERG.

bornite (peacock ore) A red-brown mineral form of copper iron sulfide, Cu_5FeS_4, with an iridescent tarnish. It occurs associated with other copper ores.

borolonite A nepheline-containing type of SYENITE.

boss A roughly circular igneous intrusion that has a diameter of less than 25 km. *Compare* stock.

botryoidal Describing a mineral consisting of spheroidal aggregates resembling a bunch of grapes. *Compare* reniform.

bottom current A current that flows close to the sea floor or within the lowermost layers of water in deep-sea areas. The term has also been used to describe the currents over shelf areas that arise owing to density-driven flow, tidal flow, or wave-induced flow. In the case of deep-water bottom currents, flow velocities have been calculated theoretically, others have been measured using current meters or special floats. Bottom flow in part of the Antarctic bottom current can attain 3 cm/sec but is usually much less. The speed of the bottom current on the W flank of the Atlantic Ocean is, theoretically at least, between 0.5 to 2 cm/sec. Bottom topography may significantly hinder the flow of bottom water, for example the transverse ridges in the South Atlantic, which the Antarctic bottom current has to negotiate. Deep gashes or gaps through such features are important in channeling bottom flow. Some sub-Arctic bottom water has its progress southward off the Labrador coast partly blocked by the Labrador submarine rise.

bottomset bed *See* delta deposit.

bottom water The lowermost layer of water in the sea, particularly the water masses that move slowly in deep ocean areas. Bottom water is relatively cold and dense. Most of the North Atlantic bottom water has a temperature around 1–2°C. The Polar Antarctic bottom water originates in the surface layers of the Antarctic and later flows as a deep-water layer. During the fall and early winter, the very cold shelf water sinks owing to convection, and slides down the continental slope to the abyssal depths. This happens especially in the Weddell Sea area. The deep water then spreads outward, especially eastward into the Southern Indian Ocean, although influenced by bottom topography. In contrast, the bulk of North Atlantic bottom water originates outside the Arctic.

boudinage When a competent bed of rock is enclosed by incompetent material and is then stretched or squeezed, the competent bed breaks in a series of short sections, because it cannot deform plastically to the same extent as the incompetent material. As the competent layer deforms, it thins locally and PINCH AND SWELL structures develop. On further stretching this thinning continues until necks are formed; finally the layer breaks. They were originally described as boudins by French geologists, to whom they resembled black sausages.

Bouguer anomaly A gravity anomaly that takes into account the effect of topography but not isostatic compensation, first observed by P. Bouguer in 1735.

boulder clay *See* till.

boundary current A fairly fast-moving current that flows near the edge of an ocean, usually along its W flank. Theoretical models of oceanic flow, particularly that of Stommel, reveal that there should be a fairly localized current flowing at depth, in a southward direction, along the W flank of the Atlantic. The necessity for such currents arises from a consideration of pressure gradients within the oceans and the forces associated with Earth rotation. Stommel's theory suggests that there should also be a current along the W flank of the Pacific Ocean, although not in this case everywhere directed southward. Field measurements have demonstrated that in the case of the Atlantic, there is a south-flowing current underneath the Gulf Stream.

boundary layer The layer of the atmosphere in which movement is closely determined by the presence of the ground surface. In practice, this layer is subdivided into the *surface boundary layer*, extending to approximately 100 m, and the *planetary boundary layer*, with an upper limit of about 600 m above the ground. In the former, the influence of the surface is paramount, whereas in the latter layer, it remains significant but not dominant.

boundary wave *See* internal wave.

bourne A stream, especially one in chalk districts such as S England. When the water table falls in summer, many bournes dry up.

Bowen ratio The ratio of the transfer of sensible heat from the ground surface to that of latent heat effected by the processes of turbulence and conduction. Over oceans, the value of this ratio can be as low as 0.1 indicating that 90% of the heat flux from the surface is in the form of latent heat. Desert areas represent the converse, with no latent heat component because all heat transfer is in the form of sensible heat.

boxwork The reticulated outward appearance of the ceiling of a cave, resulting from the deposition of hydrated iron oxides in fractures and cavities (from which material has been eroded away).

Brachiopoda A phylum of invertebrate animals having a body enclosed by two valves of a chitinous or calcareous material. Brachiopods are bilaterally symmetrical, but unlike the BIVALVIA the plane of symmetry passes across the valves, not between them. The valves are thus dissimilar but each is symmetrical about its left and right halves. The phylum is divided into two classes, the Inarticulata and the Articulata. In the latter the shells are connected by teeth and sockets along a hinge line. The shells of the Inarticulata are held together only by muscles. Brachiopods are marine benthonic animals attached to the substrate by a *pedicle*, which commonly protrudes through an opening in the larger valve (the *pedicle valve*). The other valve is called the *brachial valve*. It has an internal *brachidium*, which supports the structure responsible for maintaining a current of water for respiratory and feeding purposes. Brachiopods show variations in shell morphology and ornamentation that are valuable in taxonomic classification.

The earliest fossil brachiopods are known from Lower Cambrian rocks; they flourished throughout the Paleozoic Era and were especially abundant and diverse in the Silurian and Devonian Periods. Many became extinct at the end of the Paleozoic. There was a revival of a few groups in the Mesozoic but brachiopods are relatively insignificant today. They have been used in the stratigraphic classification of the rocks in which they occur.

brackish water Water that is partly fresh and partly saline. Areas of such water occur, for instance, behind coastal barriers that partly inhibit tidal penetration, hence the incursion of saline water, but still allow the incursion of fresh water from streams and rivers.

braided stream A stream (or river) that divides into two or more smaller channels, which further branch and separate before rejoining farther downstream. The result is a network of islands, sandbars, and channels. It occurs when the volume of flow of the stream varies widely from time to time (as in dry regions) and there is plenty of BED LOAD. It also occurs in front of a melting glacier. *See also* braiding.

braiding The division of a stream into a complex pattern of several small channels divided by mid-channel bars. After straight and meandering channels, braiding is the third major channel pattern a river can have. It is often due to lack of COMPETENCE or CAPACITY of the river to carry the load supplied to it. As a result the excess load is dumped in mid-channel to form the characteristic mid-channel bars that divide the channel into many separate subchannels. In this form, it is often seen in the periglacial environment (*see* river terrace). Easily erodible banks and variable discharge are also said to favor braiding. Meandering patterns may change to braided patterns if the stream gradient or discharge is increased.

breaker A mass of turbulent water moving landward, as a result of the breaking of a wave on approaching the coast and reaching shallower water. The decrease in water depth causes a reduction in velocity of the wave, an increase in height, and a decrease in length. The orbital paths of water particles within the wave form also change, from open circles to open ellipses, until the water under the crest is moving faster than the wave form, at which point the wave breaks. Whether a breaker results in net accumulation or erosion of beach materials depends largely upon the nature of the breaking. The crest of a *plunging breaker* falls vertically into the trough, producing little SWASH, but considerable BACKWASH. *Spilling breakers*, however, break forward and the associated stronger swash aids constructive activity. *See also* constructive wave; destructive wave.

breaker zone *See* surf zone.

breccia A RUDACEOUS sedimentary rock in which the constituent clasts or fragments are angular. *Compare* conglomerate.

brickearth Fine-grained unconsolidated deposits (sand, silt, and clay) associated with periglacial conditions in the Pleistocene. There are two types of brickearth. Aeolian loess consists of very well sorted material of predominantly silt size, picked up from glacial outwash, frost-shattered debris, or the floodplains of rivers, and transported by cold dry winds in steppelike conditions. The deposit is largely quartz, but with a mineralogy reflecting the rocks of origin, and forms a sheet over the preexisting landscape when deposited in a wedge that thins away from the source. It is commonest in E Europe, but thin veneers reach E and SE England.

Overbank material deposited on floodplains by Pleistocene rivers is also termed brickearth. This type has a lower silt content than the wind-lain type, and is less well sorted.

brine A solution of common salt (sodium chloride, NaCl) as occurs in sea water and in salt lakes. Naturally occurring brines are important sources of salts and other chemicals.

brittle failure The breaking of a rock while still within the elastic range.

brittle strength The stress applied to a rock at the point at which it breaks within the elastic range.

broad A small shallow lake in the East Anglia district of E England, formed on the site of peat extraction several centuries ago. Broads are prone to silting up and pollution.

bronzite An orthorhombic PYROXENE.

bronzitite A monomineralic ultramafic rock consisting wholly of bronzite.

brookite A brown to black orthorhombic polymorph of titanium dioxide, TiO_2, found in hydrothermal vein deposits. *See also* anatase; rutile.

brown calcareous soil A type of calcimorphic soil, typically found in the humid temperate regions of Europe and America under base-rich vegetation. Soil development is more advanced than in a rendzina, producing a soil with an A (B) C system of horizons. The soil is deeper, though rarely more than 70 cm, owing to its development over a parent material with a greater insoluble residue. The upper part of the soil has a mull humus and is dark red/brown in color with a neutral or slightly acid reaction. It has a crumb structure and is rich in organic matter. With depth the A horizon merges into an ill-defined B horizon, which is often lighter in color and contains many fragments of parent material. They are good agricultural soils and fall into the MOLLISOL order of the Seventh Approximation.

brown clay *See* red clay.

brown coal *See* lignite.

brown earth (brown forest soil) A type of soil ranging from a true brown earth, which exhibits only mild leaching, to a podzol, which exhibits extreme leaching. Brown earths are found to the south of the pozdol zone where precipitation is in excess of evaporation. True brown earths lack the distinctive horizonation of a podzol. Organic matter is rapidly formed owing to the deciduous forest vegetation being decomposed by the more abundant soil fauna. There is a thorough mixing of the mull humus into the A horizon. Carbonates are completely leached from the soil but there is no movement of the sesquioxides, as is shown by the constant silica:sesquioxide ratio down the profile. Consequently there is a weakly developed B horizon. These soils cover large areas of the middle latitudes, often forming on deposits of the Pleistocene glaciations. With their good crumb structure, mild acidity, and free drainage they are important agriculturally. *See also* acid brown soil; brown podzolic soil; sol lessivé.

brown podzolic soil A soil that is transitional in location and properties between brown earths and podzols. An Ea (eluvial) horizon is still lacking but they are more acid in nature and have a lower base status than a brown earth. An increase in free iron oxide results in a loose crumb structured B horizon. These soils are typical of the northeastern USA where they are found to the south of the true podzol zone. They fall into the SPODOSOL order of the Seventh Approximation.

brucite A white or greenish mineral with a layered structure consisting of magnesium hydroxide, $Mg(OH)_2$. It commonly occurs as a hydration product of periclase in thermally metamorphosed dolomites.

Brückner cycle A climatic cycle of about 35 years, which was deduced by Brückner in 1890 on the basis of non-instrumental data, such as harvests. He distinguished alternating warm and dry spells and cold and wet spells, but the cycle is not well defined and has too small an amplitude to have any forecasting value.

brunizem *See* prairie soil.

Bryozoa (Ectoprocta) A phylum of small aquatic colonial invertebrate animals including the moss animals and sea mats. A bryozoan colony, known as a *zooarium*, is made up of numerous polyps, each of which inhabits a small tube of calcareous or chitinous material. The colonies show great diversity in shape: many resemble seaweeds and are attached to the sea floor; others are encrusted on the hard parts of other animals or rocks. Fossils of the Bryozoa are known from rocks of the Cambrian Period onward and the group is still flourishing today. Bryozoans have a potential use in micropaleontological stratigraphy and correlation.

Buchan spell A period of the year when temperatures were anomalously cool or warm as deduced by Buchan working on 50 years of records prior to 1867 for SE Scotland. The periods of below-average temperature were Feb. 7–14, April 11–14, May 9–14, June 29–July 4, Aug. 6–11, and Nov. 6–13, and of above-average temperatures July 12–15, Aug. 12–15, and Dec. 3–14. Although there is some evidence of such periodicities from year to year they are not sufficiently persistent or clearly defined to specific dates and so Buchan's spells are largely of climatological curiosity only.

Buchan zone *See* zone (def. 1).

bulk modulus The ratio of compressive stress in rocks to the resulting change in volume.

buoyancy (in meteorology) The temperature difference between parcels of warm air and the surrounding cooler environment. Because air density is inversely proportional to air temperature, parcels or thermals of warm air have an upward acceleration through the cooler surroundings just as a cork rises through water because of its lower density. Where parts of the Earth's surface are heated differentially, the warmer parts will heat the air in contact, which will then possess buoyancy relative to its surroundings. This is the basis of natural convection within the atmosphere.

Burgess shale A bed of black Middle-Cambrian shale in British Columbia, Canada. It is the site of significant discoveries of the FOSSILS of invertebrate animals, which occur as films of carbon between the BEDDING PLANES of the rock.

buried topography A preexisting landscape that has been subsequently buried by younger strata.

bush An area of scrub in an isolated wild landscape, as occurs in parts of Australia, New Zealand, southern Africa, and the USA.

butte A small upstanding mass of rock, usually consisting of resistant capping material overlying some softer though protected rock type. It has steep sides and is produced as a result of long-continued back-wearing of a MESA.

butterfly effect Any effect in which a small change to a system results in a disproportionately large disturbance. The term comes from the idea that the Earth's atmosphere is so sensitive to initial conditions that a butterfly flapping its wings in one part of the world may be the cause of a tornado in another part of the world. *See* chaos theory.

Buys Ballot's law A law stating that if an observer stands with his or her back to the wind, then pressure will be lower on the left-hand side than to the right in the N hemisphere. For the S hemisphere the converse is true. This means that winds will blow in a counterclockwise manner around a depression in the N hemisphere, and vice versa in the S hemisphere.

bysmalith A steep-sided vertical igneous body, roughly cylindrical in form and following a steeply inclined fault. It arches up the overlying country rock or becomes exposed at the surface.

bytownite A calcic plagioclase FELDSPAR.

C

caatinga A type of thorny scrub that occurs in northeastern Brazil. The plants include acacias, cacti, and other drought-resistant vegetation (there is very little rainfall in winter).

cadastral map Usually a large-scale map, showing the boundaries of subdivisions of land. It is used for recording ownership of land.

Cainozoic *See* Cenozoic.

calamine **1.** *See* hemimorphite.
2. In the UK, another name for SMITHSONITE.

calc-alkaline Describing igneous rocks that have relatively lower sodium and potassium contents and higher calcium contents for a given silica percentage than do ALKALI rocks. The calc-alkaline volcanic suite is represented by the association of basalt, andesite, and rhyolite in orogenic regions and island arcs.

calcarenite A clastic LIMESTONE of average grain size between that of CALCISILTITE and CALCIRUDITE. It is composed of particles between 0.06 mm and 2.0 mm in diameter. Calcarenites are subdivided into: very fine, between 0.06 and 0.12 mm; fine, from 0.12 mm to 0.25 mm; medium, from 0.25 mm to 0.5 mm; coarse, from 0.5 mm to 1.0 mm; very coarse, having diameters from 1.0 to 2.0 mm.

calcareous Describing rocks or soils that contain CALCIUM CARBONATE. For example, limestone and chalk are calcareous rocks.

calcicole (calciphile) A plant that grows best on chalky (i.e. alkaline) soils. See also calcifuge.

calcifuge (acidophile) A plant that grows best on acid (i.e. lime-free) soils. See also calcicole.

calcilutite A clastic limestone of average grain size less than that of CALCISILTITES, i.e. less than 0.004 mm in diameter.

calcimorphic soil An alkaline soil occurring within broad climatic zones where parent material is the dominant factor in soil formation. The parent material is calcareous and its extreme nature resists any development toward acidity. Such soils are usually dark, organic-rich, and abundant in soil fauna. Two main soil types can be recognized belonging to this group: the rendzina and the brown calcareous soil.

calcirudite A clastic limestone of average grain size larger than that of CALCARENITES, i.e. greater than 2 mm in diameter.

calcisiltite A clastic limestone of a grain size intermediate between that of CALCILUTITES and CALCARENITES. It is formed of particles between 0.004 mm and 0.06 mm in diameter. Those of grades up to 0.03 mm are fine calcisiltites; those above this are coarse calcisiltites.

calcite The principal and most stable mineral form of calcium carbonate, $CaCO_3$. It is the chief component of limestone and marble, and occurs in the shells of mollusks. It forms colorless, white, or gray hexagonal crystals, sometimes colored by impurities. *See also* carbonate minerals.

calcium carbonate (lime) (in oceanography) Calcium carbonate is present in seawater in a dissolved state and is also locked up in the skeletons and shells of marine organisms. The factors that determine the amount of lime dissolved in seawater are numerous and quite complex, and the theories relating to lime solution have not as yet been altogether satisfactorily explained. Much of the lime in seawater originates from river discharge. An analysis of marine deposits shows that, in general, neritic sediments tend to have a far lower lime content than pelagic deposits. Also, in general, lime present in surface waters of the sea is of low concentration, perhaps because of its consumption by minute plankton, whereas average lime concentrations near the sea floor increase markedly. For example, sediments in the Atlantic Ocean that lie below 3000 m are very lime-rich. However, while the low lime content at the sea surface is very marked at low latitudes, it is far less so at high latitudes. *See also* carbonate minerals.

calcrete (caliche) Crusts and nodules of LIMESTONE precipitated at or near the surface of the ground in semiarid regions. It results from evaporation of moisture from the soil and may be associated with deposits of gravel and sand

caldera A volcanic crater whose diameter exceeds one kilometer and can reach up to 20 km, generally resulting from the collapse or explosive removal of the top of a volcano. *See also* supervolcano.

Caledonian orogeny The Lower Paleozoic orogeny in which the Caledonian mountains, extending from Ireland via Scotland to Scandinavia, were formed. It resulted from the closure of the Proto-Atlantic Ocean, between the Baltic and Canadian Shields.

calibration (in stratigraphy) The determination of a rock sequence in relation to an independent timescale. This scale may be determined radiometrically or by correlation with another continuously operating undirectional non-reversible process such as EVOLUTION, using the fossils contained in the rocks.

caliche *See* calcrete.

caliper log A subsurface logging technique, which records the variations with depth in the diameter of an uncased bore hole.

calorie A unit of heat now superseded by the joule. It was based on the amount of heat required to raise the temperature of 1 gram of water by 1°C. Because this amount depended on the initial water temperature, the standard calorie was taken as that at 15°C, raising the water temperature from 14.5°C to 15.5°C. 1 cal (15°C) = 4.1855 joules.

calving The production of icebergs by the splitting off of large slabs of ice from a glacier or ice sheet at the edge of the sea.

cambering An apparently increased dip toward valley bottoms found within horizontally bedded solid rocks located above weaker unconsolidated rocks, such as clays. No clear explanation has yet been discovered, for this phenomenon.

Cambrian The earliest period of PHANEROZOIC time and of the PALEOZOIC Era. Rocks laid down during this period are the first to show an abundance of fossils, which consist of primitive representatives of most of the invertebrate animal phyla known today. The Cambrian began about 570 million years ago, following the PRECAMBRIAN, and was succeeded about 505 million years ago by the ORDOVICIAN. The Cambrian System is often divided into Early, Middle, and Late. The name Cambrian is derived from the ancient name for Wales (Cambria), where rocks containing the earliest fossils were first studied. Cambrian rocks occur across all continents, the most complete being in North America and Siberia, and are predominantly sedimentary in origin. Many Cambrian rocks show evidence of deposition in or near shallow areas during marine transgressions into continental areas, and oceans extended over most of what is now North America during the period.

Cambrian fossils represent the animals that lived in the seas of this period. Trilobites were especially abundant and are used in the stratigraphic subdivision of the system. Other important groups include the brachiopods (inarticulate species being dominant), gastropods, primitive echinoderms, and ostracods and there is evidence from TRACE FOSSILS of a variety of worms. Bivalves and graptolites had appeared by the end of the period.

Campbell-Stokes recorder *See* sunshine recorder.

camptonite An alkaline LAMPROPHYRE.

Canadian Shield A large area of Precambrian rock that occupies 5 million sq km of Canada. Most of the shield consists of GRANITE and banded GNEISS; the remainder includes volcanic rock and some sedimentary deposits. There are extensive deposits of metal-bearing minerals, including those of copper, gold, iron, nickel, and silver.

cancrinite *See* feldspathoids.

cannel coal A fine-grained lustrous bituminous coal which burns with a bright smoky flame; the common fuel coal. Geologically, it is an example of a carbonaceous rock. *See also* coal.

canyon A deep steep-sided section of a river valley, the depth of which considerably exceeds its width, normally found in arid or semiarid regions. Canyons are usually produced where rivers have been deeply incised owing to ANTECEDENCE, or as a result of downcutting by rivers whose sources are located in areas of greater precipitation than those in which the canyons are formed. This continuous external source of water enables efficient downward erosion to proceed, while the local lack of precipitation hinders weathering of the sides and hence their degradation. Many canyons are formed by erosion of horizontally bedded rock alternations, which produce a stepped valley cross-profile, owing to their differential resistances to erosion. Others are said to have been formed by cavern collapse or by the frozen ground and blocked valley phenomena of the Pleistocene. *See also* submarine canyon.

capacity **1.** (in geology) The maximum amount of sediment of a certain size that a stream can carry as bedload. This decreases as the grain size of the sediment becomes larger, but increases as stream gradient becomes steeper or discharge becomes greater. Since the transporting surface is the stream bed, the wider it is the more it can transport; to cancel this out capacity is usually expressed as a weight of sediment per unit width of bed. Even so, capacity is still very much a function of bed width, since for a given slope and discharge, the velocity at the bed is greater in a wide shallow stream than a narrow deep one. If the grain sizes being transported are a fair mixture of large and small, capacity varies with the third power of the velocity, with a lower power if the material is mostly coarse, and with a higher power if it is mostly fine.
2. (carrying capacity) (in ecology) The maximum BIOMASS a region can support, or the maximum number of animals it can sustain during the harshest time of the year. If the capacity is exceeded, there will be insufficient resources, such as food, for the population. The number of animals must fall, through starvation, failure to reproduce, or migration.

capillarity (in soil science) The mechanism whereby capillary water moves vertically up the soil profile from the groundwater table or moist subsoil; it is a process typical of arid and semiarid zones where evaporation of water exceeds precipitation. It is more effective on claylike than sandy soils, with a maximum height of rise of 2.5 m in clays, 0.7 m in sands. As the water rises it brings with it dissolved salts and these are precipitated at the point where the capillary "current" finally dries out, which may be at the soil surface or within the profile, forming a salt accumulation, e.g. in SOLONETZ and SOLONCHAK soils. In arid zones subject to irrigation, excessive watering artificially accelerates the capillary currents, and formerly fertile areas can be ruined by the vast salt deposits that result.

capillary water (in soil science) Water held in the small pores within the soil, existing as a film around soil particles. It is this water that is mostly taken up by plant roots for plant growth, as opposed to the gravitational water that rapidly flows through the soil and removes plant nutrients during wet periods, and hygroscopic water, which is the water that remains in the soil even after air drying and is not available to plants.

capillary wave A water wave whose length is less than 2.5 cm. Its speed of propagation is largely determined by the measure of surface tension that exists in the water, a force that tends to restore the water surface to a horizontal position. As with all sea waves, the generation of capillary waves requires the transference of energy from wind flow to the surface water in the sea. Capillary waves may develop, for instance, from the action of dropping a tiny stone into a pond.

cap rock 1. A layer of SHALE or other impervious rock that overlies porous rocks containing deposits of oil or natural gas.
2. A layer of anhydrite, gypsum, calcite, and sulfur that forms a hard covering on top of a SALT DOME. In the Gulf Coast region of the USA, cap rock is a major source of sulfur.

carbonaceous chondrite A stony METEORITE that contains CHONDRULES in a claylike matrix of silicates. These meteorites get their name because of their carbon content – up to 3% mainly in the form of hydrocarbons. Some experts believe that they resemble the original material from which planets were formed.

carbonate minerals The anion $(CO_3)^{2-}$ is the fundamental unit in the structure of the carbonates. The common rock-forming carbonate minerals fall into three subgroups as follows:

1. *calcite* – $CaCO_3$, *magnesite* – $MgCO_3$, *siderite* – $FeCO_3$, *rhodocrosite* – $MnCO_3$.
2. *dolomite* – $CaMg(CO_3)_2$, *ankerite* – $CaFe(CO_3)_2$.
3. *aragonite* – $CaCO_3$, *strontianite* – $SrCO_3$, *witherite* – $BaCO_3$.

The calcite and dolomite minerals have trigonal symmetry. Dolomite results from the substitution of calcium ions in calcite by the divalent cations Mg^{2+} and Fe^{2+}, and there is continuous replacement of Mg^{2+} by Fe^{2+} between dolomite and ankerite. Calcite and aragonite are polymorphs of calcium carbonate, the latter being orthorhombic and the higher-pressure form. Calcite is usually colorless or white with a hardness 3 (*see* Mohs' scale) and cleaves into perfect rhombs. Its extreme double refraction (*see* birefringence) is apparent in crystals of the transparent variety Iceland spar. Calcite effervesces strongly in cold dilute hydrochloric acid, other carbonates reacting weakly unless the acid is warmed. Most limestones consist largely of calcite, which may be a primary precipitate or in the form of fossil shells. Calcite also occurs as a secondary cementing material in sediments. Veins and beds of fibrous calcite are called *beef*. During metamorphism, a pure limestone recrystallizes to form *marble* but if impure calcite reacts it produces such minerals as *diopside*, *wollastonite*, and *grossular garnet*. Calcite occurs in hydrothermal veins, in amygdales, and as a primary magmatic mineral in carbonatites.

Magnesite is usually white or colorless and occurs as an alteration product of magnesium-rich rocks under conditions in which carbon dioxide is available. Rhodocrosite is pink and is found in metasomatic veins and pegmatites. Siderite is usually brown and is a major constituent of bedded ironstones, an important iron ore. It is also found in hydrothermal veins.

Dolomite is colorless, white, or gray and may form as a primary precipitate. Dolomitization of calcite and aragonite takes place by reaction with magnesium-bearing sea water or by the permeation of magnesium-rich solutions along cracks and joints in limestones. During metamorphism, dolomite breaks down to produce periclase and brucite. Yellow-brown ankerite is found with dolomite in hydrothermal veins and carbonatites.

Aragonite is colorless or white and is metastable at normal temperatures and pressures. Many shells are formed of aragonite, which in time undergoes recrystallization to calcite. Aragonite occurs as a primary precipitate and in amygdales. Strontianite and witherite are usually colorless, white, or yellow, occurring in hydrothermal veins and carbonatites.

The basic copper carbonates, *malachite* ($Cu_2(OH)_2CO_3$) and *azurite* ($Cu_3(CO_3)_2(OH)_2$), are bright green and bright blue respectively. Malachite has a banded botryoidal form and occurs with crystals of azurite in the oxidized zones of copper deposits.

carbonation A process of chemical weathering that involves the dissolving of soluble rocks and minerals by weak carbonic acid, formed by the combination of water and carbon dioxide derived from the atmosphere or from soils. Carbonation is most effective on limestones, in which calcium carbonate is converted to calcium bicarbonate and is removed in solution.

carbonatite An igneous rock composed chiefly of the carbonate minerals calcite, dolomite, and ankerite. Apatite and magnetite are common accessories together with the silicate minerals, alkali feldspar, nepheline, melilite, biotite, melanite, and sodic pyroxene. Certain elements, particularly niobium, barium, strontium, and the rare earths, are concentrated in carbonatites, which are often of economic importance. Carbonatites occur as central intrusive masses, dikes, and cone sheets in ijolite complexes and are often surrounded by zones of FENITIZATION. Such alkaline complexes are confined to stable continental regions, particularly those that have been subject to rifting.

carbon cycle The series of chemical reactions that circulate carbon through the global ECOSYSTEM. Plants take up carbon dioxide during PHOTOSYNTHESIS, converting it to carbohydrates and releasing oxygen. The plants use the carbohydrates as 'food' (or animals eat the plants), oxidizing them during respiration to release carbon dioxide. The burning of FOSSIL FUELS and forest trees also releases carbon dioxide into the atmosphere, possibly contributing to the GREENHOUSE EFFECT.

carbon dioxide A gas occupying only a small proportion of the atmosphere (0.05% by weight) but having very important consequences. It has the property of absorbing radiation in wavelengths similar to those emitted by the Earth, thus it prevents the loss of much terrestrial radiation and maintains, together with water vapor, a higher temperature for the Earth than would otherwise occur. It is vitally important in plant growth for the process of PHOTOSYNTHESIS. The combustion of fossil fuel, such as coal and oil, releases carbon dioxide and so the amount in the atmosphere has been increasing since the Industrial Revolution. It is believed that this could have the effect of increasing radiation absorption and so raise the Earth's mean temperature. However, the oceans are a great store of carbon dioxide and absorb much of the increase. *See* greenhouse effect.

Carboniferous The period of the PALEOZOIC Era that followed the DEVONIAN and preceded the PERMIAN. It began about 360 million years ago and lasted until about 286 million years ago. North American geologists divide this unit of time into two periods, the MISSISSIPPIAN and PENNSYLVANIAN, corresponding approximately to the time represented by the Lower and Upper Carboniferous respectively. The lower part of the Carboniferous System, sometimes referred to as the Dinantian, is formed of two divisions, the Tournasian and Viséan Stages. The upper part, the Silesian, consists of the Namurian, Westphalian, and Stephanian Stages.

Evidence suggests that most of the continents of today formed two supercontinents in the Carboniferous Period: Gondwanaland (present-day Africa, South

America, India, Middle East, Australia, and Antarctica) was located in the S hemisphere and Laurasia (present-day North America, Greenland, and N Europe) in the N hemisphere. By the end of the period these continental plates had collided, closing the Tethys Sea; the mountain-building episodes of the Variscan and Allegheny orogenies are attributed to these movements. Large tropical swamps, from which the coal deposits of the later Carboniferous originated, extended across what is now North America, Europe, and Siberia. The rocks of the Lower Carboniferous or Mississippian are characterized by limestones formed in shallow seas of the continental shelves, although deeper-water facies of shales and sandstones were also formed. (In Britain limestone is so characteristic of this division that it is often called the Carboniferous Limestone.) The shallow seas supported a diverse fauna of Foraminifera, corals, Bryozoa, brachiopods, crinoids, blastoids, and other invertebrates, some of which contributed to reef building. The Upper Carboniferous saw the alternating transgression and regression of seas over coastal swamps and facies mark a return to more terrestrial and freshwater conditions, being composed of cyclothems of deltaic sandstones and shales with coal seams. The coal is especially prominent in the upper part of the division, being formed from vast forests of primitive land plants, such as ferns and horsetails. (In Britain the Upper Carboniferous is often divided into two units: the Millstone Grit and the Coal Measures.) The Upper Carboniferous contains fossils of freshwater bivalves, which are used stratigraphically. Fish remained abundant and amphibians became more common, some of them evolving into reptiles at the end of the period.

Carboniferous Limestone The lowest lithological division of the CARBONIFEROUS System in Britain. It corresponds to the Lower Carboniferous (or Dinantian).

carbonization The process that leads to the preservation of FOSSILS as thin films of carbon in sedimentary rock. As sediments are laid down under water, organic matter decomposes with the release of hydrogen, oxygen, and nitrogen, leaving only the carbon. Plants, arthropods, and other invertebrates, and fish have all been preserved in this manner. *See also* Burgess shale.

cardinal points The four main points or directions of the compass: north, south, east, and west.

carnallite A white mineral salt, a double chloride of potassium and magnesium, $KMgCl_3.6H_2O$. It occurs in underground salt deposits derived from sea water, and is used in fertilizers as a source of potassium.

carnegieite *See* feldspathoid.

carnelian (cornelian) A translucent red to brown type of CHALCEDONY, which gets its color from inclusions of hematite. It is used as a semiprecious gemstone.

Carnivora The order of the class MAMMALIA that includes cats, dogs, bears, weasels, etc. Carnivores have teeth modified for eating flesh and a skeletal structure suited to rapid locomotion and the capture of prey. They evolved from the INSECTIVORA during the Paleocene Epoch. Early Tertiary forms, classified as the Creodonta and now extinct, included the largest carnivores known, some having skulls up to one meter in length. Modern carnivore groups evolved during the late Eocene and Oligocene.

carnotite A yellow mineral containing uranium and vanadium, $K_2(UO_2)_2(VO_4)_2.3H_2O$. It occurs in sandstone or near petrified wood, is highly radioactive, and used as a source of uranium.

cartography The technique and science of representing spatial relationships by means of maps (including geographic maps, plans, charts, and globes). This includes working from original surveys, photography, and other COMPILED MAPS, covering every aspect of map production from carrying out surveys to final printing of the completed copies, including such problems as MAP PROJECTIONS and research into all available information. *Digital cartography*, the use of computer-based information systems in the production and use of maps, is replacing some of the more traditional methods.

cartometric testing A technique enabling a cartographer to assess the accuracy value of a map provided the map concerned is gridded. Accepted mathematical formulas are used to test the horizontal and vertical accuracy of the map against a control. The resultant answer gives the average deviation of the distances or elevations on the map from the true position or elevation on the ground.

cartouche A panel on a map giving its title, scale, etc. The cartographers of the Dutch Renaissance and the Elizabethans specialized in highly decorative cartouches.

cascade fold One of a series of minor folds developed on the limb of a major fold, as a result of gravity collapse.

cassiterite A brown to black mineral form of tin dioxide, SnO_2, found in hydrothermal veins and metasomatic deposits associated with acid igneous rocks and in alluvial deposits.

cast A type of fossil consisting of a PSEUDOMORPH in which the skeletal parts of the organism have been dissolved and the resulting space replaced by a secondary material, producing a replica of the original form. In some cases a cast may preserve features of only the outer or inner surfaces of a structure such as a shell. These are known respectively as *external* and *internal casts*. *Compare* mold.

castle kopje A type of INSELBERG developed by DEEP WEATHERING and EXHUMATION in a rock divided by evenly spaced vertical and horizontal joints. Stacks of these fairly regular blocks give a castlelike stepped appearance. The term has also been applied to heaps of joint-blocks in haphazard arrangements, which have resulted from the collapse of other landforms or from the exhumation of isolated CORESTONES. Inselberg domes may degenerate into similar features following undercutting and collapse.

cataclasis The deformation of rocks by the mechanical process of shearing and granulation. Such cataclastic rocks are said to have undergone dislocation metamorphism and range from coarsely broken breccias to intensely deformed mylonites.

catastrophism The theory, now generally thought to be false, that past geologic changes have occurred as a result of a number of sudden catastrophes (*compare* uniformitarianism). A succession of catastrophes was also invoked to explain the extinction of organisms, linked to special creations to account for the appearance of new forms.

catchment area The area from which a river and its tributaries obtain their water. It may cover a few square meters or cover a large part of a continent (such as the Amazon basin). Rainwater may run off the surface into the river or seep into the soil and flow to the river as GROUNDWATER. *See also* drainage basin.

catena The regular repetition of a characteristic sequence of soil profiles associated with a particular topography in the tropics. Relief and drainage are the dominant controlling factors of formation. Milne first used the term when mapping the soils of East Africa. There has been confusion as to whether it refers to soils on a uniform lithology or whether it can be used where another soil-forming factor besides relief and drainage varies (the term *toposequence* is used in the latter instance).

cation exchange (in soil science) An internal chemical phenomenon in soils whereby a positively charged ion (cation) held on the surface of a colloid (either a clay or humus particle) can be exchanged for another cation present in the surrounding electrolyte. Cations include the soil bases, notably calcium, potassium, sodium, and magnesium, and the acidity-inducing cations, notably hydrogen and aluminum. A typical example is for a calcium ion, held by its positive charge to the negatively charged surface of the colloid, to be replaced by hydrogen from the surrounding fluid. This makes the colloid more acid, the displaced calcium possibly being leached away. The cation exchange capacity of the soil is expressed in milliequivalents per 100 grams of clay. This varies with different colloids: in the simplest most highly weathered clays (kaolin) it is lowest, perhaps 3–15 m.e./100 grams, reaching 80–100 in the 2:1 lattice clays (e.g. montmorillonite), and then highest of all, perhaps 200+ m.e./100 g, in humus colloids.

cat's eye Any of various semiprecious gemstones that in reflected light display a bright band, like the pupil of a cat's eye. The best-known example is a green type of CHRYSOBERYL.

cauldron Any collapse structure of volcanic origin, as where a block of country rock sinks into the underlying magma. There are no limitations as to the size of the structure or the extent of subsequent erosion.

cave A hollow that is formed in rock by erosion, particularly by the action of water on LIMESTONE. Erosion starts by enlarging joints in BEDDING PLANES and at other points of weakness. Underground streams can carve subterranean caverns, and wave action at the coast can form caves in cliffs. There the erosive effect may be augmented by CORRASION. Water saturated with carbonates may drip into caves and form STALACTITES AND STALAGMITES.

cavitation One of the processes of fluvial erosion that characterizes the high-velocity parts of streams, such as waterfalls and rapids. It is relatively rare compared with the common processes of CORROSION and ABRASION, but very effective in terms of erosive power where it does occur, being largely responsible for the rapid destruction of waterfalls and rapids. Constriction of flow raises the velocity and kinetic energy of a stream, which is compensated by a decrease in the pressure of the water. This may lead to the formation of bubbles in the stream, which subsequently collapse, giving off severe shock waves when velocity decreases as the channel widens again and water pressure can rise. The stress produced by the collapsing bubbles exerts a severe stress on the channel walls, speeding up erosion.

cay (key) A small sea island. It is generally low-lying, possibly having a little vegetation, and usually formed in sand or coral. The term as applied for example to tiny islands or islets off the South Florida coast refers to quite small coral shoals.

celestite (celestine) A white, blue, or red mineral form of strontium sulfate, $SrSO_4$. It crystallizes in the orthorhombic system, and generally occurs associated with halite and gypsum, in nodules in limestone, or with sulfur (as in Sicily). It is used as a source of strontium and its compounds. *See also* barytes.

celsian A barium-containing type of FELDSPAR, $BaAl_2Si_2O_8$. It often occurs associated with manganese ores.

Celsius scale A scale of temperature named after a Swedish astronomer who was the first to divide the interval between the freezing point and boiling point of water into 100 parts. Although it was formerly known as the centigrade scale, an international agreement on units now favors the term Celsius.

cementation A rock-forming process in which fragments are bound together by a cement precipitated between the grains, making a solid sedimentary rock. These natural cements include calcite, carbonates, iron oxides, and silica (the commonest).

Cenozoic (Cainozoic, Kainozoic) The era of geologic time following the MESOZOIC and beginning about 65 million years ago. It includes the TERTIARY and QUATERNARY Periods, although it is sometimes incorrectly referred to as being synonymous with the Tertiary. Following the extinction of most of the reptiles, which were the dominant animals of the Mesozoic, the previously insignificant mammals underwent evolutionary radiation into an abundant, diverse, and dominant group. For this reason the Cenozoic is often known as the age of mammals. Birds and flowering plants also flourished and the invertebrate types characteristic of the Mesozoic were replaced by essentially modern forms. The Alpine episode of orogenic activity extends into this era. *See also* geologic timescale.

centigrade scale *See* Celsius scale.

central meridian The line of longitude that forms the axis of a map projection. Usually the central meridian is in the middle of the projection, as in the SANSON–FLAMSTEED PROJECTION.

central-vent volcano A volcano that has one central vent, around which there is a tendency for volcanic debris to accumulate into a roughly symmetrical cone.

centrifugal force The apparent force acting outward from the axis of rotation in a rotating system. Its magnitude is v^2/r where v is the linear velocity and r is the radius of curvature of the body's path, or $\omega^2 r$ where ω is the angular velocity of the body. This force is equal and opposite to the centripetal force, which tends to maintain the body in its curved path. In the atmosphere, these two forces operate when air is moving in a curved path relative to the ground surface (i.e. around a depression). It is also a part of the observed force of gravity, which is the resultant of the true gravitational attraction toward the Earth's axis and the centrifugal force resulting from the Earth's rotation. However, the centrifugal acceleration is less than 1% of the acceleration of free fall.

centripetal acceleration The acceleration of a body traveling in a curved path. It is equal and opposite to the CENTRIFUGAL FORCE per unit mass.

centroclinal Describing a basin in which the rock strata all dip toward a central low point, characteristic of cratonic areas. *See also* pericline.

Cephalopoda A class of the phylum MOLLUSCA whose members usually have an internal or external shell divided into chambers and a foot modified into a set of tentacles surrounding the mouth. All cephalopods are marine. They are usually subdivided into three groups: the NAUTILOIDEA, the extinct AMMONOIDEA and the Coleoidea (or Dibranchiata), which includes the modern squid and octopus and the extinct BELEMNOIDEA. Fossils of cephalopods are known from the Cambrian onward and have been used as zone fossils.

cerussite A mineral form of lead carbonate, $PbCO_3$. It crystallizes in the orthorhombic system, and is a secondary lead mineral, often formed by the oxidation of GALENA, and is used as a source of lead.

chabazite A colorless, white, or pink hydrated calcium aluminum silcate, with some sodium and potassium, $(Ca,Na_2)Al_2Si_4O_{12}.6H_2O$. It crystallizes in the hexagonal system and is a member of the ZEOLITE group.

chain The main equipment used in chain surveying (*see* chaining), consisting of links of thick steel wire each connected to the next by three rings. Most chains now in use have an overall length of 20 m, each individual link being 200 mm long from its center to the center of the middle connecting ring. Markers designate each whole meter, while those at five-meter intervals have a different color.

chaining A surveying method used for areas containing concentrated detail, such as urban areas, where high accuracy is required. The CHAIN is run out along a convenient line, the direction of which is obtained using a prismatic compass or theodolite. Points of detail are then fixed by measuring the perpendicular distance from the chain to the feature using a tape, these measurements being known as *offsets*. The distance along the chain and the length of the offset, for every detail point, are noted on the booking sheet. As an alternative to offsets, a point may be fixed by measuring two lines, known as *ties*, from different parts of the chain, thereby forming a triangle.

chalcanthite A bright blue mineral form of hydrated copper sulfate, $CuSO_4.5H_2O$. It crystallizes in the triclinic system and occurs in oxidized regions of copper deposits.

chalcedony A very fine-grained lustrous semitransparent type of silica, SiO_2, which may be colored by impurities (*see* silica minerals). It generally occurs as a deposit in cavities in lava and sedimentary rocks. It is used as a semiprecious gemstone.

chalcocite A dark gray or black lustrous mineral form of copper sulfide, Cu_2S. It crystallizes in the orthorhombic system but is more often found as masses. It is an important source of copper.

chalcophile An element that generally occurs in association with sulfur, especially as a sulfide. Chalcophile elements include arsenic, copper, lead, mercury, silver, sulfur, and zinc. *See also* atmophile, lithophile, siderophile.

chalcopyrite A brass-yellow copper iron sulfide, $CuFeS_2$, found in hydrothermal and metasomatic veins and associated with the upper portions of acid igneous intrusions. It is the most important source of copper.

chalk A very fine-grained pure white LIMESTONE (calcium carbonate, $CaCO_3$) formed predominantly from COCCOLITHS, but including other invertebrate skeletal fragments. It is the characteristic rock of much of the upper CRETACEOUS system in W Europe; the White Cliffs of Dover, SE England, are a famous example of exposed chalk deposits. Extensive deposits also occur in the USA.

chalybite Another name for SIDERITE.

chamosite A gray-green iron-containing mineral, a member of the CHLORITE group.

Chandler wobble The small measurable wobble of the Earth with respect to its axis of rotation, discovered in 1891 and found to have a period of 14 months and an amplitude of 0° 5′. Its origin is unknown but it is assumed to be connected with movement of materials inside the Earth.

channel Any waterway with free-flowing water, such as a natural stream or river, a connection between two lakes or seas, or an artificial canal.

chaos theory The theory of systems that exhibit apparently random unpredictable behavior. The theory originated in studies of the Earth's atmosphere and the weather. In such a system there are a number of variables involved and the equations describing them are nonlinear. As a result, the state of the system as it changes with time is extremely sensitive to the original conditions. A small difference in starting conditions may be magnified and produce a large variation in possible future states of the system. As a result, the system appears to behave in an unpredictable way and may exhibit seemingly random fluctuations (chaotic behavior). The study of such nonlinear systems has been applied in a number of fields, including studies of fluid dynamics and turbulence, random electrical oscillations, and certain types of chemical reaction. *See also* butterfly effect.

chaparral A type of low dense scrub with evergreen vegetation, such as that in California and other parts of the USA. *See also* maquis.

charnockite A rock that is mineralogically similar to leucocratic GRANULITES and contains hypersthene, quartz, and feldspar with or without garnet. Unlike granulites, the quartz does not occur in a platy flattened form. Basic varieties rich in plagioclase feldspar may be termed *enderbite*.

The formation of charnockites is controversial but it is likely that they originate either by the crystallization of granite magma under high-grade metamorphic conditions or by the high-grade metamorphism of preexisting igneous rocks.

chart A specialized form of map used for navigation and other specific purposes, e.g. synoptic (weather) charts, star charts. The most common chart is the hydrographic chart.

chatter mark **1.** A crescentic fracture seen in hard rock subjected at some time to the passage of GLACIER ice. Although the convexity of these marks can point in either direction, a study of the fracture plane from which rock material has been removed will indicate the actual direction of former ice movement. The marks are believed to be created by the impact and rolling of boulders held loosely within the base of the glacier.
2. The impact mark frequently seen on a beach cobble (*see* percussion mark).

chelation (in soil science) The combination of sesquioxides with organic acids. Water passing through leaf litter acquires an extract from the organic material that is capable of combining with the metallic cations in the soil, notably iron and aluminum, to form a *chelate* of organic and metal constituents. Experiments suggest that leaf litter is a more powerful chelating agent than peat or humus, and that the process involved is one of reduction and solution that renders the metal more mobile. The chelating agent in the litter is easily extracted by water passing over it, but its strength varies between different types of litter, pine needles being among the most powerful. This is generally thought to be an important process in podzolization.

cheluviation The combination of CHELATION and ELUVIATION. Water with organic extracts forms a chelate with sesquioxides in the soil and then moves down through the profile carrying the sesquioxides in solution. It is most effective in moving iron and aluminum.

chemical weathering The breakdown of solid rock through a number of chemical reactions, which may cause the removal of cements, resulting in weakness, or the formation of secondary minerals less resistant to erosion than those of the fresh rock. The different types of chemical action that are important in weathering are CARBONATION, HYDRATION, HYDROLYSIS, LIMESTONE SOLUTION, OXIDATION, and REDUCTION.

chenier A BEACH RIDGE on a mudflat, particularly in an estuary, consisting of sand or shell fragments. As mudflats change form through water action, the coarser material in the sediments tends to concentrate in such ridges.

chernozem (black earth) A pedocal similar to a prairie soil. It is typically found in continental interiors, such as the Russian steppes, where there is an annual precipitation of approximately 500 mm with a slight rainfall maximum in summer. The process of calcification is dominant resulting in a soil with an A Cca C system of horizons (a B horizon is absent from true chernozems). The dark base-rich A horizon, has mull humus resulting from the decomposition of the natural vegetation of grassland, incorporated to a considerable depth by rich faunal activity. This horizon may extend down to 100 cm before passing into the Cca horizon of calcium carbonate concretions. The parent material often resembles loess. These soils with their neutral to slightly acid reaction and their excellent crumb structure are agriculturally some of the most important in the world. They fall into the MOLLISOL order of the Seventh Approximation.

chert An extremely hard, white, gray, or black type of CHALCEDONY that occurs as masses or layers in limestone. The nodular form is known as flint. *See also* silica minerals.

chestnut soil A pedocal typical in the Ukraine and the Great Plains of North America. Conditions here are more arid than in the chernozem belt and consequently the natural vegetation consists of a tussocky grass cover. These soils are much shallower than chernozems and horizonation is not so distinct. The A horizon is chestnut brown in color and has a platy or prismatic rather than a crumb structure. This merges at about 25 cm into the B horizon, which is lighter in color owing to the presence of calcium carbonate and often gypsum. The calcium carbonate may form a distinct concretionary horizon (Cca) at only 50 cm depth. This lies above the parent material, which is often loess. These soils in the American West are subject to severe wind erosion and are used mainly for grazing. They fall into the MOLLISOL order of the Seventh Approximation.

chevron fold An accordian fold whose limbs are of equal length.

chiastolite A variety of andalusite in which impurities are arranged in the form of a cross. *See* aluminum silicates.

Chile saltpeter A mineral form of sodium nitrate, $NaNO_3$, that occurs as surface deposits in Chile. It is an important source of nitrates for making fertilizers and explosives.

chimney *See* smoker.

china clay *See* kaolin.

chinook A type of föhn wind found on the E slopes of the Rocky Mountains. It is named after a Native American tribe that lived near the mouth of the Columbia River and originally referred to a warm moist SW wind that blew across this area. Later it was used for a warm dry wind on the E slopes blowing from the Chinook region. The term is now only used for the dry wind bringing sudden increases of temperature and very low humidities with rapid thaw conditions when snow is on the ground.

chlorite The group of minerals having a general composition $(Mg,Al,Fe)_{12}$ $(Si,Al)_8O_{20}(OH)_{16}$. They are structurally similar to the micas and are composed of alternating layers of talc, $Y_6Z_8O_{20}(OH)_4$, and brucite, $Y_6(OH)_{12}$, where Y = Mg,Al,Fe and Z = Si,Al. Chlorites are monoclinic and typically green or white in color. They are common in low-grade metamorphic rocks, particularly those of the greenschist facies. Chlorites also occur as secondary minerals in igneous rocks owing to the hydrothermal alteration of ferromagnesian minerals such as pyriboles. Oxidized chlorites such as *chamosite*, which have a high content of iron, are found in argillaceous sediments as authigenic and detrital grains.

chloritoid A member of a group of dark green minerals with the composition $(Fe^{2+},Mg)_2Al(OH)_4Al_3O_2(SiO_4)_2$, having a structure made up of layers $(Fe^{2+},Mg)_4Al_2O_4(OH)_8$ and Al_6O_{16} linked by SiO_4 tetrahedra. Chloritoid minerals are found in aluminum- and iron-rich low-grade metamorphosed argillaceous sediments.

Chondrichthyes Cartilaginous jawed fish, including modern sharks and rays. Because of the absence of bone good fossils are rare, although teeth are often preserved. The Chondrichthyes probably evolved from the PLACODERMI and are almost exclusively marine. They first appeared in the late Devonian and continued to be important through the Mesozoic and Cenozoic to the present day. *Compare* Osteichthyes.

chondrite A stony METEORITE that contains chondrules. *Compare* achondrite.

chondrule A small rounded body usually of olivine or enstatite found in stony METEORITES.

chop The sea surface under the influence of a fairly moderate breeze, with waves having pointed crests and steep slopes, quickly developing as the breeze mounts.

Chordata The phylum of animals possessing, at least primitively, an internal skeletal rod, the *notochord*, running the length of the body. Members of this phylum include the VERTEBRATA, which contribute significantly to the fossil record. Other subphyla consist predominantly of small soft-bodied sessile marine creatures, which are rarely preserved as fossils. Some of these have an active larval stage, and the evolution of vertebrates may have begun with the retention of the characteristics of the larvae into the adult state.

C horizon The material from which soil is formed, the lower subsoil lying between the upper subsoil (B HORIZON) and the bedrock. It contains no humus and will not support plant life. *See also* horizon (def. 1).

chorochromatic map A map in which nonquantitative spatial distributions are shown by color tinting (e.g. of coalfields in an area).

chorographic map A map on which regions are delineated, usually large regions, e.g. countries or continents. Atlases and small-scale wall maps, particularly those showing political divisions, come into this category.

choropleth map A map showing quantitative spatial distributions (e.g. of population) calculated from average values per unit area.

chromite A brownish-black mineral form of iron and chromium oxide, $FeCr_2O_4$, a member of the SPINEL group. It crystallizes in the cubic system and occurs in basic igneous rocks. It is the main source of chromium.

chron The smallest interval of geologic time in the hierarchy of the Chronomeric Standard terms used in CHRONOSTRATIGRAPHY. The equivalent Stratomeric Standard term, indicating the body of rock formed during this time, is the CHRONOZONE. Chrons may be grouped together to form an AGE.

chronomere (in chronostratigraphy) Any interval of geologic time. Chronomeres are not of standard uniform duration.

chronostratigraphy The branch of STRATIGRAPHY linked to the concept of time, rather than being limited only to considerations of lithology and spatial distribution (*see* lithostratigraphy). There are two parallel hierarchies of formal terms. The *Chronomeric Standard* terms are applied to intervals of geologic time; the *Stratomeric Standard* terms are applied to the bodies of rock laid down during these time intervals. The Chronomeric Standard hierarchy is as follows: eon, era, period, epoch, age, and chron; for the major divisions of geologic time, see the table at GEOLOGIC TIMESCALE. The Stratomeric Standard hierarchy of terms has no equivalents to the eon and era; it consists of the system, series, stage, and chronozone. The individual time intervals are not of standard duration and, similarly, the rock divisions are not of uniform magnitude. Some geologists regard chronostratigraphy as essentially identical to BIOSTRATIGRAPHY whereas others insist on a formal separation of the concepts involved, regarding biostratigraphy as simply one method of calibrating chronostratigraphic scales.

chronotaxis The occurrence of units of rocks of equivalent age in separate successions. These rock units are described as *chronotaxial*. Chronotaxis can rarely be demonstrated with certainty. *Compare* homotaxis. *See also* correlation.

chronozone The smallest division of rock in the Standard Stratomeric scheme of stratigraphic classification (*see* chronostratigraphy). It indicates the body of rock that has formed during one CHRON. Chronozones are defined at a particular TYPE SECTION and should be named after geographical localities. At the TYPE LOCALITY a chronozone may correspond to a biostratigraphical ZONE but other criteria are often used in calibration and in correlation with other areas.

chrysoberyl A usually yellow to green mineral form of beryllium aluminate, $BeAl_2O_4$. It crystallizes in the orthorhombic system, often as star-shaped crystals. Some types, such as alexandrite, are used as semiprecious gemstones. *See also* cat's eye.

chrysocolla A bright blue mineral, a hydrated silicate of copper, $Cu_2H_2Si_2O_5(OH)_4$. It occurs in thin seams or as incrustations, usually in oxidized copper or copper sulfide deposits.

chrysotile A white, gray, or green fibrous SERPENTINE mineral, once the chief source of commercial ASBESTOS.

cinder A dark-colored porous volcanic fragment, generally consisting of BASALT or ANDESITE. Up to 30 mm across, cinders are semifluid when ejected from a volcano but are solid by the time they fall to the ground. *See also* pyroclastic rock.

cinnabar A bright red mercury sulfide mineral, HgS, found in veins and impregnations associated with volcanic rocks. It is the main source of mercury.

CIPW classification *See* norm.

circulation index A measure of the strength of the atmospheric circulation: used for any wind system, it commonly refers to the zonal component of the westerlies or the trade winds. *See also* general circulation of the atmosphere.

Circum-Pacific Belt A narrow belt bordering the Pacific Ocean, in which 75% of the present earthquakes occur. It consists of igneous, metamorphic, and tectonically deformed sedimentary rocks.

cirque A rounded rock basin, often containing a lake or CIRQUE GLACIER, enclosed by high headwalls and sidewalls, which are steep and frequently frost-shattered. In Scotland cirques are called *corries*; in Wales, *cwms*. The floor is usually of smoothed striated rock and frequently has a deepened basin form, with an associated convex rock lip, which may be moraine-covered. Though varying considerably in size, many cirques possess roughly the same proportions having a length to height ratio of about 3:1.

It is believed that cirques can develop from any hollow in which snow can accumulate, but in many cases they have developed from preexisting water-eroded features. The processes responsible for their creation include: FREEZE-THAW action on the headwall, which enlarges the cirque overall; abrasion by rotationally slipping ice, which deepens the rock basin; joint-block removal, which can act over the whole surface area. Cirques constitute some of the most characteristic features of glacial erosion.

cirque glacier A small glacier found in a valley head or depression on a mountain slope. Being small, they can form quickly (within 100 years) and are usually the first glaciers to form and the last to disappear during a period of glaciation. The ice occurs in distinct bands reflecting successive years' accumulation of FIRN The ice tends to move downward in the upper part (firn zone), parallel with the glacier surface in the central area, and upward in the lowest zone, i.e. a rotational movement. This is because most ac cumulation occurs in the upper zone, beneath the headwall, whereas most melting is from the lower part. This results in over-steepening of the glacier surface, and in order to return to a more balanced shape, rotation occurs when the imbalance is sufficient to overcome friction between the ice and the bedrock. *See also* cirque.

cirrocumulus cloud A cloud type indicating convection in the upper levels of the troposphere. Composed of ice crystals, the cloud has a mottled appearance consisting

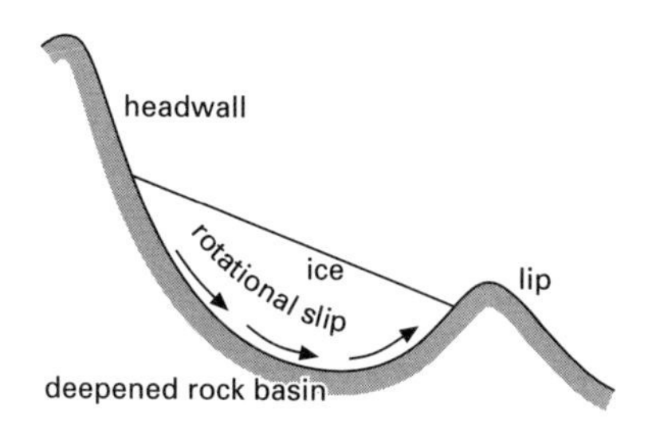

Cirque (cross-section)

of small convective elements, often regularly arranged, with clear spaces between.

cirrostratus cloud A thin high-level layer cloud composed of ice crystals. It is transparent and gives the impression of a veil spreading across the sky. During the day, cirrostratus clouds produce a halo effect, which can be seen quite clearly around the Sun. They are usually the precursor of precipitation, heralding the approach of depressions.

cirrus cloud A detached cloud in the form of a wispy streamer or filament of ice crystals. They occur at the higher levels of the troposphere, above 6000 m, and with temperatures below –25°C. When they have a hooked appearance, they are known as *mare's tails* and signify rain.

citrine A yellow or yellow-brown type of QUARTZ, used as a semiprecious gemstone. Dark samples are sometimes called false TOPAZ. *See also* silica minerals.

clapotis A STANDING WAVE that is generally formed when the incident sea waves meet waves being reflected from a vertical barrier and where the depth of water in front of the barrier exceeds the wavelength of the incident waves. An interference pattern is set up, in which there is no horizontal travel of the standing wave crests. The incident waves generally approach the barrier, perhaps a seawall, breakwater, or vertical cliff-face, more or less head on, i.e. their crests are roughly parallel to the barrier. Occasionally, the same phenomenon occurs when waves approach a very steep beach, though in this case the amount of reflection is less.

class A group in the taxonomic classification of organisms. Several related classes form a PHYLUM, and a class itself is composed of one or more ORDERS grouped together. For example, the Mammalia and Reptilia are two classes of the phylum Chordata. *See* taxonomy.

clast A small piece of rock that has been removed from a larger mass by some fragmentation process. CLASTIC rocks and sediments are composed of clasts.

clastic Describing a sediment or a sedimentary rock consisting of fragments of broken rock (*clasts*) that have been eroded, transported (usually by water), and deposited. *Clastic sediments* include a wide range of sediment particle sizes from large boulders to fine-grained sediments, such as silt. They are characteristic of coastal and shelf areas, and especially the littoral zone. The particle sizes may display wide and abrupt variations, for rarely are littoral or shelf sediments neatly sorted and graded, at least over large distances. Detrital sediments are often described as having a clastic structure. *Clastic rocks* (*fragmental rocks*), consolidated clastic sediments, include BRECCIA, CONGLOMERATE, MUDSTONE, SANDSTONE and SHALE.

clay Particles of a size less than 1/256 mm in diameter. They are usually small flaky CLAY MINERALS formed during the weathering of other rocks. The grade of particles immediately larger than clay is SILT. Together they form the ARGILLACEOUS division of clastic sediments.

clay minerals Hydrous silicates, mainly of aluminum and magnesium, occurring as platy or fibrous crystals that have a layered structure and the ability to take up ahd lose water. They are the main constituents of argillaceous rocks and are responsible for the plastic properties of clay. Four main groups may be considered:

1. *Kaolinite group*: kaolinite, $Al(Si_4O_{10})(OH)_8$, is produced during the weathering and hydrothermal alteration of feldspars and feldspathoids under acid conditions. The large-scale production of kaolinite by the alteration of granite gives rise to KAOLIN (china clay) deposits.
2. *Illite group*: illite and hydromicas are the dominant clay minerals in shales and mudstones and are derived from the alteration of micas and feldspars under alkaline conditions.
3. *Montmorillonite-smectite group*: these minerals are produced during the alteration of basic rocks under alkaline condi-

tions and are the principal components of bentonite and fuller's earth.
4. *Vermiculite*: this mineral is produced during the hydrothermal alteration of biotite and also occurs at contacts where acid magma has intruded basic rock.

clear air turbulence (CAT) Sudden severe turbulence occasionally encountered by aircraft flying at high levels of the atmosphere, which is unconnected with the vertical turbulence of convection clouds. It appears to be the result of large vertical wind shears associated with high static stability. These conditions produce the effect of two unmixed layers and on the boundary between them this wavelike motion develops, producing billowing. Clear air turbulence can be detected by high-powered radar. *See also* Kelvin-Helmholtz instability.

cleat A joint or system of joints developed within a coal seam. Generally two sets occur at right angles to each other, along which the coal fractures preferentially.

cleavage The tendency of a rock to break into closely spaced planar structures or fractures as a result of deformation or metamorphism. *See also* fan cleavage; flow cleavage; fracture cleavage; slaty cleavage.

Cleavage in minerals refers to the tendency to split along planes of weakness in the molecular framework.

cleavage plane The plane of fracture or a series of planar structures developed in a deformed or metamorphosed rock, along which it splits preferentially. Cleavage planes also occur in mineral cleavage.

cliff A high steep rock face. Coastal cliffs are attacked at their base by wave action and above by subaerial weathering and erosional processes. The steepest cliffs are those formed in resistant massive rock types, such as granite, whereas softer rocks, being more susceptible to erosion by both waves and subaerial processes, retreat and degrade comparatively rapidly. As cliffs retreat a WAVE-CUT PLATFORM is frequently left projecting out to sea.

Abandoned cliffs were eroded at their base by waves at a time when sea level was some way above that of the present, as on the melting of the Pleistocene ice sheets, so that they now stand some distance from the sea.

climate The synthesis of day-to-day WEATHER variations in a locality. The "normal" climatic conditions prevailing can be summarized, but since it is impossible to include the wide range of values in a single number and climate is not an unchanging feature, this method on its own is not a very satisfactory indicator of actual conditions.

Climate is usually taken to include the following weather elements: temperature, precipitation, humidity, sunshine, and wind velocity. Minor aspects include cloud amount, fog, snow, thunder, and gale frequencies. The major factors are summarized in terms of arithmetic averages for specific periods, usually based on at least 30 years of records. To indicate the range of values within the instrumental records, extremes are often quoted. More recently, greater emphasis has been made on the factors determining climate.

On a large scale, the climate is determined by latitude, by altitude, and by location relative to the continental margins and the main circulation belts of the Earth. On a smaller scale, more local factors can be important, such as aspect and degree of exposure to prevailing winds.

climatic change Climate is not static but constantly changes in response to variations in the factors that control it, primarily the nature of the main circulation belts. If records of temperature had been maintained for the past million years, it would be evident that throughout this period there were oscillations between warmer periods and cooler periods on a wide range of timescales. On a large scale there would be the swings between the ICE AGES and the warmer INTERGLACIAL periods, continuing through the whole time spectrum to individual years. Most of the evidence for these

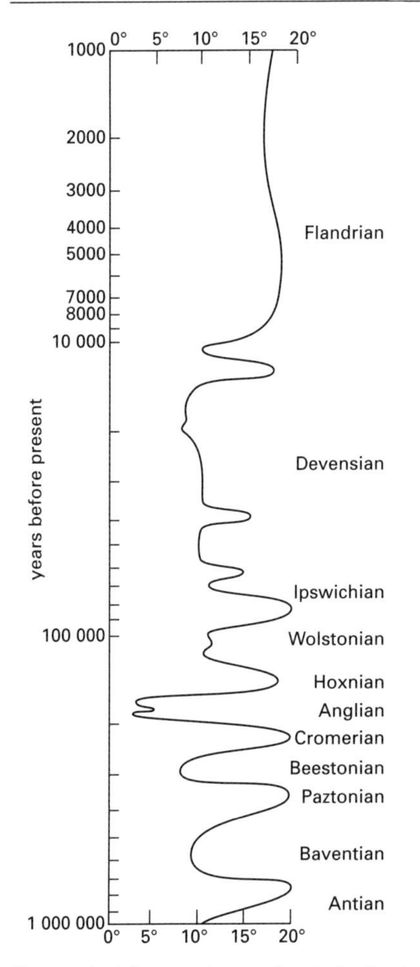

Changes in July mean temperature in lowland England

changes comes from the nature of sediments deposited and the fauna and flora they contain, because it is obviously only relatively recently that documentary evidence and finally meteorological instruments give us more precise indicators of the prevailing climate. The diagram illustrates known changes of temperature in NW Europe in this period. During this period much smaller changes became apparent, which would not have been preserved by geologic or biological evidence, so the curve is not based on consistent evidence.

The causes of climatic changes are not well understood and likely to prove very complex. The shorter-period fluctuations will probably have a different origin from the factors producing an ice age, but although mathematical models can show changes that could be produced by varying climatic controls, it is almost impossible to prove that any cause or combination of causes was the real reason for past climatic changes. Among probable factors influencing the major changes are variations in solar output, the movement of continental positions to affect the oceanic circulation, mountain building, volcanic eruptions, and the increase in the carbon dioxide content of the atmosphere.

climatic classification The climate of any individual site on the Earth's surface is unique, but there are great similarities between many records. To rationalize the large amount of data from observing points, many attempts have been made to classify climates into a small number of categories based on specific factors of similarity. There are two main bases for classification, empirical and genetic, the former being based on analysis of the observations and the latter on the atmospheric circulations from which the climate results. Most systems have used the empirical approach, with the earlier ones linking this with vegetation boundaries.

The most popular classification, which is still used in a modified form, was proposed by Köppen in 1918. He adopted 50°F (10°C) and 64°F (18°C) as critical values to produce five major climatic categories incorporating precipitation regions too. These are (A) tropical rainy climates with an average temperature for every month above 18°C, no cool season, and abundant precipitation that exceeds evaporation; (B) dry climates, where evaporation exceeds precipitation on average throughout the year; (C) humid mesothermal climates, where the coldest month has an average temperature less than 18°C but above –3°C, with at least one month having an average temperature above 10°C; (D) humid microthermal, where the warmest month is above 10°C but the coldest below –3°C; (E) ice climates where the mean temperature of the warmest month is less than 10°C. These are further subdivided into a large number of climatic types based upon seasonal temperature differences and variations in amount and distribution of precipitation. The simplicity of the system and its relationships with vegetation types have made this classification very popular although it does have limitations due to its arbitrary temperature boundaries.

A more detailed classification (*see* Thornthwaite classification) system was devised initially in 1933 and modified in 1948. This was based on two climatic factors, a moisture index and a thermal efficiency index, which are independent of any other geographical factor. However, the values were more difficult to calculate and, although frequently quoted as an example, the system is rarely used in practice. Individual aspects of the system, such as potential evapotranspiration, have been more fully developed.

Other empirical classifications have been proposed, some being based on human physiological responses to climate in terms of wind speed, temperature, and humidity combinations, others on the relationship between the actual climate of a site and a stated ideal climate.

Genetic classifications are based on the atmospheric circulations that determine climate. While being meteorologically sound, in terms of the resulting temperature and precipitation, quite varied weather conditions can be found within the same circulation belt and so such classifications have received little popular appeal.

climatic geomorphology The branch of geomorphology concerning the role of climate in landscape evolution. Landscapes are a product of processes that are conditioned by climate, geology, and base level: of these three, climate is said to be the most important, such that by its control, areas of differing climate tend also to have different landforms. The concept is particularly popular among European geomorphologists: Tricart and Cailleux have elaborated it as an alternative to the Davisian CYCLE OF EROSION concept, the difference being that the latter states that stage of evolution is the major factor determining the appearance of a landscape, whereas climatic geomorphology emphasizes that it is the processes operating, as conditioned by climate, that dictate the form of an area. The role of past climatic fluctuations in influencing the appearance of current landscapes is also involved.

climatic optimum The period during which temperatures reached their highest levels within an interglacial stage of the Pleistocene period. The climatic optimum of the present interglacial occurred about 5000 years ago and is known as the ATLANTIC PERIOD.

climatic region An area of the Earth's surface having relatively uniform climatic properties and usually determined through a climatic classification system.

climatic zone The present distribution of climates on the Earth's surface exhibits a measure of zonal symmetry with the boundaries running approximately along lines of latitude. Because of this, eight principal climatic zones have been distinguished within which there is a certain amount of homogeneity. These zones are the tropical rainy climate near the Equator, the two steppe and desert areas north and south of the Equator, the humid temperate zones, a Boreal climate in the N hemisphere, and the two polar zones.

climatograph A circular diagram showing the seasonal temperature variations of an area in graphical form. (See diagram overleaf.)

climatology The study of CLIMATE. The emphasis on particular aspects of climatology has produced many subdivisions, such as applied, regional, physical, and synoptic climatology. *See also* meteorology.

climograph (climogram) A diagram in the form of a graph in which climatic features at any one place are plotted against each other, for example wet-bulb temperatures against relative humidities.

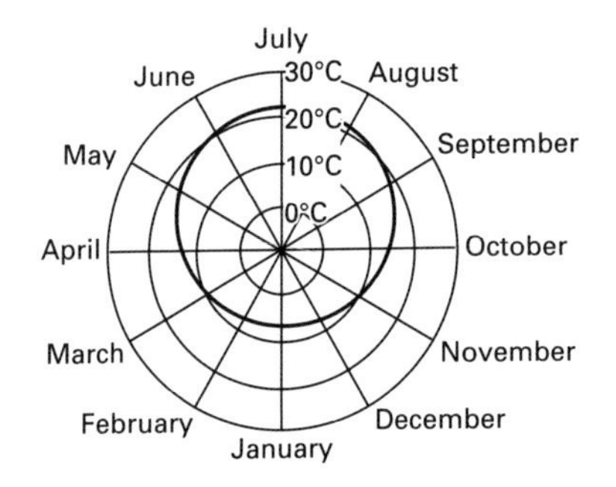

Climatograph: mean monthly maximum temperatures at Kew (London)

clinometer An instrument that measures angles of slope or inclination, such as DIP.

clinopyroxene A monoclinic type of PYROXENE.

clinozoisite A type of EPIDOTE that differs from ZOISITE in crystallizing in the monoclinic system.

clint A type of LIMESTONE PAVEMENT.

clintonite A member of the MICA group of minerals that differs optically from XANTHOPHYLLITE.

closed fold A fold that has been compressed so much that both limbs are parallel.

closure The distance between the highest point of a structure of folded rocks or structural traps and the lowest contour that totally encloses the structure. The closure determines the maximum area in which the structure can contain a fluid, e.g. oil.

cloud A mixture of minute water droplets and ice crystals produced by condensation in the atmosphere. They remain suspended within the cloud because of their small size and grow only when conditions are favorable for precipitation. Lower clouds are usually composed almost entirely of supercooled water droplets, but with cooler temperatures ice crystals become more frequent until at temperatures below –40°C the cloud is entirely ice crystals. Almost all clouds are found within the troposphere and are the result of cooling. This can be brought about by convection, uplift over mountains, or large-scale ascent in depressions.

cloud base The point at which the frequency of water droplets in the atmosphere becomes sufficient to reduce visibility to that characteristic of cloud. From the ground surface this appears to be a very rapid transition but in reality it represents a zone of transition. Cloud base indicates the level at which rising air reaches saturation and condensation results.

cloud classification Various classifications of cloud have been proposed, but the one most commonly used is based on cloud appearance and height. The major groups are divided into ten genera following plant taxonomic classification, comprising cirrus (Ci), cirrostratus (Cs), cirrocumulus (Cc), altostratus (As), altocumulus (Ac), stratus (St), stratocumulus (Sc), cumulus (Cu), cumulonimbus (Cb), and nimbostratus (Ns). The determination of cloud species is based on the appearance of the cloud, on its structure, and if possible on the physical processes involved in its formation. 14 species names are commonly used and can be referred to more than one cloud genus: *fibratus* – applied to cirrus and cirrostratus clouds in the form of filaments but without tufts or hooks; *spissatus* – denoting dense opaque cirrus cloud, often originating from the cumulonimbus anvil; *uncinus* – denoting cirrus clouds that are hooked or comma-shaped and frequently parallel; *stratiformis* – spread out into an extensive horizontal layer, applied to altocumulus and stratocumulus clouds; *nebulosus* – exhibiting a nebulous veil with no distinct details, usually cirrostratus or altostratus; *lenticularis* – lens-shaped, with sharp margins (most commonly applied to altocumulus cloud of orographic origin); *castellanus* – having a castellated or crenellated appearance connected to a common cloud base (altocumulus show this characteristic most frequently); *floccus* – having a ragged base and a small tuft or protuberance with cumuliform character above (cirrus, cirrocumulus, and altocumulus); *fractus* – denoting stratus or cumulus clouds with a broken and ragged appearance; *humilis* – denoting fair-weather cumulus clouds, showing small vertical development and sometimes flattened; *mediocris* – denoting cumulus clouds of intermediate vertical growth; *congestus* – denoting active and developing cumulus clouds, often with the appearance of a cauliflower; *calvus* – denoting an intermediate stage between cumulus and cumulonimbus in which no cirriform part can be distinguished; *capillatus* – denoting a cumulonimbus with distinct cirriform appearance frequently in the form of an anvil. If any further distinctive properties of arrangement or transparency are apparent then these are distinguished as varieties. In this way, three names could be used to classify a particular cloud. Cirriform clouds are normally above 5000 m, alto-clouds from 2000–5000 m, and stratiform (St, Sc, Ns) at lower levels. Cumulus, cumulonimbus, and to some extent nimbostratus develop vertically and so do not fit this height classification very well.

cloud forest A type of forest that grows on tropical mountains at high altitudes. The trees obtain most of their moisture as water that condenses out of the air as clouds.

cloud patterns From satellite photographs it is evident that clouds exhibit a degree of large-scale organization that produces distinctive patterns. In the analysis of these photographs, patterns such as vortices, spirals, comma-shaped masses, wave clouds, lines, and striations may be distinguished.

cloud seeding A method of trying to induce clouds to give precipitation. It attempts to simulate the natural precipitation mechanism either by adding dry ice (frozen carbon dioxide), which causes spontaneous freezing of water droplets and increases their rate of growth, or by adding silver iodide, which has a similar crystal lattice to ice and so acts as ice nuclei to assist the natural processes.

Although some success has been apparent, it is difficult to prove that seeding can increase rainfall because the clouds that were seeded may have given rain naturally and the observed increases are not sufficient to disprove this possibility.

Cnidaria A phylum of simple multicellular animals that includes the jellyfish and CORALS. They are aquatic, mainly marine, and may be colonial or solitary, attached or free-swimming. There are two types of individual: the *polyp*, which is cylindrical and sedentary, and the *medusa*, which is disk-shaped and free-living. The Cnidaria contains three classes: the HYDROZOA, which have both polyps and medusae; in their life cycle; the SCYPHOZOA (jellyfish), which have only medusae; and the ANTHOZOA (e.g. corals), which contain only polyps. The corals are geologically the most important cnidarians because many of them are colonial, forming reefs, and possess hard parts, which can be preserved.

coal A carbonaceous deposit formed from fossil plant remains. Coalification proceeds from partially decomposed vegetable matter such as peat, through lignite, subbituminous coal, bituminous coal, semibituminous coal, semianthracite, to anthracite. During this process the percentage of carbon increases and volatiles and moisture are gradually eliminated. These are all woody or humic coals. Another group are termed sapropelic coals, and these are derived from algae, spores, and finely divided plant material.

coalescence The process operating in some clouds whereby cloud droplets increase in size, eventually forming raindrops. Collision is also important in this precipitation mechanism.

Coal Measures The uppermost of the three lithological divisions of the CAR-

BONIFEROUS System in Britain. It corresponds to the upper half of the Upper Carboniferous.

coastal deposit All the deposits found at any particular coastal location are derived from one or more of the following four sources: (1) the cliffs along the coast in that vicinity; (2) inland, transported by rivers or the wind; (3) offshore, moved landward by wave action; (4) farther along the coast, transported by LONGSHORE DRIFT (this material must at one time have been derived from one of the other three sources). The nature of the materials and their quantities will depend upon the rock types found along the coast and inland and upon their relative contributions. A whole range of deposits, from massive boulders through shingle and sand to fine clays, can be found in different coastal sites.

coastal plain A gently sloping plain leading from the foot of inland upland areas down to the coast, and largely continuous with the continental shelf under the sea. Such plains are formed by the continuous processes of erosion in the inland areas, transportation of material seaward, and its deposition either on land, near the coast, or out to sea. Some coastal plains have resulted from the comparatively recent emergence of parts of a flat continental shelf due to a relative fall in sea level.

coastline The landward limit of the beach, the boundary between the coast and the shore, or the line that forms the boundary between the land and the water. Perhaps the simplest application of the term is to take the coastline as the extreme upper limit of direct wave action, which is the limit of storm wave swash during equinoctial spring tides. The term is also used to describe the type of coast as seen from a boat, for instance a rugged coastline.

coastline of emergence A shoreline that reveals evidence that the land has risen with respect to the sea. Typical features include wave-cut platforms with overhanging cliffs, and raised beach terraces.

coastline of submergence A shoreline that reveals evidence that the land has sunk with respect to the sea. Typical features include many bays and continuing erosion of cliffs and headlands by wave action.

cobble A rounded fragment of rock that is larger than a pebble but smaller than a boulder. The accepted size range is between 65 mm and 255 mm across.

coccoliths Minute round calcareous plates that formed part of the protective covering of a group of unicellular algae of the phylum Haptomonada (or Haptophyta). They have been reported from the Upper Cambrian but were more common in the Mesozoic, especially the Cretaceous, when they contributed largely to the formation of chalk.

coesite A very dense form of QUARTZ, SiO_2, stable at high pressures and occurring in and around craters caused by meteorite impact. *See also* silica minerals.

col **1.** A pattern of isobars resembling a geographical cob or saddle, i.e. a shallow dip at a high level. A col that is elongated along the high-pressure axis is known as an *anticyclonic col*, and if along the low-pressure axis is called a *cyclonic col*. Winds are light in this type of pressure system but the pattern is rarely long-lasting.
2. A depression forming a pass over a mountain ridge or other high ground, commonly formed by back-to-back cirque development or from the beheading of a dip-slope valley in a cuesta formation by scarp retreat.

cold front A concentrated thermal gradient whose movement is such that warm air is replaced by cold air. Such fronts are usually found to the rear of a depression and are associated with a sudden veering of the wind, often with a marked increase in speed, a fall of temperature, a belt of cloud, and sometimes heavy precipitation. The front slopes upward into the atmosphere at an angle of about 1 in 50. All these characteristics refer to the ideal front, and whereas most have somewhat similar properties, each individual will have slight differences. *See also* warm front.

cold glacier (polar glacier) A glacier within which the temperature remains so low that pressure-melting does not occur. In contrast to temperate glaciers these contain very little or no meltwater to act as a lubricant for movement. The glacier may remain static, being frozen to the bedrock, and requires great shear stress to induce movement, which will be very slow. These glaciers contain very little ENGLACIAL material. However, cold glaciers have been found with large bed loads, and because these can only become accumulated as a result of basal freezing and thawing (*see* glacial plucking) it is believed that the erosion responsible may have taken place during a period of warmer climate.

cold pole That location in each hemisphere where the lowest air temperatures have been measured. In the N hemisphere this is the area of NE Siberia where at both Verkojansk and Oimjakon, –68°C (–90°F) has been recorded. The mean January temperature at the former station is –51°C (–60°F), which represents the coldest spot on average. In the S hemisphere, the high-altitude research stations on the Antarctic Ice Plateau have the record with a temperature of –89.2°C (–128.6°F) at Vostok. Both cold poles are in areas where radiational cooling is extreme under clear skies and low relative humidity.

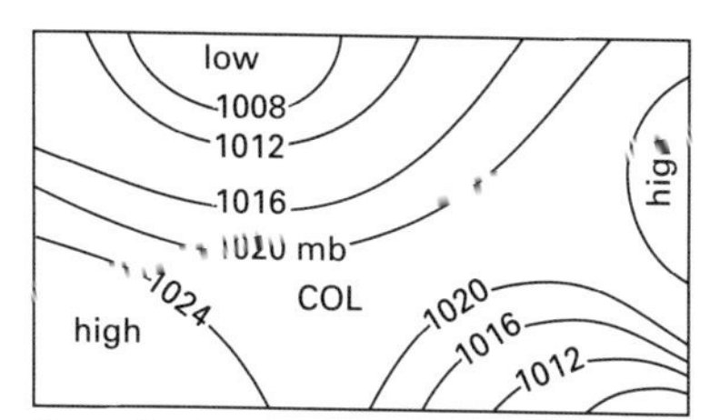

Fig. 1: Anticyclonic col

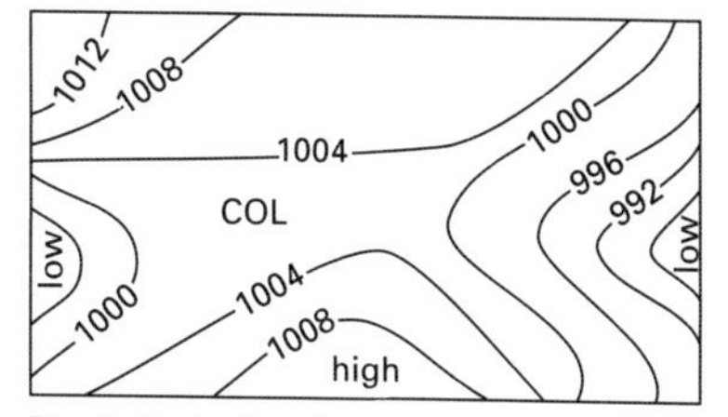

Fig. 2: Cyclonic col

Pressure patterns in a col

cold pool An area of cold air within the atmosphere surrounded by air of higher temperature at that level. Because this usually leads to a steepening of the lapse rate, it may lead to instability and the development of showers.

colemanite A white or colorless mineral form of hydrated calcium borate, $Ca_2B_6O_{11}.5H_2O$. It crystallizes in the monoclinic system and occurs as massive crystals in clay nodules originating in Tertiary lakes. It is an important source of boron.

collision zone The region where tectonic movements have caused two continents to collide, with the closure of the ocean between them.

colluvium (talus) The weathered debris (scree, mud, etc.) accumulated at the foot of a slope, which has originated from the erosion of the rockface above.

colony An association of animals of the same species having structural and physiological connections between the individuals. Examples of colonial animals are the BRYOZOA and the reef-forming corals (*see* Anthozoa).

color index An index of the relative proportions of light and dark minerals in igneous rocks. The following divisions are made according to the percentage of dark minerals:

leucocratic 0–30%
mesocratic 30–60%
melanocratic 60–90%
hypermelanic 90–100%

The terms felsic and mafic are synonymous with light and dark minerals respectively. In nearly all cases dark minerals are ferromagnesian minerals.

columbite A mineral that consists of a mixed niobate and tantalate of iron and manganese, $(Fe,Mn)(Nb,Ta)_2O_6$, in which there is more niobium than tantalum. It is so called because niobium was once known as columbium, and is found in granites and pegmatites. *See also* tantalite.

columnar joint One of a series of hexagonal or pentagonal joints resulting in columns of igneous rock, especially basalt lava. A good example is at Fingal's Cave, Staffa, Scotland.

comagmatic Describing a series of igneous rocks that are derived from the same original magma through the various processes of differentiation. *See* magmatic differentiation.

comber A wave traveling in deep water, which has a steep and high-breaking crest. The crests, much larger than whitecaps, are blown forward under powerful wind action. The term has also been applied to certain types of spilling BREAKERS.

comfort index A measure of climatic comfort experienced by humans. It is assumed that the ideal comfort state is between 40 and 70% relative humidity and air temperature between 60°F (15.5°C) and 80°F (27°C). The index is subjective in that individuals may prefer lower or higher temperatures depending upon their metabolism, but it does give an assessment of climate or weather conditions relative to an ideal state.

community (in ecology) Any group of plants and animals that occupy the same habitat and interact in various ways. Usually the dominant feature, such as a typical type of plant, gives its name to the community (e.g. beechwood).

compaction The compression of a sediment with the result that it occupies less volume (i.e., the layer becomes thinner). It is generally caused by the weight of more recent material being deposited above. It may also be caused by earth movements or, on soil, by human activity. Compacted soil is less porous, so that puddles and muddy patches may form after rainfall.

compensation level A hypothetical level within the Earth above which all columns of rock material having a unit cross-sectional area must have the same mass. *See* Pratt's hypothesis.

competence **1.** The largest size of grain that a stream can move as bed load. It is determined mostly by the stream velocity at the bed, which is controlled in turn by several factors, especially stream gradient, and the ratio between width and depth: bed velocity is higher, with other things equal, in a wide shallow river than a deep narrow one. With the larger grain sizes (coarse sand and gravel) there is a sixth-power relationship between velocity and competence. In any one river, competence varies with flood discharge over time; over space it varies with turbulence in the form of eddies rising from obstructions on the bed, which exert lift, momentarily increasing the maximum size of liftable debris.
2. The degree to which beds of rock can be folded without flowage or change in thickness. Such strata respond to deformation by fracture rather than flow.

compiled map A map that is produced from other large-scale mapping rather than original survey work. Compiled maps are usually reductions of other maps to which selected amendments have been made before reduction.

complex A complicated body of rock that cannot be designated by any of the other formal terms used in lithostratigraphy (*see* stratigraphy). A complex is usually composed of a variety of rock types and it may have distinctive structural characteristics. The term may also be used informally to refer to any complicated association of rocks.

composite log The interpreted geologic column encountered in a well, reconstructed from the evidence of geologic sampling and electrical investigations carried out in the borehole.

composite map A map, usually compiled, that brings together for ease of comparison data usually portrayed on two or more maps, e.g. land use and population distribution.

composite volcano (stratovolcano) A volcano constructed of alternating layers of lava and pyroclastic deposits, generally developed by volcanoes with fairly viscous lavas, for example, andesites.

concave slope A hill slope that is steeper at the top than at the bottom. On a map of the feature, contour lines are closer together at the top of the slope. *See also* convex slope.

concentric dike (ring dike) A roughly circular vertical dike developing when the circular fractures produced by the upward pressure of magma in a chamber are intruded by magma. The related structures known as *cone sheets* are developed by a similar stress distribution. They surround a point source (the intruding magma) and dip inward toward it, having the form of a series of concentric conical sheets infilled by magma.

conchoidal Describing a type of curving fracture marked by concentric arcuate ridges. It is characteristic of minerals such as quartz and glassy and aphanitic rocks, particularly obsidian and phonolite.

concordant coast (longitudinal coast) A coastline that parallels the hills and other features just inland of it. If the land sinks with respect to the sea (a coastline of submergence), the sea floods the valleys between the hills, which become a chain of islands.

concretion A roughly spherical body of material precipitated from a solution.

condensation The physical process of transformation from the vapor to the liquid state. In the atmosphere, condensation refers to the formation of liquid water from water vapor as a result of the atmosphere becoming saturated; for this to take place cooling must occur. If the atmosphere were absolutely pure, no condensation would take place until extreme supersaturation existed, but throughout the troposphere there are abundant nuclei on which condensation can take place on saturation. These nuclei consist of tiny crystals of sea salts, industrial and natural smoke particles, as well as volcanic, soil, or desert dusts. On condensation, latent heat is released equivalent to about 600 calories (2500 joules) per gram of water, which reduces the rate of cooling.

condensation level The level in the atmosphere at which CONDENSATION takes place. It is normally used for calculating CLOUD BASE on a thermodynamic diagram, by simulating the lifting of air of known properties of temperature and humidity.

conditional instability The state of a part of the atmosphere in which the ENVIRONMENTAL LAPSE RATE lies between the dry adiabatic and saturated adiabatic lapse rates. This means that if a parcel of air rises from the ground it will cool at the dry adiabatic lapse rate (DALR) and so remain cooler than the environment; under these conditions the atmosphere is stable. However, if the parcel reaches saturation it will then cool at a slower rate and could become warmer than the environment on further uplift, and so become unstable.

conduction The mechanism by which heat is transferred through a substance by molecular motion without overall motion of the substance itself. Air is a very poor conductor of heat and this method of transfer is relevant only when the air is absolutely calm. This may occur very close to the ground surface, but in most studies of the atmosphere, conduction can be neglected.

conduit *See* vent.

cone of depression A conical dip that occurs in the water table around a well as water is removed from the well. As a result, the water level in the well gets lower and lower.

cone sheet *See* concentric dike.

confidence limit A statistical concept used to assess the degree of confidence that can be attached to estimates of true means or standard deviations when based on only a sample of the data. For example, if the annual rainfall totals for a particular location are only known for 15 years, then the mean of this sample ($\bar{x}$) will probably differ from the true mean. By placing confidence limits it can be stated that with a 95% probability the true mean for this location will be $\bar{x} \pm 2y$, i.e. between $\bar{x} - 2y$ and $\bar{x} + 2y$, where y is the standard deviation of the data set.

confining pressure (in geophysics) The pressure exerted on a point resulting from the weight of the overlying rock or water column.

confluence The rate of convergence or approach of one streamline to the adjacent streamline relative to the direction of flow. It differs from CONVERGENCE in that it does not imply an increase of mass but simply the approach of streamlines of air flow. *Compare* diffluence.

conformable **1.** Comprising an unbroken sequence of rock strata with no angular discordance between them, marking a continuous period of uninterrupted deposition. *See also* discordant.
2. *See* accordant.

conformal projection *See* orthomorphic projection.

congelifraction The weathering of rocks through the freezing of water within pore spaces. The accompanying expansion creates stresses, which result in fracture of the rock and its consequent disintegration. The effectiveness of the process varies considerably with rock type, being dependent on the size of the pores. If these are too large, or open to the surface, the water may be forced out as temperatures fall, whereas if they are too small, the water remains in a liquid state despite supercooling. The process is most effective in schists and certain limestones, notably chalk, within which very rapid shattering is possible. *See also* frost-shattering.

congeliturbation The movement of soil materials in the PERIGLACIAL environment, as a result of freezing and thawing of water contained within the soil. Freezing is accompanied by an expansion in the soil, but this is irregular because in a heterogeneous material water will not be evenly distributed. Similarly, thawing takes place at uneven rates, and the result is that particles are transported considerable distances, even on flat ground. On slopes, material migrates downslope under the influence of gravity. The ice causing the disruption tends to be concentrated in the silt and sand fraction, coarser material being moved to the surface.

conglomerate A RUDACEOUS sedimentary rock in which the constituent clasts or fragments are more or less well rounded. *Compare* breccia.

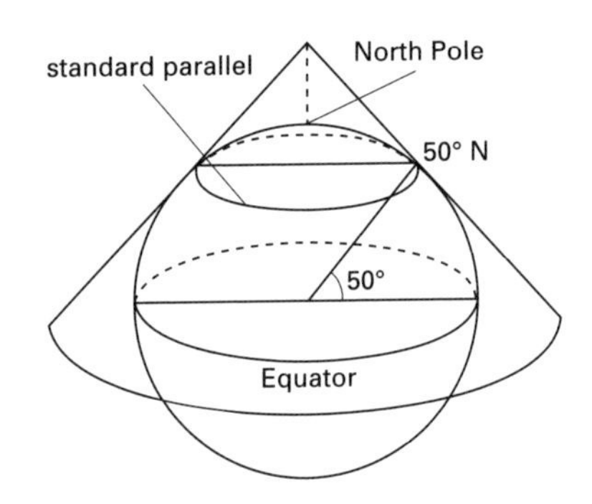

Simple conical projections

conical projection A map projection produced by projecting the meridians and parallels onto a cone (see diagram). The center of curvature of the cone is not one of the poles, although the meridians are drawn radiating from it. The parallels are concentric circles and are their true distance apart. The scale is only correct along the standard parallels; all others are too long. This projection is neither equal-area nor orthomorphic and is usually modified, e.g. BONNE'S PROJECTION and the conical with two standard parallels.

coniferous forest A forest of cone-bearing GYMNOSPERM trees. The trees are mainly evergreens (larches are an exception) with needle-shaped leaves, such as firs, pines, and spruces. Coniferous forests occur farther north than DECIDUOUS FORESTS because the trees are better able to withstand cold temperatures. Trees from the forests, which are often managed, provide timber for building and wood pulp for paper making.

conjugate joint Any of a system of joints in rocks, in which the sets of joints are of related origin. It generally consists of two sets, which intersect at 90° but share a common dip or strike.

conodont A small toothlike structure up to 2 mm long and composed of calcium phosphate. Conodonts are found in rocks of marine facies, especially shales, from Ordovician to Permian age and also in later strata, in which they are suspected as being DERIVED FOSSILS. Because they are small, change in morphology through time, and are widely distributed, conodonts are particularly useful in the correlation and identification of the rocks in which they occur. They have been variously attributed to most phyla but their zoological affinities remain unknown (*see* species); their widespread distribution suggests that they may have belonged to a group of pelagic animals.

Conrad discontinuity A seismically detectable boundary within the Earth's continental crust, separating it into a lower (basic) layer and an upper (granitic) layer. Beneath the ocean floors the upper layer is missing.

consanguinuity The common features of different igneous rocks that probably arose from the same original magma. They usually occur close together, are of about the same age, and have similar chemical compositions. *See also* comagmatic.

consequent stream A stream whose course is dictated by the slope of the land. *See also* obsequent stream; subsequent stream.

conservation (in ecology) Any measures taken to preserve resources, such as the soil, minerals, and living organisms within the environment. Methods include preserving habitats and eliminating waste of raw materials, and recycling those that are used.

conservative plate boundary *See* plate boundary.

conservative property A property of air that remains unaltered when affected by specified processes. For example, when air is cooling at the dry adiabatic lapse rate, its potential temperature remains constant and is therefore a conservative property under these conditions.

constancy of wind An index of the degree of constancy of wind direction. It is measured by determining the ratio of the vector mean wind to the scalar wind, and converting to a percentage value. It ranges from 0 if all winds are equally strong and frequent in all directions to 100 if the winds are constant or unvarying in direction. For the British Isles, values of between 15 and 50 are typical, but in the trade wind zones it can approach 90.

constant slope The scree or talus slope, i.e. the foot slope produced by the accumulation of weathered rock debris from above. The evolution of the constant slope begins with a vertical rock face at the start of the cycle and at its peak it covers the free face and buries it. At this stage the constant slope, whose angle is defined by the ANGLE OF REST of the debris of which it is composed, dominates and largely determines the hillslope form. Later on, the washing of fines from the constant slope leads to its burial by a wash slope, or alluvial toe-

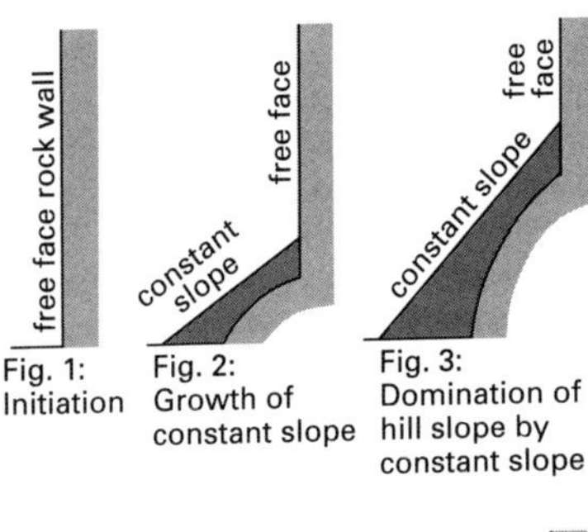

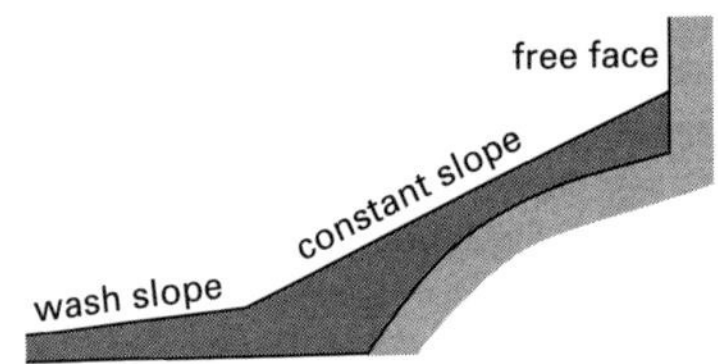

Fig. 4:
Wash of fines from constant slope evolves wash slope, which buries constant slope and ultimately dominates hill slope form

Stages of slope evolution

slope, which accumulates at the foot of the constant slope. *See also* colluvium.

constructive plate boundary *See* plate boundary.

constructive wave A wave that assists in the aggradation of a beach. For aggradation to occur the material moved landward must exceed that moved seaward, so the SWASH must be more efficient than the BACKWASH. For a powerful swash the waves must be flat, although their frequency cannot be too high, because the swash of a new wave would be interfered with by the preceding wave's backwash. The usual frequency of constructive waves is between six and eight per minute. *Compare* destructive wave.

consumer (in ecology) An organism that is above the first level in a FOOD CHAIN, which is occupied by producers. Most PRIMARY CONSUMERS are herbivores, feeding on plants. Secondary consumers are carnivores, feeding on the herbivores or other carnivores.

contact The surface forming the boundary between adjacent and distinct bodies of rock. It may be conformable within a sedimentary succession, unconformable, or produced by intrusion or faulting. The recognition of contacts is fundamental to LITHOSTRATIGRAPHY and much geologic interpretation.

contact metamorphism (thermal metamorphism) The changes resulting from the intrusion of hot magma into preexisting cold country rock. The thermal effect of the magma causes a recrystallization of minerals in the country rock and the growth of new minerals. Contact metamorphic rocks have equigranular textures, have suffered little or no deformation, and in general do not possess any foliation. Thermal effects die out rapidly with increasing distance from the contact with the igneous rock. Sedimentary rocks are converted to hornfels, quartzite, and marble. Hornfels often contain porphyroblasts or spots of contact metamorphic minerals such as biotite, andalusite, cordierite, and garnet. With the exception of the expulsion of water from the country rocks, the metamorphism is largely an isochemical process.

Four facies of contact metamorphism are recognized:

albite-epidote hornfels facies
hornblende hornfels facies
pyroxene hornfels facies
sanidinite facies.

In an ideal metamorphic aureole, pyroxene hornfels rocks occur adjacent to the intrusion and pass into hornblende hornfels facies rocks farther from the contact. Inclusions of country rock caught up in the igneous magma reach the high temperatures associated with the sanidinite facies. Such rocks may become partly fused, resulting in the formation of glass, and are known as *buchites*. *See also* metamorphism; skarn.

continent A large landmass, which rises more or less abruptly from the deep ocean floor. Associated with the exposed landmass are marginal areas (CONTINENTAL SHELVES) that are shallowly submerged. The continents occupy about 30% of the Earth's surface.

continental borderland A marine region adjacent to, or lying in place of, a continental shelf, and displaying marked topographic irregularity. A frequently quoted example lies off the coast of Southern California, a region of very diverse submarine morphology, which includes submarine ranges and basins. Some of the basins are quite deep. Often the adjacent landmass displays similar topography. The depths of continental borderlands are usually far in excess of the depths normally encountered over shelf areas.

continental climate The generalized characteristic climate of continental interiors, implying a low precipitation total, a large annual range of temperature with warm summers and cold winters, and low relative humidities.

continental crust The part of the Earth's crust constituting the continents, often referred to as the SIAL. It is composed of granite rocks and reaches a thickness in excess of 50 km under mountain regions, although it is generally only 33 km thick. The lower continental crust was formerly believed to be of basaltic composition, similar to that of the oceanic crust. It is now thought that it may simply be a more dense phase of the granitic rocks. *See also* oceanic crust.

continental drift The movement of the continental blocks relative to one another across the surface of the Earth, as a result of SEA-FLOOR SPREADING. The hypothesis of continental drift was proposed almost 100 years ago, but it is only with the advent of the PLATE TECTONICS theory that a viable mechanism was available to explain the movements of the continents. It is thought that the present continents were grouped together in pre-Mesozoic times into two large supercontinents, Pangaea and Gondwanaland. These continents subsequently broke up to form the present continents. The presence of these supercontinents was based upon the work of Alfred Wegener in 1910. His evidence was only qualitative but included distributions of rock types, flora, fauna, geologic structures, and similarity of the shape of the coastlines on either side of the Atlantic. Geophysicists of the time dismissed the hypotheses as impossible. However, in the late 1950s–1960s geophysical evidence supported the theory.

continental heat flow The flow of heat reaching the surface of the Earth in a continental as opposed to oceanic region.

continental margin (continental terrace) The combined continental shelf and slope zones, whose modes of origin are generally closely related. The depth range of the continental margin may, therefore, be taken as extending from the coast out to a depth of some 2000 m. Some continental margins are of a constructional kind, others diastrophic. Warping, faulting, and long-continued sedimentation have been shown to have played dominant roles in their evolution. Deep seismic soundings have revealed that many have thick layers of sedimentary rocks overlying older basement rocks of a crystalline nature.

continental rise The submarine surface between the abyssal floor and the base of the continental slope. Together with the slope, the rise forms a separation zone between the continental shelves and the deep ocean basins. The surface of the rise is generally smooth, with gradients between 1:100 and 1:800. The minor irregularities rarely amount to more than 20 m or so in height or depth. The rise may be up to several hundred kilometers wide, for example off part of the North African coast where it is over 600 km wide. On the other hand, it may be virtually absent, as off the Bay of Biscay. In some areas, quite large seamounts are found on the rise.

continental shelf That part of a continent that is shallowly submerged by the sea. It is a gently dipping area extending from the shoreline down to the continental slope at the SHELF EDGE, or to a depth of 200 m if the slope is absent. The average shelf width is about 70 km but shelves tend to be wide off low-lying regions and narrow off mountainous regions, varying from less than 1 to more than 1000 km wide. Average gradients tend to be steeper near the coast than over the outer shelf. Globally, shelves occupy some 7.6% of the ocean floor. Most shelves exhibit evidence of marine erosion, especially in shallow coastal waters. Some have carved within them valleys, troughs, and basins (*see* submarine canyon), often formed during low sea-level stands. Many are veneered with loose sediments that are being moved and sorted by wave and tidal action, and then molded into tidal banks and channels.

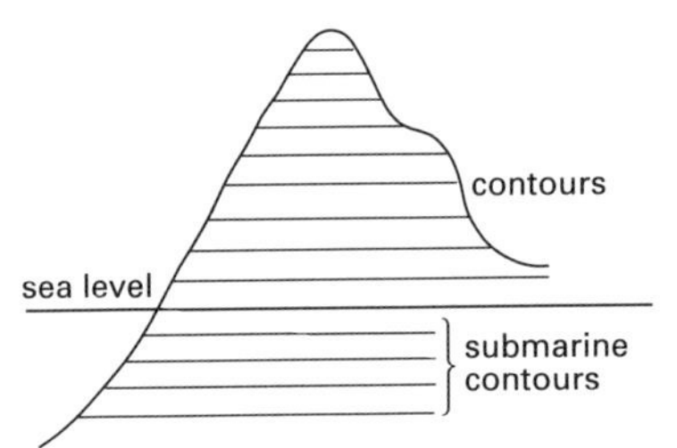

Fig. 1: Cross section of coast and hill to show contour lines

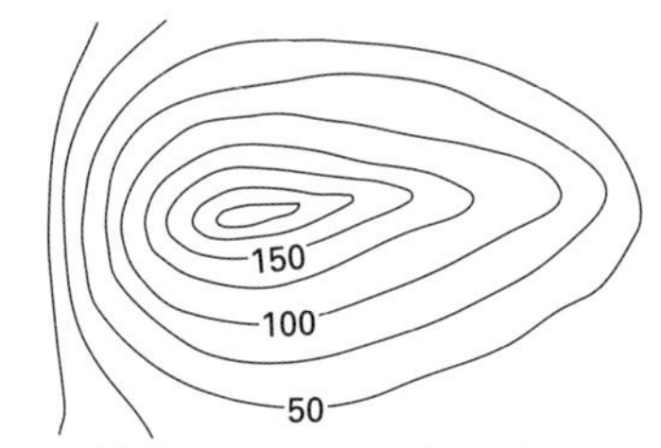

Fig. 2: Hill viewed as from above (as on maps)

Contours

continental slope A portion of the continental margin, dipping more steeply than the continental shelf, with a gradient of more than one in forty, although this varies considerably. It is bounded by the continental shelf on the landward margin, and the continental rise on the seaward side. Globally, the continental slope zone makes up about 8.5% of the total ocean floor, but apart from some SUBMARINE CANYONS that have been carefully explored and surveyed its precise form is not known. Where there is abundant sediment, especially mud, slumping and sliding is common and results in very irregular topography.

contour A line drawn on maps and charts joining points of the same elevation, above or below sea level (see diagram). This line enables the user of the map to determine the elevation and slope of the band or sea bed (contours below sea level are called submarine contours). Measurements may be in meters, feet, or fathoms (submarine). *See also* contour interval; hypsometric tinting; supplementary contour.

contour interval The difference in height between two consecutive index contours. The interval will depend on several factors: the scale of the map, the terrain the map portrays, and the purpose of the map. The scale of the map determines the maximum possible number of contours per area to retain clarity of detail; the terrain influences the number of contours required to depict accurately the true nature of the land (in level areas a few contours will show the land characteristics whereas in hilly areas more contours are needed to show the elevation variations within the area); the purpose of the map also affects the interval: plans for builders and land surveyors require the maximum number of contours per area to enable accurate measurements to be made.

control point A point on a map whose position and usually elevation are known. Control points form a network around which other map detail is plotted in its correct position, azimuth, and elevation.

Conulariidae A family of fossil animals of uncertain zoological affinities. It has been suggested that they belong to the MOLLUSCA, the ANNELIDA, or the classes Hydrozoa or Scyphozoa of the CNIDARIA. Some authorities place them, with a few other similar groups of animals, in a separate and extinct phylum, the Conulariida. They occur in sediments of marine facies and typically have a skeleton constructed from four flat triangular plates joined at the edges to form an inverted four-sided pyramid. The plates are formed of calcium phosphate and have an ornamentation of transverse ridges. Projections from the free basal edges of these plates fold over to protect the terminal opening. Conulariids are found in rocks dating from the Cambrian to the Triassic; in size they range from about 5 cm to over 20 cm.

convection (in meteorology) The most important form of heat transfer within the atmosphere. It takes place by the physical transfer of air resulting from density variations, warmer air being less dense than cool air. Two types of atmospheric convection are distinguished: (1) free or natural convection, which is essentially the transfer of warm air from lower levels of the atmosphere through the mechanism of thermals or bubbles; (2) forced convection, which is the vertical movement of air produced by mechanical rather than thermal forces, such as occurs through movement over uplands or irregular vegetation surfaces.

convection cell A continuous convection system, air that rises being replaced by air from elsewhere. This can occur on a small scale with a single thermal or convection cell rising above a plowed field heated by the Sun. At the other extreme, the meridional circulation within the Tropics has been likened to a vast convection cell with zones of rising air and the trade winds linking the vertical parts of the cell. This system is called the HADLEY CELL.

convection current A current within a fluid created by differences in temperature between layers causing differences in density. Convection currents are thought to operate in and beneath the oceans. They operate in the ocean when a portion of a water mass that is denser than the water beneath it tends to sink, thereby being replaced by less dense water. In this context, the currents caused in such regions as the Red Sea and the Mediterranean Sea by excessive surface evaporation, increase in surface salinity, and consequent sinking are claimed by some to constitute a convection current. Convection currents are also held, by some authorities, to occur within the Earth's mantle beneath the oceans. One suggestion is that the higher temperatures under the oceans at certain depths could be a consequence of convection flow that must be operating at some depth within the mantle. This could result in an increase in the temperature of certain parts of the ocean floor, such temperatures having been recorded. *See also* density current.

convective rain Rain produced by the strong but localized vertical motion in convectional clouds found in an unstable atmosphere. This is usually produced from cumulonimbus clouds, but cumulus clouds may also give slight rain. Because of the intensity of the upward movement within the clouds, convective rainfall is normally heavy but restricted in distribution at any instant.

convergence 1. A measure of the rate of inflow of a fluid into a certain volume. The reverse state is DIVERGENCE. In the atmosphere, these terms are used to indicate the horizontal components of wind velocity: convergence indicates retardation and divergence indicates acceleration. Surface convergence is usually accompanied by upward air currents. In wave refraction phenomena, convergence refers to the bunching together of the various wave orthogonals in the direction of wave advance. Wave energy tends to be concentrated in a zone of convergence. In the case of opposing currents, a convergence is the boundary or zone at which these currents meet. Usually convergence zones experience the sinking of some ocean water to greater depths.
2. *See* convergent evolution.

convergence zone An area of the atmosphere in which convergence prevails and which is characterized by a belt of convectional clouds, often with rain. Such zones can be found within otherwise uniform air streams giving a short period of heavy rain as they pass. The best known is the INTERTROPICAL CONVERGENCE ZONE, which is frequently seen on satellite photographs and can extend for several thousand kilometers.

convergent evolution (convergence) The evolution of similar characters and structural modifications in organisms that are not closely related. For example, dolphins and ichthyosaurs, although only distantly related, have a very similar external form that is presumed to have been acquired through adaptation to similar environmental conditions. Convergent evolution also occurred in marsupial and placental mammalian groups, which were isolated on different landmasses but occupied similar habitats. Such organisms are known as *homeomorphs*.

convergent plate boundary A region where tectonic plates are moving together and the crust occupies a smaller area. It may involve subduction, in which one plate dips below the other and is consumed into the mantle. Alternatively, one plate may force the crust on the other plate to pile up and form a new range of mountains. *See also* plate boundary; subduction zone.

convex slope A hill slope that is steeper at the bottom than at the top. On a map of the feature, contour lines are closer together at the bottom of the slope. *See also* concave slope.

convolute bedding Highly disturbed, folded, or crumpled laminae that are found within a single well-defined sedimentary layer, resulting from deformation of the sediment by slumping, gliding, loading, upward expulsion of water, or the effect of passing eddies in the overlying water mass, while the sediment is still in an unlithified state.

coombe A small hollow in a steep chalk hillside. It generally results from weathering, which opens and enlarges a joint in the rock. A mound of weathered material, called coombe rock, may slide down the hill and accumulate at the bottom of the slope.

coombe rock (head) A structureless deposit consisting of unweathered flints within a matrix of chalk mud and disintegrated chalk, resulting from SOLIFLUCTION movements during the PLEISTOCENE period. The name is especially applied to the deposits found in the chalk areas of S England.

coordinate A linear or angular quantity in a frame of reference such as a grid, by which positions of points on the map are defined.

coprolite The fossilized feces of an animal. Coprolites may contain other fossils, such as fragments of other animals, leaves, or pollen, that formed the diet of the creature concerned and they thus provide valuable information for paleoecology.

coral One of a group of animals belonging to the cnidarian class ANTHOZOA. They secrete a hard calcareous skeleton and are important geologically in BIOSTRATIGRAPHY and because of their reef-building activities. *See* coral reef.

coral reef A ridge of CORAL that forms offshore in mainly tropical regions. For coral animals to survive, the water should be shallow (allowing the penetration of sunlight) and temperatures should be above 21°C. The reef may be partly exposed at low tide but is always covered at high tide. Coral reefs are often very steep-sided. Close to the shore they are termed fringing reefs; barrier reefs form farther out to sea. *See also* reef.

corange line A line drawn on certain hydrographic charts joining points of equal tidal range.

cordierite An orthorhombic mineral, bluish in color and of composition $Al_3(Mg,Fe^{2+})_2Si_5AlO_{18}$. The basal section of cordierite has a hexagonal outline and often shows concentric twinning. The mineral is usually found in thermally metamorphosed argillaceous rocks, where it is associated with andalusite and biotite. It also occurs in some high-grade gneisses and rarely in gabbro that has been contaminated by argillaceous sediments.

cordillera A chain of mountains, such as the parallel chains of the Andes in South America or the Rockies, Sierra Nevada, and Coastal Ranges of North America.

core The central sphere within the Earth, separated into inner and outer units. The INNER CORE has the properties of a solid, whereas the OUTER CORE prevents the passage of S waves, suggesting that it is a liquid. It is composed of a mixture of iron and nickel, together with some dissolved silicon and sulfur, and is separated from the mantle above at a depth of 2900 km beneath the Earth's surface by the GUTENBERG DISCONTINUITY.

core sampling The collection of reasonably undisturbed sediment samples from the sea floor for oceanographic studies. The samples are corelike in form and can be split to yield a section of seabed material. A wide range of coring devices have been developed. Gravity cores drop under their own weight and drive themselves into the seabed. Piston cores are placed on the seabed and forced into it to retrieve a sample. In the case of hard-packed sand a vibro-corer may be used, the vibrator operating after the coring apparatus has reached the seabed. Some cores have been up to 20 m in length.

corestone A rounded hard block of rock within a matrix of soft decomposed material, occurring in well-jointed rocks, notably granites. When the jointed rocks are subjected to extensive weathering, it proceeds more rapidly in the joint areas, where water can penetrate, resulting in a series of corestones, but continued weathering will bring about their eventual disappearance. Individual corestones, however, may be especially resistant to weathering and these will remain as isolated unweathered blocks within a mass of incoherent weathered rock.

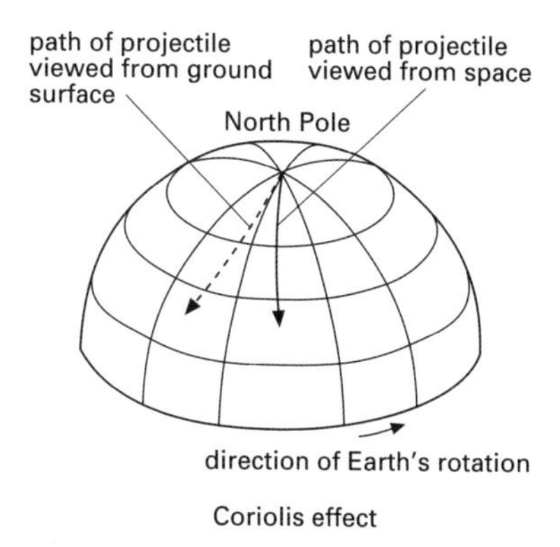

Coriolis effect

Coriolis effect A hypothetical "force" that is required to relate movement in the atmosphere to the rotating coordinate system of the Earth's surface. For example, if a projectile was launched from the North Pole with a constant southward trajectory, from space it would appear to maintain this track. However, as the Earth is rotating from west to east, to an observer on the ground surface it would appear to be constantly deviating to the right (see diagram). In Newtonian mechanics, a body will maintain its course unless acted on by other forces, so to explain this deflection the Coriolis "force" is introduced. Its magnitude is obtained from the equation $2\omega v \sin\phi$, where ω is the value of the Earth's angular velocity, v is air velocity, and ϕ the latitude. It therefore varies from zero at the Equator to a maximum at the poles. Without the Coriolis effect, wind flow would be at right angles to the pressure gradient, as found in equatorial areas, instead of being parallel to it. It is named after C. G. Coriolis, a French mathematician.

corona structure The growth of a concentric mineral zone around a core mineral. The second mineral may be a REACTION RIM or the result of EPITAXIAL GROWTH upon the primary mineral.

corrasion A type of erosion involving abrasion of a rock surface by small fragments of rock carried along by a river or glacier. *See also* abrasion.

correlation A statistical technique that determines the degree of association between two sets of data. The most common method in climatology is to use the product moment correlation coefficient whose value ranges from +1 for a perfect direct relationship through zero for no association whatsoever to –1 for a perfect inverse correlation. The data must have a normal distribution for this coefficient, but others that do not have this constraint are available. The product moment coefficient is most frequently used for comparing rainfall values at different locations. The best known was the work of Walker on Indian monsoon rainfall, who found significant associations between rainfall in India and sea-level pressure at various locations around the world.

In stratigraphy, correlation refers to the process of equating parts of isolated geologic successions, usually in relation to time (*see* chronotaxis; homotaxis). Strata may be correlated on the basis of similarities in lithology or biostratigraphically on the basis of similar contained fossils. Other methods employ radiometric age determinations or paleomagnetic information (*see* paleomagnetic correlation).

corrie *See* cirque.

corrosion The process of erosion by chemical solution. All common rock-forming minerals are to some extent soluble in water, even quartz, generally considered to be highly resistant, and so removal in solution is a significant process. In humid areas, corrosion renders limestone one of the most easily eroded of rocks, whereas in arid zones, where water is lacking, limestone is highly resistant, demonstrating the effectiveness of corrosion. Rivers draining areas of organic materials, especially bogs, swamps, and marshes, are rich in organic acids, and these accelerate corrosive action. *See also* abrasion; corrasion.

corundum A mineral consisting essentially of aluminum oxide, Al_2O_3, but minor amounts of other ions give rise to a variety of colors. The blue of sapphire is due to the presence of iron and titanium whereas the variety ruby contains chromium. Corundum occurs in some silica-poor igneous rocks such as nepheline-syenites. It is also found in hornfels.

cotidal line A line drawn on certain hydrographic charts joining points at which average high water (or low water) occur simultaneously. These lines are often given as differences from the times of high (or low) water at a "standard port", or they may be expressed as time intervals after the time of the moon's transit. A *cotidal chart* depicts both CORANGE LINES and cotidal lines.

coulee A series of deep branching channels that are formed as MELTWATER is suddenly released from a lake dammed by ice at the foot of a glacier. Well-known coulees occur on the Columbia plateau in Washington, USA.

country rock The preexisting envelope of rocks into which an igneous magma is intruded. The country rock often shows the marked thermal effects of CONTACT METAMORPHISM. Xenoliths contained in the solidified magma are usually derived from the country rock.

covellite (covelline) An indigo-blue iridescent mineral form of copper sulfide, CuS. It usually occurs as compact or slaty masses in metasomatic zones or hydrothermal veins, associated with other copper minerals.

crag A rocky outcrop on a hillside, characterized by very steep edges. Most crags occur in hard-rock regions, resulting from erosion and weathering.

crag and tail A glacial erosion feature found in areas of relatively low relief, formed where a knob of particularly resistant rock exists on the floor of a valley beneath a glacier. This knob will cause deflection of the ice, protecting any less resistant rock behind it. The knob, which is invariably striated and fluted on the stoss end (facing the oncoming ice) is known as the crag, while the protected rock behind, which may also receive a covering of glacier-derived material, is known as the tail. Together they form a semistreamlined form. These features become visible only on the disappearance of the ice.

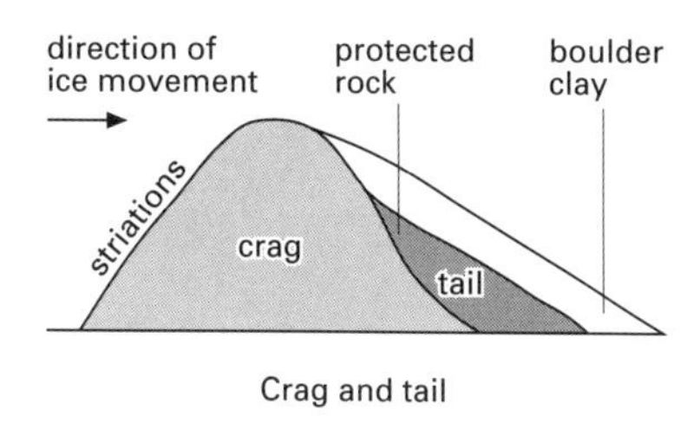

Crag and tail

crater A large bowl-shaped depression in the Earth's surface, resulting from volcanic activity, out-gassing, or the impact of a meteorite.

crater lake A lake that forms in a volcanic crater or CALDERA. The water comes from direct PRECIPITATION and runoff from the crater sides. The water may be forcibly ejected if the volcano becomes active.

craton Any of those parts of the Earth's crust that have been relatively stable (i.e. unaffected by orogenic activity) during the last 1000 million years, e.g. the Zambia nucleus of Africa. Frequently they represent the roots of deeply eroded ancient mountain chains.

creep The slow movement of soil and rock debris down gentle slopes under the influence of gravity: it is the slowest of all the types of MASS MOVEMENT, and it is not certain whether it is a flow, which is movement by internal deformation and overturning, or a slide whereby the whole mantle slips over the bedrock beneath. Current evidence rather supports the latter explanation. Creep occurs in all environments but is most important in the maritime periglacial areas, where it is of a type known as SOLIFLUCTION. It is said to occur on all nonvegetated slopes of more than 5°, but in the areas subject to solifluction it can occur on much gentler slopes.

Evidence of creep is commonly seen when structures such as tombstones and telephone poles lean upslope, as the material in which they stand has moved downslope. Many processes are involved, including freeze-thaw, wetting and drying, needle ice, and the preferential filling of cavities from upslope. It is suggested that creep is the dominant process in shaping the convex parts of slopes. *See also* slope convexity.

crenulation cleavage *See* strain-slip cleavage.

crepuscular ray An optical phenomenon of the lower atmosphere produced by the alternations of sunlight and shadow radiating from the Sun, made visible by haze. The shadowing may be produced by gaps in low-level layer cloud, or by the irregular profile of convective cloud. A similar effect can sometimes be seen at sunset when the Sun is just below the horizon and distinct rays are seen radiating from it, the dark bands indicating parts cut off from the Sun's rays by clouds or hills.

crest The highest point of an ANTICLINE. A line joining the highest points of a given bed is the *crest line*; a plane including all the crest lines of successive folds is the *crest surface*.

Cretaceous The final period of the MESOZOIC Era. Beginning about 135 million years ago and lasting about 70 million years, it followed the JURASSIC and preceded the TERTIARY (which marked the start of the Cenozoic Era). The name derives from *creta*, the Latin word for chalk, the characteristic rock of the period; extensive deposits form the white cliffs along the English Channel. The lower part of the Cretaceous System consists of six stages: the Berriasian, Valanginian, Hauterivian, Barremian, Aptian, and Albian. The six stages forming the Upper Cretaceous are the Cenomanian, Turonian, Coniacian, Santonian, Campanian, and Maastrichtian.

During the period South America began to move westward away from Africa; India was moving north, and Australia and Antarctica were also breaking away. North America was moving away from Eurasia, and by the end of the Cretaceous the Atlantic Ocean extended as far as the Arctic Ocean. Sea levels were higher than at any other time in the Earth's history. Cretaceous rocks are widespread, the Early Cretaceous characteristically deltaic and lacustrine, and the Late Cretaceous including marine deposits, such as sandstone and the characteristic chalk, a pure fine-grained white limestone formed largely of planktonic COCCOLITHS. In the seas in which the coccoliths were deposited, bivalves, ammonites, belemnites, echinoids, and bony fish also flourished. On land angiosperm plants appeared, and dinosaurs and other reptiles reached their peak of development. At the end of the period the ammonites and many other invertebrate groups and most of the reptiles became extinct. *See* K/T boundary event.

crevasse A crack of variable width in the surface ice of a glacier, caused by shear stresses set up by differential movements within the ice. These movements occur in a VALLEY GLACIER, the edges of which move more slowly than the center, owing to the frictional effect of the valley sides. In this case crevasses form pointing up the glacier. Crevasses also occur where the ice moves over a steeper section of ground, in which case they are transverse to the valley. Despite the fact that their depths rarely exceed 30 m, crevasses are very important because they assist in the downward penetration of both rock debris and meltwater.

crinanite A type of ALKALI GABBRO.

Crinoidea The class of the phylum ECHINODERMATA that contains the sea lilies. They are typically sessile benthonic marine animals, consisting of a stem, formed of ossicles of calcite, which supports a cup (*calyx*) of calcite plates. On the upper side of this is a centrally placed mouth surrounded by five feathery arms that bear pinnae. These are used in food collection. Fossils are known from the Cambrian Period onward and the class is still in existence today. Crinoids formed an important group in the Lower Carboniferous and contributed to the formation of bioherms and rocks. Most became extinct at the end of the Paleozoic. Crinoids have been used biostratigraphically as ZONE FOSSILS.

cristobalite A white microcrystalline type of silica, SiO_2, which occurs in cavities in volcanic rocks and in some thermally metamorphosed rocks. *See* silica minerals.

critical angle The angle of incidence of a light ray which, if exceeded, will lead to total internal reflection.

critical distance (in seismic refraction surveying) The distance between the source of seismic waves and that point in an upper rock horizon at which the arrival time of a direct wave is matched by the arrival time of a higher-velocity wave that has been refracted from a lower horizon.

Cromwell Current *See* equatorial current.

cross-bedding (cross stratification) The development of internal laminations within a stratum inclined at an oblique angle to the main bedding planes, resulting from changes in the direction of water or wind currents during deposition. It is most commonly developed in sandstones.

Crossopterygii A group of bony fish (*see* Osteichthyes) whose sole present-day representative is the coelacanth. The group is important for having given rise to the AMPHIBIA and hence to all terrestrial vertebrates. Unlike most modern fish, the Crossopterygii had fleshy paired fins. They were common from the Devonian to the end of the Paleozoic. *Compare* Actinopterygii.

cross profile A transverse section of a river's channel or valley. A stream can adjust these parameters as well as its long profile in response to environmental changes. Valley cross profile can show such features as valley-in-valley forms, river terraces, extent of floodplains, symmetry or asymmetry of the valley-side slopes, angle of valley-side slopes, etc., all of which contain clues to the geomorphic history of the river and the region through which it flows. *See also* long profile.

crude oil Petroleum as it occurs in the ground. It consists of a mixture of impure hydrocarbons which, after purification, can be separated to yield various fuels and compounds used by the petrochemical industry.

crust The outermost shell of the Earth, varying in thickness from 5 km under the oceans to 60 km under mountain ranges. The lower boundary is marked by the Mohorovičić discontinuity. The crust is composed of two units, the CONTINENTAL CRUST (also known by the acronym SIAL) and the OCEANIC CRUST (SIMA).

Crustacea A class of invertebrate animals of the phylum ARTHROPODA, characterized by the presence of two pairs of antennae and one pair of mandibles. Crustaceans are almost exclusively aquatic and range in size from the large lobsters to almost microscopic forms. Included in the six subclasses are the OSTRACODA, important geologically in stratigraphic correlation, the Malacostraca (crabs and lobsters), and the Cirrepedia (barnacles). Fossil crustaceans, mostly ostracods, are known from the CAMBRIAN Period onward.

cryolite (Greenland spar) A mineral form of sodium aluminum fluoride, Na_3AlF_6. Rarely crystalline, it occurs as colorless or white masses. It is used as a flux in the electrolytic extraction of aluminum and in making various ceramics.

cryoturbation The modification of the soil through the processes of CONGELITURBATION and the effects of NEEDLE ICE in a periglacial environment.

cryptocrystalline Describing material that is crystalline yet so fine-grained that individual crystals cannot be resolved even under the microscope.

crystal A chemical substance with a definite geometrical form, having plane faces at regular angles to each other, which is the expression of the regular arrangement of atoms or ions composing the substance. *See* crystal symmetry; crystal system.

crystal axes Imaginary lines that pass through the center of a CRYSTAL. There are generally three axes, but four in the hexagonal system. The angles between the axes and their relative lengths characterize the structure of any crystal. *See* crystallographic index.

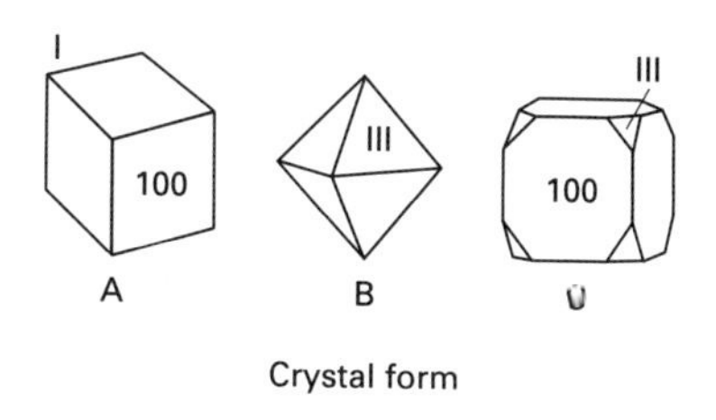

Crystal form

crystal form A complete set of regular planes forming a crystal. The diagram illustrates different forms found in the cubic CRYSTAL SYSTEM: A and B are simple crystal forms and C is a combination of the two forms cube and octahedron. The form exhibited by a crystalline mineral may be a great aid to identification.

crystal habit The shape of a crystal arising from the shape and size of faces and the development of different forms. In the diagram the two crystals possess the same combination of crystal forms but A has a prismatic habit and B a pyramidal habit owing to the relative development of prism and pyramid faces.

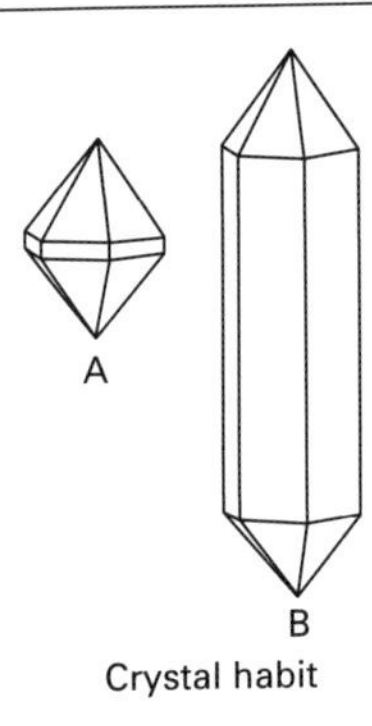

Crystal habit

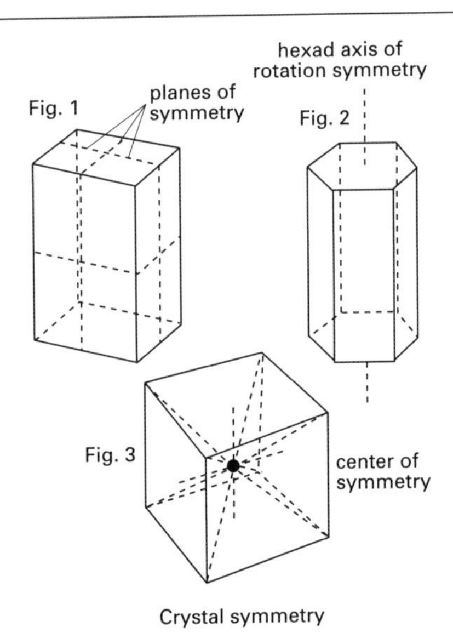

Crystal symmetry

crystalline rock Rock formed by the process of crystallization, i.e. metamorphic and igneous rock.

crystallite A microscopic embryonic crystal found in glassy rocks as a result of DEVITRIFICATION. Crystallites have rodlike, stellate, or feathery forms.

crystallographic index A crystal face may be indexed according to a notation that depends upon the intercepts made by the face with the crystallographic axes. The distances along the axes from the origin to the points where the plane cuts are termed the intercepts. The ratios of the intercepts (referred to a standard parametral plane or axial ratio) are termed the parameters of that crystal face. According to the system of Miller, the reciprocals of the parameters of a crystal face are called the indices.

crystal symmetry The crystal faces of minerals are symmetrically arranged and reflect the internal regularity of the atomic structure. Crystal symmetry is described by reference to the following elements:
1. plane of symmetry – a plane along which a crystal may be cut into two equal halves, one being the mirror image of the other (fig. 1).
2. axis of symmetry – a line about which a crystal may be rotated through 360° so that the crystal assumes a congruent position every 180°, 120°, 90° or 60°. Such axes are termed *diad* (2-fold), *triad* (3-fold), *tetrad* (4-fold), and *hexad* (6-fold) axes respectively (fig. 2).
3. center of symmetry – a central point about which every face and edge of a crystal is matched by one parallel to it on the opposite side of the crystal (fig. 3).

On the basis of these symmetry elements, 32 different crystal classes are recognized together with 7 CRYSTAL SYSTEMS:
cubic – 4 triad axes
tetragonal – 1 tetrad axis
hexagonal – 1 hexad axis
trigonal – 1 triad axis
orthorhombic – 3 diad axes
monoclinic – 1 diad axis
triclinic no axes.

Other symmetry elements may be possessed by crystals in addition to those listed. For example, a cube has three tetrads, four triads, six diads, nine symmetry planes, and one center of symmetry.

crystal system A category of crystal with reference to the position of the crystal faces in relation to the intercepts that the planes containing the faces make with three (or four) axes, which intersect at an origin. All crystals of the 32 different symmetry classes can be referred to seven crystal systems as follows (see diagram):
1. *cubic (isometric)* – 3 orthogonal tetrad axes of equal length, a_1, a_2, a_3.

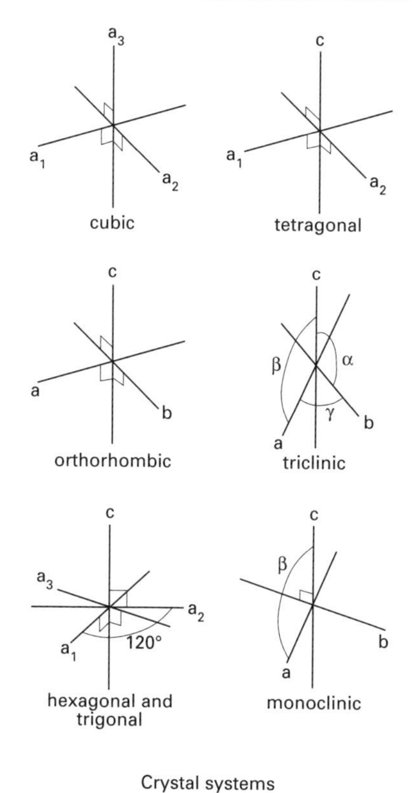

Crystal systems

2. *tetragonal* – 3 orthogonal axes, 2 horizontal diads of equal length, and one vertical tetrad, a_1, a_2, c.
3. *orthorhombic* – 3 orthogonal diad axes of unequal lengths, a, b, c.
4. *hexagonal* – 4 axes, 3 horizontal diads of equal length at 120° apart and one vertical hexad at right angles, a_1, a_2, a_3, c.
5. *trigonal* – 4 axes, 3 horizontal diads of equal length at 120° apart and one vertical triad at right angles, a_1, a_2, a_3, c.
6. *monoclinic* – 3 unequal axes, one vertical, one horizontal diad, and a third making an oblique angle with the plane containing the other two, a, b, c.
7. *triclinic* – 3 unequal axes, none at right angles, a, b, c.

Different CRYSTAL FORMS can be referred to the same set of crystal axes and hence belong to the same crystal system. For example the cube, rhombdodecahedron, and octahedron are different crystal forms of the same cubic system.

crystal tuff A type of tuff that contains many broken crystals. *See* pyroclastic rock.

crystal zoning Many crystals, especially in igneous rocks, are zoned such that their composition changes systematically and gradually from the core of the crystal to the rim. The succession of zones reflects changes in the composition of the melt in which the crystal is growing or changes in the physical conditions as crystallization proceeds.

cubic (isometric) *See* crystal system.

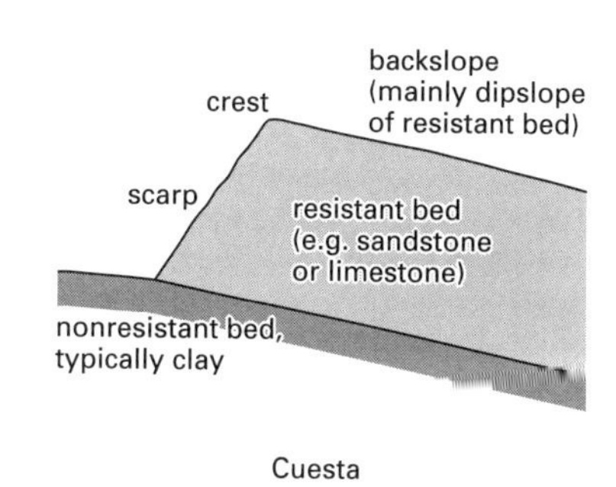

Cuesta

cuesta An asymmetrical ridge produced by differential erosion of gently dipping strata. A cuesta is characterized by its steep scarp face, well-defined crest, and gentle backslope, which largely conforms to the dip of the strata. It is constructed of a more resistant geology (e.g. sandstone or limestone) than the etched out foot (e.g. clay).

culmination The highest point on a structural feature. As the axis of a fold often undulates up and down, a series of culminations separated by depressions can develop. (See diagram at FOLD.)

cummingtonite A monoclinic type of AMPHIBOLE.

cumulate rock A type of igneous rock that is formed largely by the accumulation of early-formed PRIMARY MINERALS that crystallize from a magma and sink under the influence of gravity to settle in layers on the floor of the magma chamber. Often the cumulus crystals trap small quantities of liquid in the interstices and this liquid subsequently crystallizes to produce intercumulus minerals. Cumulate rocks occur in large basic layered intrusions, often of lopolithic form. Monomineralic and bimineralic rocks are characteristic and cumulate dunites, peridotites, pyroxenites, anorthosites, and gabbros commonly occur.

cumulonimbus cloud The main rain cloud of convective origin. It is heavy and dense with considerable vertical extent to give the appearance of a vast cauliflower, each protuberance being a tower of rapidly rising saturated air with velocities of up to 30 meters per second. At the top of the cloud there is often a more fibrous flattened cloud composed of ice crystals: this is the ANVIL. Most thunderstorms are associated with this type of cloud.

cumulophyric *See* glomerophyric.

cumulostratus cloud *See* stratocumulus cloud.

cumulus cloud A type of cloud indicating convectional activity extending above the CONDENSATION LEVEL. Cumulus clouds range in vertical extent from shallow fair-weather cumulus clouds indicating weak convection topped by an inversion, to the much more extensive cumulus congestus, which almost approaches a cumulonimbus cloud in size.

cupola A slender upward projection from a large igneous body into the overlying country rock. *See also* roof pendant.

cuprite (ruby copper) A red to black mineral form of copper oxide, Cu_2O. It crystallizes in the cubic system and occurs in oxidized copper deposits, sometimes associated with limonite. It is used as a source of copper.

Curie point The temperature above which permanent magnetism disappears. Each element has its own specific Curie point; in the case of iron it is 760°C.

current (in oceanography) A horizontal flow of water, the movement affecting the whole water column or only a part or parts of it. The speed at which currents flow, and their direction, may vary markedly, or the current velocity may be remarkably constant in a temporal sense. Often variations are seasonal, or they may be due to changing meteorological factors. Permanent currents operate independently of the tides and weather, although they are indirectly affected by these. These include the general circulatory current systems in the oceans. Other currents flow on account of freshwater discharge from rivers, the action of waves and tidal motion, differences of seawater density, and wind-drag. All or certain of these currents may be superimposed, one upon the other. *See also* density current; ocean current.

current bedding A sedimentary structure resulting from the action of either wind or water currents. It includes cross-bedding and ripple bedding.

cusp A crescent-shaped mass of beach material, which may range from sand to quite large shingle or cobbles. The coarser material accumulates on the promontories or horns between the bays and the finer material in the bays. Cusps are regularly spaced and generally display coarser material than is found over the remainder of the beach surface. Those formed purely in sandy material are less common than those developed in shingle or in sand and shingle mixtures. They point down the beach toward the sea and the margin between coarser and finer material has a scalloped shape. They vary greatly in size, ranging between several centimeters in height to giant cusps or megacusps, such as those found on parts of the coast of West Africa. It seems that there exists a relationship between cusp spacing and swash length, and that a type of cellular water flow together with swash periodicity are important additional factors in their development, but their origin has yet to be explained satisfactorily. Waves breaking perpendicular to the beach are thought to be the most conducive to cusp formation.

cuspate delta A DELTA within which material is evenly deposited on either side of the river mouth. It is usually found on straight coasts where wave action is fairly strong.

cuspate foreland An approximately triangular accumulation of beach materials, usually shingle, the apex of which extends out to sea and produces an irregularity in the coastal outline. Such forelands are created as a result of the combination of SPITS or BEACH RIDGES approaching each other from opposite directions, owing to the action of two major wave sets in the area, each being more active on one side of the foreland. The growth of these features can often be traced by the existence of many parallel shingle ridges. The maintenance of a sharp projection out to sea largely depends upon the presence of a nearby island or coastline, which provides shelter from direct frontal attack by destructive waves.

cutoff *See* oxbow lake.

cut-off high A warm anticyclone that has moved poleward from the main subtropical high-pressure belt, often producing BLOCKING in the westerlies. This system usually forms as a ridge in the upper westerly circulation and this intensifies into a cellular form becoming detached from the main flow like a meander (or oxbow lake) in a river. Its influence then extends downward toward the surface. (See diagram.)

cut-off low A low-pressure system of similar origin to the CUT-OFF HIGH but extending toward the Equator after developing from a trough in the upper westerly circulation. Cut-off lows frequently drift into the Mediterranean Sea area after cutoff has taken place farther north.

cwm *See* cirque.

cycle of erosion A concept explaining the evolution of dissected land surfaces from uplifted areas of little relief to dissected landscapes and then to level surfaces (peneplains) via a series of stages (youth, maturity, and old age). The process is said to be cyclic because it will begin again on further uplift (*see* rejuvenation). The concept of strict erosional cycles is not now adhered to, because climatic changes mean that processes in any one area are not constant, and most landscapes have evidence of evolution under a series of cycles or parts of cycles of different climatic types as follows:

Arid

The key to the cycle of arid erosion is the process of PEDIMENT formation. Penck and King developed this idea, and as a result arid lands are said to evolve by a process of slope retreat (*see* parallel retreat) not decline, such that the trend is not the increase of relief to a maximum in maturity and thereafter increasingly subdued relief but decrease of the area of plateau and increase of the area of pediment. In youth the rivers incise themselves, and firstly retreat produces valley-side pediments; by maturity the divides have shrunk and the initial topography is nearly lost, by old age retreat of slopes has reached the point where divides are lost altogether, and pediments of

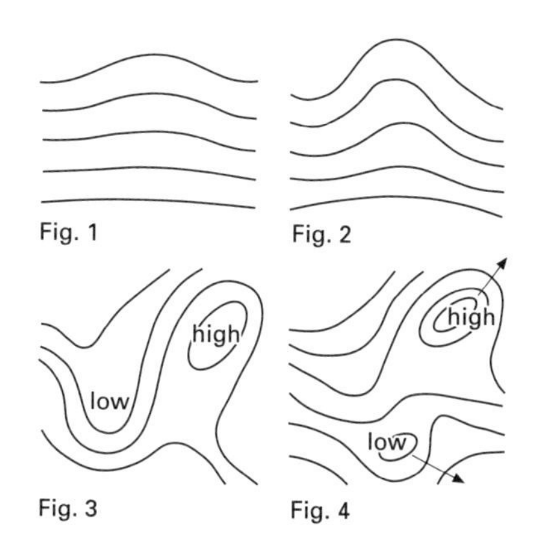

Cut-off high and low

individual basins coalesce to give the ultimate PEDIPLAIN. *See also* river.

Coastal

This cycle, evolved by Johnson in 1919, differs on emergent (recently uplifted) and submergent (recently drowned) coasts. On both types, the general pattern is a simplification of coastal outline. In the emergent case, the coastline will be initially straight, with a shallow offshore slope. Waves break offshore and build up a barrier island, with a lagoon behind. As the offshore profile is deepened, the island is attacked by larger waves which push it onshore, migrating over the lagoon, until at maturity the barrier is coterminous with the original coast. It then continues to migrate, eroding the shore, until at theoretical old age a plain of marine planation is created. On a submergent coast, the initial outline is very complex, as existing drainage patterns are drowned. The headlands are subject to concentrated attack by refraction of waves, and as they are cut back spits begin to build out across the bays, eventually cutting them off and leading to their infill with waste. At submaturity the headlands are cut back and the bays closed; continued erosion pushes the coast to maturity, at which point the shoreline lies behind the heads of the initial drowned valleys. Thereafter the course is as in the emergent case.

Glacial

The major process involved is the creation of cirques and the retreat of headwalls, destroying the preglacial landscape. Initiation involves the collection of snow fields and the creation of the cirques by nivation. Once created, headwall retreat eats into the original surface, eventually leaving only horned peaks and arêtes. Such features typify the passage from youth to maturity. By full maturity the main regional valleys are filled with ice, only the highest peaks projecting above the ice surface as NUNATAKS. The retreat of the headwall of the glacial valleys extending from the cirques eventually exceeds the rate of retreat of cirque headwalls, and the merging of the two marks the end of maturity. The old age phase has never been observed: as with the coastal cycle, only the phases of youth and maturity are well defined.

Normal (Davisian)

Developed by W. M. Davis in the latter part of the nineteenth century, this was an attempt to devise an orderly sequence for the evolution of landscapes under a humid temperate climate. The sequence involved is as follows:

1. Uplift initiates the stage of youth, when rivers begin to incise themselves into the landscape. Steep V-shaped valleys are separated by broad flat interfluves.
2. Maturity of the landscape, when widening of the river valleys has completely destroyed the original landscape and valleys meet at sharp interfluves. In the rivers, this stage occurs when the valleys are graded, and down-cutting is replaced mostly by side-cutting and floodplain creation.
3. By old age the landscape is subdued owing to SLOPE DECLINE. The streams meander widely on broad floodplains, losing their adaptation to structure, a feature of maturity, due to its masking beneath debris. Only MONADNOCKS on rocks of great resistance in high divides rise above a peneplain.

Periglacial

This scheme revolves around the role of frost shattering and solifluction. Originating on a dissected landscape, frost shattering attacks the valley-side slopes, producing a frost-shattered cliff, which migrates into the divide. The debris from the frost shattering is transported downslope by solifluction and meltwater to the valleys. As this process continues, the frost-shattered cliffs retreat until no solid rock remains. The scree slopes then dominate the landscape, being progressively flattened until by old age a landscape of faint relief is formed.

cyclogenesis The initiation or development of a depression or cyclone.

cycloidal wave (trochoidal wave) A wave with a flatter trough and a sharper crest than the typically smooth sinusoidal wave. It is a steep symmetrical wave with a crest angle of some 120°. The wave form is that of a cycloid or trochoid, i.e. a curve that would be described by a point rotating within a circle that itself was being rolled along a straight path. With an increase in wave steepness, the cycloidal wave sharpens its crest and increases its asymmetry. Because the flow is of a rotational type, no mass transport is possible (*see* mass transport current).

cyclolysis The process of decay or weakening of the cyclonic circulation around a low-pressure center.

cyclone An area of low pressure with a series of closed isobars, usually of circular or oval form, around its center. In the N hemisphere, it is surrounded by a counterclockwise wind circulation and in the S hemisphere by a clockwise rotation, but both are known as cyclonic circulation for each respective hemisphere. *Compare* anticyclone.

In mid- and high-latitudes, the cyclone is usually referred to as a DEPRESSION. In tropical areas, it is a storm system of great intensity with wind stronger than 64 knots (120 km per hour) and is synonymous with a hurricane or typhoon, although the nomenclature is not fully standardized. *See* tropical cyclone.

cyclostrophic winds A class of winds in which there is extremely strong curvature of the airflow, such as a tropical cyclone or tornado. Under these conditions, the centripetal acceleration becomes the major control of the gradient wind.

cyclothem A series of sedimentary beds deposited as part of a single cycle. The term is generally applied to Carboniferous strata, where the sequence begins with a sandstone layer, followed by shale and freshwater limestone. Above the limestone is a clay layer, above which rests a coal seam. Following the coal there is a return to marine conditions, with the deposition of a marine shale, followed by a marine limestone. Each cyclothem is separated from the next by a disconformity. The cycle represents an episode of emergence and erosion.

cylindrical equal-area projection *See* Lambert's cylindrical equal-area projection.

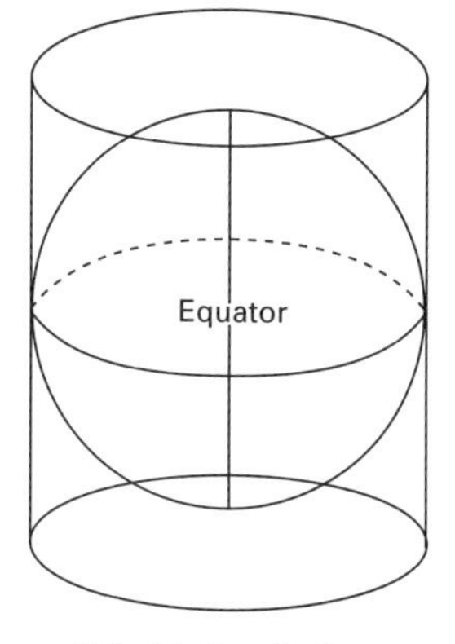

Cylindrical projection

cylindrical projection A map projection constructed as though a cylinder were placed around or cutting through the Earth, onto which the relevant details are projected (see diagram), meridians and parallels being drawn as straight parallel lines. There are three basic types: the simple cylindrical, LAMBERT'S CYLINDRICAL EQUAL-AREA PROJECTION, and the MERCATOR PROJECTION. In the simple cylindrical, the cylinder touches the Earth at the Equator, which has the effect of making the poles the same length as the Equator and therefore causes considerable distortion in the higher latitudes. The scale along the meridians is correct, but directions are not true. It is a projection that is seldom used owing to the existence of more practical modified versions.

D

dacite A dark igneous rock consisting mainly of quartz and plagioclase. *See* rhyolite.

darcy A unit used to measure the porosity of a rock (the permeability coefficient). It represents the resistance to flow through the rock of a fluid (gas or liquid). The usual practical unit is the millidarcy (md).

Darcy's law The rate of flow of a fluid through a porous material (suck as a rock) is proportional to the product of the permeability of the material and the pressure causing the flow.

datolite A mineral form of hydrated calcium borosilicate, $CaBSiO_4(OH)$. It crystallizes in the monoclinic system as colorless or white prismatic crystals. It occurs in veins and in AMYGDALES.

datum The vertical datum is a horizontal line or a point used as the origin (zero) from which heights and depths are measured. On most maps and charts it is that of mean sea level, e.g. in Britain the *Ordnance Datum* (*O.D.*) of Ordnance Survey maps is mean sea level at Newlyn, Cornwall, calculated from tidal observations between 1915 and 1921. Horizontal datum is generally used in connection with geodetic information where it forms the basis for horizontal control. Horizontal datum is determined, using the reference spheroid, from bearings (azimuths) and longitude and latitude readings.

daughter element An element that forms as a result of the radioactive decay of another element (e.g. radon is a daughter element of radium, which is in turn a daughter element of thorium).

day Astronomically, a *solar day* is the period of time between successive occasions on which the Sun is in the meridian of any fixed place; the *sidereal day* is the time between successive transits of a fixed distant star. As the Earth varies in its rate of movement during its orbit round the Sun, the solar day is not constant. The climatological day extends for a 24 hour period from 9 a.m. GMT. This is the time at which most climatological observations are made and so rainfall for April 21, for example, would in reality be the amount that fell between 9 a.m. April 21 to 9 a.m. April 22.

debris An accumulation on the surface of soil or rock fragments. It may result from glacial action (and is found when the ice melts) or from processes (such as frost action) that cause rock to break into fragments.

debris load That part of a river's total load carried as solid material. It includes material moving by suspension wholly above the bed, by SALTATION, and by rolling and sliding along the bed. The total debris load is the sediment discharge of a river. *See also* load.

decalcification (in soil science) The removal of lime from a soil. The process involved is carbonation, whereby rainwater percolating through humus becomes enriched with carbon dioxide, forming an acid, which combines with calcium carbonate in the soil and carries it away in solution as calcium bicarbonate. It operates best at low temperatures, because the solubility of calcium carbonate decreases as temperature rises. If the content of carbon dioxide in the rainwater falls, then its ability to dissolve calcium also falls. Decalcification is said by some pedologists to precede LESSIVAGE and PODZOLIZATION: in a given soil it influences the upper profile first, and so there is a point at increasing depth beyond which decalcification has not proceeded, called the *acid point*.

deciduous forest Areas occupied by trees and shrubs that periodically shed their leaves, usually in winter or during the dry season. Deciduous trees are also called broadleaved trees.

decile If a series of values are ranked in order of magnitude they can be subdivided into any number of groups depending upon the amount of data. A decile represents a tenth part of the series and would normally be quoted as the first or second decile, etc.

declination *See* angle of declination.

décollement The plane marking the boundary between two different types of deformation. The rocks above are generally deformed, whereas those below may be unaffected. It is caused by the upper rock series sliding over the lower during folding.

decomposer (in ecology) Any organism that breaks down dead plants and dead animals or their excreta. They are important in the FOOD CHAIN because they set free such substances as nitrogen compounds, which pass into the soil or atmosphere. Bacteria and fungi are the main decomposers, but some invertebrate animals (such as earthworms) also fulfill this function.

decomposition The weakening and break-up of a rock mass through CHEMICAL WEATHERING. A decomposed rock will lack its former cohesion owing to the washing away of cementing materials or the alteration of the constituent minerals.

deep (abyss) A depression in the floor of the ocean, generally at depths in excess of 5000 m.

deep-focus earthquake An earthquake whose focus is at a depth of more than 300 km. Most are found along the BENIOFF ZONES.

deep ocean One of the fundamental divisions or zones of the Earth's surface. The total volume of deep oceans far exceeds the total volume of land lying above sea level. Also, the oceans are, in general, markedly deeper than the continents are high. The Pacific Ocean is the largest ocean, occupying, together with its adjacent seas, some 50% of the total ocean area, which is 361 million sq km. The Atlantic Ocean occupies roughly 106 million sq km and the Indian Ocean 75 million sq km. The average depth of the deep oceans is 3800 m, with a large percentage of the oceans varying in depth between 3000 and 6000 m below sea level.

deep-sea trench *See* trench.

deep weathering Weathering at great depths, occurring when suitably porous, permeable, or chemically reactive rocks are subjected to humid tropical or subtropical weathering in areas where erosion is very slight. REGOLITHS over 100 m deep have frequently been reported and others up to and over 300 m are known. There is a transition in all profiles from solid rock (at depth), through partly weathered rock (containing CORESTONES), to totally weathered material that has lost the former structures of the parent material. Evidence of deep weathering occurs in areas that are now totally unsuitable for its development, in which cases weathering is assumed to have occurred during former periods of warmer climate.

deflation Wind removal of fine material (silt and clay) from areas of increased weathering. Typically, a chance depression in the desert surface increases moistness at that point, producing increased chemical weathering. This leads to clay formation. On drying, the clay is susceptible to wind action, and when picked up can be carried long distances in suspension, often in the form of sandstorms. This fine material is dropped elsewhere, leaving behind a concentration of coarse material as a *deflation surface*. Deflation surfaces usually take the form of gravel-strewn areas, although this process can also produce large depressions, and sometimes OASES may result. *See also* aeolian erosion.

deforestation The permanent removal of trees and shrubs, usually as a result of human activity for obtaining timber or clearing land for agriculture or mining. The exposed soil is easily washed away by rain or blown away by the wind, resulting in an eroded infertile landscape and loss of natural habitats. *See also* badlands; desertification.

deformation The alteration of rock formations that generally results from tectonic plate movements. The consequences include compression, extension, faulting, and folding.

deglaciation The processes operating during an INTERGLACIAL period once climatic conditions begin to improve and the ice sheets that may have developed in the previous GLACIAL PHASE start to shrink. Although considerable melting takes place at the margins, accumulation can still be maintained in the source zones, thereby continuing the transportation of debris by moving ice. Most melting takes place during the summer when the ice margins may recede toward the source or alternatively the whole ice mass may become shallower, resulting in the landforms characteristically created by stagnant ice. During the winter an ice sheet may remain static or even advance again slightly, causing deposition of TERMINAL MORAINES or PUSH MORAINES. Successive moraines may be distinguished in order to trace the recessional pattern of the ice sheet. The complete disappearance of an ice sheet may take thousands of years, and crustal adjustments and sea-level changes may continue long afterward.

degradation Although largely synonymous with EROSION degradation is most usually applied to the lowering of stream beds over a time period measurable in years. It is not applied to the short-term cut and fill of stream beds extending over a period of a year or less, which is an integral part of the equilibrium regime of streams. Degradation can arise through a fall of sea level, causing REJUVENATION, or a change of flow conditions giving the river increased erosive power, such as decreased load or increased runoff. Evidence of degradation can be found in river terraces and knickpoints. *See also* aggradation.

degree **1.** A unit of measurement along the lines of latitude and longitude. One degree is equal to l/360 of the Earth's circumference on the lines of *longitude*. However, a degree of *latitude* decreases in length toward the poles. Each degree is subdivided into 60 minutes, and each minute into 60 seconds.
2. A unit of temperature. *See also* absolute temperature; Celsius scale; kelvin.

degree-day The basic unit used for ACCUMULATED TEMPERATURE. It can refer to either degrees Celsius or Fahrenheit and is the number of degrees above (or below) a specified datum during the day.

delayed runoff Rainwater that flows into streams (usually from springs) after having first been absorbed into the soil. Delay may also be caused if rainwater is temporarily locked up in snow or ice.

delta A large accumulation of sediment deposited at the mouth of a river where it discharges into the sea (or a lake). Deltas are formed as a result of the decrease in load-carrying capacity following the deceleration of river water on entering the comparatively static sea (or lake). Fine clay material, normally carried in suspension, is also deposited because the very small particles coagulate and sink in the presence of saline water. For the sediment to accumulate the amount of material deposited by the river must exceed that removed by coastal erosional processes. Deltas grow in size because the river tends to bifurcate once a certain amount of deposition has occurred; the smaller streams (distributaries) then deposit material over a wider area. They may become abandoned later, and that part of the delta no longer receiving sediment will become eroded. *See also* arcuate delta; bird's-foot delta; cuspate delta; delta deposit.

delta deposit Prior to the formation of a DELTA proper, fine material is deposited on the seabed at the river mouth. These materials (*bottomset beds*) are laid down horizontally and extend a considerable way out to sea. Above these, and making up the major volume of the delta, are materials (*foreset beds*) deposited with an inclination from top to bottom of up to 35°; progressive deposits extend farther out to sea and thereby cause enlargement of the delta. These beds are overlain by the *topset beds*, which are again horizontal and are a seaward extension of the river's alluvial floodplain.

demersal Describing a fish that lives on or near the sea floor. Flatfish, including the flounder, halibut, and sole, are specially adapted for life on the bottom, with special forms of camouflage. The other main types of demersal fish are the round fish, which include haddock, cod, and hake.

demography The use of statistics and mathematics to study human populations geographically to determine such factors as composition, distribution, and size.

dendrite A fernlike branching pattern of material deposited within a rock or mineral, often resembling a plant fossil. Most dendrites are black, composed of such minerals as pyrite (iron sulfide) and pyrolusite (manganese dioxide). *See* moss agate.

dendritic drainage The drainage pattern that develops where structural controls of slope, variable lithology, or fault and joint patterns are absent; as a result the drainage net is entirely random, with equal probability of stream flow in all directions. It characterizes areas of flat rocks with uniform lithology, notably plains, plateaus, and massive crystalline rocks. One example is the US Great Plains, notably the Badlands of South Dakota.

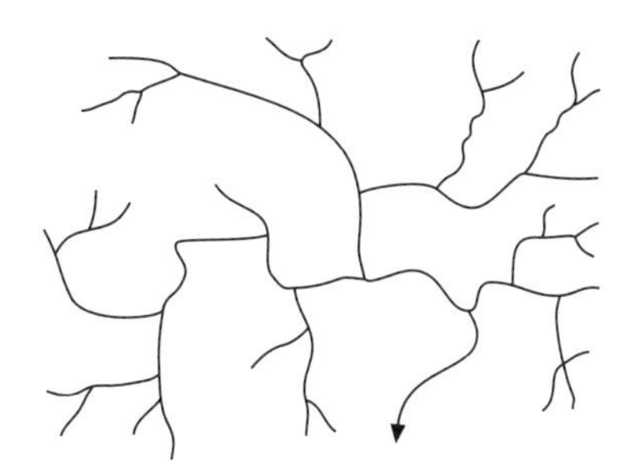

Dendritic drainage

dendrochronology The interpretation of former climates from changes in width of annual growth rings of certain tree species. By using the bristlecone pine, a long-lived conifer from the southwestern USA, and preserved specimens, it is possible to go back as far as approximately 5000 BC. However, the width of the tree ring does not bear a simple linear relationship with one climatic factor, being related to an amalgam of many, of which temperature and amount of precipitation during its growing season are the most important. This limits the reliability of the method.

denitrification The process by which bacteria in the soil break down nitrates to produce nitrogen gas, which passes into the atmosphere. Denitrification is an important part of the NITROGEN CYCLE.

density The mass of unit volume of a substance at a specified temperature and pressure. In the case of the atmosphere, density is not normally measured directly but is calculated from its relationships with temperature, pressure, and humidity through the GAS LAWS. For dry air at 1000 mb and 290 K, density is about 1.2 kg/m^3 (see diagram at ATMOSPHERE).

density current A CURRENT caused by differences in density. The density of a mass of water may become different from that of the surrounding water; if it is greater, the density-driven flow is usually a

BOTTOM CURRENT or *underflow*. On the other hand, the density of a mass of water may be intermediate between that of the water above and beneath it, in which case the density-driven flow may develop as an *interflow*. A common cause of density-driven flow arises when relatively light fresh water discharging from a river rides out across and above denser saline water entering from the sea; in this case, a saline wedge tends to drive upstream along the bed of the river or estuary, often transporting appreciable amounts of fine sediment.

density log A subsurface logging technique, which records the variations with depth in the density of strata in an uncased bore hole.

denudation The weathering of rocks, the entrainment of debris, and its subsequent transport and deposition. Denudation is highest in areas of high relief, heat, humidity, steep slopes, and rocks with abundant sediment yield. The highest rates of denudation recorded, reaching 3000 mm/1000 years, occur in glaciated areas; lowest rates, about 1.2 mm/1000 years, occur in hot dry lowlands. *Compare* erosion.

denudation chronology The study of how a landscape has evolved through time to its current state, by arranging in sequence the pieces of evidence discovered, to obtain a series of pictures of the evolution of the relief through a series of stages. For each major stage in the evolution of a regional landscape the following broad groups of features are researched. Climate, geology, and base level (the independent variables) dictate the geomorphological system, consisting of erosion, transport, and deposition. This leads to the creation of landforms, the remains of which constitute the basic research material. Partly reflecting the above factors and in turn influencing them are the flora and fauna. As flora and fauna reflect so many environmental variables, especially climate, their study in the form of pollen, insect, and mammalian remains is significant in environmental reconstruction. Once the type of environment is known, inference can often be made of the geomorphological processes operating.

Dating is done by various methods: organic matter of 60 000 years old or less can be dated by radiocarbon methods, and for greater time range other methods are available. Paleobotany has been used to establish pollen zones through time, with similar changes occurring over wide areas allowing organic remains to be correlated between regions and assigned to a relative dating system. Pedological methods can be used to study superficial deposits of past climates still preserved.

deposition The laying down of material subsequent to EROSION and TRANSPORT. Deposition could be described as the creative part of the geomorphological system.

In the fluid transport media of sea, rivers, and wind, deposition occurs when the forward movement and turbulence in the transporting medium falls below the settling velocity of the load, i.e. when the transporting medium lacks the COMPETENCE or CAPACITY to carry the load any farther. Low-energy environments are therefore zones of deposition; in rivers, these occur on the inside of meander bends; on coasts, in bays and estuaries; in an air stream, in the lee of obstructions. All these environments are characterized by lack of strong upward eddies and divergent streamlines of flow.

In the solid transport medium of ice movement, most deposition occurs where changes in the climatic regime stop the advance of ice and lead to its stagnation and melting, rendering it incapable of transporting any farther.

depression **1.** (in meteorology) A mid-latitude CYCLONE (*low*) or a weak TROPICAL CYCLONE with wind speeds of less than Force 6 (BEAUFORT SCALE). The term may also be used for smaller extratropical depressions not associated with surface fronts such as lee lows, thermal lows, or polar lows. However, its most common usage is for the main synoptic disturbances of mid-latitudes. These are areas of low pressure surrounded by several closed isobars, frequently accompanied by surface FRONTS and moving toward the northeast in the N hemisphere and the southeast in the S hemisphere. They appear to undergo evolution during their movement eastward, commencing as shallow lows with widely separated fronts (see diagram) and only moderate winds. With time the pressure at the center of the depression falls with a strengthening of the winds and a narrowing of the warm sector. Eventually the warm sector disappears from the surface (the process being known as OCCLUSION), the central pressure starts rising, and the system begins to decay. This is the general pattern for depressions, but they vary appreciably in their intensity, size, direction of movement, and time of existence.

Depressions are the main source of precipitation in most lowland parts of the mid-latitudes, the amount falling being determined by distance to the depression center, intensity of the depression, and the time taken for the depression to pass. As a result, the area of highest precipitation totals is about 50–60° latitude where depressions are most frequent and at their most intense stage. In the early stages of the depression, the troposphere is warm and the cyclonic circulation is shallow, but as it develops the troposphere cools and the cyclonic rotation extends to much greater heights. The origins of the depression have been disputed, but they appear to be due to DIVERGENCE in the upper atmosphere above suitable thermal gradients near the ground. **2.** (in geology) A structurally low area in the Earth's crust, lying between culminations along a fold axis. *See* fold.

deranged A type of drainage pattern characteristic of recently glaciated areas, such as the Canadian Shield and Siberian tundra, where the drainage pattern has not yet adjusted to the structures in the relatively recent glacially deposited surface, which masks the previously adjusted drainage pattern developed on the solid geology beneath. As a result, drainage is not coordinated, and is characterized by many small local drainage basins and lakes.

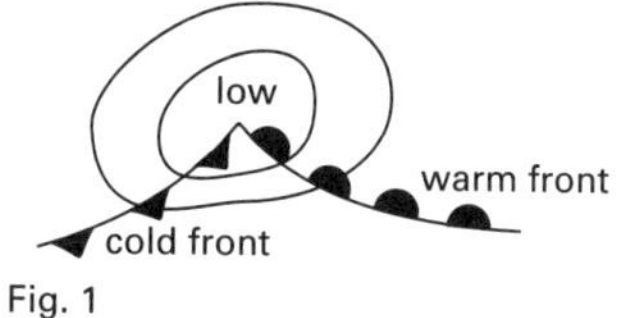

Fig. 1

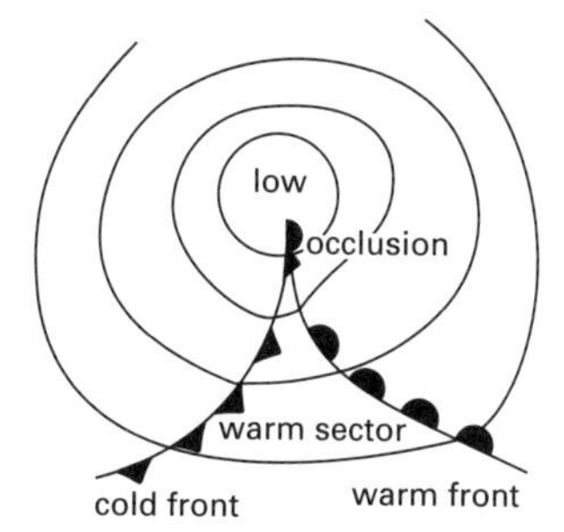

Fig. 2

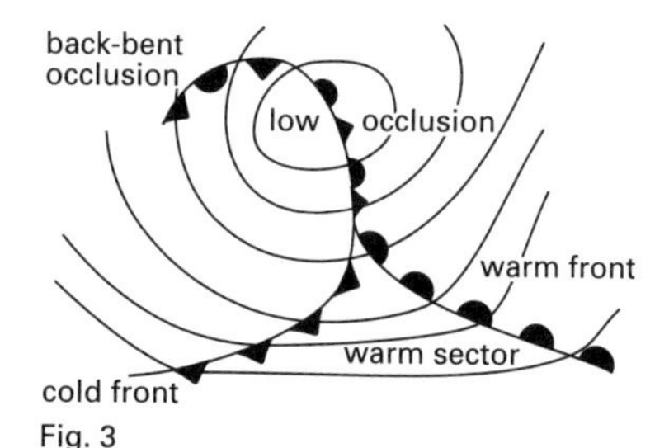

Fig. 3

Depression evolution

derived fossil A fossil that was originally preserved in a stratum older than that in which it is ultimately found, having become incorporated in the more recent stratum by processes of erosion and deposition. If its true nature is not suspected, misleading conclusions regarding age and stratigraphical relations of the deposit may be made.

desert An area of the Earth's surface where precipitation (usually taken as below 250 mm mean annual rainfall) is too low to compensate for evaporation throughout the year. This reduces the types of vegetation that can survive and the absence of surface runoff produces distinctive types of landform. The atmospheric state producing deserts is either a persistent high-pressure area, such as the subtropical anticyclones in the Sahara, or areas in which natural atmospheric stability is emphasized by surface cooling due to cold water ocean currents, such as the Humboldt and Benguela currents. Deserts are also found in continental interiors where mountain barriers prevent the passage of moisture-bearing winds; examples are the Gobi Desert and the deserts of the southwestern USA. On the margins of a desert there are areas of climatic variability that experience true desert conditions in some years but not in others.

Soils are usually very poorly developed, stony, or saline (*see* sierozem). Weathering debris accumulates close to its source because surface drainage cannot be maintained. When rain does fall it is usually in the form of sudden downpours, which can cause brief surface runoff, moving considerable amounts of debris. Mechanical weathering is more important than chemical owing to the lack of water, and wind is a very active agent of erosion and transportation of fine material. Vast areas of unconsolidated sand and dunes are less typical of most deserts than stony scrublands with occasional resistant rock uplands.

desertification The process by which a desert is formed by either changes in climate or human intervention. Natural causes also include fires that destroy vegetation. Most often, however, the cause is overgrazing by farm animals or DEFORESTATION, resulting in erosion and infertility (*see* badlands). The process can sometimes be reversed by planting special grasses or incorporating into the soil water-absorbent grains of plastic. *See also* salinization.

desert pavement An area of gravelly desert plain or REG, over which the abrasive effect of wind-blown sand has created a closely packed level surface. The individual stones may be cemented together by precipitated salts drawn to the surface in solution by capillarity.

desert rose A flowerlike arrangement of platelike mineral crystals, often found among the sand of deserts. The most common types consist of BARYTES or GYPSUM.

desert soil A characteristic type of soil that has poorly developed horizons (*see* profile). There is little vegetation and only a thin organic layer, because it is too dry for the formation of humus.

desert varnish A thin coating, varying in color from pale yellow to very dark red, found on the surfaces of pebbles and blocks in stony deserts. It is thought to be caused by the deposition of iron and manganese oxides from solutions drawn to the surface by capillarity and then evaporated. Highly polished surfaces of this type can occur through the extremely abrasive effect of fine sand carried by strong winds.

desiccation crack (mud crack, sun crack) A type of crack that develops in fine-grained deposits as a result of shrinkage owing to the evaporation of the water they contain, producing polygonal patterns. These patterns are usually irregular and polygons can vary from a few millimeters up to a meter in diameter. Subsequent wetting of a dried-up sediment will cause swelling and the disappearance of the cracks. PATTERNED GROUND may be maintained, however, if long-lasting cracks become infilled by some extraneous material, such as wind-blown sand. When preserved they act as useful indicators of paleoenvironments and orientation.

desilication (in soil science) The removal of silica from a soil profile by intense weathering and leaching. The process is typical of tropical areas and leads to the development of latosol profiles (*see* ferrallitic soil).

destructive plate boundary *See* plate boundary.

destructive wave A wave that moves more beach material seaward than landward, resulting in a diminution in the size of the beach. Characteristic destructive waves are steep, so that on breaking their BACKWASH is more active than their SWASH. They occur at high frequency, usually between 13 and 15 per minute, which means that the backwash of a preceding wave can interfere with the swash of the next, reducing the potential landward movement of material. Destructive waves frequently occur in association with local onshore winds, which cause the setting up of a seaward current on the sea floor, assisting in transporting the material stirred up by the breaking waves away from the beach. *Compare* constructive wave.

detritus 1. Rock or mineral waste produced by the breaking up and wearing away of rock surfaces; debris.
2. Organic debris from dead or decaying organisms, particularly the remains of aquatic creatures that fall to the bottom of a lake or the sea.

detrivore An animal that feeds on organic detritus, the small particles of matter formed when dead plants and animals decompose. Detrivores are most common at the bottom of the sea, where they feed on matter that drifts down from above.

deuteric changes Small-scale textural and mineralogical changes brought about by a residual hot volatile phase during the final stages of crystallization of a magma. These include the alteration of feldspars to albite, analcime, and zeolites and the alteration of mafic minerals to chlorite. Deuteric changes are exceedingly difficult to distinguish from METASOMATISM when material is introduced from outside. *Compare* hydrothermal process; pneumatolysis.

development equation In the atmosphere there exists a high degree of compensation, so that the change in pressure observed at the surface is often only a small net effect between strong DIVERGENCE or CONVERGENCE in the surface layers and the reverse flow in the upper atmosphere. This will result in vertical motion to compensate for the changes. If there is divergence in the upper levels and convergence at the surface, the vertical motion will be upward with the likelihood of precipitation.

The development equation, deduced by R. C. Sutcliffe, relates the difference in divergence between high and low levels to the thermal wind, vorticity, and the Coriolis effect, the first two of which can be measured from charts and the last easily calculated. If the result is positive it implies cyclonic development and if negative anticyclonic development.

devitrification The slow crystallization of natural volcanic glasses such as obsidian and pitchstone, which are metastable, often accompanied by secondary hydration. Devitrification is indicated by the presence of incipient CRYSTALLITES and SPHERULITIC growth of quartz and feldspar. The process often goes to completion and many rocks show little or no evidence of a former vitreous state.

Devonian The first period of the Upper PALEOZOIC. Beginning about 408 million years ago and lasting for some 45 million years, it followed the SILURIAN and was succeeded by the CARBONIFEROUS. The period is named for the county of Devon, England, where these rocks were first recognized as a major group.The Devonian System is divided into seven stages: the Lochkovian, Pragian, and Emsian form the Lower Devonian, the Eifelian and Givetian the Middle, and the Frasnian and Famennian the Upper Devonian. These are divided on the basis of fossils from rocks of the shallow-water marine facies, where invertebrates including corals, brachiopods, ammonoids, and crinoids flourished. Graptolites became completely extinct and the trilobites declined.

Outcrops of Devonian rocks occur in all continents with extensive deposits underlying areas of North America, South America, Europe, and Russia. A giant landmass, Gondwana, was located in the S hemisphere with smaller landmasses in equatorial areas. At the close of the Silurian and during the early Devonian the collision of the continents of what is now North America and Europe was accompanied by extensive volcanic activity and mountain uplift, especially in a belt that included New England, Nova Scotia, Newfoundland, Scotland, Scandinavia, and E Greenland. Extensive continental deposits accumulated, consisting of conglomerates, red silts, and sandstones. This facies is known in Europe as the Old Red Sandstone and contains remains of a large variety of ostracoderm fish.

Other Devonian sedimentary rocks include the carbonate reef deposits of Western Australia, Europe, and Canada. Black shale deposits formed locally and there were widespread evaporite deposits. Pelagic limestones rich in fossil cephalopods occur in Europe and the Urals. Gnathostome (jawed) fish, including placoderms, were also common. By the end of the period primitive amphibians had evolved from certain crossopterygian fishes. There is also evidence of land plants, such as ferns and horsetails, and of associated insects and spiders.

dew Water droplets deposited on the ground after radiational cooling has reduced the temperature of the ground surface below the dew-point temperature of the air in contact with the surface. The source of the moisture may be either dewfall from the atmosphere during conditions of light wind and a downward transfer of water vapor to the ground, or diffusion of water vapor from the soil and condensation onto vegetation, which is also being cooled by long-wave radiation losses. The latter process takes place only when the air near the surface is calm, but is one of the most frequent sources of dew.

dew-point The temperature of air at which saturation will take place if the air is cooled while remaining at a constant pressure and moisture content. Although it can be measured directly, dew-point is normally determined indirectly from tables based on dry- and wet-bulb temperatures.

dew-pond A shallow artificial depression lined with clay that collects water and is used in fields without running or piped water. The ponds are particularly common on chalk or limestone rock. It was originally believed that their main source of water was dew because they appeared to be well maintained with water even during dry periods. However, later investigation showed conclusively that natural precipitation was most important and that dew was of minor significance.

dextral fault A transcurrent (wrench or tear) fault in which the rocks on the opposite side of the fault plane are offset to the right. See diagram at FAULT. *Compare* sinistral fault.

diabase (dolerite, microgabbro) A rock that differs little mineralogically from gabbro, of which it is the medium-grained equivalent. Calcic plagioclase and augite are essential; in addition, diabase may contain olivine, hypersthene, quartz, or feldspathoids. The characteristic texture is ophitic but many examples are intergranular and porphyritic. The term *epidiorite* is sometimes applied to altered diabase. Diabase occurs mainly as dikes, plugs, and sills. *See also* alkali basalt; basalt. In the UK the rock is known as dolerite and the term diabase has been used to refer to altered dolerite.

diabatic Describing a thermodynamic process in which heat enters or leaves a system. There are many examples of this in the atmosphere, such as evaporation, turbulent mixing, and radiation absorption. *Compare* adiabatic.

diachronism The condition of a lithological unit whose base is not a time plane, i.e. whose age is different in different successions. Diachronism occurs when the boundaries of facies move in time. It can often be detected only if suitable zone fossils are available, and failure to recognize diachronism can lead to false impressions of past events and geography.

diagenesis A collection of processes by which loose accumulated sedimentary material becomes sedimentary rock. Diagenetic processes are postdepositional. In time, the pressure on a sediment increases owing to the increasing load of superposed material, and compaction results, involving a reduction of pore space. Chemical reactions take place between the sediment and entrapped and circulating fluids, leading to the cementation of grains by materials precipitated from the fluids. Calcite, silica, and hydrated iron oxides are common cementing materials. The diagenetic replacement of calcite by dolomite may take place in calcareous marine sediments. Such processes taking place near the Earth's surface at low temperatures and pressures ultimately lead to the induration of loose aggregates and to lithification. Diagenetic changes grade into those taking place at higher temperatures within the domain of METAMORPHISM.

diallage A mineral name for diopside and augite when displaying SCHILLER.

diamagnetic Describing a substance that has a magnetic susceptibility of slightly less that 1. When a diamagnetic substance is placed in a magnetic field, its induced magnetization is directed opposite to that of the applied field. The most diamagnetic substance is bismuth.

diamond A crystalline form of the element carbon, which occurs mainly in pipes of KIMBERLITE and in alluvial deposits. It crystallizes in the cubic system, forming colorless or colored crystals (tinted by impurities), and is the hardest known mineral. It has long been valued as a precious gemstone; nongem varieties are used as industrial abrasives.

diapir A vertical body of IGNEOUS ROCK that rises into the Earth's crust because it is less dense that the surrounding rocks. SALT DOMES also rise for the same reason. *See* diapirism.

diapirism The upward intrusion of a less dense rock mass through overlying more dense rock. It was originally applied to SALT DOMES, but is also an important mechanism for the inplacement of granitic rock types.

diaspore A hydrated form of alumina, AlO(OH), which occurs in ALUMINA and some BAUXITES. It crystallizes in the orthorhombic system as gray, green, or pinkish aggregates.

diastem A depositional break of a very short time period, with or without erosion. The beds above and below the bed have the same dip and strike. The absence of beds can be determined only by paleontological evidence.

diastrophism Movement within the lithosphere, including folding, faulting, orogenesis, and the formation of new ocean floor, causing large-scale deformation of the Earth's crust.

diatom A microscopic marine or freshwater alga (*see* algae). Diatoms form an important constituent of plankton, providing food for a great variety of aquatic animals. Their geologic importance derives from the fact that many diatoms possess a case of silica, which may become fossilized, and fossilized diatoms often form extensive deposits (*see* diatomite). The earliest diatoms are found in rocks of Cretaceous age.

diatomite A very fine-grained siliceous rock consisting of the skeletal remains of DIATOMS. They are formed under both freshwater and marine conditions.

diatom ooze A deep-sea siliceous ooze (*see* pelagic ooze), containing over 30% organisms. It is a cold water deposit, especially prominent in an elongated belt in the N Pacific Ocean and flanking Antarctica. On a global scale, it occupies some 9% of the total ocean floor. Living diatoms consist of siliceous algae belonging to the phytoplankton. They thrive in zones of upwelling water, where nutrients are abundant.

diatreme A volcanic vent, often filled by brecciated material, that has been cut from the sides of the conduit by high-pressure gas charged with particles. The best-known examples are the diamond-bearing kimberlite pipes of South Africa.

differential compaction The reduction in volume of sediments during compaction to different degrees, depending on their porosity, grain size, and the rigidity of the particles that compose the rock. For example, shales are more compressed than sandstones.

differential erosion The erosion of rocks subject to differential weathering. Well-weathered rocks are obviously more susceptible to subsequent erosion than resistant types. After a period of active erosion, easily weathered rocks will form lowland areas, whereas resistant strata remain as upstanding blocks. Even within a single rock type certain bands may be more or less resistant than the main mass, resulting in ridges and depressions respectively. TORS and INSELBERGS represent good examples of differential weathering and erosion within homogeneous rocks, different joint spacings accounting for the varied susceptibility.

differential shear A type of rock deformation in which movement takes place throughout the whole rock, just as in a flow, but it occurs in distinct laminae or planes.

differential weathering The more intensive weathering of certain parts of a rock mass even when the same weathering processes have been acting on the whole mass for the same length of time. Within many theoretically homogeneous rocks there may be variations in mineral composition or grain size, which can explain such differences. Joints play an important part, especially in the case of rocks with low porosity, as the percolation of water through them causes preferential weathering around the joints. There will be differential weathering between different rock types subjected to the same weathering processes.

differentiation *See* magmatic differentiation; metamorphic differentiation.

diffluence The rate of separation of adjacent streamlines in the direction of airflow. It is the reverse of CONFLUENCE.

diffusion The slow process by which different fluids or fluids having different densities mix together as a result of molecular movements. It obeys similar laws to thermal conduction but is too slow to be of importance in the atmosphere. Far more important is mixing achieved by eddy transfer (EDDY DIFFUSION) in turbulent air.

digital cartography (digital mapping) *See* cartography.

dike (dyke) A tabular body of igneous rock that is intruded vertically and discordantly to the structure of the rocks through which it passes. *See also* dilation dike; neptunian dike; radial dike; sill.

dike swarm A large number of dikes, often arranged in either a radial or parallel pattern.

dilatancy An increase in the volume of a rock deformed by pressure, caused by the expansion and extension of small cracks within it. The effect can be detected in strained rocks just before an earthquake, and is the basis of one type of earthquake prediction.

dilation dike A discordant igneous intrusion that causes the walls on either side of a fracture to move apart. The term is also applied to swarms of dikes that have filled fractures in the Earth's crust when the area has been subjected to tensional forces.

dimorphism (in mineralogy) The existence of an element or compound in two different crystal forms, such as marcasite and pyrite (both forms of iron sulfide, FeS_2) or diamond and graphite (both forms of carbon).

dinosaur A member of a group of extinct archosaur reptiles that were the dominant form of terrestrial life from the end of the Triassic period to the Cretaceous. They evolved into a great variety of both carnivorous and herbivorous species, some of very large size. They are classified in two orders: the ORNITHISCHIA and the SAURISCHIA.

diopside A clinopyroxene with composition $CaMgSi_2O_6$. *See* pyroxene.

diorite A coarse-grained intermediate igneous rock containing plagioclase of oligoclase-andesine composition and mafic minerals. The An_{50} (anorthite) composition of plagioclase feldspar marks the division between diorite and GABBRO. In diorites, the most common mafic minerals are hornblende and biotite although some rocks contain pyroxene. A little alkali feldspar may be present together with accessory magnetite, apatite, and sphene. More acid varieties of diorite contain quartz up to 10% and may be called *tonalites*. With a further increase in quartz, tonalites pass into granodiorites. With an increase in the amount of alkali feldspar, diorites pass into syenodiorites. In the USA, the term tonalite is applied to all rocks containing quartz, sodic plagioclase, and mafic minerals and is equivalent to the British tonalite (quartz-diorite) and granodiorite (in part).

Both leucocratic and melanocratic diorites occur. Meladiorites containing euhedral hornblende crystals are termed *appinites* and often appear to be a pegmatitic facies of more normal diorite. Diorites usually have equigranular textures. The medium-grained equivalents, microdiorites, are often porphyritic. *Markfieldite* is an oversaturated microdiorite consisting of a groundmass of graphically intergrown alkali feldspar and quartz with phenocrysts of plagioclase and hornblende. The volcanic equivalents of diorites are andesites. Dioritic rocks tend to occur in small intrusive masses associated with granodiorite and gabbro bodies and many diorites are thought to be hybrid rocks. *Compare* syenite; syenodiorite.

dip **1.** The angle made between the horizontal plane and that of the bedding plane, measured perpendicularly to the strike, in a stratified rock or any planar structure. *See also* apparent dip. *Compare* hade.
2. (magnetic dip inclination) The angle between the Earth's magnetic field at any point on the Earth's surface and the horizontal. It is 90° at the MAGNETIC POLES and 0° at the Equator. It is measured using a *dip circle*, an instrument in which a magnetic needle is free to rotate, in the vertical plane, around a circular scale.

dipole field That portion of the Earth's magnetic field that can best be described as originating from a dipole magnet in the Earth's interior, inclined at 11° to the Earth's axis of rotation. *See also* nondipole field.

dip slope The slope of the surface of the land that more or less mirrors the slope of the rocks beneath.

direct circulation (in meteorology) Circulation in which potential energy is converted into kinetic energy through the rising and sinking of juxtaposed lighter and denser air, respectively. This occurs at scales ranging from land and sea breezes to the meridional cells of the GENERAL CIRCULATION OF THE ATMOSPHERE.

direct wave A wave that travels from one point to another along the path of shortest distance.

discharge The volume of water flowing through a cross section of a stream. It is measured by gauging mean velocity in the cross section, which is multiplied by the cross-sectional area to give an expression of flow in m^3 per second. Within a river discharge increases downstream, except in arid areas where volume is lost by evaporation. By far the greatest discharge of all the world's rivers is that of the Amazon, with a flood-stage discharge of 180 000 m^3 per second, which is at least as great as the combined discharges of its nearest rivals, the Congo, Yangtze, Mississippi-Missouri, Yenisei, and Lena.

disconformity (paraunconformity) An unconformity that marks a considerable time gap but where the beds of rock above and below the plane of the unconformity have a similar dip and strike (*compare* angular unconformity), the absence of beds being detectable only by paleontological means. It is of greater magnitude than a DIASTEM. *Compare* unconformity.

discontinuity 1. (in meteorology) A sharp change in the value of a meteorological variable, for example at a cold front, where there is usually a discontinuity in the temperature, humidity, and wind velocity fields across the frontal surface. Most meteorological properties are continuous functions of space and time and their values can therefore be mapped in the forms of pressure charts or isotherm maps.
2. (in geophysics) A marked change with depth in one or more of the physical properties of the materials constituting the Earth's interior. For example, a boundary at which the velocity of earthquake waves changes is a seismic discontinuity.

discordant 1. Decribing a rock unit that cuts across the bedding or foliation of adjacent rocks. Intrusive igneous rocks, such as dikes, show discordant relationships.
2. Denoting a drainage pattern characterized by streams cutting indiscriminately across structures. Discordance at a local level can be a product of various detailed factors, but at a regional level it can be due to glacial blocking (*see* deranged drainage), tectonic activity (*see* antecedence), or to superimposition of the drainage from a preexisting cover of rocks. It is also referred to as *unconformable* drainage. *Compare* accordant.

discordant coastline A coastline running at right angles to structural features shaping the landscape immediately inland of it. It is most obvious where lines of hills run inland, giving an undulating coastline with alternating cliffs and level beaches. If the land sinks relative to the sea, the valleys between the hills become deep inlets (*see* fiord; ria). Such coastlines are common on the western side of the Atlantic Ocean.

disharmonic folding Folding in which differences in the COMPETENCE of the various beds of rock result in variation in types of folding within the fold, the less competent beds forming numerous folds that are smaller than those formed in the more competent beds enclosing them.

dispersion The separation of SEISMIC WAVES into groups with different frequency, resulting from variations in the velocity of the waves. Such differences, more noticeable in surface waves than body waves, are caused by variations with depth in the density and elasticity of the rocks through which they travel.

disphotic zone Ocean depths at which photosynthesis is not effective because of the small amount of light that penetrates to this layer, which extends from about 80 m to 200 m (the edge of the CONTINENTAL SHELF) below sea level. *See also* aphotic zone; euphotic zone.

disseminated deposits Deposits of minerals that are formed when hydrothermal fluids fill small fissures and pores in a rock (*see* hydrothermal process). They are usually found with igneous INTRUSIONS, and often comprise useful metallic minerals.

dissolved load That part of a river's load carried in chemical solution. Rainwater, being mildly acidic, can dissolve rocks, especially limestone, and then feed that dissolved content into a river via groundwater flow. Rivers may also be mildly acidic, especially if they pass through areas of bogs or marshes, where organic acids are produced, and they may dissolve minerals of their beds. In the Mississippi 29% of the load by weight is carried in solution, but this proportion varies in other streams according to climate and according to what proportion of runoff is contributed via groundwater flow. The major ions carried are bicarbonate, sulfate, chloride, calcium, and sodium, although in small basins there is variation according to the nature of the local rocks, and in populated areas pollution can significantly alter chemical content. *See also* debris load; load.

distributary A branch of a river that flows from the main course and does not rejoin it; the distributary makes its own way to the sea. Most are shallow and narrow. *See also* tributary.

disturbance (in meteorology) Any small-scale synoptic feature causing a disturbance from the normal gradient wind flow, especially a shallow DEPRESSION.

diurnal variation The changes in magnitude of any climatological property recorded during the solar or climatological DAY. Most elements exhibit some variation. Temperature and relative humidity have an inverse cyclic fluctuation reaching peak and trough respectively about 1400 hrs. Atmospheric pressure also shows a daily variation, but it is based on a 12-hourly oscillation that proceeds according to local solar time. Maxima occur about 10.00 and 22.00 and minima at 16.00 and 04.00 hours. Because the synoptic variations in mid-latitudes are large, it is only within the tropical areas that the diurnal variation of pressure is immediately apparent. Some proposals have been made for a nocturnal maximum of precipitation but this has never been proved.

divergence A measure of the rate of outflow of a fluid from a certain volume. It is the opposite state from CONVERGENCE and the mathematical term used to describe both, convergence being negative divergence. It has important implications in the atmosphere as the cause of pressure changes and vertical motion, but is very difficult to measure directly or accurately. Values of divergence in the free atmosphere range up to 10^{-5} per second, although locally, as in fronts, it can reach higher values.

In wave refraction phenomena, divergence refers to the spreading out of the wave orthogonals in the direction of wave advance. Wave energy and wave height tend to decrease in areas of divergence.

In the case of ocean currents, a divergence is the zone in which currents flow away from each other, for example at roughly latitude 10°N where the North Equatorial Current and the countercurrent associated with it separate.

divergent plate boundary A region in which two of the Earth's tectonic plates are diverging, always occurring on the ocean floor. As the plates move apart, material wells up from the mantle beneath to form new oceanic crust. Such boundaries are often associated with MID-OCEAN RIDGES and active volcanic action.

divide (watershed) The area of higher ground that lies between two separate drainage systems. *See also* interfluve.

division **1.** An informal word for any unit in any scheme of stratigraphical classification, including LITHOSTRATIGRAPHY, BIOSTRATIGRAPHY, and CHRONOSTRATIGRAPHY. *See also* stratigraphy.
2. In traditional plant classifications, a taxonomic grouping corresponding to a phylum.

doldrums Equatorial oceanic areas in which winds are light and variable and where navigation by sailing ships was difficult. It coincides with the intertropical convergence zone or thermal equator with frequent thunderstorms, heavy rains, and squalls. It is now only used as a graphic term for this belt of variable winds anywhere near the Equator between the trade wind zones.

dolerite *See* diabase.

doline (dolina) A large SINKHOLE formed by solution in KARST country.

dolomite A magnesium-containing CARBONATE MINERAL of composition CaMg$(CO_3)_2$. The term dolomite is also used to denote a rock with a high ratio of magnesium to calcium carbonate.

dolomitization The process by which a calcium carbonate rock is transformed into a calcium-magnesium carbonate rock through the partial or complete replacement of calcite or aragonite by dolomite. *See* carbonate minerals.

dome An anticlinal fold in which the beds dip in all directions away from the central point of folding. A dome may be merely structural or it may constitute an exposed landform. *See also* salt dome.

Doppler effect A change in the measured frequency of a wave due to the relative motion of the source and the recorder. As the source moves toward the recording station the wavelength decreases, whereas when the source moves away from the recorder the wavelength increases.

Doppler radar A type of radar that relies on the Doppler effect. If the radar target is approaching or moving away from the transmitted signal, it will affect the frequency of the returning signal, being less than that transmitted if moving away or greater if approaching. In meteorology, precipitation drops are the usual target and indicate the horizontal air motion, the fall speed of the particles, and, with certain assumptions, the vertical air motion. This technique is being increasingly used for investigations of subsynoptic levels, such as at frontal zones.

dormant volcano A volcano that has erupted within recorded history but is apparently not active at present. Because nobody can predict if it will one day become active again, it is best described as an inactive volcano.

dot map (in ecology and demographics) A map on which the distribution of a particular measurable quantity is denoted by same-sized dots (each representing a particular number). For example, population densities of plants, animals, or even people can be represented graphically in this way (with each dot standing for, say, 50 trees or 10 elephants).

double refraction *See* anisotropic; birefringence.

downland An area of pasture on hilly terrain. The term is used mainly in Australia and New Zealand.

downs Open rolling uplands, usually on chalky soil. The soil is thin and there are few trees; the chief vegetation is grassland, often used as pasture. Plowed areas of the downs are used to grow cereal crops. The term is used mainly in S England.

downthrow side The side of a fault that has moved downward relative to the other, i.e. the side that has younger beds brought down against the older beds of the other side. (See diagram at FAULT.)

drag fold A small fold associated with a larger folding structure, often found adjacent to faults where the rock strata have been bent as a result of the movement along the fault plane.

drainage Any process that removes rainwater from the land, whether by means of streams or by artificial means (such as drainpipes and channels. *See also* drainage basin; drainage pattern.

drainage basin The area that supplies water to a particular network of drainage channels. At the edge of the basin a divide, usually in the form of hills, mountains, or a plateau, separates it from the adjacent basin. The drainage basin constitutes the unit of study for drainage pattern, rates of denudation, relation between precipitation and runoff, and various other geomorphological factors. Little is known of the controls on basin size and shape, but regional geologic structure and tectonic history are clearly important.

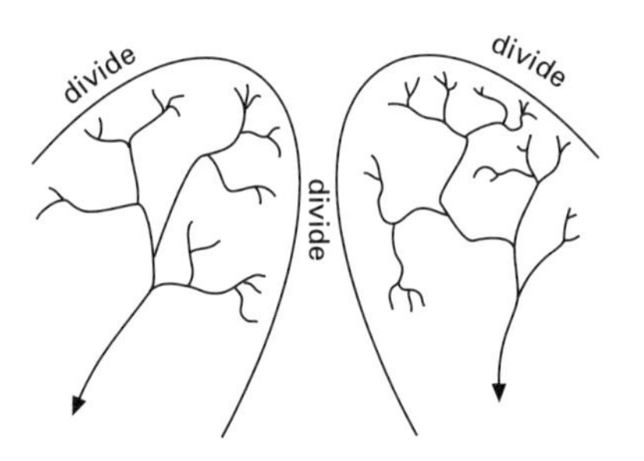

Drainage basins

drainage density The ratio between the total length of stream channel in a basin, and the area that the channel occupies. Closely related is *stream frequency*, which is the number of streams divided by the total channel area. Drainage density is high on rocks yielding impermeable soils, e.g. shales and clays; it is also high where sediment yield is high, on easily erodible rocks, e.g. clays.

Other things being equal, drainage density is governed by climate, through its control on runoff via rainfall and vegetation. Relief is also significant: high-relief areas tend to have high drainage densities. Drainage density is an integral part of drainage pattern description: in particular it shows the degree of dissection of the drainage basin.

drainage divide *See* divide.

drainage pattern The spatial arrangement of streams within a basin. It is a reflection mainly of the area's geologic structure, i.e. the nature and arrangement of faults, folds, rocks (especially lithologies), and relative relief. Past and current climate and the tectonic and geomorphic history of the basin are also important. These variables control the rate and nature of stream incision and headward erosion, the basic processes by which the pattern develops. Field studies show the pattern evolves very rapidly at first, then reaches a steady state of little further change, contrary to W. M. Davis's original idea that the rate of development was constant through time. *See also* accordant; angulate drainage; barbed drainage; dendritic drainage; deranged drainage; discordant (def. 2); drainage density; parallel drainage; radial drainage; rectangular drainage; stream order; trellis drainage.

dravite A brown variety of TOURMALINE that is rich in magnesium. It generally occurs in metamorphic rocks, and is sometimes used as a semiprecious gemstone.

dredging The deepening of a river, port-approach channel, or other such area by excavating loose sediment or in-situ rock from the river or seabed. The dredger or dredge is a vessel specially designed to undertake this. Some dredgers actually dig up the bed using a combination of buckets or grabs; others suck up loose sediments mechanically using suction devices, as in a trailing suction-dredger. Dredging may be performed to improve navigable depths or to remove obstructions such as shoals or banks that are dangerous to navigation. Other dredgers operate farther offshore and recover useful minerals, such as sand, gravel, tin, ore, and diamonds from the seabed. Dredging can have far-reaching environmental consequences, as when it increases turbidity, to the detriment of fish and shellfish production.

dreikanter A pebble with three facets, formed by the erosive action of windblown sand in desert regions. The wind moves the pebbles back and forth, because they are too heavy to be lifted entirely.

drift Glacial and fluvioglacial deposits. Great thicknesses of this drift accumulated during the PLEISTOCENE period, although much has subsequently been removed by erosion. An area of glacial deposition may be referred to as one of *drift topography*. A *drift map* is a geologic map on which glacial drift and other superficial deposits are shown, as contrasted with a *solid* edition showing only the underlying rocks.

drizzle Water droplets ranging in diameter from 0.2 to 0.5 micrometers. It forms by coalescence in stratiform cloud with only weak vertical motion, otherwise the droplets would be unable to fall. It also requires a high relative humidity between cloud base and ground surface or only a short distance between cloud and surface. In both cases, evaporation of the droplets would occur without these constraints.

drought (drouth) A period of dryness due to the absence of significant precipitation. It is an unsatisfactory term unless very carefully specified because it implies some effect on agriculture and vegetation, but this is dependent upon factors other than rainfall alone. In purely climatic terms, it can be used to indicate lack of rain. In the USA a drought is defined as a period of 21 days or more when rainfall is 30% below average for the time and place.

drowned coastline A strip of land along the coast that has been submerged under the sea, either because the sea level has risen or because the land has sunk relative to the sea. Valleys are flooded, forming FIORDS or RIAS, and hills may become islands.

drowned valley A river valley cut at a time when sea level was at a much lower level than it now is, usually during the PLEISTOCENE Period, and subsequently drowned as sea level rose again on melting of the large ice sheets. These former inland valleys now exist as indentations in the coastline, their internal shapes being largely hidden underwater. Differently shaped coasts are produced by the flooding of geologically different areas. If the former drainage system was perpendicular to the general outline of the present coast a series of inlets separated by headlands will exist; if the drowned valleys lay parallel with the present coastline numerous elongate islands, the former interfiuves, will be present, with narrow flooded areas between them. *See also* fiord; ria.

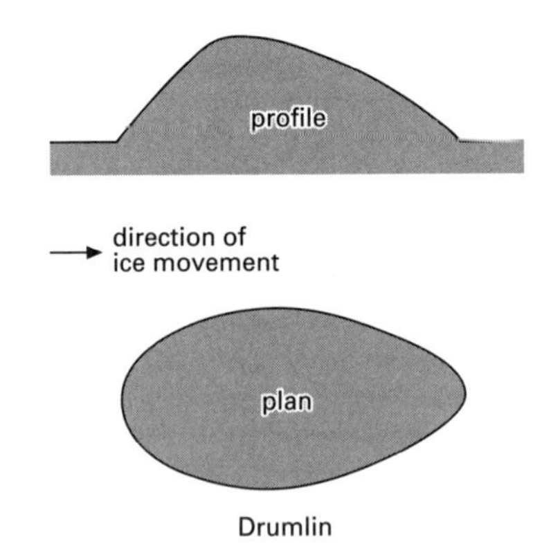

Drumlin

drumlin A streamlined glacial form tending to occur in groups, all with their long axes parallel with the direction of ice movement. They may be composed of TILL, which can vary considerably in composition, of preexisting DRIFT, or of rock. Some contain stratified deposits. Since these forms are so variable a number of explanations of their origins have been suggested: there are two principal explanations. Those composed of rock or preexisting drift materials must have involved erosion, although if erosion were the only process, one would expect shapes similar to those of ROCHES MOUTONNÉES, which is not the case. Drumlins composed of newly deposited till must involve deposition, and it has been suggested that they form around a nucleus of frozen till, or rock. Despite their variability, all drumlins are linked by their common shape, and the fact that their formation is associated with moving ice.

druse (drusy) A cavity in a rock or mineral vein into which large euhedral crystals project. A rounded cavity containing crystals that project toward the center is called a *geode*. *See also* miarolitic cavity; vugh.

dry adiabatic lapse rate The constant rate at which air will cool upon uplift, prior to saturation being reached. Its value of 0.98°C/100 m depends upon the properties of the gases in the atmosphere and the gravitational force. *See also* environmental lapse rate; saturated adiabatic lapse rate.

dry site A small hill in a wetland or marshy region, often occupied by human settlement.

dry valley A typical subaerial river valley in all respects except that it contains no stream or river. Such valleys are mostly found in areas of soluble limestone where the waters responsible for the cutting of the valley, at an early stage in the erosional history of the area, have subsequently disappeared underground through enlarged joints and subsurface cavern systems, developed as a result of LIMESTONE SOLUTION. At times of exceptional discharge the underground systems may be incapable of carrying all the water, and surface drainage may return temporarily.

Some dry valleys may have been formed in former periglacial conditions, when permafrost rendered the ground impermeable, or in times of wetter climate, when the water table was higher.

dumortierite A blue or green fibrous mineral form of hydrated aluminum borosilicate, $Al_8BSi_3O_{19}(OH)$.

dune A mound or hill of sand. Dunes are characteristic landforms of deserts, in an unvegetated form, and of certain coastal and riverside areas, where they are usually vegetated. Desert dunes are simpler in form owing to the lack of moisture and plant growth, the two major forms being the BARCHAN (crescentic dune) and the SEIF DUNE (longitudinal dune). Coastal dunes show a development inland from the young FOREDUNE through the main MOBILE DUNE to the STABILIZED DUNE.

The major requirement for dune-building is a sand supply; in the deserts this is transported by wind from areas of erosion, notably scarps and areas of DEFLATION, whereas in coastal areas it is transported by the longshore drift from a source that may be a preexisting supply or an area of erosion creating a supply, and then left on the foreshore ready for wind action to take it inland and build the dune.

Coastal dunes are hence due to a three-way relationship between littoral drift of sand, blowing of sand onshore, and vegetation growth trapping the sand and building a dune, whereas desert dunes may be wave patterns resulting from the interaction of air flow and land surface, the dunes accumulating in areas between turbulent eddies.

dune-bedding A type of CROSS-BEDDING that occurs in DUNES, resulting from variable deposition of sand by the wind.

dunite A monomineralic ultramafic rock consisting wholly of OLIVINE.

Dust Bowl An area in the western USA that has suffered extensive wind erosion, which has removed the fertile topsoil. It lies mainly within W Kansas, Oklahoma, and Texas and extends into Colorado and New Mexico. A smaller dust bowl occurs in central Nebraska, North Dakota, and South Dakota. *See also* badlands.

dust devil A whirlwind or small tornado in which sand and dust rise into the atmosphere and can reach a height of up to 30 m. It forms by strong convection above an intensely heated sandy surface and is distinguished as a dust devil only if the surface is sufficiently sandy for the surface material to be drawn into the rotating column.

dust storm A cloud of fast-moving windblown dust, which occurs when the wind is strong enough to lift the dust particles from the ground. The dampness, density, shape, and size of the particles determines how strong the wind has to be. Local dust storms are often associated with rain and thunder; extensive dust storms usually occur in regions of low atmospheric pressure. In desert areas, very small occurrences are known as DUST DEVILS.

dyke *See* dike.

dynamical meteorology The branch of meteorology concerned with the causes and nature of motion within the atmosphere.

dynamic equilibrium (in geomorphology) The state of balance between erosion and deposition to which rivers seek to adjust. The emphasis on *dynamic* is because rivers never actually achieve a balance of no erosion or deposition, but are always actively adjusting to the constant changes of load and discharge, etc., by some minor erosion and deposition. *See also* equilibrium profile; river.

E

earthflow A mass movement of soil on a steep slope, well mixed with water. Such flows are commonest in areas with little vegetation, where rainfall comes in sharp bursts, saturating the soil, and where the soil has a high content of fines. The difference between earthflows and mudflows is purely of degree: earthflows occur on gentler slopes and move at a lesser velocity than mudflows. The soil eventually comes to rest as a tongue at the slope foot, the thickness of the tongue becoming less as the velocity of flow increases. Like the other mass movements, flows act on unstable slopes, reducing them to a more stable angle.

earth hummock A rounded mound of frozen soil, up to 20 cm high, that has fine material at its core and is covered with vegetation. Common in alpine and arctic regions, earth hummocks create a type of PATTERNED GROUND.

earth movement Any movement of the Earth's crust caused by processes occurring beneath it. Sudden earth movements accompany earthquakes and volcanic eruptions; slow movements cause folding and uplifting of rock strata.

earthquake A series of shocks, subdivided into FORESHOCKS, PRINCIPAL SHOCKS, and AFTERSHOCKS, which generate seismic waves within the Earth, as a result of the fracturing of brittle rocks within the lithosphere. They result from the accumulation of forces within the rocks until they are strained to a point beyond which they fracture. The *magnitude* of an earthquake is the amount of energy involved. Depending on the quantity of energy liberated when the overstrained rocks fracture, the earthquake may vary from mild quiverings to violent oscillations of the land surface.

earthquake focus *See* focus.

earthquake intensity The degree of violence of an earthquake at a particular point on the Earth's surface, expressed on a descriptive scale divided into twelve points. Point one on the scale is only detected by seismograms, whereas at point twelve objects are thrown into the air and there is total destruction.

earthquake zone *See* seismic zone.

earth science Any one of the sciences that study the Earth. The earth sciences include geology, geomorphology, meteorology, and oceanography.

earthslide The movement of the soil mantle over a shear plane. For the process to operate there must be an unstable slope and a dry soil. If the soil becomes wet, it may turn into an EARTHFLOW. An important part may be played by water, however, in lubricating the shear plane. *See also* mass movement.

earth temperature The temperature of the ground surface as determined by the characteristic wavelengths of long-wave radiation emission.

earthworm The most important of the macrofauna in the soil. Under favorable conditions, as when the soil is moist and rich in lime and organic matter, there may be up to a million earthworms per acre. They may pass as much as 10 tonnes per acre of soil through their bodies each year. This material is humified and excreted in the form of casts, which contain more humus, nitrates, and exchangeable bases than the surrounding soil. Besides their function in the chemical and physical breakdown of plants they are also important in improving the texture, aeration, and drainage of the soil.

easterlies Any winds in which the zonal component of air flow is from the east. They are subdivided into the tropical easterlies or trade winds and the polar easterlies. The presence of easterly winds at the ground surface is an essential requirement to counteract the effect of the mid-latitude westerlies on the Earth's rate of rotation.

easterly wave A shallow trough or disturbance in the trade-wind flow associated with an increase of cloudiness and precipitation to its rear. After initial debate about the status of the easterly wave, satellite photographs have confirmed characteristic cloud patterns associated with the trough although they do exhibit greater variety than was first thought. In some instances they probably act as initial disturbances for the development of hurricanes.

easting Any of the north-south grid lines on a map, quoted before the NORTHING when coordinates are being given, showing distance east from the origin of the grid.

ebb tide The outgoing of the tidal stream; the retreating tide, i.e. that part of a tide cycle following the high-water stage and preceding the low-water stage. *Compare* flood tide.

Echinodermata The phylum of marine invertebrate animals that includes the starfish, sea urchins, and crinoids. The body of an echinoderm usually has five radii, along which are grouped hydrostatic *tube feet*, functioning in locomotion and feeding and operated by the *water vascular system*, an internal system of fluid-filled canals. These radii are known as *ambulacra* and the areas between as *interambulacra*. An external skeletal system (*test*) composed of plates of single calcite crystals is usually present and sometimes bears spines. There are five living classes: the Asteroidea (starfish), Ophiuroidea (brittle stars), ECHINOIDEA (sea urchins), Holothuroidea (sea cucumbers), and the CRINOIDEA (sea lilies). In addition there are a number of important extinct classes, such as the Cystoidea. Echinoderms extend from the early Cambrian Period to the present day.

Echinoidea The class of the phylum ECHINODERMATA to which the sea urchins belong. They are protected by a hemispherical and usually spiny test of interlocking calcite plates, which in some later forms is flattened. Typically the anus is situated in the center of the upper surface and the mouth is diametrically opposite. In later forms the mouth and anus may have become displaced, the anus sometimes opening on the oral surface. Echinoids are marine, benthonic, and free living and are known as fossils from the Ordovician Period onward. The group was affected by the widespread extinctions at the end of the Paleozoic Era but revived in the Mesozoic and extends to the present day. Echinoids are used as biostratigraphic ZONE FOSSILS in the Cretaceous System.

eclogite A coarse-grained granular rock consisting essentially of bright green omphacite and deep red almandinepyrope garnet. Diopside, quartz, and kyanite may also be present in small amounts. Eclogite has the chemical composition of basalt and may be considered to be the high-pressure high-temperature metamorphic equivalent. Eclogite is found as lenses and bands in regionally metamorphosed rocks of the highest grades and as inclusions in basalt and kimberlite.

ecology The scientific study of how living organisms affect, or are affected by, their natural environments and by other organisms. Both living and nonliving components of the environment are considered (including plants, animals, soil, climate, temperature, etc.). Ecology may be concerned with individual organisms, populations, or even whole communities. *See also* ecosystem.

economic basement Strata below which there is little chance of finding economic mineral resources, particularly oil.

ecosystem An ecological unit that consists of the physical environment (both living and nonliving) and the organisms that occupy it. It may be as small as a pond or tidal pool, or as large as a tropical rainforest or even the whole global system. Nutrients and energy pass through an ecosystem in a certain way (*see* food chain). The organisms living in it may occupy various TROPHIC LEVELS, from plants (producers) at the bottom to animals (consumers) at the top. Left alone, an ecosystem achieves a balance, with all the organisms living successfully together. It will even usually recover from a disaster such as a drought or flood. Human intervention, however, may cause changes from which the ecosystem cannot recover, for example the large-scale cutting down of forests or extensive pollution of the air, land, or oceans. *See also* carbon cycle; nitrogen cycle.

ecoulement The downward sliding of large masses of rock as a result of gravity.

edaphic Describing those factors of the soil (chemical, biological, and physical) that affect the growth of plants. Examples of such factors include moisture, mineral content, and texture.

eddy A rotational feature of a fluid, which retains its identity for a limited time while moving within the main body of the fluid and eventually amalgamating with it. Such eddies can frequently be seen in river flow after the current has been disturbed by a bridge or shearing. *See also* whirlpool.

In the atmosphere, eddies are found in a wide range of scales. Near the ground surface they are important in transferring momentum, heat, and moisture. On the large scale, eddies in the form of depressions are a necessary mechanism for much of the meridional transfer in the GENERAL CIRCULATION OF THE ATMOSPHERE.

eddy diffusion The exchange of atmospheric matter and properties brought about by eddies. *See also* diffusion.

edenite A monoclinic AMPHIBOLE.

edge dislocation A defect in a crystal lattice resulting from the insertion of an extra plane of atoms.

effluent (in geology) A stream that flows out of a lake or other stream. In general usage, the term describes waste material discharged as a liquid, and a potential source of pollution.

ejecta 1. Material that is thrown out of an erupting volcano. *See* pyroclastic rock.
2. Material that is thrown out when a METEORITE impacts with the ground. It consists mainly of rock fragments, but may also include glassy particles that result from the melting of the meteorite or the rock it hits.

ejecta blanket A deposit of EJECTA that surrounds an impact crater after a METEORITE hits the ground.

ejectamenta *See* tephra.

Ekman flow The movement of surface sea water in the direction of a wind, which, as it blows, exerts a force on the sea surface in the direction of the wind and, because of frictional forces operating within the water column, causes a certain thickness of water to flow. The ocean's response is highly complicated, especially because of Earth rotation and the fact that water is fluid, but also because of such factors as land-ocean configuration. In the N hemisphere, average Ekman flow is some 45° to the right of the wind, and the speed decreases and the direction swings increasingly to the right of the wind direction with increasing depth. Ekman flow has several far-reaching effects; for instance, off the coasts of California and Peru, Ekman flow tends to be in an offshore direction, thereby causing zones of upwelling of fertile deep water near the coast, which provides nutrients for fish populations.

Ekman layer In general terms, the circulation of ocean water, viewed in the vertical column, occurs in two layers, the uppermost layer being relatively thin, about 100 m deep, and beneath this and extending down to the ocean floor, the deep-water circulation involving deep water masses. The upper water masses, in the shallow surface layer, move in what is termed the Ekman layer, the thickness of which has been calculated theoretically. Flow within this layer is termed EKMAN FLOW.

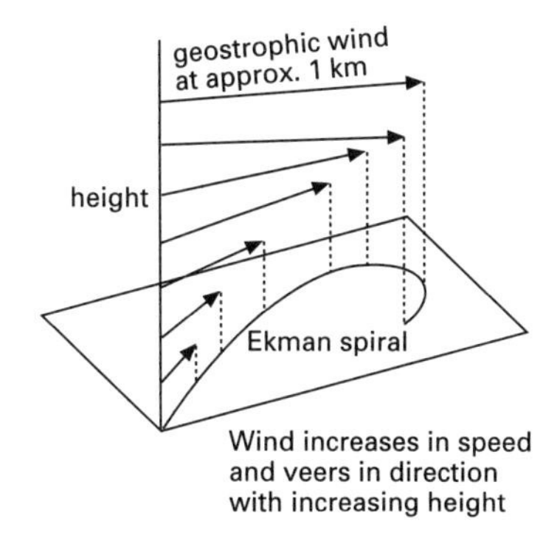

Ekman spiral

Ekman spiral The ground surface has a frictional interaction with the atmosphere and its effect decreases with height until, between 500 and 1000 m, the GEOSTROPHIC WIND is observed. The spiral illustrates in vector form how the wind velocity changes with height (see diagram), blowing across the isobars at low speed near the surface and parallel at higher speeds in the free atmosphere.

elastic wave A seismic wave that is propagated through a medium by elastic deformation.

elbow of capture A right-angled bend in a river downstream of which is the capturing river and upstream of which is the captured river. *See* river capture.

electrical logging A subsurface logging technique in which an electrode or series of electrodes are lowered into an uncased borehole. As the sonde is raised a continuous record of the electrical properties of the rocks through which it passes is recorded. By an examination of the variations in several properties an assessment of the rock types present in the borehole can be made.

electromagnetic radiation The form of energy that travels in waves from its source without the necessity of an intervening medium. The waves move with the speed of light (3×10^8 meters per second) in space and only slightly slower through the atmosphere. The total spectrum of this form of radiation includes wavelengths as short as gamma rays (10^{-13} m) to long radio waves (up to 10^5 m), as shown on the diagram overleaf. Within this range is visible light between 0.4 and 0.7 micrometers, to which our eyes are sensitive. The behavior of electromagnetic radiation is explained by the following physical laws:

1. All matter with a temperature above absolute zero (–273°C) emits radiation.
2. Some substances emit radiation of certain wavelengths only.
3. A substance that emits the maximum amount of radiation for a given temperature in all wavelengths is known as a black body (*see* black-body radiation). The amount of radiation emitted is then proportional to the fourth power of the substance's absolute temperature.
4. Substances will only absorb radiation of wavelengths that they can also emit.
5. The hotter a substance, the shorter will be the wavelengths at which most of the radiant energy is emitted.
6. The amount of radiation received at a point is inversely proportional to the square of the distance of that point from the radiation source.

element A substance that cannot be broken into simpler substances by chemical means. All the atoms in an element have the same atomic number although, in isotopes, some of an element's atoms have a different mass (because they contain different numbers of neutrons). There are 92 elements that occur naturally on Earth.

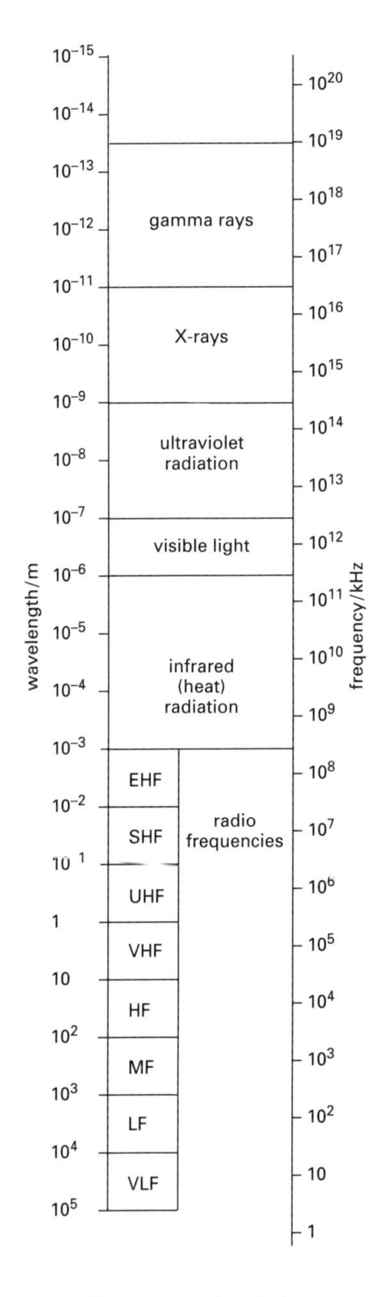

Electromagnetic radiation

elevation The vertical distance of any point on the Earth's surface above the DATUM level, measured in meters or feet.

El Niño The periodic extension along the Peruvian coast of the warm equatorial current. This replaces the cold HUMBOLDT CURRENT and sea surface temperatures may increase by 10°C. Heavy rains then fall in the normally desert areas of Peru and there is a fall in the anchovy catch, on which the local economy depends. It is called El Niño (the baby boy) in an allusion to the Christ child, because it appears just after Christmas.

The mechanism of El Niño is well understood. The east–west trade winds push large volumes of water west toward the coast of Indonesia. Every few years there is a change in the wind pattern, which allows warm water to flow back across the Pacific toward the coast of South America. The phenomenon is now known to have an effect on a wider scale; it has been associated with drought in southeastern Asia and in Australia and with flooding in North America.

El Niño is connected with an oscillation in the atmosphere across tropical regions, called the *Southern Oscillation.* The coupled air–water flow is called *ENSO* (El Niño–Southern Oscillation). As part of this cycle there is, in some years, formation of a cold region in the eastern Pacific. This is called *La Niña* (the little girl).

elutriation The natural sorting of rock fragments into finer and coarser particles. It most commonly occurs when the fragments are transported by water, but may also happen during pyroclastic flow, when volcanic material flows down the side of an erupting volcano.

eluviation The washing out of fine material from a soil, especially from the upper part. The eluvial horizon is the A horizon, which has less clay than the rest of the profile as a result.

emerald A bright green transparent variety of BERYL (the color is caused by chromium impurities), valued as a precious gemstone. It crystallizes in the hexagonal system and occurs mainly in mica schists and veins of calcite.

emery A black or dark gray impure form of the mineral CORUNDUM. It occurs as fine granules, often with hematite or magnetite impurities. It is used as an abrasive.

emissivity The ratio of emission of radiation from a substance to the emission from a black body at the same temperature and wavelength (*see* black body radiation). Values range from slightly below 1.0 to 0.85 for most substances.

enclave An inclusion or XENOLITH.

enderbite A rock of the CHARNOCKITE group, rich in plagioclase feldspar.

endomorphism The change in composition of igneous rock derived from magma as it assimilates material from the COUNTRY ROCK surrounding it. *See also* metamorphism.

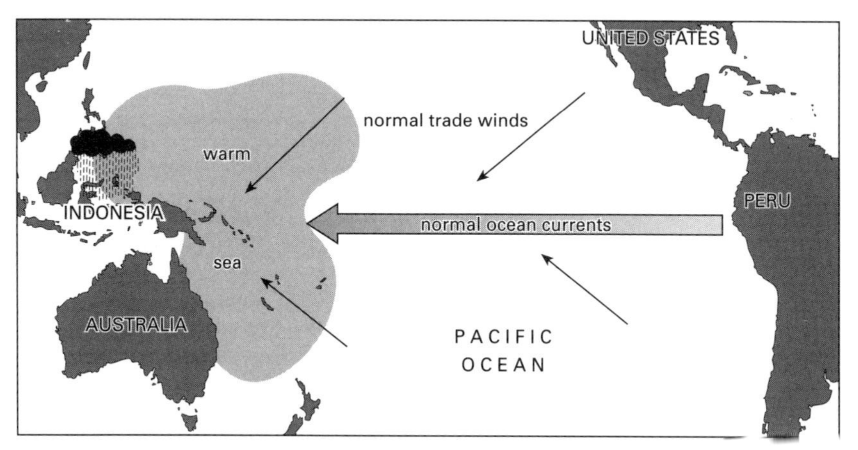

Under normal conditions the trade winds push water from east to west and warm water accumulates around Indonesia

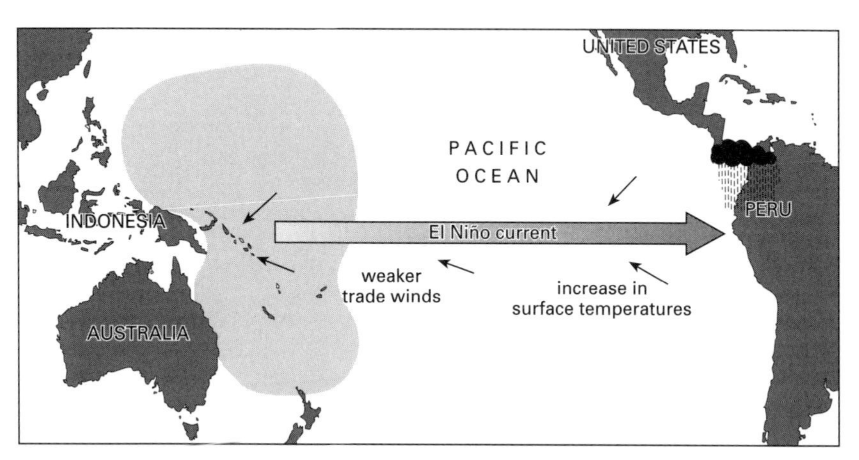

In El Niño years, the trade winds weaken and the warm water flows back toward the coast of Latin America

en echelon Denoting overlapping or offset geologic structures.

energy balance A concept applied to the static energy of the Earth-atmosphere system, rather than its dynamic energy. It relates the net radiation flux at a surface to the utilization of this available energy. This is consumed by the conduction of heat below the ground surface, the transfer of sensible heat to the atmosphere by turbulence and convection, and the transfer of latent heat by similar methods. All these processes are reversible.

The energy balance equation is often called the *heat balance equation* and is expressed in the form $R = H + LE + G$ where R is net radiation, H is sensible heat, LE the latent heat transfer, and G the component transferred into the ground.

englacial Describing materials or meltwater contained within an ice mass. Such debris or water can reach the interior ice either by movement upward from the bed or downward from the surface.

enrichment The natural increase in the proportion of one of the elements or minerals that constitute a given type of rock. It may result when additional amounts of a constituent are introduced from outside sources, or when one constituent is selectively removed from the rock. It can be caused by chemical processes, as when mineral-laden water filters through a porous rock, or by mechanical processes, as when relatively light quartz is transported away from heavier metallic minerals.

ENSO *See* El Niño.

enstatite An orthopyroxene. *See* pyroxene.

entisol One of the ten soil orders of the SEVENTH APPROXIMATION classification, approximately equivalent to the azonal category of the old classification. These are recent soils without natural genetic horizons and they include lithosols (shallow stony soils), regosols (thin soils on unconsolidated drift), and alluvial soils or fluvisols (soils that are constantly being added to because of their site on active floodplains). These soils are not extensively used for agriculture but may become productive when managed properly.

entrainment (in meteorology) The process of mixing between the environment and a rising thermal in the updraft of a cumulus or cumulonimbus cloud. It has the effect of diluting the warmer rising air and so slightly reduces its buoyancy. This prevents the rate of cooling in the thermal from being adiabatic, but in practice the difference is slight.

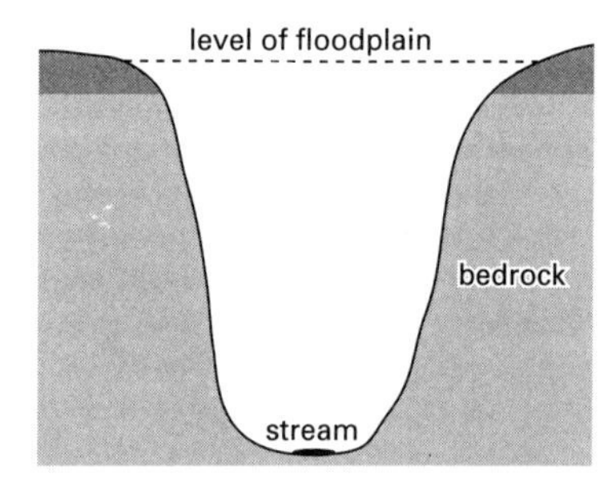

Section through an entrenched meander

entrenched meander The type of INCISED MEANDER that results from the REJUVENATION of floodplain meanders and subsequent incision into the bedrock below. They are therefore the product of a two-stage process, unlike the one-stage origin of INGROWN MEANDERS. Entrenched meanders are often distinguished from ingrown meanders on the basis of their symmetrical valley cross profile, contrasted to the asymmetrical cross profile of the ingrown type. However, some floodplain meanders do not incise themselves vertically into their floodplains, but at an angle, producing an asymmetric cross profile.

environmental lapse rate The actual rate of change in temperature with height of the atmosphere, at a given time, used particularly when investigating the stability of the lower atmosphere. The differences between the DRY ADIABATIC LAPSE RATE, SATURATED ADIABATIC LAPSE RATE, and environmental lapse rate are especially important in this respect when drawn on a thermodynamic diagram. The lapse rate is steep in the lower layers during daytime heating, decreasing with height, so that by the upper atmosphere it approximates to the saturated adiabatic lapse rate. Inversions are frequently present indicating a warming of the air with increasing height, but within the troposphere they are rarely very deep.

Eocene The epoch of the TERTIARY Period that extended from the end of the PALEOCENE, 57.8 million years ago, to the beginning of the OLIGOCENE Epoch, about 36.6 million years ago. Some classifications dispense with the Paleocene, thus adding another 8.6 million years to the Eocene, which then becomes the first epoch of the Tertiary. During this epoch mammals were abundant, perissodactyls, including large and small horses, and artiodactyls appeared and the modern carnivore families became established. The first evidence of bats and whales also comes from this epoch.

eon The largest interval of geologic time in the Chronomeric Standard scheme of chronostratigraphic nomenclature (*see* chronostratigraphy), formed of several ERAS grouped together.

Eötvös unit The unit employed to express gravitational curvature or gradient. 1 Eötvös unit equals 10^{-6} mgal cm^{-1}.

epeiric sea A shallow INLAND SEA.

epeirogenesis Uplift or subsidence of large areas of continents or ocean basins. *Compare* orogenesis.

ephemeral stream A stream that contains water only immediately after rainfall, found mainly in arid and semiarid areas. For most of the year, its channel is dry (*see* gully).

epicenter The point on the Earth's surface situated directly above the focus of an earthquake.

epicentral angle The angular distance between the epicenter of an earthquake and a recording station, expressed in terms of the angle subtended at the center of the Earth between verticals from these two points.

epidiorite A rock with a dioritic mineralogy derived from the low- to medium-grade metamorphism of labradorite-pyroxene assemblages of basic igneous rocks (gabbro and dolerite), which results in the formation of hornblende and a plagioclase of more sodic composition.

epidote A group of minerals having the general formula $X_2Y_3Si_3O_{12}(OH)$, where X is mainly Ca but also Mn^{2+}, Ce^{3+}, and other rare earths, and Y = Al, Fe^{3+}, Mn^{3+}, and Fe^{2+}. The following compositions occur:

zoisite $Ca_2Al_3Si_3O_{12}(OH)$
clinozoisite $Ca_2Al_3Si_3O_{12}(OH)$
epidote $Ca_2Fe^{3+}Al_2Si_3O_{12}(OH)$
piemontite $Ca_2(Mn^{3+}Fe^{3+},Al)_3Si_3O_{12}(OH)$
allanite (orthite) $(Ca,Mn^{2+},Ce)_2(Fe^{2+},Fe^{3+}Al)_3Si_3O_{12}(OH)$

The epidote minerals are monoclinic except for zoisite, which is orthorhombic. Zoisite may be gray, green, or brown in color; clinozoisite is colorless to green; epidote is green to yellow-green; piemontite is typically red-brown; allanite is brown to black. A pink manganiferous variety of zoisite is called *thulite*. Minerals of the epidote group are found in the medium-grade regionally metamorphosed rocks of the greenschist and amphibolite facies. Epidote is produced during retrograde metamorphism and may be found on joint surfaces and along fractures. Plagioclase feldspar, clinopyroxenes, and hornblendes can all be replaced by epidote minerals and basic igneous rocks in particular may suffer extensive epidotization. Allanite is found as an accessory mineral in some acid igneous rocks.

epilimnion The warmer upper layer of water in a lake or shallow sea. Photosynthesis may occur and green plants grow within the epilimnion because light can penetrate it. *See also* hypolimnion.

epipedon A diagnostic surface horizon, constituting that part of the soil with organic matter or the upper eluvial horizon or both. Introduced by the SEVENTH APPROXIMATION, there are six common epipedons: mollic, histic, plaggen, anthropic, umbric, and ochric, each being indicative of a certain class of soil. For example, MOLLISOLS are identified by the mollic epipedon; soils influenced by human use are identified by the anthropic epipedon (>250 ppm acid soluble salt phosphate due to farming) and the plaggen epipedon (a layer >50 cm deep produced by manuring).

epitaxial growth The parallel or orientated overgrowth of one mineral on a crystal of another such that there is some form of continuity maintained between the crystal structures of both minerals. Examples include the overgrowth of augite on orthopyroxene and idocrase on garnet.

epithermal *See* hydrothermal process.

epoch An interval of geologic time in the Chronomeric Standard scale of chronostratigraphic classification (*see* chronostratigraphy). The equivalent Stratomeric Standard term, indicating the body of rock formed during this time, is the SERIES. Several epochs together form a PERIOD and are themselves formed of a number of AGES.

equal-area projection *See* homolographic projection.

Equator An imaginary line that girdles the Earth at latitude 0°. It is a GREAT CIRCLE 40 076 km long. North of it is the N hemisphere; to the south is the S hemisphere.

equatorial current Among the prominent currents that flow in the oceans are those in the equatorial regions. The currents flowing at or close to the Equator in the Pacific Ocean are, in general, similar in pattern to those flowing in equatorial regions of the Atlantic Ocean, there being a *South Equatorial Current* flowing astride the Equator, which is separated from the westward-flowing *North Equatorial Current* by an eastward-flowing *Equatorial Countercurrent*. However, this countercurrent is much more strongly developed in the Pacific than is its counterpart in the Atlantic. There are also currents that flow at some depth; for instance, the *Cromwell Current* (also known as the *Pacific Equatorial Undercurrent*) flows as a narrow swift current in an easterly direction and at a fairly shallow depth beneath the South Equatorial Current. The flow rate in the Equatorial Countercurrent probably attains some 25 million cubic meters per second, with speeds of up to 2 knots.

equatorial rainforest *See* tropical rainforest.

equatorial trough The belt of low pressure that occurs in equatorial areas and oscillates in its mean location depending upon the position of the overhead Sun. It marks the convergence zone of the trade winds blowing from each hemisphere and is most clearly observed over the ocean areas where continental modifications of airflow are least.

equatorial westerlies A zone of rather variable winds, found when the INTERTROPICAL CONVERGENCE ZONE extends more than 5° from the Equator. They probably represent trade-wind air that has acquired a westerly component on crossing the Equator but their origins may vary depending upon the precise location of formation.

equatorial zenithal gnomonic projection A map projection in which the tangent touches the surface of the Earth at the Equator. All the great circles are straight lines but the angle of intersection of the meridians and the parallels (shown by curves) becomes increasingly more acute poleward, and therefore the distortion is greater. Places best represented on this pro-

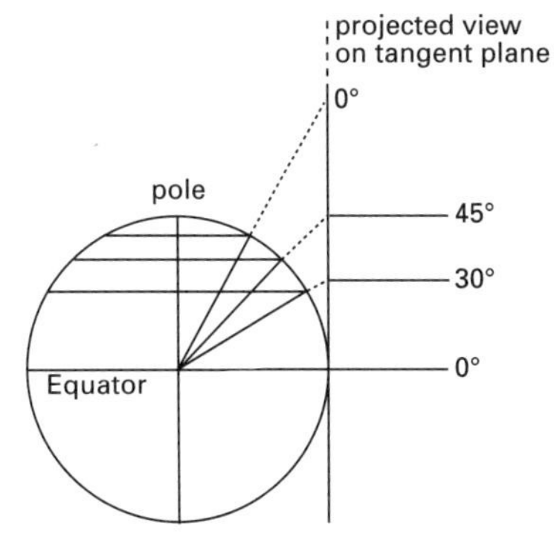

Equatorial zenithal

jection are those near the Equator, e.g. Africa; beyond about 30° north or south the distortion is too large for the projection to be of any use.

equigranular Describing a rock that has crystals of a uniform grain size.

equilibrium profile (graded profile) A long profile of a river in which there is a delicate balance between erosion and deposition. Slope of the equilibrium profile is adjusted so that the sediment load provided to the river can just be transported by the available discharge and channel characteristics. Traditionally this is thought of as being a concave profile, but in practice it need not be perfectly smooth. Streams with uneven long profiles have been found to be in equilibrium in the sense that in no part of their profile is active erosion or deposition taking place. *See also* dynamic equilibrium; equilibrium regime.

equilibrium regime The regime of a natural system, such as a river channel, coastline, slope profile, etc., when it is in a state of balance considered over time, neither eroding nor depositing. The term is most often used in connection with rivers, where it is largely synonymous with GRADE.

equinox The period of the year when the Sun is overhead at local solar time at the Equator. This happens twice a year, about March 21 and Sept. 22, when day and night are equal time periods of twelve hours. Because of refraction through the atmosphere, actual day length is slightly longer than the true astronomical day. In NW Europe, the first and second equinoxes are taken to mark the beginning of spring and of fall respectively.

era An interval of geologic time in the Chronomeric Standard scheme of chronostratigraphic nomenclature (*see* chronostratigraphy). It is formed of several PERIODS grouped together and a number of eras may be compounded to form an EON. For example, the Paleozoic Era is composed of six periods (the Cambrian, Ordovician, Silurian, Devonian, Carboniferous, and Permian) and lasted for some 325 million years, from about 570 to 250 million years ago.

erg **1.** A type of arid desert landscape consisting of a very extensive sand cover, especially in the Sahara. *See also* hammada; reg.
2. A unit of work or energy equal to the work done by a force of 1 dyne acting through a distance of 1 cm. 1 erg is equal to 10^{-7} joules.

erosion The lowering of the land surface by agents that involve the transport of rock debris. Unlike degradation, erosion tends to be an episodic process, varying greatly over time and space (*see* cycle of erosion). For erosion to occur, the eroding medium (gravity, river flow, waves, currents, wind, ice, etc.) must exert a force on the land surface greater than its shear strength. The elastic limit of the material then being exceeded, fractures form and the material moves in the direction of the force applied to it. The movement may be sharp and sudden, as in LANDSLIDES and ROCKFALLS, or slow and imperceptible, as with CREEP. *Compare* denudation. *See also* differential erosion.

Soil erosion is a great problem because the surface part is the most fertile and plant nutrients have to be replaced artificially. Wind erosion of soil can be severe in semiarid regions (*see* Dust Bowl). Water erosion (sheet wash, rill action, gullying, etc.) is usually most prominent in humid lands, e.g. in the Tennessee Valley. *See also* abnormal erosion.

erratic A rock that has been transported by ice and deposited in an area of dissimilar rock type. Volcanic and metamorphic rocks frequently make the most impressive erratics because they are often distinctively colored and their high resistance to erosion permits their continued existence despite considerable transportation. By linking source outcrop with final resting place, some insight into directions of ice movement can be gained, although one erratic may, in turn, be moved by several ice masses in different directions. The existence of some erratics hundreds of meters in extent indicates the phenomenal power available in glacial transportation.

eruption plume A cloud of gas, rock fragments, and molten magma that rises into the air above an erupting volcano. Its height depends on its density and the pressure that forced it upwards.

escape velocity The initial velocity that a particle, space probe, etc., on the surface of a body such as the Earth would require to overcome the gravitational influence of that body and move away from it.

escarpment *See* scarp.

esker An elongate ridge of rounded stratified FLUVIOGLACIAL deposits, consisting primarily of sands and gravels with some finer and coarser materials; they can be only tens of meters long but some extend for hundreds of kilometers. Eskers form in contact with stagnant or very slow moving ice from materials deposited by meltwater streams, which may be underneath, within, or above the ice mass. The cross-sectional shape of an esker will depend largely upon the position of the meltwater stream. If it is above the ice, the esker will have an ice core and melting will cause collapse; if it is in tunnels within the ice, collapse will also occur when the walls melt. If the tunnels are high and narrow, sharp-crested ridges are produced, whereas low wide tunnels result in flat-topped and comparatively undisturbed eskers. *See also* beaded esker.

essential mineral Any of certain diagnostic minerals on the presence or absence of which is based the classification and subdivisions of igneous rocks. For example, the presence of calcic plagioclase and pyroxene is implied in the term gabbro.

essexite A type of ALKALI GABBRO.

estuarine (in ecology) Describing an environment at a river ESTUARY, where there are tidal effects and fresh and salt water mix.

estuary The part of a river mouth within which tides have an effect and therefore where fresh and saline water are mixed. Most present-day estuaries are DROWNED VALLEYS, owing their existence to the postglacial rise in sea level, and for this reason they usually contain much deposited sediment. Deposition will continue if the river introduces more sediment than can be removed by tidal streams and by what little wave action is possible within the estuary. Many estuaries exhibit intricate patterns of channels, largely the result of erosion by both incoming and outgoing tidal streams, which give flood and ebb channels respectively. *See also* braiding; delta.

etesian wind A wind blowing from a northerly direction in the Aegean Sea or E Mediterranean during the summer period. Such winds are a response to the shallow thermal low pressure over the Sahara and the northward movement of the Azores anticyclone to be centered over the W Mediterranean.

eucrite A variety of coarse-grained basic igneous rock containing clinopyroxenes and orthopyroxenes, olivine, and plagioclase of bytownite-anorthite composition.

euhedral Describing crystals that are well developed and have good crystal faces. *Compare* anhedral; subhedral.

Euler's theorem A theorem that provides a mathematical explanation for the distribution of conservative PLATE BOUNDARIES. Plate tectonics requires that all conservative plate boundaries (transform faults) lie on small circles, the axes of which form the axis of rotation for the relative motion of the plates on each side. It also indicates that the velocity of relative motion across a destructive or constructive plate boundary is proportional to the angular distance of the particular point from the axis of rotation and to the angular velocity about the axis of rotation for the motion of the plates. Thus velocities will vary along plate boundaries.

eulittoral zone *See* littoral zone.

euphotic zone Ocean depths to which abundant sunlight for photosynthesis penetrates, extending from the surface to about 80 m. *See also* aphotic zone; disphotic zone.

eustasy Worldwide movements of sea level. The origin of these movements is most commonly attributed to the withdrawal and release of water due to the growth and decay of ice masses in the Quaternary (GLACIO-EUSTATISM). But eustatic movements can also be caused by tectonic movements of the sea floors or landmasses. It may be difficult to distinguish between the two, but most effects can be attributed to glacio-eustasy. Since the last ice age there has been a gradual rise in sea level as conditions have become relatively warmer. This is a positive eustatic change. *See also* base level.

eutaxitic Denoting the streaky or banded appearance common in ignimbrites. Individual streaks or fiamme are discontinuous flattened bodies. The term *parataxitic* is applicable when the fiamme are extremely elongated. *See* pyroclastic rock.

Eutheria The placental mammals. Their young are born at a late stage in development, having been nourished inside the body of the mother by means of a connection called the *placenta*. Fossils suggest that the Eutheria diverged from other mammalian groups in the Cretaceous, but their main evolutionary radiation took place during the Cenozoic, after the extinction of the dinosaurs. They rapidly became the dominant land animals and now include such diverse forms as the flying bats and the marine whales (which are the largest living animals known). *Compare* Marsupialia.

eutrophic (in ecology) Describing a freshwater HABITAT in which the water is rich in plant nutrients. *See* eutrophication.

eutrophication The process by which a freshwater HABITAT becomes excessively enriched with nutrients. These nutrients include nitrates and phosphates (generally from artificial fertilizers washed off farmland by rainwater) and possibly sewage. As a result, there is rapid growth of algae (an algal bloom), some of which die because insufficient light can penetrate the overgrown water. The dead algae decompose, using up dissolved oxygen and leading in turn to the death of fish and other aquatic animals.

euxinic environment A marine environment in which the bottom water is poorly ventilated, deficient in oxygen, and (in extreme cases) characterized by the presence of hydrogen sulfide, i.e. chemically reducing conditions. Such conditions apply, for example, to parts of the Black Sea and to certain of the Norwegian fiords, where sea-floor basins occur at different depths. One effect is that the floor of the Black Sea is sparsely populated with living organisms. The seabed deposits in euxinic environments are characteristically fine-grained ones containing a high percentage of decomposable organic material.

evaporation The process by which a liquid is changed into a gas or vapor. The rate of evaporation in the atmosphere represents a net effect between the removal of water molecules from the surface layers and replacement by other water molecules in the lower layer of the atmosphere, which through continual motion strike the surface and become absorbed. As the molecules are being transferred from a lower to a higher energy state through evaporation, a supply of energy must be provided for the process to be maintained. This is a very important aspect of the Earth's energy balance; by this process surplus energy in tropical areas is transferred by the flux of water vapor to the radiation-deficit areas nearer the poles.

If evaporation continued into still air, the overlying layer would very soon reach saturation and prevent further evaporation. However, air movement usually mixes lower layers by turbulence, reducing the water vapor content and enabling further evaporation to take place. Clearly, the stronger the wind, the greater will be the rate of evaporation as long as the air is not saturated. This introduces the second factor influencing evaporation, that is the requirement for the air to be capable of absorbing moisture. An index of this is the saturation deficit, which is the difference between the saturation vapor pressure and actual vapor pressure. If the saturation deficit is large, as in warm dry air, the gradient between the moist surface and the atmosphere will be high, and so the rate of transfer will be large. With moist air, the humidity gradient will be less and the rate of evaporation correspondingly smaller.

The measurement of evaporation is difficult because of the problem of adequately simulating the evaporation surface. The standard method is to measure the loss of water from an evaporation pan surface, but it is accepted that this does not equate very well with true atmospheric evaporation. Measurements of the changes in weight of soil samples are a closer approach to reality but still differ from the natural situation.

evaporite A rock or deposit formed as a precipitate from a saturated solution. Evaporation from virtually closed bodies of saline water is the most common process involved. Evaporites are classified according to their chemical composition. Great thicknesses of evaporite deposits occur, with minerals layered in a sequence depending upon their relative solubilities.

evapotranspiration The combined system of vapor transfer by evaporation and transpiration from the ground surface and its vegetation layer. Except in areas of sparse vegetation, such as deserts, transpiration by plants is the dominant factor in the total loss of water from the land surface. *See* hydrologic cycle.

evergreen trees Trees that do not shed their leaves in winter or the dry season. Typical evergreens include conifers and trees of the tropical rainforest. *Compare* deciduous forest.

evolution The gradual change in organisms through time, resulting in the origin of new SPECIES. In 1859 Charles Darwin suggested the process of NATURAL SELECTION as the means by which this could come about. The diversity of extinct and modern organisms has resulted from their becoming adapted to particular habitats. Evolution is generally a progressive change, with organisms becoming more able to deal with ever more extreme environments. This theory of modification and change is opposed to the doctrine of Special Creation, which holds that species are immutable and are replaced at intervals by newly created forms. Much fundamental evidence for the various pathways of evolutionary change comes from paleontology. *See also* convergent evolution; parallel evolution.

exfoliation The splitting of rocks into a series of concentric shells by a number of different weathering processes: UNLOADING, SPHEROIDAL WEATHERING, and FLAKING.

exhumation The exposing by denudation of a surface or feature that had previously been buried by deposition. Theories of exhumation have been used to explain the creation of a number of landforms, such as TORS and INSELBERGS. A first stage consists of progressive DEEP WEATHERING on a rock type embodying considerable variation in its susceptibility to decomposition, which largely reflects variation in its jointing pattern. A highly irregular WEATHERING FRONT develops and may remain for a considerable length of time. Subsequent erosion of the REGOLITH, initiated by earth movements or sea-level changes, will eventually exhume the least weathered rock masses.

exosphere The upper layer of the Earth's atmosphere, extending upwards from a height of about 600 km. It consists mainly of hydrogen and helium in varying proportions. *See also* atmosphere.

exotic A boulder or large rock body that has been transported by one of a variety of processes into an area of unrelated rock types. Where the transporting mechanism is tectonic in origin the rock body is said to be *allochthonous*, whereas if ice is the transporting mechanism such blocks are called ERRATICS.

exposure The degree of openness of a site and the amount of interference to airflow by natural and artificial obstacles. For climatological instruments to give representative values, they must be sited correctly with a uniformity of exposure. The correct exposure varies from element to element, but is most important for rain gauges and anemometers, where airflow can affect the recorded values. It is never measured quantitatively although attempts to do so have been made.

exsolution (unmixing) Several minerals, notably feldspars and pyroxenes, are homogeneous solid solutions of two or more chemical end-members at high temperatures. During cooling, the homogeneous mix becomes unstable and such a mineral unmixes, producing two distinct mineral phases. For example, an alkali feldspar rich in potassium may undergo exsolution to produce a perthite intergrowth consisting of separate structural units of sodium and potassium feldspar, indicated by the occurrence of blebs of albite within an orthoclase host.

external magnetic field That part of the Earth's magnetic field that results from effects produced above the Earth's surface.

extinction The disappearance of a SPECIES or other group of organisms representing a particular evolutionary line. It may occur when an organism that has become highly specialized for living in a particular environment is unable to adapt to extreme changes in its HABITAT. Some extinctions affect a wide variety of habitats, as occurred at the end of the PALEOZOIC and at the end of the MESOZOIC Eras. The latter extinction affected the dinosaurs on land, pterodactyls in the air, and ichthyosaurs in the sea; many invertebrate groups, such as the ammonites, also disappeared (*see* K/T boundary event). The causes of these widespread extinctions are not clearly understood. Major climatic changes combined with unsuccessful competition with other groups of organisms have been suggested as likely causes: the extinction of one group is often broadly contemporaneous with the evolutionary radiation of another unspecialized and often insignificant group.

extrusive rock A rock that forms after igneous materials cool and solidify on the surface of the Earth. The rapid cooling generally results in a fine-grained type of rock. *Compare* intrusive rock.

eye The central area of a hurricane or typhoon, where wind speeds are light, breaks in the main cloud sheets appear, and the driving rain stops. The diameter of the eye varies, reaching up to 80 km in the larger storms, although its duration depends upon the speed of movement of the storm. The weather in the eye feels particularly oppressive after the strong winds preceding it, but it is soon followed by even stronger winds in the opposite direction.

F

face 1. One of the surfaces of a CRYSTAL. 2. The chief or most obvious surface of a landform (e.g. the face of a CLIFF).

facies 1. A collection of metamorphic rocks that have formed over the same range of physical conditions. Differences in mineralogy are attributed to the variations in chemical composition of the original rock types. For example, at the amphibolite facies, a shale may be represented by the assemblage staurolite + garnet + biotite + plagioclase + quartz, a basalt by hornblende + plagioclase + quartz, and an impure limestone by diopside + calcite + quartz. A number of metamorphic facies and subfacies are recognized in both regional and contact metamorphism. Metamorphic facies and their relative temperatures (T) and pressures (P) are as follows:
Contact metamorphic facies – low P, low to high T.
Zeolite facies – low T, low to intermediate P, gradational from DIAGENESIS.
Greenschist facies – low to intermediate P and T.
Amphibolite facies – intermediate to high P and T.
Glaucophane schist facies – low T, very high P.
Granulite facies – high P and T.
Eclogite facies – high T and very high P.
See also grade (def. 1); zone (def. 1).
2. A rock unit or group of associated rock units having particular features, lithological, sedimentological, and faunal, that reflect some specific environmental conditions. Thus, rocks deposited at the same time in different regions may be dissimilar because they are of different facies, as is the case, for example, with the PERMO-TRIASSIC rocks of marine facies and continental facies in Britain. Correlation of one facies with another may be difficult, especially if the fossils they contain are restricted to each particular facies. If this is the case they are known as *facies fossils*. A unit of a particular facies may be diachronous, that is, it is present in different areas at different times.

facing direction The direction in which younger beds of rock lie in relation to the rocks in a FOLD. If the younger rocks lie above the fold, the facing direction is upward; if they lie below, it is downward.

Fahrenheit scale A scale of temperature in which the freezing point of water is at 32° and the boiling point at 212°.

fair-weather cumulus cloud A cumulus cloud with limited vertical extent. The clouds are prevented from rising farther by an inversion, usually formed by subsidence in an area of high pressure. Because high-pressure areas rarely give rain, clouds formed in this way are known as fair-weather cumulus.

false bedding An obsolete term for CROSS-BEDDING.

family A group in the taxonomic classification of organisms. A number of related families together form an ORDER, and a family itself is composed of one or more genera (*see* genus). For example, the Hominidae, to which MAN belongs, is a family of the order Primates. *See* taxonomy.

fan cleavage CLEAVAGE in which the planes are not parallel but form a fanlike arrangement with either an upward or downward convergence. This arrangement often results from rotation of beds in the limbs of a developing fold.

fathom A unit of depth equal to 1.83 meters (6 feet).

fault A fracture in the Earth's crust along the plane of which there has been displacement of rock on one side relative to the other, either in a horizontal, vertical, or oblique sense. See illustration overleaf.

fault basin A depression separated from the enclosing higher ground by faults.

fault block 1. An area of the Earth's crust that behaves as a single unit during faulting.
2. A rock body bounded by at least two faults, which can be either elevated or depressed relative to the surrounding region.

fault breccia The mass of broken rock fragments along a fault plane produced during dislocation.

fault line The line along which a FAULT intersects the surface of the ground.

fault-line scarp A type of cliff, formed originally by a FAULT, whose face has retreated because of erosion. The original fault line lies in front of the scarp under accumulated rock fragments. *See also* fault scarp.

fault plane The surface along which a FAULT forms.

fault scarp A type of cliff formed when a block of rocks is forced upward by a FAULT. It is located on the FAULT LINE. *See also* fault-line scarp.

fayalite A mineral of the OLIVINE group.

feldspar The most important type of rock-forming minerals and the most abundant minerals in igneous rocks. Feldspars have a framework structure in which $(Si,Al)O_4$ tetrahedra are linked together with calcium, sodium, potassium, and barium ions occupying the large spaces in the framework. Feldspars may be considered to be mixtures of the four components:

$CaAl_2Si_2O_8$ – anorthite (An)
$NaAlSi_3O_8$ – albite (Ab)
$KAlSi_3O_8$ – orthoclase (Or)
$BaAl_2Si_2O_8$ – celsian (Ce).

Barium feldspars are rare and most feldspars are members of the ternary system An-Ab-Or. Members of the series between $NaAlSi_3O_8$ and $CaAl_2Si_2O_8$ are plagioclase feldspars and those between $NaAlSi_3O_8$ and $KAlSi_3O_8$ are alkali feldspars.

Plagioclase feldspars. The replacement NaSi↔CaAl results in a complete gradation between the two end members of the series, calcic (anorthite 100–90% An, 0–10% Ab; bytownite 90–70% An, 10–30% Ab; labradorite 70–50% An, 30–50% Ab) and sodic (andesine 50–30% An, 50–70% Ab; oligoclase 30–10% An, 70–90% Ab; albite 0–10% An, 100–90% Ab).

Plagioclases are triclinic and exhibit multiple lamellar twinning. They are milky white or colorless. Plagioclase occurs both as phenocrysts and in the groundmass of most basic and intermediate igneous rocks. Calcic plagioclases are characteristic of basic rocks and anorthosites. Sodic plagioclases are found in intermediate rocks. Albite occurs in some granites and also in spilites. Plagioclase is also found in a wide variety of metamorphic rocks.

Alkali feldspars. At high temperatures the replacement Na↔K results in a continuous chemical series between $NaAlSi_3O_8$ and $KAlSi_3O_8$. In most volcanic rocks, the feldspar has crystallized at a high temperature and has been quenched, thus retaining its high-temperature crystal structure. Slow cooling under plutonic conditions or crystallization at relatively lower temperatures results in the formation of two separate feldspars, one rich in $NaAlSi_3O_8$, the other in $KAlSi_3O_8$. The homogeneous high-temperature structural state of an alkali feldspar may become unstable at a lower temperature and unmixing occurs by the rearrangement of the alkali ions in the solid state. The resulting intergrowth of the two components is called *perthite.*

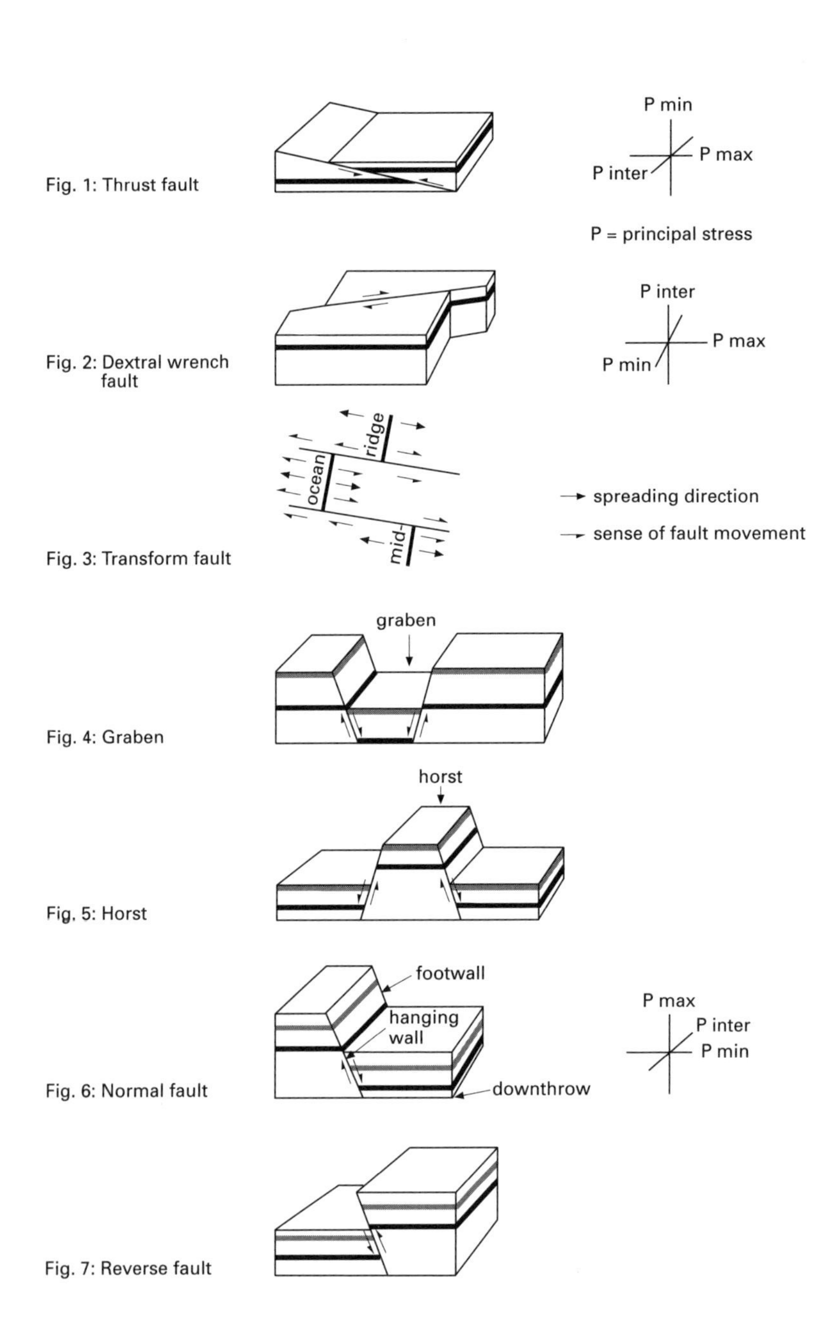

Fig. 1: Thrust fault

Fig. 2: Dextral wrench fault

Fig. 3: Transform fault

Fig. 4: Graben

Fig. 5: Horst

Fig. 6: Normal fault

Fig. 7: Reverse fault

Types of fault

Sanidine is a high-temperature monoclinic potassium feldspar found in volcanic rocks, such as trachytes and rhyolites. *Anorthoclase* is a high-temperature sodium-potassium feldspar. *Orthoclase* (also monoclinic) and *microcline* (triclinic) are intermediate- to low-temperature potassium feldspars found in plutonic acid- and alkali-rich rocks. *Adularia* is a very low-temperature monoclinic potassium feldspar found in pegmatites. Alkali feldspars also occur in a wide variety of metamorphic rocks. Sanidine and orthoclase exhibit simple twinning but microcline and anorthoclase have multiple twins in two directions at right angles, producing a tartan effect. Alkali feldspars are colorless, white, pink, or red.

feldspathic Describing a rock or mineral aggregate that contains FELDSPAR, usually in a specified amount. *See also* feldspathoid.

feldspathoid A member of a group of undersaturated minerals in which the alkali cations sodium and potassium are located in cavities within the framework structure of $(Si,Al)O_4$ tetrahedra. Feldspathoids appear in place of feldspar in rocks deficient in silica. They do not occur with free quartz because they react with silica to form feldspar.

Nepheline has an ideal composition $NaAlSiO_4$ but always contains some *kalsilite* ($KAlSiO_4$), the amount of solid solution increasing with increasing temperature. Both nepheline and kalsilite are white or colorless and hexagonal. *Carnegieite* is a high-temperature cubic form of nepheline and *kaliophilite* is a rare form of kalsilite. Nepheline is a primary mineral in many alkaline igneous rocks and is produced during the metasomatic alteration of rocks through the process of FENITIZATION. Nepheline also occurs as a result of the reaction between basic magma and limestone. Kalsilite is found in potassium-rich volcanic rocks.

Leucite is a white pseudocubic mineral of composition $KAlSi_2O_6$ and often shows complex twinning. It occurs as euhedral icositetrahedral crystals in potassium-rich basic lavas. *Analcime (analcite)* is a white or pink cubic mineral of composition $NaAlSi_2O_6.H_2O$ with some replacement K↔Na. It occurs as a late-stage mineral in basic and intermediate igneous rocks.

The minerals of the sodalite subgroup are as follows:

sodalite $3(NaAlSiO_4).NaCl$
nosean $3(NaAlSiO_4).Na_2SO_4$
haüyne $3(NaAlSiO_4).CaSO_4$

These minerals are cubic, commonly blue in color, and are found in alkaline igneous rocks such as nepheline syenites and phonolites. *Lazurite* is the major constituent of lapis-lazuli, formed by the metamorphism of limestone. *Cancrinite* is a hexagonal white or colorless mineral of approximate composition $Na_8(AlSiO_4)_6$-$(HCO_3)_2$. It is formed as a late-stage or secondary mineral as a result of the reaction between nepheline and carbon dioxide.

fell An upland in northern England, usually covered by heathland or rough pasture for the summer grazing of sheep.

felsic Denoting light-colored minerals such as feldspar, feldspathoids, and quartz. *Compare* mafic.

felsite Any crystalline acid volcanic rock.

femic Denoting the iron- and magnesium-rich minerals of the CIPW normative classification. *Compare* salic.

fen A flat marshy low-lying region, often located near the coast where rivers and lakes have silted up. Such terrain is named after the Fens in E England, where the marshes have been drained and turned into fertile farmland.

fenite A type of igneous rock formed by METASOMATISM through contact with a CARBONATITE intrusion. It is an alkaline SYENITE.

fenitization The alkaline metasomatism of country rock, typically gneissose, surrounding ijolite-carbonatite complexes.

Quartzo-feldspathic rocks may be completely made over to syenitic alkali feldspar-nepheline-aegirine assemblages, termed *fenites*. Fenitization precedes or accompanies the emplacement of the carbonatite magma.

fen soil *See* organic soil.

fenster *See* window.

ferrallitic soil (latosol, lateritic soil, red earth) A soil that is typically found on the old planation surfaces of the humid tropics. The chief soil-forming process is ferrallitization where weathering and leaching are intense. Rapid formation results in a deep profile rich in clay and hydrated oxides of iron, aluminum, and manganese. Free drainage gives a red coloration to the soils but they may be yellowish where drainage is poorer. Any plant remains are rapidly broken down and the released nutrients are immediately used again by the plants so that little organic matter accumulates. This factor plus the kaolinitic clay results in a low cation exchange capacity, which makes these soils of little agricultural value. They fall into the OXISOL order of the Seventh Approximation.

ferrimagnetism A type of weak FERROMAGNETISM in which the magnetic domains of a substance are aligned in opposite directions, although there are slightly more of one alignment than the other. The mineral MAGNETITE is a ferrimagnetic material.

fersiallitic soil (ferrisiallitic soil) A tropical soil formed in a less humid climate than FERRALLITIC SOILS. The profile differs in that weathering and leaching are not so dominant, producing a soil that is not so deep. Where the parent material is particularly resistant only an A C profile occurs, although sometimes an A Bt C profile develops. The whole profile is richer in weatherable minerals and has a higher cation exchange capacity than ferrallitic soils.

ferrisol A soil that has developed similarly to FERRALLITIC and FERSIALLITIC SOILS, its profile being less intensely weathered and leached than ferrallitic soils but more so than fersiallitic soils. Unlike ferrallitic soils these are found on sloping sites where surface erosion is an important process. The profile is similar to a ferrallitic soil, being red in color and having a similar structure, but often clay movement results in a textural B horizon. These soils have a low cation exchange capacity but are more fertile than ferrallitic soils. They fall into the ULTISOL order of the Seventh Approximation.

ferroaugite A monoclinic PYROXENE.

ferromagnesian mineral A mineral that is rich in iron and/or magnesium and usually dark in color, for example olivines, pyroxenes, amphiboles, and biotite.

ferromagnetism A type of magnetism exhibited by substances (usually metals) that have a magnetic permeability very much larger than 1. Cobalt, iron, nickel, and many of their alloys are ferromagnetic materials, used for making permanent magnets. They show such behavior because, in a magnetic field, their magnetic domains become permanently aligned in one direction.

ferruginous Describing a substance that contains iron, such as a sedimentary rock (e.g. types of sandstone) whose constituent grains are cemented together with iron oxide (*see* cementation).

fertilizer Any organic substance or inorganic salt applied to the soil to improve crop production. The nutrient elements most important for plant growth are nitrogen, phosphorus, and potassium.

fetch The distance of the stretch of open water over which the frictional effect of a wind that is blowing in a constant direction is actually generating waves. In small sea areas, fetch distance is usually limited by the dimensions of that sea area; in the open ocean, however, the limit is usually imposed by the constantly changing meteorological situation. There is also a relation between fetch distance, wind strength, wind duration, and the maximum size of waves produced. For a given wind speed, the largest possible waves require a minimum fetch distance; and for a given fetch distance, there is an optimum size wave that can be generated, however hard or long the wind blows. Very large ocean waves may require fetch distances of the order of 1000 km or more.

fiamme Flattened disk-shaped glassy bodies found in ignimbrites and imparting a eutaxitic structure. *See* pyroclastic rock.

fibrolite A fibrous or acicular form of SILLIMANITE. *See* aluminum silicates.

field completion A check of a map under compilation carried out in the field to ensure that all the details are shown and to make any corrections or amendments that may be necessary. Details usually affected by the field check are names and civil boundaries.

field intensity The strength of a magnetic field at a particular point on the Earth's surface.

field reversal *See* magnetic reversal.

filtering (in climatology) A method of separating components of a selected time period from all other possible periodic components. It is most frequently used for investigations into changes of climate.

filter pressing *See* magmatic differentiation.

finger lake A deep long narrow lake, usually occupying a valley carved by glacial erosion. The Finger Lakes in New York State are well-known examples.

fiord (fjord) A long narrow sea inlet lying between steep mountain slopes, which often reach up to several hundred meters above sea level. Fiords are glaciated valleys which, owing to a relative rise of sea level after the melting of the Pleistocene ice sheets, have become flooded by the sea. Many fiords have very considerable depths, some well in excess of 1000 m. Because of glacial action, they are characterized by the existence of gouged basins separated from the open sea by sills of solid rock, often capped by moraine, marking the point where the glacier that cut the valley lost much of its erosive power, either through melting or slowing down. The cross section of a fiord is often U-shaped.

fireclay A type of clay that can be baked to high temperatures without crumbling. Fireclays are composed mainly of hydrous aluminum silicate, as in illite, kaolin, or kaolinite (*see* clay minerals), and often occur beneath beds of coal.

firn (nevé) A half-way stage in the transformation of fresh snow into glacier ice. Fresh snow is very loosely packed and has a density between 0.1 and 0.2. Compaction and recrystallization reduce the pore space between grains, increasing the density. Firn exists when the density reaches 0.5. The word means "of last year", and in this case refers to snow that has survived one summer. Further compaction continues until glacier ice of density 0.89 to 0.90 is formed, although this may take many years.

firth (in Scotland) A lengthy estuary or arm of the sea. Firths bear some similarity to FIORDS and some have the closed and deep depressions and rock sills typical of fiords. They developed from river or glacial valleys that later experienced the postglacial transgression, becoming partly flooded by the sea. Firths on Scotland's E coast are not classed as the fiord type. The lochs and firths of the W coast of Scotland are fiords but have rather flatter hillside slopes than, for example, the fiords of Norway and Alaska.

fish One of a variety of aquatic vertebrates belonging to any of the following classes: AGNATHA, PLACODERMI, CHONDRICHTHYES, and OSTEICHTHYES.

fissure Any extensive break, cleft, or fracture in the Earth's surface, generally caused by earth movements (earthquakes and faulting) or volcanic action. *See also* crevasse.

fissure eruption A volcanic eruption through a FISSURE rather than a central vent. The largest occur along MID-OCEAN RIDGES, where molten magma rises through fissures in the oceanic crust.

fjord *See* fiord.

flaking The splitting of rocks around their margins into curved flakes, which vary in thickness from a few millimeters to a few centimeters, depending on the process involved. The most common process is the chemical decay of minerals, especially HYDRATION, causing expansion. The growth of salt crystals from solution beneath the surface of porous rocks can give rise to flaking, the flakes in this case being frequently backed by a thin layer of salt. In both these cases there tends to be a series of concentric layers of flakes around one boulder. Flaking results in a rounding of the block, because there is a preferential attack on edges and corners. The process can take place only above ground level, and the final development is a flat rock surface. Flaking on concave surfaces produces hollows.

flame structure A sedimentary structure in which an underlying bed has been squeezed up into an overlying horizon in flamelike tongues as a result of loading. Generally there is some degree of horizontal displacement.

flap structure A structure resulting from gravity collapse, consisting of a bed originally on the limb of an anticline that has bent as it slipped downward and come to rest in an inverted position.

flash flood A flood of sudden occurrence that results from rapid runoff after heavy rainfall. Flash floods are commonest in desert areas, where there is no vegetation to prevent fast runoff, although they also occur in wetter regions. The water may flow along gullies and cause a flood many kilometers away from the site of the rainfall.

flexure fold A type of fold best developed in rocks where there is good layering. As the fold develops the beds slip over one another.

flint A nodular variety of CHALCEDONY formed in chalk. *See also* silica minerals.

flocculation (in soil science) The process by which colloidal (i.e. clay-size) particles join together forming groups, called *floccules*. Colloidal material tends to adopt this habit in the presence of neutral salts, notably salts of calcium; in the presence of alkaline salts, especially sodium, the particles adopt the opposite habit and separate as independent units. Calcium therefore improves the structure of the soil by helping aggregation, and this in turn improves its agricultural value; flocculation of otherwise intractable clays is often encouraged by the addition of lime to make them more productive.

flood The state of a river when the volume of water flowing in it exceeds BANKFULL, and water commences to spread away from the channel over the FLOODPLAIN. Floods are produced by discharge increases due to exceptional rainfall and runoff. In hydrology, much effort has been devoted to the study of floods because of their great geomorphic significance and their cost to life and property on floodplains. Flood frequency curves have been produced for many rivers, showing the likely recurrence interval of floods of varying magnitudes, based on the analysis of past DISCHARGE records from gauging stations. Geomorphologically, floods can accomplish great landscape modifications in short periods, because the increased velocity of flood rivers produces an increase in erosional capacity proportional to the square of that velocity: thus if velocity doubles, erosional capacity increases four times. *See also* geomorphic process.

flood basalt Very fluid basaltic LAVA (usually from a FISSURE ERUPTION) that flows over large areas; there may be a series of such flows one after the other. The largest known area of flood basalt covers 250 000 sq km of the Deccan plateau on the Indian subcontinent; smaller examples occur in southern Africa and the north-western USA (Columbia River Plateau).

floodplain A relatively level area bordering a river, subject to periodic flooding and made up of sediments deposited by the river, which bury the rock-cut valley to a variable depth. In the larger rivers it may contain the following features: the river channel itself, OXBOW LAKES, POINT BARS, scars of former MEANDERS resulting from their migration downvalley, areas of stagnant sedimenting water (*sloughs*), LEVÉES, backswamp deposits of fine sediments in slack water away from the river, and sandy flood debris. In smaller streams all these features will not be present, but the Mississippi and Mekong, for example, display them all.

Floodplains originate from deposits within the river channel on the point bars and from overbank deposits in time of flooding. It is now thought that deposits within the river channel are the major source, constituting possibly 60–80% of floodplain material, but actively aggrading rivers probably owe more to overbank deposits.

The existence of a floodplain does not mean a river is actively aggrading: meanders migrate downstream, depositing material on the inside of bends and eroding material on the outside. Once deposited, a particle will remain static in the floodplain for a considerable period (about 1000 years in one stream studied) until the next meander upstream had eroded its way down to the position of that grain and re-entrains it. Floodplains in streams in equilibrium therefore represent a temporary storage for sediment on its passage through the valley. *See also* lateral erosion; river.

flood tide The incoming of the tidal stream; the rising tide, i.e. that part of a tide cycle following the low-water stage and preceding the high-water stage. *Compare* ebb tide.

Florida Current A major surface current on the W periphery of the Atlantic Ocean. It forms one of two branches of the North Equatorial Current. At about longitude 60° W, the North Equatorial Current splits up into two important currents: one flows into the Caribbean Sea but then re-enters the Atlantic Ocean via the Gulf of Mexico, and thereafter flows as the Florida Current. The Florida Current extends from the Straits of Florida to roughly the latitude of Cape Hatteras. As such, it constitutes the S portion of the Gulf Stream system. The discharge rate of the Florida Current is probably of the order of some 26 million cubic meters per second, but this figure varies significantly from one period to another, and almost certainly seasonally. It is a fast-flowing current that is particularly confined to the surface waters (reaching down to depths of between 600 and 800 m), probably attaining speeds in excess of 150 cm per second close to the coasts of Florida. An important factor in the generation of this current is the significant difference in water level that exists between the Gulf of Mexico coast and the Atlantic coast of Florida.

flowage Rock deformation resulting from the stressing of the rock beyond its limit of elasticity, but without fracture.

flow cleavage A type of cleavage resulting from solid-state flowage in a rock, accompanied by the regrowth of minerals. Traces of bedding are almost destroyed. *See also* slaty cleavage.

flume A deep narrow gorge or ravine containing a fast-flowing stream or series of cascades.

fluorescence The emission of light as a result of previously absorbing radiation. Many minerals fluoresce on exposure to ultraviolet light, and this property may be used as an aid to identification.

fluorite (fluorspar) A cubic mineral form of calcium fluoride, CaF_2. The purple color of the familiar variety Blue John is due to the presence of colloidal calcium; other varieties are white, green, and yellow. Fluorite is found in hydrothermal vein deposits and as a late-stage mineral in some alkaline and acid igneous rocks.

fluorspar *See* fluorite.

flute A sedimentary structure in the form of a hollow on the surface of a poorly consolidated bed of sediment, resulting from the scouring action of passing water. When this is infilled by an overlying bed of silt or sand the form of this hollow is preserved, and is exposed as a *flute cast* on the undersurface of the overlying bed.

fluvial (fluviatile) Of or relating to a river.

fluviatile *See* fluvial.

fluvioglacial Describing any process or resultant landform involving the presence of water derived from a glacier or ice sheet. Glacial meltwater can transport and erode material just as a normal river, but if it lies within or beneath the associated ice, it may be capable of extremely efficient erosion due to increased pressures or velocities. Fluvioglacial deposits can generally be distinguished from those of purely glacial origin because they exhibit rounding, sorting, and stratification.

flux (flux density) A rate of flow of some quantity, usually some form of energy, through a unit area.

flysch Thinly bedded marine sandstones, marls, shales, clays, conglomerates, and graywackes, which fill a trough adjacent to a rapidly rising mountain chain and represent rapid erosion and deposition prior to the main period of orogeny. The term was originally used by Alpine geologists and applies strictly to the sediments associated with the Alpine orogeny. *Compare* molasse.

focal depth The shortest distance between the focus of an earthquake and the Earth's surface.

focus The center of an earthquake, where strain energy is converted into elastic waves. It is the source of seismic waves produced during an earthquake. For convenience it is thought of as being a point source, although strictly speaking earthquake waves originate from a nonfinite source, such as a fault plane.

fog A cloud near the ground surface composed of minute water droplets in suspension. It is sufficiently dense for the drops to produce a reduction in visibility to less than 1 km. Condensation of water to form the droplets in a fog can occur in many ways, and fogs are often classified by their method of formation – radiation fog, advection fog, etc. The most frequent method is by some form of cooling, such as radiational cooling at night; in advective fog the lower layers of the atmosphere are cooled through turbulent contact with the ground. More rarely, condensation can be achieved through evaporation from the surface.

In industrial areas, the greater supply of condensation nuclei and hygroscopic particles enable fogs to form even when relative humidity is less than 100%. Where smoke is abundant, this can produce a thick and particularly unpleasant type of fog, known as SMOG (an abbreviation of *smoke fog*).

föhn A warm dry wind that descends the leeward side of mountain ranges to give extremely rapid thaws of lying snow in winter. Such winds occur whenever rapid descent favors adiabatic warming, whether or not there has been cooling at the SATURATED ADIABATIC LAPSE RATE on the upwind side of the mountains. However, they are of climatological significance only where the descent is sufficient to produce a marked warming, as in the Alps, Rockies, or Andes, where airflow is at right angles to the mountain barrier and the maximum rate of descent achieved.

fold A buckling of bedded sedimentary rocks due to deformation processes or the effect of gravity. *See also* orogenesis.

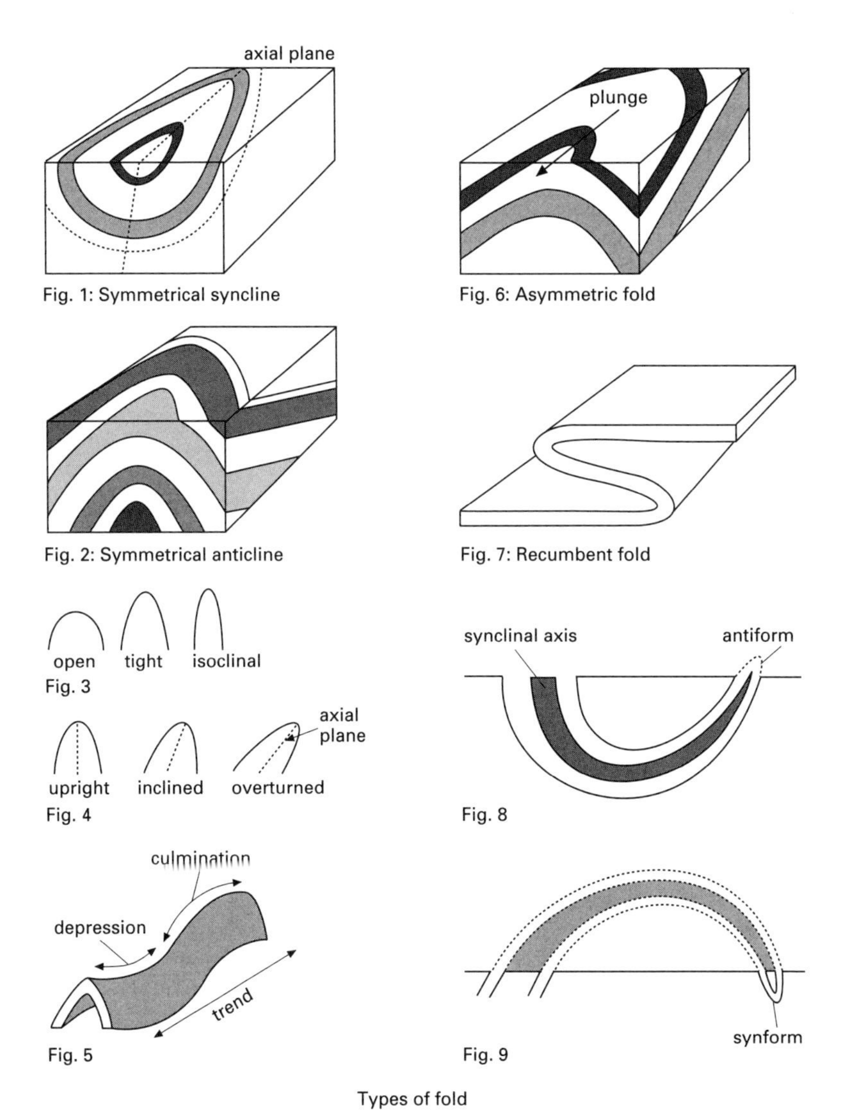

Types of fold

fold axis An imaginary line parallel to the HINGE LINE of a fold, from which the rock strata dip away in different directions (downward from a crest; upward from a trough). See diagram.

fold belt *See* mobile belt.

folding The action of forming a FOLD in rock. *See* orogenesis.

fold limb Either of the two flanks on either side of the axis of a fold. The angle between the limbs is called the *interlimb angle*.

fold mountain A mountain that is formed by large-scale folding. *See* orogenesis.

foliation A banded or laminated structure within a metamorphic rock, resulting from the metamorphic segregation of minerals into compositionally different layers, parallel to the schistosity. The term is also commonly used to describe any parallel planar element in a metamorphic rock, such as schistosity and cleavage.

food chain A series of linked feeding relationships between organisms. Each organism depends for its food on the one next lowest in the hierarchy (called a TROPHIC LEVEL. The lowest level is occupied by PRIMARY PRODUCERS (autotrophs), such as plants, which provide food for the herbivores (PRIMARY CONSUMERS) in the level above them. These, in turn, become food for carnivores (secondary consumers). Consumers also include decomposers (DETRIVORES), which break down the waste products and dead bodies of all the other organisms. A better picture of complex feeding interactions between many organisms is provided by a FOOD WEB.

food web A sequence of complex interlinked FOOD CHAINS that demonstrates the feeding relationships between all the organisms in a natural COMMUNITY.

fool's gold A common name for pyrite (iron pyrites, FeS_2), which is so called because of its bright golden color when first exposed to the air.

foot wall The surface of rock beneath a fault plane or ore body. *Compare* hanging wall.

Foraminifera An order of planktonic organisms (*see* protozoa). Most species display jellylike bodies surrounded by a shell or casing of calcium carbonate. The casing is often flask- or sphere-shaped. Foraminiferans are most abundant in the surface layers of the sea, and tend to reach their maximum development near the Equator. The tests or casings of dead foraminiferans make up a large proportion of the organic ooze deposits, for example, GLOBIGERINA OOZE, which contains some 30% foraminiferal tests.

forecast (in meteorology) A statement about the probable weather events for specific areas for certain time periods. Forecasts are normally given at three levels of detail: (i) short-range, dealing in depth with the probable events for the next 24–48 hours; (ii) medium range, covering about one week, in which the expected movements of the main high and low pressure systems are covered; (iii) the long-range forecast for a month or even a season, in which the only details refer to climatological averages and whether the weather is expected to be warmer or cooler, wetter or drier.

The methods required for working out these forecasts differ for the time period involved. For the short-range forecast, satellite images show the major cloud belts, radar detects the areas of precipitation, and over land areas there are large numbers of stations continuously recording the weather elements. Together they provide a fair assessment of likely events supplemented by computer calculations of pressure fields for 24–48 hours ahead (*see* numerical weather forecasting). Interpretations can then be made on the basis of these forecast maps.

Medium-range forecasting for the USA and Europe is largely based on experience of previous weather sequences together with a knowledge of the upper atmosphere, particularly the Rossby waves (*see* long wave) of the mid-latitude westerlies. As these to some degree control the direction of movement of low-pressure systems, information about the state of the upper atmosphere aids forecasting for this timescale. However, as even quite small initial disturbances can dramatically and rapidly change the weather, success is only moderate.

For long-range forecasts, different techniques are required. In Britain, the ANALOG method is often used. In this method former weather charts are compared with present charts and it is assumed that similarities in the charts can be used to predict weather events. Sea-surface temperatures and their anomalies are also included as these are important in affecting depression formation and energy transfer to the atmosphere. In the USA, long-range forecasts are based on computer simulations of the atmosphere. Details of the present state of the atmosphere at many different levels are fed into powerful computer systems and then run for a time period equivalent to one month in real time to forecast pressure movements. Neither method is entirely satisfactory and there is considerable debate as to whether it is theoretically possible to forecast movement in the atmosphere accurately for such long time periods.

foredeep *See* trench.

foredune A coastal dune growing in the newest line of dunes nearest to the sea. Some authorities put a maximum size of about three meters on foredunes. The usual distinguishing characteristics are dominance by the grass *Agropyron junceum*, which is capable of tolerating the occasional drenching with salt water to which the foredune is subject, and a very open nature with low cover values and large areas of bare sand. Since the foredune is produced by the coalescence of individual grass and sand mounds, at a very early stage it will appear broken with an uneven crest line.

foreland A stable area of older cratonic rocks marginal to an orogenic belt, generally part of the continental crust. When two continental plates collide the intervening wedge of sediments is squeezed between two forelands, onto which these sediments are thrust and overfolded.

foreset bed *See* delta deposit.

foreshock One of a series of seismic waves recorded before the principal shock of an earthquake, resulting from small slips or fractures in brittle rocks as they reach their YIELD POINT, as a result of a buildup in stress.

foreshore The part of a beach that becomes covered and uncovered by water during the process of tidal rise and fall. It extends from low-water spring tide level to normal high-water springs level, and its width will naturally depend upon the tidal range and gradient of the beach.

foresight *See* leveling.

formation The fundamental unit in the lithostratigraphical classification of bodies of rock (*see* lithostratigraphy; stratigraphy). A formation is formally defined by a TYPE SECTION at a type locality, on the basis of readily observable lithological features that distinguish it from bodies of adjacent rock. It should preferably be marked off from these by a distinct lithological change, such as might be found at an unconformity, but it may have more arbitrary boundaries in a gradually changing rock sequence. Thickness is not a criterion in its definition; a single formation will often differ in thickness at different points. Stratigraphically adjacent and related formations may be associated to constitute a GROUP. A formation may contain MEMBERS but need not be entirely composed of them.

form line A line drawn on a map to give an impression of the terrain where there is insufficient information to construct accurate contours. Form lines are usually represented by dashed lines rather than unbroken ones. In areas mapped by aerial photography there may be parts obscured by cloud and these are often covered by form lines in the absence of alternative heighting data.

forsterite A mineral of the OLIVINE group.

fosse A long narrow depression between the side of a glacier and the valley that contains it. The depression forms as some ice melts because of heat absorbed by or reflected from the walls of the valley.

fossil The remains of any organism that lived in the past. The term was originally applied loosely to anything ancient and dug up and as an adjective may still be used in this sense (as in FOSSIL FUEL). A fossil can be the whole or any part of an animal or plant, usually chemically altered (i.e. fossilized); it may be an impression of the shape of an organism that has been preserved in some way (*see* cast; mold) or it can be simply the remains of the effects of an organism in the past, such as fossil tracks, excrement, or burrows (*see* trace fossil). *See also* derived fossil; zone fossil.

fossil fuel Any naturally occurring hydrocarbon fuel such as coal, natural gas, oil (petroleum), and peat. The fuels formed slowly underground by the action of pressure on the remains of dead plants and marine animals. Fossil fuels represent a nonrenewable resource.

Fourier analysis (harmonic analysis) A mathematical method of analyzing a data series that contains periodic variations, such as seasonal precipitation. These can be represented by a series of sine or cosine functions.

foyaite A type of alkali SYENITE.

fractocumulus cloud Broken or fragmented cloud rising only slightly above the CONDENSATION LEVEL. Such clouds are most frequently seen on early mornings when there has been sufficient convection for the clouds to approach the condensation level.

fractostratus cloud (scud) Cloud that is broken and ragged, similar in appearance to fractocumulus but found beneath altostratus or nimbostratus clouds, frequently preceding a warm front. It is the result of condensation taking place in the moist layer beneath the main cloud base, either by turbulent mixing or through the addition of moisture by precipitation falling from higher levels.

fracture cleavage A series of closely spaced fractures and parallel joints produced in deformed rocks that have been subjected to only minor metamorphism.

fracture zone A region that lies astride a line along which faulting has occurred. It is often the site of earthquakes and volcanic activity, and may run across a continent or along the seabed.

franklinite A rare black crystalline mineral of the SPINEL group consisting of oxides of iron, manganese, and zinc, $(Fe^{2+},Zn,Mn^{2+})(Fe^{3+},Mn^{3+})_2O_4$. It crystallizes in the cubic system and is used as a source of zinc.

frazil ice Small crystals of ice that form at the edges of fast-flowing streams or in moving seawater.

free air anomaly A gravity anomaly that takes account of altitude but does not take account of attraction effects resulting from topogaphy and isostatic compensation.

free face A vertical or near vertical facet of slope cut in bare rock. It is rare to find faces actually at 90°, and various authorities have adopted different critical values for the minimum allowable slope; slopes of above 40–45° are usually termed free faces. They occur in consolidated materials, which can maintain their verticality when the removal of debris from the footslope exceeds the rate of debris production from the face above. Wave erosion at the foot of a sea cliff, or lateral planation by a stream at the side of its valley, are situations producing free faces. Owing to this removal of debris from its foot, the slope becomes constantly oversteepened, and slips and slides lead to stripping of material from the whole face equally, thereby maintaining the face by PARALLEL RETREAT. If the agency removing the debris at the foot ceases to do so, for example the accumulation of a beach at a cliff foot cutting off direct wave action, retreat at the foot of the free face slows down compared with its upper part, and the face will tend to be gradually obliterated by decline to a gentle angle. *See also* basal sapping.

freeze-thaw Describing weathering processes involving the freezing and thawing of water within preexisting rock fissures. When water expands on freezing, it tends to enlarge cracks; on thawing, the enlarged cracks can contain more water than before, which will cause increased enlargement on renewed freezing. Eventually, particles become broken away from the main mass of rock. Freeze-thaw processes weaken rocks before they are subjected to other agents of weathering and erosion. A notable example is GLACIAL PLUCKING.

freezing level The height above sea level of the layer of the atmosphere that is at a temperature of 0°C. More frequently, its height is quoted in terms of pressure level, e.g. 850 mb. As water does not automatically freeze at 0°C (*see* freezing nucleus), it would be more accurately called the melting level, representing the point at which solid precipitation would melt. If the ground temperature is below freezing, the freezing level would be stated as being at station level, although if there is an inversion aloft the situation may be complex.

freezing nucleus Water droplets in clouds do not automatically freeze when the temperatures fall below freezing point. Being very small, the droplets require some form of NUCLEUS on which the crystalline growth can develop but such nuclei are not as effective or numerous as condensation nuclei and do not produce total freezing at 0°C. Instead a few ice crystals will appear at between –5°C and –10°C and become more frequent with lower temperature. The nuclei are composed of volcanic dust or clay particles with a similar crystalline structure to that of ice, and in some cases fractures or splinters of ice crystals may themselves act as nuclei. At temperatures below –40°C, spontaneous freezing takes place even in the absence of nuclei.

freezing point The temperature at which the liquid and solid phases of a substance are in equilibrium at standard pressure (1013.25 mb); for water this is 0°C. Because nuclei (*see* freezing nucleus) are required for cloud droplets to freeze, water can exist at temperatures far below its true freezing point. The term is therefore primarily used for ground-surface temperature observations and forecasts, particularly in connection with road surface conditions.

freezing rain Rain or drizzle that falls onto a ground surface that is below 0°C, the water freezing to give a coating of ice. It can occur when a layer of rising warm air, such as at a warm front, is moving slowly over a cold surface where the temperature remains below freezing point. Because the warm air eventually reaches the ground surface this state is normally of short duration.

fresh water Water that contains less than 2% of dissolved minerals. It is not necessarily safe to drink.

friable Describing a mineral or rock that crumbles easily (using the pressure of the fingers).

friction The mechanical force of resistance operating between two substances in contact. In the atmosphere, it is the resistance between the ground surface and the layer of air above it, although the thickness of this layer varies appreciably. The term can also be used when referring to wind shear.

The effects of friction on airflow are numerous. Wind speeds are reduced near the ground and there is a movement of wind across the isobars from higher to lower pressure, compared with the GEOSTROPHIC WIND when friction is excluded. The rate of increase of speed with height away from the surface is affected by the roughness of the ground, extremes being represented by oceans and urban areas.

frigid zone A cold area of the Earth's surface, either in polar regions or at high altitude.

fringing reef A REEF that grows directly on bedrock at the seashore, especially in tropical regions. It forms a shelf whose rough surface appears above the water at low tide. *See also* barrier reef.

front A transition zone between air of different thermal characteristics and origins. On surface pressure charts, fronts are depicted in two-dimensional form along the line of maximum thermal gradient, where this can be determined. In cross section they represent a surface of separation between the two AIR MASSes, sloping gently away from the surface warm air. Distinctive weather phenomena are associated with fronts and hence they are very important in short-period weather forecasting. Problems of precise definition of a front occur because it represents a thermal gradient, the strength of which is not necessarily related to weather activity, so the same gradient may have quite different degrees of vertical motion. In forecasting, a front is often called a trough of low pressure. *See also* depression.

frontal rainfall Heavy rainfall that originates in a DEPRESSION, where there is rapid uplift of large air masses at the juncture of cold and warm fronts.

frontogenesis The process of intensification of the thermal gradient along a frontal zone: part of the formation of a DEPRESSION. It can take place whenever the isotherms are suitably orientated during conditions of CONFLUENCE and CONVERGENCE in the lower atmosphere.

frontolysis The weakening of the thermal gradient at a frontal zone. This is produced under the reverse conditions of frontogenesis, i.e. with surface DIVERGENCE of air and often subsidence. When frontolysis is taking place, precipitation along the front stops and the cloud sheet breaks up and eventually disappears.

frost Frost occurs when the temperature falls below 0°C. Subdivisions are made into ground frost and air frost based on thermometer readings at grass level and screen level (1.2 m) respectively. It is normally recognized by the icy deposit that usually forms under such temperature conditions, but if the air is very dry, frost may occur without forming icy deposits. For weather-forecasting purposes, grades of severity of frost are distinguished with –0.1 to –3.5°C slight, –3.6 to –6.4°C moderate, –6.5°C to –11.5°C severe, and below –11.5°C very severe. Allowances are made for wind speed and wind-chill index, which influence the sensation of cold.

frost action A type of MECHANICAL WEATHERING that results in the breakup of rock through repeated freezing and thawing of water in its crevices.

frost heaving The swelling of the ground surface through a combination of CONGELITURBATION and the effects of NEEDLE ICE.

frost hollow An area that has a greater liability to frost incidence than its surroundings. Because cold air has a slightly greater density than warm air, it tends to collect in hollows or low ground and so areas of high frost incidence are usually in hollows where free drainage of air is prevented. The most famous frost hollow is Gstettneralm in the Austrian Alps where temperatures as low as –52°C have been recorded.

frost-shattering The breaking apart of masses of rock by the continued enlargement of cracks within them through FREEZE-THAW action, in areas where the temperature fluctuates for considerable periods around freezing point. The resultant outcrops and debris are all highly angular in appearance, and this is the major characteristic of frost-shattered forms. *See also* congelifraction.

fugitive *See* volatile.

fulgurite A tubular or dendritic mass of fused SILICA originating when lightning strikes sand.

full A shingle BEACH RIDGE or a BERM composed of sand. When ridges are called fulls, the separating parallel depressions are known as *swales*.

fuller's earth A clay rich in montmorillonite. *See* clay minerals.

fumarole A vent in the ground in a volcanic region which emits steam and other hot gases. Temperatures may be as high as 1000°C within fumaroles, which occur mainly on lava flows and in the calderas and craters of active volcanoes. If sulfurous gases are also present, it is called a *solfatara.*

fumigation (in meteorology) The result of the influx of heavily polluted air down to the ground surface. It most frequently occurs after night-time cooling has given a temperature inversion a few hundred meters above the ground surface, beneath which industrial pollution has accumulated. When convection starts after solar heating next morning, rising air motion will resume and compensating downward flow will take the polluted air to the ground, giving sudden high values of pollution.

Fungi A kingdom of heterotrophic organisms, important in soil. They are less abundant in numbers than bacteria (up to 1 000 000 per gram of soil) but greater in bulk, because they have a filamentous nature. They require moisture and may be found in alkaline, neutral, or acid soils. Thus in the latter type they number more than bacteria. Although, unlike bacteria, they cannot oxidize ammonium compounds or fix atmospheric nitrogen, they are more important than bacteria in the formation of humus and in the promotion of soil stability. Fungi are capable of decomposing even the most resistant plant materials such as cellulose and lignin. The mycelia of certain mushroom fungi may form associations with plants called mycorrhizas. This association is of mutual benefit to plant and fungus and where present there is a marked increase in the uptake of nutrients.

G

gabbro A coarse-grained basic igneous rock, the product of plutonic crystallization of basalt magma. The essential constituents are plagioclase that is more calcic than anorthite (usually labradorite) and clinopyroxene (usually augite or titanaugite). Many gabbros contain olivine in addition. Accessory minerals include magnetite, ilmenite, apatite, biotite, and hornblende.

Rocks in which orthopyroxene, usually hypersthene, predominates over clinopyroxene are called *norites*. With an increase in the proportion of plagioclase, gabbros and norites pass via leucocratic varieties into anorthosites; with an increase in olivine and pyroxene they pass via melagabbros into ultramafic rocks. With an increase in the proportion of olivine at the expense of pyroxene, gabbro passes into troctolite and allivalite. With a change to plagioclase of more sodic composition, gabbros pass into diorites. Eucrite is a variety of gabbro that contains olivine, two pyroxenes, and plagioclase of bytownite-anorthite composition. Gabbroic rocks containing pyroxenes and olivines rich in iron are termed *ferrogabbros*. Alkali gabbros (syenogabbros) are marked by the introduction of feldspathoids or alkali feldspar.

Most gabbros have equigranular textures; others are ophitic or, rarely, orbicular. Reaction rims are common features and rarely multiple rims involving olivine-pyroxene-hornblende-biotite may be preserved. The medium-grained equivalent of gabbro is termed microgabbro or DIABASE. Gabbroic rocks occur in stratiform layered bodies, ophiolite complexes, and as nodules in basalts.

Gaia hypothesis An idea championed in the 1970s by the British scientist James Lovelock which suggests that all the living and nonliving systems on Earth form a unity that is regulated, and kept suitable for life, by the organisms themselves. The whole planet can therefore be regarded as a huge single organism. The hypothesis stresses the interdependence of living things and the environment.

gal A unit of acceleration equal to one cm per second per second (named after Galileo). In practice this unit is too large and measurements of the strength of the Earth's gravitational field at a point are made in milligals (one thousandth of a gal); 1 milligal is equivalent to 9.8×10^{-3} N kg^{-1}. *See* gravity unit.

gale A wind speed averaged over a 10 minute period of between 50 and 102 km per hour. On the BEAUFORT SCALE Force 7 is a moderate gale; Force 10 a whole gale (or storm). Gales most frequently occur during the passage of an intense depression.

galena The chief lead ore (lead sulfide, PbS), found as gray metallic cubes of high density and often associated with sphalerite, barytes, fluorspar, and calcite in hydrothermal veins. It is sometimes mined for its silver content.

Gall's projection A cylindrical map projection with the cylinder cutting the Earth's surface at 45° north and south (*see* cylindrical projection). The meridians have true scale only at the points of intersection, the scale increasing poleward and decreasing toward the Equator. It is neither a homolographic nor orthomorphic projection, and is not often used.

gamma-ray log A logging technique run in wells that records the natural radioactivity of the formations through which it passes. High values are given by shales, whereas sandstones give low readings. It can be run in cased boreholes, because it does not rely upon the electrical properties of the formation.

gangue Minerals found in an ore deposit that have no commercial value and must be removed during the refining process.

gap An opening in a ridge, usually formed by the action of a glacier or river. Gaps provided routes through mountain ranges and often became the sites of human settlement.

garnet A member of a group of cubic minerals with a general formula $R_3^{2+}R_2^{3+}Si_3O_{12}$ where the divalent metals are magnesium, iron, manganese, or calcium and the trivalent metals are aluminum, iron, or chromium. The following six end-members are recognized:

pyrope – $Mg_3Al_2Si_3O_{12}$
almandine – $Fe_3^{2+}Al_2Si_3O_{12}$
spessartite – $Mn_3Al_2Si_3O_{12}$
grossular – $Ca_3Al_2Si_3O_{12}$
andradite – $Ca_3(Fe^{3+},Ti)_2Si_3O_{12}$
uvarovite – $Ca_3Cr_2Si_3O_{12}$

Compositions corresponding to pure end-members are rare and garnets are divided into two series, *pyralspite* (*pyr*ope, *al*mandine, *sp*essartite) and *ugrandite* (*u*varovite, *gr*ossular, *an*dradite). Little chemical variation between pyralspite and ugrandite garnets occurs.

Garnets commonly develop dodecahedral and icositetrahedral forms and are a variety of colors. Pyralspite garnets are typically pink, red, or brown. *Melanite* is a dark brown to black variety of andradite, rich in titanium and occurring in ijolites and nepheline syenites. In general, garnets are characteristic of metamorphic rocks. Almandine is found in medium- to high-grade regionally metamorphosed argillaceous sediments. Pyrope is found in ultrabasic rocks and eclogites, and grossular occurs in metamorphosed limestones. Many types of garnet are used as semi-precious gemstones.

garnierite A bright green amorphous form of hydrated nickel magnesium silicate. It occurs as a mineral in SERPENTINITE and is used as a source of nickel.

gas cap An underground pocket of natural gas that occurs above a reservoir of oil. It may be tapped for use as a fuel.

gas laws The basic physical equations that relate the pressure (p), volume (V), temperature (T), and density (ρ) of a perfect gas. These include the equation of state ($p = R\rho T/M$, where R is the universal gas constant and M is the molecular weight of the gas) and Boyle's law, which states that at constant temperature, the volume of a given mass of gas is inversely proportional to the pressure of the gas. This law controls the cooling of air as it rises and expands into regions of lower atmospheric pressure.

gas sand A type of sandstone or sand deposit that contains useful quantities of natural gas, which may be extracted.

gastrolith A stone or pebble ingested by reptiles and birds to assist in the breakdown of food in the gizzard. Fossil structures associated with the remains of extinct reptiles are thought to be gastroliths.

Gastropoda The class of the phylum MOLLUSCA that includes the slugs and snails. Typically they have a shell that is often coiled in a helical spiral; unlike that of the CEPHALOPODA it is not divided into chambers. Most gastropods are marine but some live in fresh water and others are terrestrial. Fossils are known from the Cambrian Period onward and the class has continued to flourish and diversify to the present day.

geanticline A very large anticlinal structure that has developed within geosynclinal sediments, as a result of lateral compression.

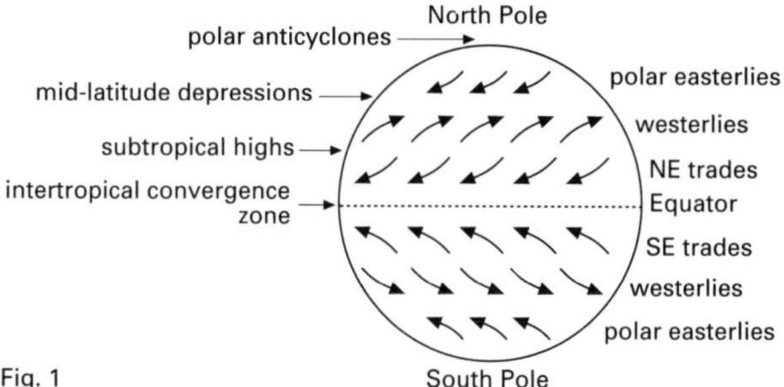

Fig. 1

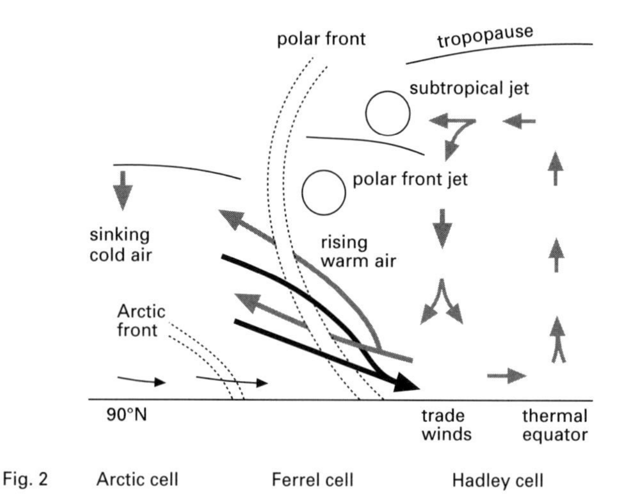

Fig. 2 Arctic cell Ferrel cell Hadley cell

General circulation of the atmosphere

gedrite An orthorhombic mineral of the AMPHIBOLE group.

general circulation of the atmosphere The mechanism by which energy is transferred from surplus to deficit areas of the world. It is constrained by a number of processes that interact with the main driving force of the circulation to result in the observed system of winds and disturbances. This circulation is basic to an understanding of the climates of the world because it determines the direction and strength of winds in any latitudinal position.

The atmospheric circulation is very complex, but it can be simplified into a series of belts (see Fig. 1). Within the tropics there are the trade winds converging toward the intertropical convergence zone (ITCZ). Poleward of the subtropical anticyclones are the main westerly winds with a small area of easterlies near the poles. These mean winds are essentially zonal in direction and do not appear to be an effective system in transferring surplus radiational energy toward the temperate and polar latitudes. However, long-term studies have shown that meridional exchange takes place as shown in Fig. 2. In the tropics, air rises at the ITCZ, releasing latent heat and gaining potential energy. This is moved toward the subtropics where it descends and maintains the trade winds and where much moisture is evaporated from the sea surface to continue the convection cell toward the ITCZ. This cellular circulation has been called the HADLEY CELL. Poleward, the mean charts show a dominance of zonal winds, but exchange here is effected by depressions, which advect warm and cold air and aid their mixing. Because this process takes place at different longitudes for only short periods of time, it is averaged out on the mean maps.

The driving force for the atmospheric circulation is the differential pattern of net radiation received at the Earth's surface. If the Earth were uniform and stationary this would give rise to a simple thermal cell between equatorial (surplus) and polar (deficit) areas. However, the effects of the Earth's rotation, the presence of mountain barriers, the distribution of land and sea, and the positions of the ocean currents all help to modify the simple thermal cell into the complex general circulation of the atmosphere.

Increasingly realistic models of the general circulation of the atmosphere are being obtained now that more is known about the controlling factors and powerful computers with large memory stores are available. Over the years the models or simulations have become more sophisticated, starting by trying to account for the primary processes responsible for maintaining the large-scale features of the circulation, using the primitive equations of motion. It is now possible to construct models of the circulation during an ice age from what we know about the controlling factors, but the large scale on which the models work makes them unsuitable for very long-range forecasting. Depressions can be initiated or affected by small-scale processes and greatly change the expected weather.

genus (*pl* genera) A group in the taxonomic classification of organisms that consists of one or more related SPECIES. One or several genera form a FAMILY. For example, the genus *Homo*, to which man belongs, is placed in the family Hominidae. A generic name always begins with a capital letter and is italicized. It is always included with the specific name in naming a species, when it is often abbreviated to the initial letter. Thus, the species to which modern man belongs is *Homo sapiens* or *H. sapiens*. A genus is defined on the basis of the species it contains, the most typical of which is designated as the *type species*. *See* taxonomy.

geocentric At or relating to the center of the Earth.

geochemical cycle (rock cycle) The route followed by an element or group of elements as they circulate through the Earth's crustal and subcrustal rocks. Elements may combine or separate during the cycle. Weathering of igneous rocks produces particles that settle as sediments, which may become sedimentary rock or undergo metamorphism into metamorphic rocks. These may melt and form new magma, which rises to the surface and crystallizes as new igneous rock.

geochemistry The branch of geology that deals with the chemical elements and compounds in the atmosphere, water, soil, and rocks of the Earth, particularly their composition and how they are formed and distributed.

geochronology The science of age determination of parts of the Earth. It has two branches: absolute dating and relative dating. The former is more modern, and involves techniques such as carbon–14, potassium-argon, and radioactive decay methods to put actual ages in years BP (before present) on organic remains, rocks, or sediments. Relative dating includes fossil correlations, pollen analysis and correlation, matching of tills from different areas on the basis of weathering or geologic content, archeological indicators, and several lesser used techniques that help establish stratigraphy. Relative dating seeks to put events in order; absolute dating tries to give precise dates of occurrence. In geomorphology both are used, especially in studies of the Pleistocene and post-Pleistocene periods.

geode *See* druse.

geodesy The science concerned with determining the exact shape and size of the Earth and the exact position of points on the Earth's surface. In addition it investigates the Earth's gravitational field, variations in tides, and the Earth's rotation. It provides the reference surface from which astronomical observations are made.

geodetic surveying The large-scale surveying of the Earth's surface, taking into account its curvature. Precise measurements from geodetic surveys provide the horizontal control points for more detailed surveying.

geodimeter A modern surveying instrument used for the accurate measurement of distance. The measurement is indirect, being obtained by timing the period required for a beam of light to travel from one end of a line to the other and return. Since the velocity of light is known, the distance can then be calculated. The instrument sends out a light beam, often from a laser, which is reflected by a mirror at the other end of the line to be measured. Suggested maximum ranges are 5 to 10 km in daylight and 15 to 25 km in darkness, depending on the light source. The geodimeter is ideal for rapid baseline measurement.

Geographical Information System (GIS) A computer system that includes the hardware, software, and data for capturing, storing, updating, manipulating, analyzing, and displaying geographical and spatial data. Within the earth sciences GIS systems have been applied, for example, to create maps displaying specific information, such as the geology of an area, ocean circulation, topography, the watershed for a stream system, and hazard risks from natural disasters for an area.

geography The scientific study of the features of the Earth's surface. *Physical geography* includes CLIMATOLOGY, GEOMORPHOLOGY, METEOROLOGY, and PEDOLOGY; human activity is covered by political or socioeconomic geography.

geoid The Earth pictured as a smooth oblate spheroid, coinciding with sea level and taken to continue across continents at the same level. At all points on Earth the geoid surface is at right-angles to the downward direction of the pull of gravity. *Geodetic leveling* employs the geoid as its reference surface.

geologic timescale (geologic column) A chronological series of events in the geologic history of the Earth, beginning in the PRECAMBRIAN and extending to the present (the HOLOCENE epoch of the QUATERNARY period). It is divided into eons, eras, periods, epochs, and ages, each ascribed dates in terms of millions of years ago (see table). *See also* chronostratigraphy.

geology The study of the structure and composition of the Earth. *See* geomorphology; geophysics; petrology. *See also* chronostratigraphy; geologic timescale.

geomagnetic equator A great circle equidistant between the geomagnetic poles. It connects all points of zero geomagnetic latitude and its plane is at right angles to the Earth's magnetic axis.

geomagnetic field The Earth's magnetic field, which causes a compass needle to align north-south whatever its location on the Earth's surface.

geomorphic process The chemical and physical interaction between natural forces (e.g. running water, waves, ice, gravity, biological phenomena, etc.) and the Earth's surface, which act to modify the landscape. An increasing trend in geomorphology is toward the study of processes, inferring from these the landforms being created, as contrasted with more traditional geomorphology, in which processes were merely inferred from landforms. Processes are determined by the environmental variables of climate, geology, and base level; a change in one or some of these will change the nature or rate of processes. Processes fall into the broad categories of weathering, erosion. transport, or deposition, within each of which many different types can be rec-

Era		*Period*	*Epoch*	*millions of years*
			HOLOCENE	0.01
		QUATERNARY	PLEISTOCENE	1.6
			PLIOCENE	5.3
			MIOCENE	23.7
			OLIGOCENE	36.6
			EOCENE	57.8
CENOZOIC		TERTIARY	PALEOCENE	66.4
		CRETACEOUS		144
		JURASSIC		208
MESOZOIC		TRIASSIC		245
		PERMIAN		286
	(UPPER)	CARBON-IFEROUS	PENNSYLVANIAN	320
			MISSISSIPPIAN	360
		DEVONIAN		408
		SILURIAN		438
	(LOWER)	ORDOVICIAN		505
PALEOZOIC		CAMBRIAN		570
		PRECAMBRIAN Origin of Earth's crust		about 4600

Geologic timescale

ognized. Rates of a process, as shown by measurements or inference, are characterized by great spatial and temporal variation.

A central concept in geomorphology is the relative significance, in terms of work done in landscape formation, of infrequent but highly destructive events such as storms, floods, and gales and the very frequent but far less effective events of average conditions. In storms, whole beaches can come or go overnight; rivers can change their courses and expand their floodplains during floods; sand dunes can be built and eroded in a single severe gale. This is because capacity for work increases as a power function of the increase of the agency concerned; erosional capacity of rivers is proportional not to velocity, but the square of velocity; wave energy is proportional to the square of wave height; the transporting power of wind increases as the cube of its velocity. Major landscape features are largely shaped by the infrequent severe events, but Leopold, Wolman, and Miller (1964) concluded that moderate events of moderate frequency are responsible for most work done.

A better understanding comes from actually measuring the processes: this shows the rates at which the same process operates in different environments, and the comparative effectiveness of different processes in the same environment. It also shows the spatial and temporal distribution of changes, and once this is understood, prediction to the future is easier and more reliable, and the formation of existing landforms is more easily appreciated. Measurements of processes are taken as they operate, by the use of rods inserted into slopes and river banks for example, recording losses, or by the use of map evidence recording, for example, the successive growth of a spit across a river mouth, measurements from which can show variation in rates at different periods.

geomorphology The study of the evolution of landforms, excluding the major forms of the Earth's surface, such as mountain chains and ocean basins. It includes the study of the relationships between structures and landforms at the basic level, but then has two major branches: the inductive study of existing landforms, from which the processes of their evolution are inferred, and the deductive study and measurement of actual processes operating, inferring their ultimate influence on the landscapes on which they are acting. These two are complementary, because they approach the cause and effect relationships between landforms and processes from two contrasted viewpoints. The inductive branch is older than the more scientific deductive branch.

geopetal cavity (spirit level structure) A WAY-UP STRUCTURE produced within a closed cavity in a rock, generally within a fossil. On death some mud-sized material entered the fossil and settled out with its upper surface parallel to the horizontal. Once the cavity was closed no further sediment could enter. The trapped sediment hardens and leaves a record of the past horizontal level. The empty space above the sediment is generally later infilled, usually by coarse calcite crystals.

geophysics The science that combines the principles of mathematics and physics with the use of sophisticated equipment to examine the Earth's interior and the properties of its various internal constituents.

geosphere The nonliving part of the Earth, as opposed to the living BIOSPHERE. It includes the Earth's crust (lithosphere), all bodies of water (hydrosphere), and the air (atmosphere).

geostrophic wind A theoretical wind representing a balance of forces between the pressure gradient and the CORIOLIS EFFECT. This produces a wind flow parallel to the isobars. Because the effects of FRICTION are ignored, the real wind only approximates to the geostrophic wind above the BOUNDARY LAYER. At these levels the geostrophic wind can easily be calculated from the pressure gradient.

geosyncline (geotectocline) A large linear trough developed along the margin of a continent, within which a considerable thickness of stratified sediments including turbidites accumulates, with occasional extrusive volcanic rocks. After a long period of accumulation the trough is folded and uplifted to form mountain chains. In the light of the theory of plate tectonics and sea-floor spreading, this term is falling into disuse. *See also* trench.

geotectocline *See* geosyncline.

geothermal Describing heat that originates in the Earth's interior. Evidence of geothermal activity includes fumaroles, geysers, hot springs, and boiling mud volcanoes.

geothermal gradient The increase in the Earth's temperature with depth from its surface, expressed either in depth units per degree, or more generally degrees per unit depth.

geyser A type of hot spring that throws up jets of boiling water and steam, which are formed when groundwater comes into contact with hot rock underground. High pressure increases the boiling temperature of water, but after it is forced out as steam the pressure falls and water boils at lower temperatures. This causes an intermittent cycle of events, with eruptions taking place at intervals after periods of apparent inactivity.

GIS *See* Geographical Information System.

glacial *See* glacial phase.

glacial deposition The accumulation of rocky material that has been transported by a moving glacier or ice sheet. The deposits, collectively known as DRIFT, are usually left stranded and include drumlins, erratics, eskers, and moraines. *See also* till.

glacial drift *See* drift.

glacial erosion (glacial scouring) The wearing away of rock by the action of a moving glacier. The process is accelerated by the presence of meltwater streams and rock fragments beneath the ice. *See also* glacial plucking.

glacial maximum The greatest extent of PLEISTOCENE ice existed about 200 000 years ago, during the penultimate glacial period. Ice extended across the Arctic Ocean and joined with the ice sheets that covered much of North America, Greenland, N Russia, and NW Europe (the Baltic, Norway, Sweden, Denmark, and Finland were completely covered). This ice is believed to have extended across the North Sea to engulf all but the very south of the British Isles. Iceland was covered by an isolated ice cap. Antarctica was concealed by ice, as today, and mountainous areas in both hemispheres underwent glaciation. At this maximum an area of some 46 million sq km was beneath ice, over three times that of the present day.

glacial phase (glacial) The period during an ice age when ice sheets expand to lower latitudes. This would normally be associated with a global drop of temperature and changes in position of the main circulation belts. *See also* deglaciation; glacial maximum; interglacial; interstadial.

glacial plucking The erosive process by which rock material can be included within the ice of a glacier, later becoming a tool of glacial ABRASION. It entails the freezing of the glacier ice onto bedrock and the subsequent plucking out of blocks on movement. For this to occur FREEZE-THAW activity must be taking place at the ice-rock interface, thereby restricting the process to TEMPERATE GLACIERS.

A prerequisite for plucking is the presence of already loosened blocks, because the tensile strength of consolidated rock is far greater than that of ice. Well-jointed rocks are most susceptible (an optimum mean joint separation of from one to seven meters has been suggested). Freeze-thaw activity both before and after the appearance of the glacier will assist, while UNLOADING joints can also form as a result of the replacement of eroded rock by less dense ice, providing more susceptible layers.

glacial scouring *See* glacial erosion.

glacial wastage A reduction in thickness or areal extent of an ice mass resulting from more material being lost by ablation than is added by accumulation.

glaciated sea floor One of the most characteristic types of CONTINENTAL SHELF is found off coasts that have been heavily glaciated. Such shelves tend to have highly irregular surfaces and in their deepest parts may have 200 m or more of water covering them. Glaciated shelves are, in general, very wide, averaging 160 km; the average shelf-edge depth is around 220 m. Shelf areas that have been heavily glaciated, such as that off the Norwegian coast, often display enclosed basins and trough features. Such submarine topographic forms are analogous to lake basins found in many glaciated valleys on the continents. Not all of the sea-floor features testify to erosion; there may also be accumulations of morainic material and other deposits, built into or retaining a constructional form. Drumlins have been found off the coast of Maine, while concentric stone deposits on part of the floor of the North Sea are probably of morainic origin. Many glacially-eroded features have become buried by shifting sediments and can be found only by seismic measurements. Many of the banks and channels formed in glacial debris have been shaped by a combination of wave and tidal action, for example, in the S North Sea.

glacier An accumulation of ice of limited width moving downslope, under the influence of gravity, from a source area. Glaciers are formed by the accumulation and compaction of snow (*see* firn). They may be contained within basins (*see* cirque glacier) or within preexisting valleys (*see* valley glacier). *See also* cold glacier; temperate glacier.

glacier regime The pattern of relative gaining or losing of ice from a glacier or ice sheet. A glacier having a positive regime is gaining ice and will therefore advance and thicken, whereas one with a negative regime loses ice (*see* ablation) and will be reduced in size.

glacio-eustatism Worldwide change in sea level due to the effect of ice. During the PLEISTOCENE glacial periods sea level was much lower than at present, owing to the presence of much water on the continents in the form of ice. Estimates give values for this lowering of around 100 m. During the INTERGLACIAL periods the ice sheets melted and sea level rose again, sometimes to higher levels than those of today. Such changes would have a worldwide effect on erosional processes, because of the lowering of BASE LEVEL. However, these changes were not exclusively the result of the reduction in the volume of sea water, because deformation of the Earth's crust at the same time also had an appreciable effect.

glass (in geology) An amorphous (noncrystalline) material that is formed by the rapid cooling of lava or magma, such as OBSIDIAN.

glauconite A bright green authigenic mica-like mineral with an approximate composition $(K,Na)_{1.2\text{-}2}(Fe,Al,Mg)_4(Si_{7\text{–}7.6}\text{-}Al_{1\text{–}0.4}O_{20})(OH)_4.n(H_2O)$, which is found only in marine sediments, particularly greensands.

glaucophane A monoclinic AMPHIBOLE common in certain kinds of metamorphic rock.

glazed frost *See* black ice.

gleying The permanent or seasonal presence of either perched water or groundwater within a soil profile. This creates anaerobic conditions, leading to a dominant process of REDUCTION of ferric iron to its ferrous form, giving the soil a blue-gray color. The water may either be perched on a heavier impermeable horizon below, as in the PEATY GLEY PODZOL, or the result of a high groundwater table in areas near drainage lines. If the gleying is seasonal, the profile may have ochreous mottlings due to periodic oxidizing conditions, changing the iron back to its ferric state. The lack of oxygen inhibits microbiological breakdown of humus, which accumulates as acid peat. Anaerobic bacteria tend to produce gases toxic to large plants, such as hydrogen sulfide, ammonia, and marsh gas (methane). Together, the effects of hydromorphism are sufficient to completely change the direction of development of the soil profile.

gley soil A type of intrazonal hydromorphic soil that may be found in tundra, temperate, or tropical regions where there is excessive moisture in the soil profile. The characteristic poor drainage may be a result of the profile characteristics, i.e. a fine-textured parent material, or the site of the soil, i.e. a low-lying area. Two main types of gley can be recognized: the surface-water gley and the groundwater gley. The former is found where, because of a poorly drained subsoil, water is held within the upper part of the soil, giving waterlogged conditions for part of the year. A typical profile would consist of a very dark AI horizon rich in humus. Below this there would be Eg and Bg horizons with typical gleyed characteristics (*see* gleying). The C horizon would not show any gleying and would be similar to the parent material. The groundwater gley often occurs where there are permeable sands overlying impervious clay. The high groundwater table that is formed results in increased gleying with depth. The surface horizons are typically aerobic, gray or brown in color, and rich in humus. With depth a mottled Bg and a gray/blue G horizon are common. Most gley soils are under grassland vegetation and are used mainly for grazing.

glide twinning A complex form of crystal TWINNING resulting from deformation. It results from one layer of atoms changing its position so that the lattice is reversed. This results in broad lamellae within the crystal. It is commonly shown in the calcite crystals of marble.

Global Positioning System (GPS) A satellite-based coordinate positioning tool that can rapidly and accurately determine the latitude, longitude, and altitude of a point on or above the Earth's surface. It is based on a constellation of 24 satellites orbiting the Earth at a very high altitude and was originally developed by the US for military purposes. It has become an important tool in cartography, increasing accuracy and speeding production. Among its applications in earth sciences it has been used in studies to measure movements of the Earth's crustal plates, providing important evidence to support the theory of plate tectonics.

global warming The supposed gradual increase in the Earth's average air temperature, generally attributed to an increasing concentration of carbon dioxide in the atmosphere (and the consequent GREENHOUSE EFFECT). It has been calculated (using computer modeling) that a 1.5°C rise in temperature would result in melting of ice in the polar ice caps, causing a 20-cm rise in sea levels worldwide, and widespread flooding. Actual evidence of global warming is not yet conclusive but during the 20th century average temperatures on Earth rose by 0.5°C.

globigerina ooze A calcareous deep-sea ooze (*see* pelagic ooze) containing over 30% organisms. It is very widespread on the deep-sea floor. The organisms largely comprise the calcareous skeletons or tests of minute FORAMINIFERA. Globally, globigerina ooze represents some 130 million sq km covering something like half of the Atlantic Ocean floor, being widespread in the Indian Ocean, particularly on the W side, and very conspicuous in the S Pacific Ocean.

glomerophyric (glomeroporphyritic cumulophyric) Describing a PORPHYRITIC texture indicated by the occurrence of phenocrysts in aggregates or clumps set in the groundmass of an igneous rock.

GMT *See* Greenwich Mean Time.

Gnathostomata Vertebrates possessing jaws. The Gnathostomata include most fish and all higher vertebrates. *Compare* Agnatha.

gneiss A coarse-grained metamorphic rock in which quartz and feldspar predominate over micas. The schistosity is poorly defined and segregation banding is irregular and discontinuous. Coarse granular bands of quartz and feldspar alternate with thin often undulating schistose bands in which micas and amphiboles are concentrated.

Paragneiss has an unambiguous sedimentary composition. Orthogneisses are metamorphosed igneous rocks. Rocks of granitic composition containing a planar or linear orientation of minerals, imparted by stress, are termed granite-gneiss. Gneisses are formed during high-grade regional metamorphism.

gnomonic projection A form of azimuthal map projection in which the great circles of the Earth are shown as straight lines. Great circle routes can thus be plotted for navigation purposes. This projection is not often used because of its limited application.

goethite A yellow-brown mineral, FeO.OH, formed by the oxidation and hydration of iron minerals or as a direct precipitate. *See* limonite.

Gondwanaland The S hemisphere supercontinent, which takes its name from the Gondwana system of India, thought to have existed over 200 million years ago. Since then it has been fragmented as a result of SEA-FLOOR SPREADING into the present continents of Africa, Australasia, India, South America, and Antartica. *See also* Pangaea.

goniatite One of the earlier mollusks of the subclass AMMONOIDEA, whose shells had simply folded angular suture lines. Goniatites are used as ZONE FOSSILS in the Devonian and Carboniferous System but were replaced by the AMMONITES in the PERMIAN.

goniometer An instrument that measures the angle between crystal FACES. Because such angles are characteristic of the crystal, they provide a useful means of identification.

gorge A narrow steep-sided valley, generally formed in hard rocks by the erosive action of a flowing stream or river. A large gorge is called a CANYON.

gossan A mass of quartz and hydrated iron oxides that often marks the outcrop of a sulfide-bearing vein, the sulfides having been oxidized to soluble sulfates. Metalliferous material removed in solution may be redeposited at depth in a zone of secondary enrichment. *See also* hydrothermal process.

GPS *See* Global Positioning System.

graben A generally elongated block of rock that has been downthrown between two parallel faults relative to the surrounding area. It differs from a RIFT VALLEY in that it is a structural feature and not necessarily also a topographical one. *Compare* horst. (See diagram at FAULT.)

grade **1.** The degree of metamorphic change undergone by a rock. When such changes can be traced from unmetamorphosed rocks to highly metamorphosed rocks, the grade of metamorphism is said to increase. High-grade rocks are produced under conditions of high temperatures or pressures; likewise low-grade metamorphic rocks form at low temperatures and pressures. *See* zone.
2. The condition reached in the later stages of a normal CYCLE OF EROSION when a river or slope has an EQUILIBRIUM REGIME, with a smooth concave long profile. Since the theory was proposed by Davis, other factors besides the modification of the long profile have been taken into account, especially adaptation to changes in the controlling environmental factors of base level, rainfall, and the quantity and type of the load supplied to the stream. A river that is at grade is able to maintain its COMPETENCE so that it is just able to transport its load at

GRANITE-GRANODIORITE ROCKS			
Dominant feldspar	*Coarse-plutonic*	*Medium*	*Fine-volcanic*
alkali feldspar	granite	microgranite	rhyolite
alkali feldspar= plagioclase feldspar	adamellite	microadamellite	rhyodacite
plagioclase feldspar	granodiorite	microadamellite	dacite

a given discharge. It does this by adjusting its long profile, cross section, channel pattern, and bed roughness: any environmental changes will be met by changes in some or all of these factors, so that the river uses up its available energy and no more.

graded bedding Sedimentary bedding in which particles show a size distribution. The coarsest material forms the base and then the sequence becomes progressively finer upward. It is often present in turbidity deposits. On a larger scale such upward fining sequences are developed in braided river channels, as they are infilled. *See* braided stream.

graded profile *See* equilibrium profile.

gradient wind An extension to the concept of the GEOSTROPHIC WIND. Isobars in the upper atmosphere are rarely straight, so to account for the curved flow, following Newton's laws of motion, another force must be included. This is the CENTRIPETAL ACCELERATION. The equilibrium wind for these three forces is known as the gradient wind. Curvature of the isobars can be either in the cyclonic or anticyclonic sense.

granite A member of a family of acid coarse-grained plutonic igneous rocks containing essential quartz, alkali, and plagioclase feldspars and small quantities of mafic minerals. The rocks of the granite-granodiorite suite and their volcanic equivalents, rhyolites-dacites, are divided according to the proportions and compositions of the feldspars.

Granites contain on average about 25% quartz and contain either a single feldspar, a perthitic intergrowth (*hypersolvus granites*), or two kinds, oligoclase and potassium feldspar (*subsolvus granites*). All are leucocratic rocks (*see* color index), the only common dark mineral being biotite. Accessory minerals include muscovite, zircon, apatite, and tourmaline. The typically granular texture in which most of the crystals are subhedral is often termed granitic. Porphyritic varieties containing large zoned phenocrysts of white, gray, or pink feldspar are common. Many granites exhibit graphic, orbicular, and rapakivi textures and contain miarolitic cavities. Some strongly alkaline granites are chemically distinctive, having low aluminum and calcium contents and characterized by the presence of the soda pyriboles, aegirine and riebeckite, and an abundance of otherwise rare accessory minerals. With a decrease in the amount of quartz, granites pass into quartz-syenites and syenites. Adamellites, of which Shap granite is an example, contain oligoclase and potassium feldspars, quartz, and possibly hornblende in addition to biotite. With a decrease in the amount of quartz, adamellites grade into monzonites; the term quartz-monzonite is sometimes used as a synonym for adamellite. Granodiorites that are almost devoid of alkali feldspar are called *trondhjemites*. Granitic rocks are particularly susceptible to the pneumatolytic (*see* pneumatolysis) processes of alteration, greisening, tourmalinization, and kaolinization.

Microgranites are the medium-grained equivalents of granites. Porphyritic varieties have been termed quartz-porphyries or granite-porphyries. Granophyre is a micrographic rock exhibiting a micrographic (granophyric) texture. Granodiorites are volumetrically the most abundant plutonic rocks and constitute the bulk of the batholiths in orogenic regions. Granodioritic magma may be produced by anatexis at deep crustal levels. When such sites of magma generation become exposed at the surface, the association of granites, migmatites, and high-grade regionally metamorphosed rocks is apparent. Once generated, granitic liquids rise up in the crust and undergo differentiation to produce the more chemically extreme members of the suite. At high crustal levels, intrusive granite plutons produce marked metamorphic aureoles. Although most granites crystallize from a magma, some granitic rocks may be produced by granitization, the pervasive metasomatism of preexisting rocks. *See also* aplites; pegmatite.

granitization The process of metasomatic transformation of preexisting rocks into granite due to the action of granitic fluids (*ichors*) arising from depth.

granodiorite A plutonic igneous rock containing plagioclase feldspar of oligoclase-andesine composition, subordinate potassium feldspar, biotite, or hornblende. *See* granite.

granophyre A microgranite exhibiting a micrographic (granophyric) texture. *See* granite.

granular disintegration The separation of individual grains from the main mass of a rock. It most frequently occurs in coarse-grained rocks, notably granite, and both mechanical and chemical weathering processes can be responsible. Mechanically, the differential expansion and contraction of different mineral grains due to INSOLATION may cause disintegration, while crystallization of salt from solution, within the surface pores of a rock, may prise grains apart. Chemically, HYDRATION usually involves a considerable volume change, and granular disintegration may occur where there is differential expansion of various mineral grains. The usual product of this type of weathering is a coarse sand.

granule A small fragment of rock up to 5 mm across. It is thus larger than a coarse grain of sand but smaller than a pebble.

granulite A granular textured metamorphic rock occurring in areas of regional metamorphism of the highest grade. Granulites lack the hydrous minerals, micas and amphiboles, but may contain pyroxene, garnet, kyanite, or sillimanite. Characteristically the quartz and feldspar crystals are flattened and their parallel alignment gives rise to a foliation. *See also* charnockite; metamorphic facies.

graphic intergrowth An intergrowth of alkali feldspar and quartz commonly found in pegmatites and granitic rocks. The quartz is orientated along preferred directions within the feldspar and on a flat surface resembles hieroglyphic or runic writing.

graphite (plumbago) A soft iron-gray massive or laminar form of pure carbon, which is characteristically greasy to the touch. Graphite is found mainly in metamorphic rocks; it has many commercial uses.

Graptolithina A primitive group of invertebrates of the phylum Hemichordata. Graptolites are totally extinct, fossils being confined to the Paleozoic. The body consists of a number of minute cups (*thecae*) of chitinous material, which house individual polyps. The thecae are arranged along a stem (*stipe*) and graptolites show great variation in the arrangement and orientation of the thecae and the disposition of the stipes. Biseriate forms have thecae on both sides of the stipe, uniseriate forms along only one, and there may be a single stipe or many.

Graptolites were benthonic and planktonic marine organisms and the planktonic forms attained a wide distribution. This, coupled with their great morphological variation and rapid evolution, makes them of considerable importance in biostratigraphy and they are used as ZONE FOSSILS throughout much of the Ordovician and Silurian Systems. Graptolites are first known from rocks of the Cambrian System; they suffered widespread extinctions at the end of the Silurian but some groups continued to the early Carboniferous.

grassland Any region where grass is the natural vegetation, generally where there is insufficient rainfall to support woodland or forest trees but too much for a desert to form. Tropical grasslands, also called *savannas*, occur mainly in E Africa, N Australia, and parts of N South America. Temperate grasslands include the PAMPAS of South America, PRAIRIES of central North America, STEPPES of central Asia and Russia, and VELD of southern Africa and Australia.

grass minimum thermometer A thermometer used to record the minimum air temperature just above (20–30 mm above) ground level, where it is often lowest during the night.

graticule The network of lines on a map projection representing the lines of latitude and longitude. On some maps the graticule may not be shown over the entire face of the map; in these cases the lines shown have short graticule ticks representing lesser lines not shown.

gravel According to some particle size classifications, gravel is unconsolidated material (usually rock fragments rather than individual mineral crystals) in the 2–60 mm size range. It is the size of material lying immediately above sand, and below cobbles. In practice, however, the term tends to be loosely used in geomorphological literature for all material of an unconsolidated nature of greater than sand size and less than boulder size (i.e. from 2–200 mm in diameter).

gravimeter (gravity meter) An instrument used to measure slight variations in the Earth's gravitational field, commonly employed in prospecting for oil and other minerals (whose deposits cause local anomalies in the gravitational field).

gravitational constant The constant (G) used in the law of universal gravitation. It has the value 6.670×10^{-11} N m^2 kg^{-2}.

gravitational gradient A regional shift in the measured value of gravity across an area.

gravity *See* acceleration of free fall.

gravity anomaly On the assumption of uniform density, calculations can be made to predict the value of gravity for a particular point on the Earth's surface. Where this value differs from that actually measured an anomaly exists. A *negative gravity anomaly* is an area in which gravity is less than that predicted; it is an area out of isostatic balance. An area of greater gravity than that predicted is a *positive gravity anomaly* and usually results from a large mass of basic rocks near the Earth's surface.

gravity collapse structure A tectonic structure produced mainly as a result of downward movement under the influence of gravity.

gravity fault *See* normal fault.

gravity fold A fold structure resulting from sliding in response to gravity, as distinct from vertical or horizontal compressional forces operating during an orogeny.

gravity gliding The movement of a large rock mass downhill as a result of gravity.

gravity meter *See* gravimeter.

gravity unit One gravity unit (gu) equals 10^{-6} m s^{-2} and is often used in place of the milligal (*see* gal). One gravity unit is roughly equivalent to one ten millionth of the value of gravity at the Earth's surface.

gravity wave A wind-generated wave at sea whose length exceeds 50 mm and whose speed of propagation is influenced mainly by gravitational forces.

gray-brown podzolic soils Soils found covering large areas of the northeastern USA to the south of the brown podzolic zone. They are more mildly podzolized soils, less acid conditions resulting in more abundant soil fauna incorporating humus deep into the AI horizon. An Ea horizon is present but it is darker (gray-brown) in color than the typical eluvial horizon of a podzol. Clay is illuviated in the B horizon, which normally exhibits a blocky structure.

gray desert soil *See* sierozem.

gray forest soil A type of soil found in the transition zone between forest and steppe in North America and Russia. The typical profile consists of a thin litter below which is a dark gray A horizon with a crumb structure. Organic and mineral matter have been well mixed. Below is the Ea horizon, with its nutty structure, which is light gray in color due to silica coating the ped faces. Clay and humus have been removed from this horizon and deposited in the illuvial Bt horizon, which has a prismatic structure. The gradation to parent material, which is often loess, is usually characterized by calcium carbonate concretions. Thus the soil exhibits the leaching characteristics of a podzol and the calcification of a chernozem. From the profile characteristics it has been suggested that these soils originally developed under steppe vegetation, which was replaced by forest when there was a change from a warm dry to a cool humid climate. These soils may be subdivided zonally. Westward from E Russia to European Russia these soils change from dark gray forest soils to normal gray forest soils and finally to light grayish-brown forest soils. The humus content decreases westward from 10% to 3% owing to more podzolic conditions. They fall into the ALFISOL order of the Seventh Approximation.

graywacke An ARENACEOUS sedimentary rock in which fairly angular particles of SAND grade, mainly lithic fragments, are suspended in a matrix of much finer material. *See also* turbidite.

gray wooded soil A soil belonging to the podzolic group, morphologically resembling a podzol but chemically resembling a gray-brown podzolic soil. It is typically found in the mid-continental states of Canada on a base-rich parent material under boreal forest. It has a well-developed Ea horizon with a platy structure and a Bt horizon with a blocky structure. Leaching is less intense than in a podzol, resulting in less movement of sesquioxides in the soil profile and a soil with a higher pH value. Gray wooded soils fall into the ALFISOL order of the Seventh Approximation.

Great Barrier Reef A CORAL REEF lying off the NE coast of Australia, extending for some 1900 km and varying from 30–160 km in width. It is an example of a barrier reef found off a continental coast in contrast to those reefs that circumscribe volcanic islands. It is separated from the mainland by a discontinuous channel 40–90 m deep and therefore too deep for coral growth. On the seaward side of the reef, the shelf attains depths of at least 420 m. The coral composing it covers about half of the Queensland shelf region. The reef is considered one of the best coral features in the world, and fears are being expressed concerning serious damage to the reef caused by certain species of starfish. Borings have indicated that recent coral extends down to depths of between 120 and 140 m, and that sand and mud intercalations occur within the coral. The coral tends to be of two quite different types, the dividing line between the two being the *Trinity Opening*. The reef may be based on a step-fault platform.

great circle A circle round the Earth's surface whose plane passes through the center of the Earth. Great circle routes, broken down into a series of RHUMB LINES, are often used in navigation because they represent the shortest distance between two points on the Earth's surface.

green flash When the Sun sinks below the horizon, the last rays may be seen to change from pale yellow, to orange, then to green. The rapidity of the process has given this phenomenon the name of the green flash. It occurs only when the air is exceptionally clear and it is caused by differential refraction of the spectral colors by the atmosphere. Normally the colors complement each other when the Sun is high in the sky, but as the Sun disappears the individual colors may very occasionally be identified.

greenhouse effect The process by which the Earth's surface is warmed by heat reradiated back to Earth by gases, especially carbon dioxide but also including methane, chlorofluorocarbons, nitrous oxide, water vapor, and other gases, in the atmosphere. The presence of the atmosphere maintains the temperature of the Earth's surface at a much higher level than would be expected on the basis of equilibrium with solar radiation input. This is because the gases are transparent to short-wave radiation but almost opaque to long-wave radiation. Glass has similar properties and the effect of the Earth's atmosphere has been likened to that of a greenhouse. Increases in carbon dioxide levels resulting from the combustion of fossil fuels may intensify the greenhouse effect. *See also* global warming.

greensand A greenish sandstone or sand deposit that consists mainly of GLAUCONITE.

greenschist A type of green metamorphic rock whose color is caused by the presence of ACTINOLITE, CHLORITE, or EPIDOTE. It should not be confused with GREENSTONE.

greenstone Basic volcanic rock that has suffered low-grade METAMORPHISM or alteration.

greenstone belt A group of very ancient basic and ultrabasic volcanic and sedimentary rocks that occur on most continents. The composition of KOMATIITE lavas within such sequences suggests that the temperature of the mantle was much higher at the time of their formation.

Greenwich Mean Time (GMT) Local time at Greenwich, London, which is located on the 0° meridian and from which the standard times of different areas of the globe are calculated. 15° longitude represents one hour in time from the Greenwich meridian.

greisening *See* pneumatolysis.

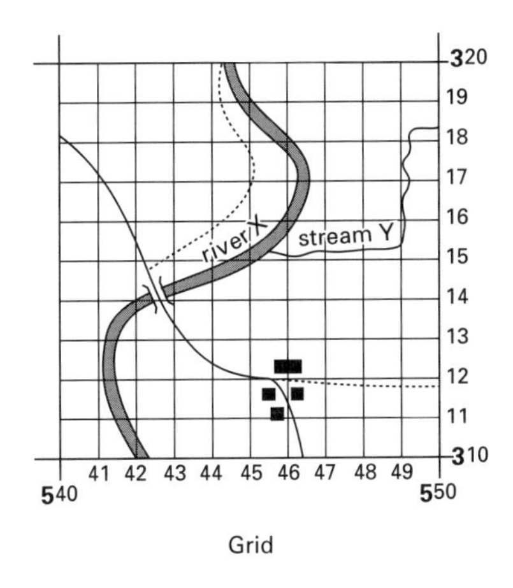

Grid

grid A network of parallel lines intersecting at right angles and forming reference squares over the face of a map. Grid lines are numbered, therefore each square has unique easting and northing readings. Grids are used to enable identification of points on the map with respect to other points, or distances between points. (In the diagram the junction of stream Y and river X has a grid reference of 545 315). They are also used to determine the accuracy of a map: by knowing the true position of a point in relation to the grid and the position as shown on a map the discrepancy between the two can be determined.

gridiron drainage A type of drainage system characterized by many parallel streams that each have tributaries running at right angles to them, resulting in a rectangular network of drainage channels. *See also* trellis drainage.

grid zone The Earth's surface can be arbitrarily divided into grid zones, which are areas of different grid systems, each with its own origin. Grid zones enable the identification of points on the Earth's surface without the use of geographical coordinates, i.e. longitude and latitude.

grike (gryke) *See* limestone pavement.

grit A type of coarse SANDSTONE with a large grain size, typical of the MILLSTONE GRIT of England.

groin (groyne) An artificial barrier constructed on beaches and extending seaward perpendicular to the coastline in order to trap beach materials moving along the shoreline. Groins are usually built in groups, their length, height, and separation varying with type of beach material and desired results. Although very effective at building up the beach in their immediate vicinity, by halting or at least slowing down longshore movement of material, they may cause considerable erosion of that part of the beach down-drift of the groins.

grossular A white, yellow, or greenish member of the GARNET group of minerals, $Ca_3Al_2(SiO_4)_3$. It occurs in limestone, usually as a product of contact metamorphism. It is used as a semiprecious gemstone.

Grosswetterlagen (large-scale weather patterns) A concept devised by Baur to identify the major trends in atmospheric events over a particular area. It has subsequently been modified and became used in Germany as a method of assisting long-range weather forecasting. In the classification the main anticyclone or depression center is located and this is used to predict the likely wind and weather systems.

ground frost The outcome of a fall in temperature below 0°C, measured by a thermometer at ground level on a grass surface. In climatological statistics, note is made of whether a ground frost occurred, not its intensity. Data prior to January 1 1961 took 30°F as the base for ground frost.

ground ice Ice that is locked within the pores of the soil and rock fragments in the REGOLITH. It represents frozen GROUNDWATER.

groundmass The finer-grained bulk of crystalline or glassy material that composes an igneous rock and into which the coarser components (PHENOCRYSTS, XENOLITHS) are set.

ground moraine A sheet of TILL, usually some tens of meters thick, with a surface characterized by very low relief. Ground moraine was deposited from the base of an ice sheet or glacier on the melting of the ice, and forms a cover that tends to mask former bedrock features, because where hollows formerly existed the greatest thicknesses of till were deposited.

ground swell **1.** Swell waves as they enter shallow water, i.e. water with a depth approximately half that of the wavelength. This causes the swell waves to decrease in length and increase in height. In this sense, the waves are beginning to be influenced by the ground (seabed).
2. Swell waves in deep water that possess considerable height and length.

groundwater Water precipitated from the atmosphere that has percolated into the ground and become trapped within pores, cracks, and fissures. Its presence is essential for practically all weathering processes and for this reason more weathering takes place in well-jointed rocks than in massive types. In the upper layers of permeable rock, groundwater can flow through the strata quite easily; however, at depth a level is reached below which all pore spaces are water-filled (the *water table*). Weathering still occurs below the water table, in the form of reduction and hydrolysis.

group A division in the hierarchy of the lithostratigraphic classification of bodies of rock (*see* lithostratigraphy; stratigraphy). It is formed of two or more adjacent and related FORMATIONS; adjacent groups may be aggregated to form *supergroups*.

growing season The period of the year during which plant growth can proceed without temperature restriction. By temperature latitude convention, a daily mean temperature at screen level of 6°C is the critical lower limit; the growing season can then be determined from the monthly variation of mean temperature. It is not a very precise concept because a mean temperature of 6°C could be produced by a variety of diurnal ranges, and different plants respond to different critical temperatures.

groyne *See* groin.

grumusol *See* tropical black soil.

grunerite A monoclinic AMPHIBOLE.

gryke (grike) *See* limestone pavement.

guano Seabird excrement found in large amounts on certain of the Pacific Islands and along part of the coast of South America. It is of great economic importance since, being rich in nitrogenous matter and phosphates, it serves as a valuable land fertilizer.

guide fossil A fossil of known age that, because of its location, helps to establish the age of the stratum in which it is found.

Gulf Stream The largest and most important permanent ocean current in the N hemisphere. It is most prominent in its southern portion, beyond the meeting-point of the Florida Current and the Antilles Current. In this region, it attains its greatest depth, probably some 1000 m, and achieves a maximum width of some 50–70 km. Farther north, it becomes less effective at depth, while to the south of the Newfoundland Banks it begins to split into several branches. It is not a very warm current, differing only little in its thermal characteristics from the Sargasso Sea region which lies close to its path. Indeed, as it spreads out northward, it tends to mix with colder surrounding waters and is fed also by the rather cold Labrador Current. Where strongly developed, its sharply defined left-hand flank contrasts markedly with its more diffuse right-hand flank, where it is common for large eddies to be thrown off. *See also* North Atlantic Drift.

gully A narrow channel on a hillside lacking vegetation, generally formed by the rapid runoff of surface water following heavy rainfall. A gully is dry most of the time. *See also* badlands.

gully erosion A type of erosion in which gullies develop from RILLS when sheet flow of water becomes concentrated into distinct channels. They originate naturally near the head of the drainage pattern, but they are better known as the most widespread and obvious signs of human influence on geomorphology due to the removal of vegetation cover in farming. Plowing down the slopes aids the concentration of sheet flow into gullies, which then erode their headwalls and advance upslope, as well as widening their valleys and developing tributaries. Gullies remove great volumes of sediment, especially topsoil, and are known and feared in many currently settled areas, such as central southwestern USA and southern Australia. There is evidence that they were also once widespread in Europe. It is now known that the simple conservation practice of plowing parallel to contours will retard gully development. *See also* abnormal erosion; arroyo.

gust A sudden increase of wind speed of short duration. Gusts are due to mechanical interference by the ground surface, so that where the surface is aerodynamically rough, as in cities, gustiness is increased. Conversely, winds in exposed coastal sites tend to experience fewer gusts, although their mean wind speed can be high.

Gutenberg discontinuity The boundary between the Earth's core and mantle, at a depth of about 2900 km, at which seismic primary (P) waves slow down and secondary (S) waves disappear. The behavior of the waves is thought to be caused by the change from solid rock in the mantle to molten rock in the core.

guyot A relatively smooth flat-topped SEAMOUNT or tablemount. Guyots are best known in the Pacific and Atlantic Oceans. Their flat tops are interpreted as being planation surfaces eroded by a combination of marine and subaerial processes. The tablemount is more or less conical in shape, and the summit surface usually lies in fairly deep water. The summit depths often range between 1000 and 2000 m; others have been sounded between 2000 and 3500 m. Few guyots with summit depths less than some 1000 m have been located. Guyots may be veneered with sediments such as volcanic sand, gravel, and globigerina ooze.

gymnosperms A group of primitive seed-bearing plants that differ from the ANGIOSPERMS in having seeds that are not protected by an outer covering. They include the conifers, cycads, and ginkgos. Gymnosperms are known as fossils from the Carboniferous Period and flourished during the early Mesozoic. Most gymnosperms became extinct during the Cretaceous, presumably because of unsuccessful competition with the angiosperms.

gypsum A common and widespread colorless or white mineral form of hydrated calcium sulfate, $CaSO_4.2H_2O$. It crystallizes in the monoclinic system and is an evapotite mineral that occurs in limestones and shales, often associated with anhydrite and halite. The fine-grained massive white form is alabaster, tabular colorless crystals are selenite, and fibrous crystals form satin spar. Gypsum is used in making plaster of Paris and cement.

gyre The gross circulation of water that occurs in each of the Earth's major ocean basins. It is caused by the convection of warm water from the surface, the effects of prevailing winds, and the rotation of the Earth. *See also* ocean current.

H

haar A local term for advection fog on the east coast of Britain. It is also called *sea-fret* or *sea-roke*. In spring and early summer the North Sea is relatively cold and quickly cools moist air to saturation point giving low stratus cloud. Whenever winds are onshore, the low cloud drifts inland where it is gradually dispersed by heating. It is most frequent under anticyclonic conditions when vertical motion and mechanical turbulence are weak.

habitat The place where an organism lives, as defined by those aspects of its total environment that affect it and to which it is adapted. For example, sponges have a benthonic habitat. *See also* niche.

haboob An Arabic word applied to any dust storm in the Sudan, occurring chiefly in the summer, raised by strong winds, without reference to its origin. Many of them result from the downdrafts of cumulonimbus clouds with wind speeds reaching 90 knots in extreme cases, and they are often followed by heavy rain or thunderstorms.

hachures Short lines drawn on a map to show the relief of an area without the use of contours. They point downhill and are thicker and closer together where the gradient is steepest. Before cartographic techniques were sufficiently advanced to enable the construction of contour lines this method was extensively used. It can be of particular effect in mountainous areas.

hade The angle measured between the vertical plane and that of the incline of a structural surface, generally applied to faults but applicable to any structural surface. *Compare* dip.

Hadean *See* Precambrian.

Hadley cell A cellular wind circulation in tropical latitudes with surface winds blowing from the subtropics at about 30° to the Equator and winds in the upper atmosphere blowing in the reverse direction. This corresponds with the surface temperature gradient from the thermal Equator toward the poles. This simple circulation was first suggested by Hadley in the 18th century. Although it is modified by the effects of the Earth's rotation, it is still believed to be a reasonable approximation to reality. *See also* general circulation of the atmosphere.

hail Approximately spherical particles of ice that fall from cumulonimbus clouds. The pieces of ice can be irregular in shape and have been known to weigh up to almost 1 kg, although they are normally far less. They often have a concentric structure with alternating layers of clear and opaque ice, providing some evidence of their evolution within the cloud. It is believed that this structure is the result of the stone being transported into different parts of the cloud during its development. Beginning

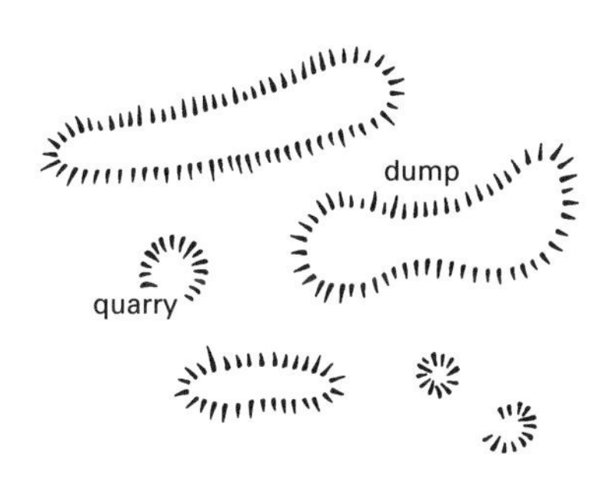

Hachures

with an ice nucleus, opaque ice will form around it when it is in the colder parts of the cloud. Small cloud droplets freeze rapidly to produce a spherical aggregation of ice with many air enclosures to give the characteristic white and opaque appearance. In the lower parts of the cloud with temperatures only slightly cooler than 0°C, the relatively larger droplets there tend to spread over the hail surface before freezing so that little air is trapped and the ice is transparent. The distribution of hail in relation to the main path of the storm will depend upon the nature of updrafts and the cloud structure.

Hailstorms occur most frequently in the continental interiors of temperate latitudes and decrease toward the poles and Equator, and over the sea. This is because conditions most favorable to cumulonimbus development are found here (intense surface heating, sufficient concentrations of cloud water, and strong thermal gradients).

half-life The time taken for one half of a sample of radioactive element to decay (to another element). It has a constant value for any particular radioisotope, and its measurement is the basis of RADIOMETRIC DATING.

halide A compound (a salt) whose negatively charged element is one of the halogens: fluorine, chlorine, bromine, or iodine. The halides are thus fluorides, chlorides, bromides, or iodides.

halite (rock salt) The common mineral form of common salt, sodium chloride, NaCl. It may be colorless or white, and is often colored pink by impurities. It crystallizes in the cubic system and occurs in underground evaporite deposits, derived from ancient seawater. It is used as a source of sodium compounds and chlorine. *See also* salt dome.

halomorphic soil A soil dominated by sodium salts. There are three main types: the SOLONCHAKS or saline soils, which are dominated by chlorides and sulfates of sodium; the SOLONETZ or alkali soils, dominated by carbonates of sodium; and the leached or degraded SOLOD. There are also many arid zone soils subject to some SALINIZATION. These halomorphic soils are not true INTRAZONAL SOILS, but tend to occur in certain climates, e.g. the arid to semiarid transition zone; they are often related to the beds of former lakes or seas, such as the area north of the Caspian Sea, or to depressed areas where a water table is near enough to the surface to feed salts upward by CAPILLARITY. In highly irrigated but poorly drained areas, e.g. the San Joaquin Valley of California, the water table may be artificially raised by irrigation, leading to salinization of formerly fertile soils through increased capillary rise of water bringing up dissolved salts.

hammada (hamada) A type of arid desert plain, consisting of an extensive almost bare rock surface, especially in the Sahara. *See also* erg; reg.

hanging valley A tributary valley in a glaciated area, the gradient of which becomes much steeper on entering the main valley. A large glacier will cause greater deepening of a preexisting valley than will a smaller glacier, and therefore where glaciers invade a major valley and its tributaries the major valley will be deepened to a greater extent. On the disappearance of the ice, the floor of the major valley will be at a lower altitude than those of the tributaries, which will hang above the main valley. Subsequent rivers will reach the main valley from these hanging valleys as waterfalls.

hanging wall The surface of rock above a fault plane or ore body. *Compare* foot wall.

harbor A sheltered stretch of water where ships may anchor or tie up to buoys or jetties, in order to be safe from storms or other adverse conditions, or to load or unload their cargoes. Some are formed by natural features, including sheltered bays and estuaries; others are created by the construction of jetties, breakwaters, moles, and other artificial structures. Artificial harbors are usually built, sometimes after careful scale hydraulic model tests, so as to minimize wave action, SEICHE action, and undesirable currents. Many harbors, both natural and artificial, have to be dredged (*see* dredging) in order to maintain or to improve limiting depths.

hardness *See* Mohs' scale.

hardpan A layer of hard material just below the surface of the ground, usually deposits of carbonates, hydroxides, oxides, silica or, most often, clay. It may also contain some organic material. It is common in sandstone regions and is impervious to water, thus preventing good drainage. Marshy patches and puddles may form on the surface after rain.

hard water Natural water that contains dissolved ions of calcium and magnesium. These ions form a scum with soap. Water with *temporary hardness* contains dissolved calcium hydrogencarbonate, which is formed by the action of dissolved carbon dioxide on chalk or limestone. When temporarily hard water is boiled, calcium carbonate precipitates out, which forms limescale on boilers, pipework, and kettles. Water with *permanent hardness* generally contains dissolved calcium sulfate or calcium fluoride. These are not removed by boiling.

harmatome A white or gray hydrated barium aluminosilicate, $Ba(Al_2Si_6O_{16}).6H_2O$. It occurs in basic igneous rocks and is a member of the ZEOLITE group of minerals.

harmattan A dry wind blowing from between north and east over West Africa. Because the air has had a long trajectory across the Sahara, it is very dry, cool by night but warm by day, and is laden with dust. It represents the normal dry season state of the area, oscillating in its extent with the seasons, occasionally reaching the Gulf of Guinea in January but rarely south of 15°N in July.

harmonic analysis *See* Fourier analysis.

harzburgite An ultramafic rock consisting largely of olivine and orthopyroxene.

hastingsite A monoclinic AMPHIBOLE.

haüyne A member of the sodalite subgroup of FELDSPATHOIDS.

hawaiite A type of alkali basalt. *See* trachybasalt.

haze Atmospheric obscurity due to minute suspended solid matter, such as dust or smoke particles, in the sky. It is not a very precise term; when visibility reduces to 1 km, the phenomenon is called mist, especially when it is produced by water droplets.

head *See* coombe rock.

headward erosion (headwater erosion) The increase in length of a gully or valley, or a stream within it, caused by erosion at the upper end (head) of the valley.

heat The form of energy to which all other forms will eventually revert. Heat can be transferred by conduction, convection, or radiation and may lead to an increase of temperature, but it can cause a change of state from liquid to vapor without any corresponding alteration in temperature. Heat is normally measured in joules, although the former unit, the calorie, is still sometimes used.

heat balance The static ENERGY BALANCE between the Earth and atmosphere. It indicates the utilization of net radiation by evaporation, sensible heat, and advection.

heat island An urban area that has a higher temperature than its rural surroundings: on calm clear nights excesses of up to 10°C have been recorded. They are the result of the different nature of the surfaces in a city, which in turn affects the heat balance and airflow. *See also* urban climate.

heave The horizontal displacement along a fault.

hectare (ha) A metric unit of area equal to 10 000 m^2 (100 ares), and equivalent to 2.47 acres.

hedenbergite A monoclinic PYROXENE.

helical flow (helicoidal flow) The most significant type of turbulence in streams. Superimposed on the primary downstream flow, a secondary flow moves across the surface of the stream toward the outside of the meander bends, compensated by a reverse flow along the bed toward the inside of the meander bends. This gives the streamflow a net corkscrew movement, concentrating erosion on the outside of the meander bends, with deposition on the inside, causing downstream propagation of the meanders.

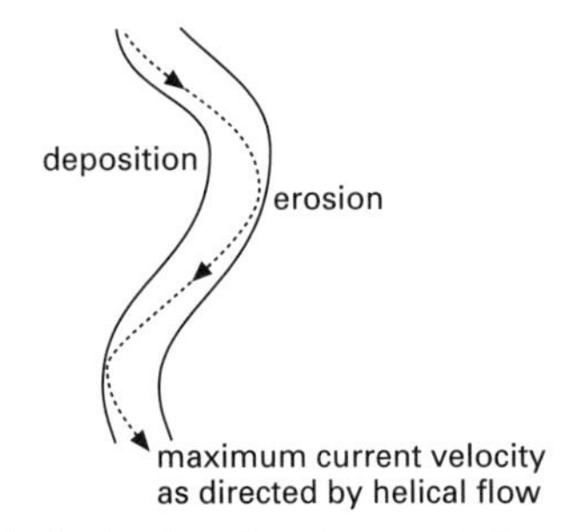

Fig. 1: Distribution of maximum velocity in a meandering stream due to helical flow

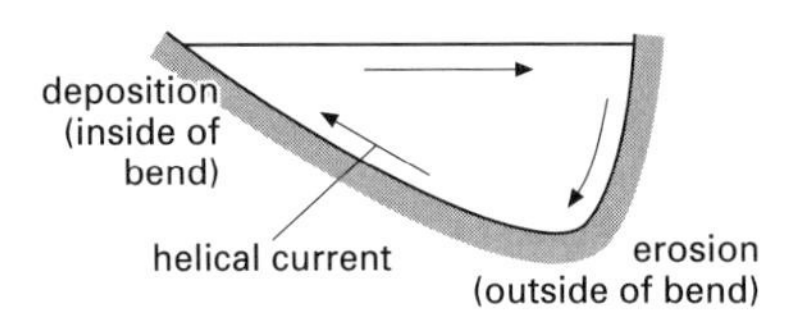

Fig. 2: Section through the outside of a meander bend

Helical flow

helicitic structure Curved or S-shaped trails of inclusions occurring in porphyroblasts (*see* porphyroblastic), especially garnets, and representing an earlier foliation.

hematite The mineral form of ferric iron oxide (Fe_2O_3), an important iron ore. It occurs in two main types, as a massive red botryoidal or reniform ore (kidney ore) and as metallic crystals known as *specular iron ore*. Hematite often occurs as a cement in sandstones producing a red coloration. Most ore deposits result from the alteration of iron carbonates and silicates in sedimentary rocks.

hemimorphite (calamine) A mineral form of hydrated zinc silicate, $Zn_4Si_2O_7(OH)_2.H_2O$. It crystallizes in the orthorhombic system and generally occurs as fibrous crusts or masses. It is used as a source of zinc.

herbivore An animal that eats green plants or algae. In ecological terms, it is a primary CONSUMER and is generally the source of food for secondary consumers (carnivores). *See also* food chain.

Hercynian orogeny A phase in the Variscan orogeny affecting Europe and characterized by a northwest fold trend. The term is used without time significance and covers the Carboniferous and Permian.

hercynite *See* spinel.

heterogeneous strain The changes in shape or dimensions occurring when a body of rock does not deform equally in all directions. A linear structure would become curved and parallel structures would diverge after deformation.

heterolithic unconformity *See* nonconformity.

heteromorphism The existence of rocks of very similar chemical composition but with different mineral aggregates, resulting from different rates of cooling of the magmas from which they formed.

heterotrophic Describing an organism that is a CONSUMER, i.e. one that feeds on plants or other organisms because it cannot synthesize its own food (unlike AUTOTROPHIC organisms). Such organisms include some algae and bacteria, most protozoa, and all fungi and animals.

heulandite A white, gray, or brown hydrated calcium sodium aluminum silicate, $(Ca,Na_2)Al_2Si_7O_{18}.6H_2O$. It crystallizes in the monoclinic system and occurs in basic igneous rocks. It is a member of the ZEOLITE group of minerals.

hexagonal *See* crystal system.

hiatus A break in a stratigraphic sequence, either as a result of nondeposition or erosion. It represents the period of time missing between beds above and below an unconformity.

high An area in which atmospheric pressure is higher than surrounding areas. It is used as a more general term than an ANTICYCLONE, which is centered on a closed isobar. Areas of high pressure are normally associated with dry weather, although the skies may be either cloudy or clear.

high index circulation (in meteorology) The phase of the INDEX CYCLE when the westerly circulation reaches its maximum intensity in specified areas. This represents a strong surface westerly flow with Rossby waves of long wavelength and small amplitude. Meridional exchange is at a minimum value.

high-resolution satellite imagery *See* satellite.

hill fog Persistent low cloud, associated with low atmospheric pressure. It does not affect low-lying regions. *See also* fog.

hill shading (plastic shading) A means of showing relief on maps, shading the east- and south-facing slopes (the steeper the slope, the darker the shading) to give the effect of an oblique light shining from the northwest over a relief model.

hinge line **1.** (in folding) An imaginary line joining points of maximum curvature of the folded strata.
2. The boundary between a stable region of the Earth's crust and one undergoing changes in elevation, generally as a result of isostatic readjustment but also including areas of tectonic deformation.

hinge fault (pivot fault) A normal fault that dies out along its trend by a gradual decrease in throw. (See diagram at FAULT.)

histosol One of the ten soil orders of the SEVENTH APPROXIMATION classification, which includes bog soils such as peat. Histosols are characterized by accumulations of organic matter, which remains more or less undecomposed because of the waterlogged conditions. The histic EPIPEDON (surface horizon) has at least 20% of its weight in organic matter, or over 30% if half the weight of the horizon is clay. Histosols are divided into four suborders on the basis of the degree of decomposition of the organic matter.

Although found throughout the world, histosols occupy the smallest area of all the ten orders of the Seventh Approximation (under 1% of all soils). They tend to develop particularly in cooler, poorly drained, more humid areas.

hoar frost A silvery-white deposit of ice crystals formed on surfaces cooled below freezing point by radiation. It may result from the freezing of dew or by the direct sublimation of ice crystals from water vapor in the atmosphere when temperatures are less than 0°C.

hodograph A diagram used to analyze the wind field above an observing station. Winds at different pressure levels are drawn as vectors from the origin and so give an indication of the changes in wind speed and direction with height. It is then possible to make deductions about the thermal patterns at these upper levels and about the advection of warm or cold air. (See diagram overleaf.)

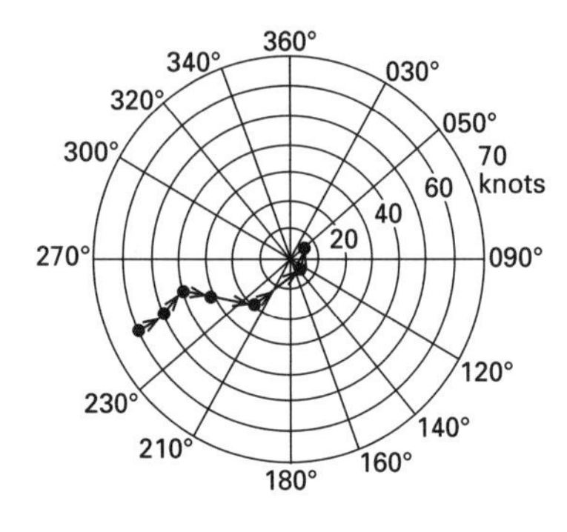

Hodograph

hogback A steep-sided narrow ridge that stands above the surrounding plain. It generally results from the folding or faulting of strata, followed by differential erosion that removes softer rocks from alongside it.

Holocene (Recent) The present epoch, covering the last 10 000 years or so of geologic time from the end of the PLEISTOCENE. In Britain it follows all the major Pleistocene glacial episodes and is therefore often known as the *Post-Glacial.* However, some authorities consider it to be no more than an interglacial phase of the Pleistocene. At the beginning of the Holocene, Britain was isolated from the rest of the continent of Europe by a general rise in sea level. Descendants of most fossil species of this epoch are still alive today.

holocrystalline Describing a rock composed wholly of crystals.

holozoic Describing an organism that feeds on complex organic matter (as opposed to taking in simple soluble materials). The term thus includes most animals.

homeomorphism The existence of crystals that have the same form and habit but different chemical composition. *See* crystal form; crystal habit.

homeomorphy Similarity in structure between organisms that are not closely related as a result of evolutionary adaptation to similar habitats (*see* convergent evolution). The term is often restricted to similarities between more closely related species, in which homeomorphy leads to confusions in taxonomy (*see* parallel evolution).

homocline A series of beds of rock all of which have a similar dip and strike, including monoclinal structures and isoclinally folded beds. The term is applied only to small areas of the Earth's crust.

homogeneous strain The changes in shape and dimensions that occur when a body of rock deforms equally in all directions. It results in straight-line and planar structures.

homolographic projection A map projection possessing the property of equal area, i.e. the area bounded by adjoining meridians and parallels on the map is equal in area, by ratio, to that area bounded by the same meridians and parallels on the ground, e.g. MOLLWEIDE'S PROJECTION and BONNE'S PROJECTION.

homotaxis The occurrence of divisions of rock occupying the same position in the stratigraphic sequence in separate successions. Such divisions are said to be *homotaxial.* Although the divisions may be of the same age, the concept does not take account of considerations of time and simply reflects the order of deposition. *Compare* chronotaxis. *See also* correlation.

honeycomb weathering Weathering within jointed rocks in which joint infillings are more resistant to weathering than the main mass of the rock, so that after erosion they project and surround recesses in the rock. Typical infilling materials are iron oxides, secondary silica, calcite, and manganese oxides. Migration of rock cements into the joints may produce a corresponding decrease in the resistance of the rock itself, thereby increasing the effectiveness of honeycomb weathering.

Hooke's law A stressed body deforms to an extent that is proportional to the force applied. Materials that obey Hooke's law are said to be elastic.

horizon **1.** (in soil science) One of a number of layers of soil arranged in a vertical sequence in the PROFILE. Each layer is reasonably uniform. Most soil-forming processes result in this layered arrangement but some, e.g. the action of earthworms, are destructive in this respect.
2. (in stratigraphy) A plane within a rock sequence that is assumed to be a time plane. In practice, a horizon is not a sharp plane but a thin bed characterized by some distinctive feature of lithology or fossil content. A particular horizon is sometimes merely theoretical, and it may be regarded as extending through a contemporaneous succession that carries no internal evidence of its presence.

horizontal equivalent The distance between two points on a slope projected onto a horizontal plane, as they would be represented on a map.

hornblende A monoclinic mineral of the AMPHIBOLE group, common in both igneous and metamorphic rocks.

hornfels A fine- to medium-grained nonfoliated rock composed of equidimensional mineral grains showing no preferred orientation. Porphyroblasts of minerals such as biotite, andalusite, and cordierite may occur (*see* porphyroblastic) and their incipient development is often indicated by a spotted appearance to the rock. Hornfels are the product of contact metamorphism and are found in aureoles bordering intrusive plutonic igneous rocks. *See also* contact metamorphism.

horn (horn peak) *See* pyramidal peak.

hornito A small volcanic structure resulting from the accumulation of small blebs of lava thrown out around a parasitic vent; a spatter cone.

hornstone A very fine-grained volcanic ash. *See* pyroclastic rock.

horse latitudes The zone of light and variable winds between 30° and 40° N and S where the subtropical anticyclones are dominant. The name arose during the days of sailing ships when unfavorable winds could prolong the journey; when food became short, horses would be thrown overboard or eaten.

horst An elongated block of rocks bounded by faults. The sense of movement of the block is upward relative to the surrounding rocks. *Compare* graben. (See diagram at FAULT.)

hot spot An area on the Earth's surface that has a higher than average heat flow, e.g. a volcano. Hot spots are also thought to develop within the mantle, giving rise to MANTLE PLUMES. These plumes are thought to be important features in the mechanism of plate tectonics.

hot spring *See* thermal spring.

Hudson canyon One of numerous submarine canyons that cut into or across the wide continental shelf lying off the E coast of the USA. It is one of the best-known submarine canyons in the world and has been carefully surveyed several times. It cuts well into the shelf opposite the Hudson River and has been traced seaward as far as a wide submarine fan. The cross section of the canyon tends to be V-shaped, and flanking it are what appear to be small tributary valleys. Core samples have yielded a variety of deposits (rock, gravel, and clay), certain of which suggest that some sort of powerful turbidity flow may have operated, perhaps when sea level was significantly lower.

human influence on geomorphology In the 8000 years of the Holocene geologic period during which humans have become established, their activities have had considerable geomorphological impact, principally in accelerating existing processes but to a lesser extent in creating new landforms. The major effects are those consequent on destruction and removal of the vegetation cover. Clearing the land for agriculture can cause gullying and soil erosion, leading to destruction of valuable farmland, and wind erosion of bare soil can create dust bowls (*see also* badlands). In coastal areas, human use of sand dunes leads to vegetation destruction and creation of BLOWOUTS. River basins can have higher sediment yields than those not so much used, owing to the increased erosion from farming, etc. Brown (1970) found erosion rates could locally increase a hundredfold because of human influence.

Landforms created by human action include lakes resulting from subsidence of the land surface in mining areas, the Norfolk Broads, created by peat-cutting in medieval times, and reclaimed land from the sea, severely changing the coastline, as with the Zuider Zee in the Netherlands. Spoil heaps, railroad cuttings, and the blowing out of hillsides to build towns are other deliberate results. Sometimes human action can produce quite unexpected results, especially in the coastal zone where the erection of groins, harbor arms, and other artificial structures severs the littoral drift leading to the decay or growth of beaches in quite unnatural situations. Large-scale recharge schemes may be used to overcome these problems, with convoys of trucks shuttling beach materials that can no longer be moved by natural agencies. *See also* arroyo.

Humboldt Current (Peru Current) A current of cold ocean water that flows from the Antarctic northward along the western coast of South America. It veers westward as it reaches the Equator. It cools winds blowing onto the coast from the Pacific Ocean, reducing rainfall on the western slopes of the Andes. *See also* El Niño.

humic acid An acid derived from resynthesized organic materials that dominate the soil humus along with the fulvic acids. Humic acids are polysaccharides and contain more carbon and nitrogen and less oxygen than fulvic acids. Fulvic acids tend to be more important in newly formed humus but eventually they are overtaken by the humic acids.

humidity The amount of water vapor in the atmosphere. It can be measured or calculated in a variety of ways. The most frequently used measure is the *relative humidity*, which is the mass of water vapor in a given volume of air expressed as a percentage of the mass of water vapor in an equal volume of saturated air at the same temperature and pressure. As its name implies, this is only a relative index and it varies in opposite phase to temperature, even though the absolute amount of moisture in the air may not change. A better system from the meteorological point of view is the use of VAPOR PRESSURE, which indicates the water vapor content of the atmosphere irrespective of temperature. The *absolute humidity* is the mass of water vapor per unit volume of air, given in grams per cubic meter.

humification The formation of HUMUS from organic matter.

humite A member of a group of orthorhombic and monoclinic minerals consisting of layers of Mg_2SiO_4, which have an olivine structure, alternating with layers of composition $Mg(OH,F)_2$. Humite minerals are found in metamorphosed limestones.

hummock A mound or knoll that rises above a generally level surface, or a small mound of soil or turf in alpine areas (*see* earth hummock).

humus Amorphous colloidal material in the soil, dark in color and composed of resistant plant tissues, such as lignin, and new compounds, such as polysaccharides, synthesized by microorganisms. It is important to the soil both physically and chemically (its cation exchange capacity far exceeds that of the clays). *See also* moder humus; mor humus; mull humus.

hurricane A TROPICAL CYCLONE with surface wind speeds in excess of 117 km per hour (64 knots), found in the Caribbean and W Atlantic, the E Pacific, and the S Indian Ocean. They are identical to similar cyclones in the W Pacific, N Indian Ocean, and near Australia, which have their own local names. Hurricanes are scaled according to their strength ranging from category 1 (minimal damage) up to category 5 (catastrophic). At the center of a hurricane is an area of light winds and higher temperature known as the EYE, about which clouds and rain bands spiral with the associated winds.

The term is also used in the BEAUFORT SCALE for wind speeds above 117 km per hour (64 knots), whether or not they are linked to a tropical cyclone. Such speeds rarely occur over land.

Huygen's principle A principle stating that each wave front acts as a source of secondary wavelets.

hyaline Describing a rock that is glassy.

hyaloclastic rock *See* palagonite.

hyalocrystalline Describing a rock composed of both crystals and glass.

hybridization *See* assimilation.

hydrate **1.** A chemical compound that has water in its composition (as *water of crystallization*).
2. A chemical compound produced by HYDRATION.

hydration The chemical addition of water to a substance. If the substance is a mineral, it usually involves a fairly considerable expansion in the mineral grains. This may be important in producing subsequent mechanical weathering, in the form of EXFOLIATION and GRANULAR DISINTEGRATION. Minerals affected by hydration are prepared for further chemical weathering processes, such as OXIDATION and CARBONATION, while the process is also very important in the formation of CLAY MINERALS.

hydraulic action (hydraulicking) (in geomorphology) The removal of loose, incoherent, or weathered material by flowing water, assisting in stream-bed and bank erosion. Material carried can range from single minerals or aggregates of crystals to pieces of layered bed strata, e.g. slate or shale, if the materials dip downstream. *See* abrasion.

hydraulic jump A sudden increase in mean water level in the direction of flow, which is of a rapidly varying type and leads to a form of stationary wave. In the simplest terms, a surface flow at high velocity confines a deeper flow, thereby resulting in the dissipation of energy through turbulence.

hydrographic chart A map of the seabed, showing depths of water, heights of underwater features, and, sometimes, geologic information. *See* hydrography (def. 2).

hydrography **1.** The study of the oceans, seas, rivers, and other water bodies, together with the strips of land bordering these. It involves description and measurement and the subsequent presentation of this information on hydrographic charts.
2. The shape of the sea floor and the deposits of which it is composed, including navigational information.

hydrolith A type of rock that was formed by chemical precipitation from water, such as gypsum or halite.

hydrologic cycle (water cycle) The continuous circulation of water between the oceans, atmosphere, and land. Water evaporates from the oceans (and lakes and rivers) as water vapor in the atmosphere, where it may condense into clouds. Clouds release precipitation (rain, snow, or hail), which falls on the land. Some evaporates, some is taken up by plants and released into the atmosphere when they transpire, and some runs off to form streams and rivers that flow to the oceans. Any that penetrates the ground forms GROUNDWATER.

hydrolysis A chemical reaction involving water. In geology, it is a form of chemical weathering involving a reaction between the H and OH ions formed by the decomposition of water and the ions of rock minerals; it occurs wherever rocks and water are in contact. Hydrolysis is a particularly important process in the weathering of FELDSPARS; when water comes into contact with orthoclase feldspar, potassium hydroxide and alumino-silicic acid are produced. These react with atmospheric carbon dioxide giving potassium carbonate and water, while the acid breaks down into clay minerals and easily removable soluble colloidal silica. Kaolin (china clay) is produced by weathering in this way.

hydromorphic soil A type of soil found in a wide range of environments where drainage is poor, e.g. alluvial flats, and therefore they are classified as intrazonal. Reduction is a more dominant process than oxidation, resulting in a blue/gray subsoil, with red/brown mottles. The most important soil of this group is gley but also included are peaty gley podzols, peaty gleys, meadow soils, planosols, groundwater podzols, and groundwater latosols. With increasingly poor drainage these soils grade into the organic soils.

hydrosphere All the waters of the Earth, as opposed to gases (atmosphere), rocks (lithosphere) and living organisms (biosphere). In all, water covers about 74% of the Earth's surface and has a total mass of 10^{21} kg.

hydrostatic equation An equation relating the variation of pressure with height to the density of air and the force of gravity: $dp/dz = -\rho g$, where dp/dz is the rate of change of pressure with height, ρ is the density of air, and g is the acceleration of free fall. It is a useful approximation when considering horizontal movement relative to the Earth and is used as one of the basic equations in models of the GENERAL CIRCULATION OF THE ATMOSPHERE.

hydrostatic pressure **1.** Pressure that is exerted by a liquid, often water. In a body of water, the pressure increases with depth. **2.** (in geology) The pressure exerted on rock in the Earth's crust by the rocks that overlie it. It is also termed confining pressure.

hydrothermal process Following the pegmatitic stage during the crystallization of an igneous melt is the hydrothermal stage, when the residual fluid is a relatively low-temperature aqueous solution. Such a fluid may effect considerable alteration of the crystallized portion of the magma by such processes as saussuritization, kaolinization, and the alteration of mafic minerals to serpentine and chlorite.

Many mineral deposits are formed by precipitation from hydrothermal solutions. A characteristic hydrothermal deposit is the sulfide-bearing vein filling rock fissures. Hydrothermal deposits may be divided into three groups: hypothermal deposits formed at temperatures of 300–500°C at considerable depth and including cassiterite, wolfram, and molybdenite veins; mesothermal deposits including sulfides of iron, lead, copper, and zinc and formed at temperatures of 200–300°C; epithermal deposits formed at the low temperatures of 50–200°C and including stibnite, cinnabar, sulfur, silver, and gold. *See also* pneumatolysis.

Hydrozoa A class of the phylum CNIDARIA. As few hydrozoans possess hard parts they are not common as fossils, although the Paleozoic STROMATOPOROIDS may belong to this class.

hyetograph A type of rain gauge that records the amount and duration of rain as it falls. Rain is collected in a cylinder with a float on the water surface. As the water level rises during precipitation, the float rises in proportion to the rate of rainfall and this is recorded by a pen on a rotating chart.

hygrometer An instrument that measures HUMIDITY (the water vapor content of the air).

hygroscopic nuclei Atmospheric particles that have an attraction for water vapor and help in the initiation of precipitation processes.

hypabyssal rock *See* plutonic rock.

hypersthene An orthorhombic PYROXENE.

hypervelocity impact The impact of an object (such as a METEORITE) that is moving so fast that the rocks it hits are not strong enough to withstand the shock waves created. The result is generally a large crater, surrounded by IMPACT BRECCIA and other EJECTA.

hypidiomorphic Describing a rock consisting of EUHEDRAL and ANHEDRAL crystals or in which the majority of crystals are SUBHEDRAL. *Compare* allotriomorphic; idiomorphic.

hypolimnion The cool lower layer of water in a lake or shallow sea. Insufficient light penetrates for photosynthesis to take place, so there are no green plants and there is little dissolved oxygen in the water. *See also* epilimnion.

hypothermal *See* hydrothermal process.

hypsographic curve A curve showing the proportion of a landscape lying at, above, or below particular elevations. The percentage of the landscape between successive contours is calculated from a relief map, and the results expressed either as simple proportions between different heights, or as a cumulative measure showing the percentage of area lying above each level, starting with 100% at sea level and working up to nil at the absolute peaks. It has been used to look for development of erosion surfaces, e.g. terraces and former peneplains, which appear on the curve as extensive areas at their particular elevations. It is also used to display the proportions of the Earth's surface at particular elevations above and depressions below sea level.

hypsometer An instrument for determining atmospheric pressure or altitude by measuring the boiling point of water. The air pressure can be determined directly from the temperature at which boiling occurs, but altitude must then be obtained from the altimeter equation. Accuracy is not very high: the boiling point must be known to within 0.01 K to obtain the height to within 3 meters.

hypsometric tinting The differential coloring of elevation bands on a map to enable the user to determine quickly the higher and lower areas of the land portrayed. Greens usually portray lowland areas; as the land rises so the colors change from yellows to browns to purples. This technique is common in small-scale maps, e.g. atlases and wall maps.

I

ice The solid state of water. It melts at 0°C, requiring approximately 340 000 joules per kilogram (80 calories per gram) of ice (latent heat of fusion) or 2.8×10^6 J/kg (677 cal/g) if converted directly from ice to vapor (latent heat of sublimation). Pure water does not necessarily freeze to ice when the temperature falls below 0°C, especially if it is in the form of small droplets such as those found within clouds. This has very important implications in the BERGERON-FINDEISEN THEORY of precipitation formation. Even when FREEZING NUCLEI are present, droplets may still remain in the liquid phase to temperatures of –30°C or less. By –40°C the water will freeze spontaneously irrespective of the presence or absence of nuclei.

At the ground surface, ice can take many forms: rime, hail, black ice, and, on a vast scale, glaciers. Because ice reflects solar radiation, and requires heat for melting, it is a very important aspect of the heat balance in high latitudes.

Ice Age A period in the Earth's history when ice spread toward the Equator accompanied by a general lowering of surface temperatures, especially in temperate latitudes. The PLEISTOCENE period, ending from about 10 000 years ago, experienced at least four major ice advances (*see* glacial maximum), with the margin reaching about 52°N over NW Europe and about 45°N in NE America. With this change in location of ice surfaces the whole atmospheric circulation altered, the main climatic belts being compressed and pushed toward the Equator. At present we appear to be in an interglacial circulation but it is assumed that further ice ages will be experienced in the future; as the true cause or causes of ice ages are not known it is impossible to predict when this will be. It is possible that changes in landmass altitude and variations in solar radiation are at least partly responsible.

From geologic records, it has been deduced that ice ages also existed in Permo-Carboniferous times about 250 million years ago, one in late Precambrian times some 500 million years ago, and possibly several earlier. These former ice-age deposits have even been found in present tropical latitudes, because the continents have since changed their relative positions.

Between 1550 and 1850 temperatures in much of the N hemisphere fell to their lowest since the last ice age, and this period has been called the *Little Ice Age*. Alpine glaciers advanced and settlement in many northern areas, such as Greenland, Iceland, and N Norway, had to be abandoned.

iceberg A mass of ice in the sea that has moved from a land area to the coast, and from there into the sea, and which exceeds 5 m in height. These ice masses, many of which result from the calving off of fragments of glacier ice, float partly above but largely beneath the surface of the sea. They vary greatly in shape one from the other: some are dome-shaped, some are sloping or pinnacled, while others have a more tabular form. Glacier bergs display very irregular shapes.

ice cap A permanent mass of ice such as might cover a highland area; a small-scale ICE SHEET.

icefall A section of a glacier at which the ice moves down a steep slope. Many transverse crevasses form across the glacier, and there may be piled up SÉRACS at the foot of the icefall.

ice floe An area of sea ice that measures 20 m or more across and is essentially flat. Small floes are 20 to 100 m across, big floes are 500 to 2000 m across, and giant floes are over 10 km.

ice fog When visibility falls to less than 1 km as a result of ice crystals, an ice fog is said to exist. Such fogs have increased in frequency in recent years as human activities extend to colder latitudes. When the air is very cold, only a little water is required before saturation takes place. If the temperature is below –40°C the ice crystals will form directly on saturation. Normally little evaporation takes place, but the release of water vapor from motor-vehicle exhausts or power station cooling systems can quickly saturate calm air, as often happens in the Fairbanks area of Alaska.

Icelandic Low The mean low pressure system of the North Atlantic. Depressions form in the latitude of Newfoundland, move northeastward as they intensify, then gradually decay. Although their tracks vary quite considerably, on average they are most frequent and at their most intense in the Iceland area. As the circulation of the atmosphere varies with time, the precise location of the Icelandic Low changes.

Iceland spar A transparent variety of CALCITE of optical quality. *See also* carbonate minerals.

ice-rafting The transport of debris in sea ice. Widespread at high latitudes are sediment deposits that have been rafted by ice, either icebergs or ice floes. They reached the ice from meltwater streams or by being frozen into the base of ice that later became waterborne. These materials range from very large boulders to clays and silts. They may be transported over considerable distances to be deposited among material of an entirely different character. Icebergs and floes reach nearer to the Equator in the N hemisphere than they do in the S hemisphere. Ice-rafted pebbles and boulders seem to be capable of making pits or dents in soft sea-floor deposits, presumably as they fall to the sea floor, as was shown by underwater photography in the Arctic.

ice sheet An accumulation of ice hundreds of meters thick and covering vast areas. Many developed during the PLEISTOCENE period, and are represented by the present Greenland and Antarctic ice sheets. The ice develops at high altitudes, spreading outward and submerging all the underlying topography. Being unrestricted by rock walls, unlike CIRQUE and VALLEY GLACIERS, the directions of ice movement within these ice sheets reflect the slopes of the ice surface, which depend upon relative accumulation rates within the various parts of the sheet.

ice shelf An area of very thick ice floating on the sea, as around much of Antarctica. It is attached to the land and built up by accumulated snow and outward moving glaciers.

ichor *See* granitization.

Ichthyosauria An order of extinct reptiles that became secondarily adapted to a marine life. In shape they were similar to dolphins (mammals): they had fishlike tails and the limbs were modified as paddles. Ichthyosaurs were totally aquatic and the young were probably born alive. They were prominent throughout the Mesozoic, especially in the Jurassic period, becoming extinct at the end of the Cretaceous. *Compare* Plesiosauria.

iddingsite The red-brown alteration products of forsteritic olivine consisting largely of iron oxides and clay minerals.

idioblastic Describing a metamorphic rock consisting largely of crystals having euhedral form.

idiomorphic Describing a rock in which the majority of crystals are anhedral. *Compare* allotriomorphic; hypidiomorphic.

idocrase (vesuvianite) A yellow, green, or brown tetragonal mineral with an ideal formula $Ca_{10}(Mg,Fe)_2Al_4Si_9O_{34}(OH,F)_4$, found in thermally metamorphosed limestones.

igneous rock Any of the major rocks that have crystallized from a high-temperature molten silicate (or rarely carbonate) liquid or magma. Igneous rocks include PLUTONIC, hypabyssal, VOLCANIC, and PYROCLASTIC ROCKS. The compositional range of igneous rocks is described by the terms acid, intermediate, basic, and ultrabasic. Classification may also be based upon the SILICA SATURATION principle. *See* andesite; basalt; carbonatite; diorite; gabbro; granite; ijolite; lamprophyre; layered igneous rock; rhyolite; syenite; syenodiorite; trachyte.

ignimbrite *See* pyroclastic rock.

ijolite A suite of strongly alkaline undersaturated plutonic rocks often associated with carbonatites. They are characterized by varying proportions of nepheline and clinopyroxene, an abundance of accessory minerals, and the virtual absence of feldspar. Members of the suite are classified in terms of COLOR INDEX as follows:

urtite <30 (% dark minerals)
ijolite 30–70
melteigite 70–90
pyroxenite >90

Thus an ijolite may consist of about equal proportions of nepheline and aegirine-augite. Accessory minerals include apatite, melanite, phlogopite, sphene, and calcite. *Uncompahgrite* is a melitite pyroxenite and *jacupirangite* consists largely of titanium-rich augite, magnetite, and biotite.

illite A micalike clay mineral, a hydrated silicate of aluminum and potassium. It occurs in mudstone and shale. *See also* clay minerals.

illuviation The accumulation of fine material in a soil, or a part of a soil, washed down from above. The *illuvial horizon* is the B horizon: the degree of development of the B is therefore a guide to the time and effectiveness of operation of the illuvial processes. The materials involved are sesquioxides, bases, clays, and colloidal organic matter.

ilmenite A brown or black mineral form of iron titanium oxide, $FeTiO_3$. It crystallizes in the trigonal system, and occurs as an accessory mineral as aggregates in igneous and metamorphic rocks, as well as in some mineral sands. It is a major source of titanium.

imbricate structure A series of thrust sheets all dipping in the same direction and displaced by roughly the same amount.

impact basin A large IMPACT CRATER with a diameter in excess of 200 km. It is formed when a meteorite strikes the ground. There are usually terraced walls, a sunken floor, and perhaps a hill at the center. The best examples can be seen on the Moon and on Mercury.

impact breccia Fragments of rock produced when a meteorite hits the ground. Small pieces, including droplets of impact melt, may fall back into the crater. Larger fragments are mixed with hot gases and flow out across the surface of the ground, or are hurled into the air as EJECTA. The preexisting rocks around the crater may exhibit SHATTER CONES.

impact crater A crater formed when a meteorite hits the ground. *See* impact basin.

impermeable Describing a type of rock that is nonporous and therefore does not absorb water. Igneous rocks such as granite are good examples (although joints in the rock may allow water through, making it pervious). Clay is impermeable after it has become saturated with water. *See also* impervious.

impervious Describing a type of rock that, because it contains no cracks or fissures, does not allow water to pass through it. Metamorphic rocks such as shale and slate are good examples.

inceptisol One of the ten soil orders of the SEVENTH APPROXIMATION, covering soils that are better developed than entisols but not so advanced in formation as the alfisols, i.e. there is no horizon of iron and aluminum accumulation. They are young soils in which the horizons present have developed quickly. They are equivalent to the humic gley soils, acid brown soils, and andosols of the old classification system.

incised meander A meander of a stream that is cut deeply into bedrock. This incision could arise from original development, because meanders can develop before a stream has completed downcutting (*see* ingrown meander). Alternatively, incision may result from REJUVENATION of typical floodplain meanders causing them to cut down and incise themselves into their floodplain (*see* entrenched meander).

inclination *See* dip (defs. 1 and 2).

inclusion The occurrence of one substance enclosed within another, e.g. a preexisting rock fragment caught up in a later rock (xenolith) or crystals of one mineral contained in another. Small volumes of volatile constituents trapped during the growth of crystals are termed *fluid inclusions*. Inclusions of gabbro and peridotite within alkali basalt are often called *nodules*.

incompetent Describing beds of rock that deform plastically and flow when stressed during folding, or folds in which the thickness of the beds changes during folding. *See also* competence.

index contour A contour line that is accentuated by its width, in order to facilitate the reading of elevations on a map. For example, where the contour interval is 10 m the 50 m and 100 m contours may be accentuated.

index cycle (in meteorology) The ZONAL INDEX is believed to undergo a cyclic variation from a high-index circulation, with its strong zonal westerlies, to the low-index state, with much meridional transfer, then back to the high-index phase, with a wavelength of about 28 days. However, further work has cast doubt on the validity of a true cycle.

index fossil *See* zone fossil.

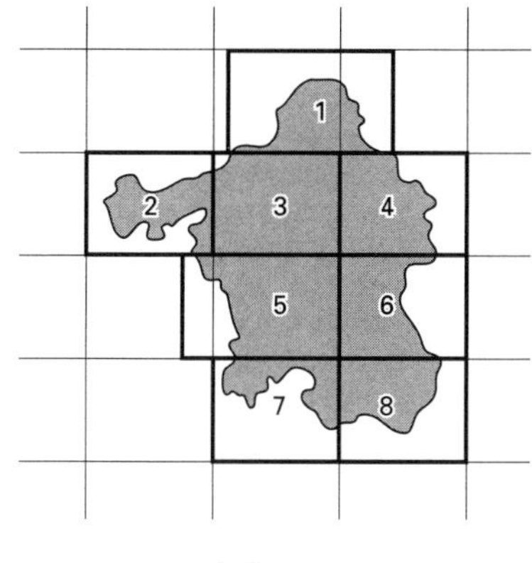

Index map

index map Usually a small-scale map forming a base upon which certain information concerning larger scale maps, or other data, may be portrayed for easy reference. In series mapping the individual sheet lines are drawn on a small-scale outline map and such information as sheet numbers, dates of publication, and edition numbers are shown for each sheet. This enables the user to find out immediately which sheets cover any part of the area.

index mineral *See* zone.

Indian clinometer A surveying instrument used in PLANE TABLING for determining the heights of points fixed by intersection or radiation methods. The height of the plane table and of the point over which it is set up must be known. The clinometer is placed on the plane table and an adjustable sight is used to view the points for which heights are required. This adjustable sight indicates a tangent value, positive for an elevation, negative for a depression, which is multiplied by the distance between the points, measured from the table, to find the height difference.

Indian summer Any period of warm settled weather in October or early November in the N hemisphere. The term was first used in New England when under such weather conditions the Native Americans made preparations for the coming winter by storing food and repairing tents. Most years do have at least one spell that could be called an Indian Summer but it has no precise meaning.

induration The hardening of porous rocks or of soils through the deposition of minerals, which act as a cement, on or within the surface layers. Chemical weathering usually causes the weakening of rocks through the removal of elements in solution. Where strong evaporation occurs, as for instance in the semiarid regions, these elements are brought to the surface by capillarity, where they combine to cement the soil or weathered rock. There are three main types of induration rock forms. The cement in calcrete consists mainly of calcite ($CaCO_3$), that in ferricrete is of hematite (Fe_2O_3), while silcrete is cemented by silica (SiO_2).

infiltration The movement of water downward into the ground. It is affected by many factors: the amount of vegetation cover, the compactness of the surface layers, the porosity of the underlying rock, and the amount of rainfall. The presence of water already in the ground also affects the rate of infiltration.

infrared satellite imagery *See* satellite.

infrastructure A structure produced deep within the Earth's crust under conditions of high temperature and pressure. It is characterized by plastic folding, migmatites, and granites. Overlying this area are the less highly deformed rocks of the superstructure.

ingrown meander The type of INCISED MEANDER that develops from origin. If a winding course develops in a stream before

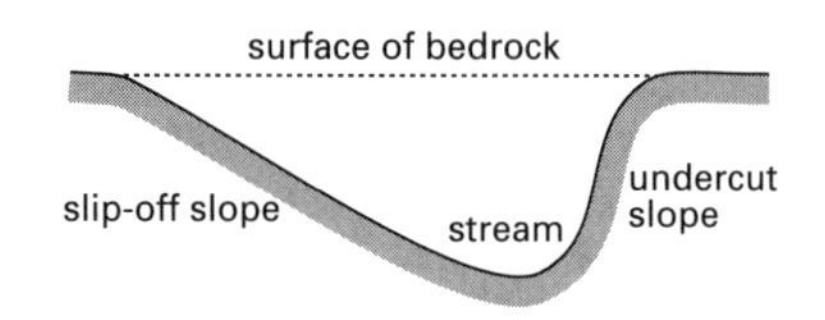

Cross-section through an ingrown meander

it has completed downcutting, it will meander and cut down simultaneously, lowering its course into bedrock. As with floodplain meanders, lateral erosion on the outside of meander bends, complemented by deposition on the inside of bends, produces an asymmetric cross profile of the valley and downstream migration of the meanders. *Compare* entrenched meander.

inland basin (bolson, playa) A closed depression within an upland desert region toward which material is moved whenever rain falls in sufficient amounts. The material is moved both down wadis and as sheet wash on broad slopes. The finest grains and material in solution are moved farthest, so that there is a gradation from vast accumulations of unsorted material at the basin margin, through sand and loam, to clay and salts in the center. Many of the mineral grains also exhibit salt or calcareous coatings. The basins have a very flat surface when dry, although extensive cracking may be produced by the heat. Salt lakes or salt marshes may form during periods of rain.

inland sea A category of ADJACENT SEA (*compare* marginal sea) consisting of an extensive water body lying in the interior of a landmass, often remote from the ocean. Inland seas may or may not have access to the open coast; where they do have access it is usually restricted to a single or a few narrow sea passages. Inland seas display a large depth range. Those with depths generally less than 250 m are called *shallow inland* (or *epeiric*) *seas*, e.g. the Baltic Sea. Far greater depths are encountered in the deep inland seas, for example the Mediterranean, where depths range from 2000 to 5000 m. Very large lakes have some of the characteristics of inland seas. All inland seas and lakes experience wave action, SEICHE movements, and wind-induced flow. Waves on a large lake such as Lake Kariba can be more than 3 m high, and very much larger waves affect some inland sea areas. The shores display many of the morphological forms typical of sea coasts in general.

inlier An area of exposed rock that is surrounded by stratigraphically younger strata, usually a result of folding, faulting, and erosion (such as the erosion of the crest of an anticline). *Compare* outlier.

inner core The Earth's innermost shell, believed on geophysical evidence to consist of solid iron and nickel and separated from the OUTER CORE at a depth of 5000 km. It has a diameter of 2600 km.

Insecta The largest class of the phylum ARTHROPODA, comprising the only invertebrate animals adapted for flight. They are air-breathing, have three pairs of legs in the adult, and all but the most primitive have one or two pairs of wings. This highly successful and diverse group occupies an enormous range of habitats. Many fossil species are known, despite the rarity of suitable environments for preservation. Fossil insects have been found in the Devonian, and some Carboniferous dragonflies attained a wing span of nearly 80 cm.

Insectivora The order of the class MAMMALIA that includes the shrews, hedgehogs, and moles. Modern insectivores are small and usually very unspecialized in their anatomy and feeding habits. In these primitive features they resemble the first Mesozoic placental mammals, which are generally regarded as early insectivores and the ancestors of most modern placental groups.

inselberg (bornhardt) A large domelike residual hill of hard rock, characterized by very steep slopes rising abruptly from the surrounding surface. Inselbergs were formerly thought to be semiarid landforms, but it is now widely believed that they are more typical of humid tropical climates. The main reason for their occurrence is differential DEEP WEATHERING followed by subsequent EXHUMATION, which exposes the more massive, less weathered parts of the weathering profile as inselbergs. Some large forms extend to a height of more than 300 m and are possibly formed by successive stages of weathering, which proceed at the same time as REGOLITH removal. As an inselberg is exposed at the surface, UNLOADING joints usually develop, and movement of rock slabs along such partings can bring about the destruction of these domed forms.

insequent stream A stream that has a random course, which cannot be predicted by reference to a slope, rock type, or faults or depressions in the surface. As a result, it adopts a DENDRITIC DRAINAGE pattern.

insolation (from *in*coming *sol*ar radi*ation*) The intensity of either direct or global (direct and diffuse) solar radiation on a unit area at a specified time on a specified surface. Its value is dependent upon the SOLAR CONSTANT, the time of year, the latitude of the receiving surface, the slope and aspect, if any, of the surface, and the transparency of the atmosphere.

insolation weathering Weathering resulting from the expansion and contraction produced by temperature changes. It was formerly believed that in arid areas practically all rock fragmentation was caused by the extreme variations in temperature and consequent differential expansion between the outside and inside of blocks, which set up stresses producing cracks. However, experimental work has suggested that the process was extremely overrated, although there are many examples of fragmented rocks for which this process seems to be the best explanation. In addition to large-scale cracking, GRANULAR DISINTEGRATION may be produced by the differential expansion and contraction of several mineral types in one rock.

instability The state of the atmosphere when thermals are able to rise freely by their own buoyancy forces. This occurs when the ENVIRONMENTAL LAPSE RATE is steeper than the DRY ADIABATIC LAPSE RATE and is known as *absolute instability*. This normally is found only near the ground surface. Other states of instability depend on the relative rates of cooling of the environment and dry and saturated thermals. Once condensation has taken place and the atmosphere remains unstable, cloud development will be extensive and heavy rainfall probable.

insular shelf The shelf zone surrounding a large island. The shelf extends from low-tide level out to the shelf-edge zone where, as with all shelves, there occurs a marked and steep descent to the deep-sea floor. Insular shelves, like continental shelves, may be broad or narrow, but they tend to contrast with the zone surrounding coral reefs and oceanic islands, for in these cases the sea floor plunges steeply to the deep-sea floor from close to the shore. The shelf edge of insular shelves usually lies at a depth of some 200 m.

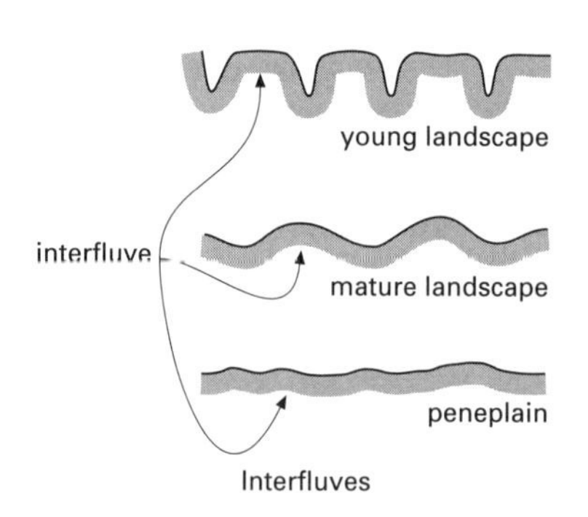

Interfluves

interfluve The higher ground between one river and the next in the same drainage system: an upstanding area not yet subject to fluvial denudation. Interfluves become lowered and narrowed as denudation proceeds, until at the stage of peneplanation they are almost insignificant. *See also* divide.

interglacial A period of warmer climatic conditions separating GLACIAL PHASES of the PLEISTOCENE. Temperatures were probably similar to those of today, or perhaps a little warmer. They are believed to have lasted longer than the glacials and very much longer than the 10 000 years or so since the last glacial phase.

intergranular Describing a texture common in basic lavas in which the spaces between plagioclase crystals are occupied by relatively small grains of pyroxene and olivine. *Compare* ophitic.

intergrowth An interlocking growth of two minerals that crystallize simultaneously from a melt. Intergrowths may also result from the EXSOLUTION of a homogeneous mineral to form two discrete phases, one included in the other. *See* graphic intergrowth; myrmekite; perthite; symplectite.

interlocking spur One of a series of spurs on alternate sides of a river valley. They are strips of high ground that remain after a meandering river has eroded away the ground between them, and generally occur mainly in the upper and middle reaches of the river.

intermediate contour A contour line that falls between the INDEX CONTOURS.

intermediate rock Strictly, a type of rock containing 55–66% silica (by weight); in current usage intermediate rocks contain less than 10% free quartz, sodic plagioclase, and/or alkali feldspar. Typical intermediate rocks are syenite, diorite, trachyte, and phonolite. *See also* acid rock; basic rock; ultrabasic rock.

intermittent stream A stream that flows at only certain times of the year, as after heavy rain or when it receives water from springs. Such streams dry up in the dry season, usually because the water table becomes lower.

intermontane A basin of deposition enclosed by mountains.

internal magnetic field That portion of the Earth's magnetic field that originates within the Earth itself.

internal wave (boundary wave) A wave that occurs within a water mass, which may be stratified or in which density may vary with depth. Internal waves have quite different characteristics from those of waves that develop at the sea's surface. The density variations sometimes occur abruptly, at sharp surfaces of discontinuity called *interfaces*, and sometimes quite gradually. The amplitude of internal waves often far exceeds that of waves on the sea's surface; for example, one measurement recorded in the Atlantic Ocean amounted to some 60 m.

International Date Line A line corresponding to the 180° meridian, deviating to some extent around island groups. The date immediately to the east of the line is one day earlier than to the west, because 180°E is 12 hours ahead of GREENWICH MEAN TIME and 180°W is 12 hours behind.

International Map of the World projection A modified polyconic map projection having two standard parallels and a scale of exactly 1:1 000 000. Straight lines joining points on the truly divided top and bottom parallels of the projection represent the geographic meridians. Arcs of the same circle, rather than concentric circles, form the parallels.

intersection A type of PLANE TABLING in which points of detail are fixed by setting up the plane table at two or more stations for each detail point and drawing lines of sight to it, the point of intersection of these lines representing the location of the point.

intersertal Describing a texture common in basic rocks in which interstices are filled by a glassy MESOSTASIS or a fine-grained aggregate of quartz, alkali feldspar, and deuteric minerals.

interstadial A warmer stage within a GLACIAL PHASE, in which the ice receded, to be followed by a readvance on the return of colder conditions. *Compare* interglacial.

interstice The angular space inside a meshwork of lathlike crystals. In basalts, interstices between feldspar crystals may be occupied by grains of pyroxene or glass. *See also* intergranular; intersertal.

intertidal Describing the part of a seashore between the high- and low-water marks. Organisms that live there have to be able to withstand periods of drying out and periods of being completely covered by the sea.

intertropical convergence zone (ITCZ) The zone of convergence between the airstreams from the N and S hemispheres. It oscillates in position with the thermal equator, reaching its maximum northward extent in July and maximum southward extent in January. Over the ocean areas it marks the meeting point of the trade winds, but over the continents it becomes affected by the monsoonal circulation and has somewhat different properties. The ITCZ can be clearly seen over ocean areas from satellite photographs as a narrow band of enhanced convectional activity indicating upward motion and cloudy showery weather at the surface. Sometimes two bands are found separated by about 3–5° latitude, each exhibiting the usual ITCZ characteristics. Many theories have been proposed to account for this feature, but as yet the real origin remains in doubt. As the dual convergence zone is found only over the oceans, it is presumed that it must be connected with sea-surface temperatures and upwelling.

Reference to this zone as the intertropical front still remain, but as it has dissimilar features to temperate latitude fronts and is not really frontal, the use of this term is discouraged.

intraclast An ALLOCHEM that is a penecontemporaneously eroded fragment of calcareous material. Intraclasts are not derived from preexisting limestones.

intraformational Existing or formed with a formation. For example, an intraformational conglomerate is one in which the clasts are eroded fragments of newly consolidated sediment. These clasts are then incorporated immediately into the accumulating sediment.

intrazonal soil A type of soil differing from surrounding ZONAL SOILS because of the dominance of local drainage or parent material over the zonal influence of climate and vegetation. These soils include hydromorphic soils (poor drainge), halomorphic soils (salinization), calcimorphic soils (calcareous parent material), soils dominated by human influences, soils on organic matter, and various horizons in zonal soils regarded as intrazonal. They differ from AZONAL SOILS in that they have more or less well developed horizons, whereas azonal soils still have little horizonation. They are often linked to zonal soils in catenary sequences, with poorly drained soils occurring downslope from the well-drained zonal types. In the SEVENTH APPROXIMATION, these soils are linked with the nearest zonal soils as aquic or halic variants, for example there is no separate group of gleys.

intrusion The process by which INTRUSIVE ROCKS are formed.

intrusive rock An igneous rock that forms by injection of magma or other plastic material into the Earth's crust (the process of *intrusion*) and has solidified beneath the Earth's surface. *Compare* extrusive rock.

inversion An increase in temperature with height in the atmosphere. This can act as an upper limit to convection, hence inversions are associated with atmospheric stability and usually dry weather. They can be formed by cooling beneath the inversion or warming at the inversion level and above. The former method occurs most frequently as a result of radiational cooling at night from the ground surface, an extreme case being over the Antarctic ice cap, and the latter by subsidence of air under anticyclonic conditions. In these circumstances the air may be extremely dry, as occurs above the trade wind inversion, and quickly evaporates any saturated thermals reaching this level. It acts therefore as a very effective ceiling to convectional cloud.

inversion layer A layer of air next to the Earth's surface that is cooler than the air above it. This unusual condition generally occurs in basins or valleys, and when it happens over cities it may cause SMOG.

invertebrates Animals that do not possess a jointed backbone formed of vertebrae. The members of all animal phyla except the CHORDATA are invertebrates. *Compare* Vertebrata.

inverted relief Topographic relief that is the opposite of the geologic structure, as for example where anticlines form valleys and synclines form high ground. It is the result of prolonged denudation.

involution **1.** A lobe of one type of material projecting upward or downward into material of another type, produced by the effects of CRYOTURBATION. These involutions are seen in the upper sections of bedded deposits and are a clear indication of former PERIGLACIAL conditions. The exact cause of this type of movement is not yet fully clear.
2. The refolding of a nappe structure resulting in very complex fold patterns.

ionosphere The layer of the atmosphere above the stratopause in which free ions and electrons occur as a result of ionization of gas molecules by solar ultraviolet and X-radiation. It consists of several layers, indicated by the letters D to G, which reflect electromagnetic waves back to Earth and thus it is very important in radio communications.

iridescence A series of rainbow colors produced by a mineral caused by the interference of light within its crystals or at its surface. The colors change as the angle of incident light changes.

iron bacteria Bacteria that feed on soluble iron compounds and as a result precipitate insoluble iron(III) hydroxide, $Fe(OH)_3$.

iron meteorite *See* meteorite.

iron pan A type of HARDPAN that is rich in iron oxide.

iron pyrites Another name for PYRITE.

ironstone A type of sedimentary rock that is rich in iron minerals, found in beds, layers, or nodules. The iron may take various forms, including CHAMOSITE, HEMATITE, PYRITE, and SIDERITE.

irradiance The quantity of radiant energy received in unit time on unit area of surface. It is expressed in watts per square meter.

isallobar A line on a chart connecting places of equal pressure tendency. Such lines are used in forecasting the movements of pressure systems, particularly depressions.

isarithm A line on a map joining points of equal value. There are many forms of isarithm, depending on the values concerned (*see* isobar; isogon). A contour is also a form of isarithm joining points of equal elevation.

island arc A long chain of islands, usually arcuate or bow-shaped. Many island arcs are bordered on their ocean flank by deep slits or TRENCHES in the deep-sea floor. The E seaboard of Asia, for example, has a whole series of island arcs – the Aleutians, the Kurils, the Japanese, and the Marianas – flanked on their ocean side by a series of trenches: the Aleutian Trench, Kuril Trench, Japanese Trench, Ryukyus Trench, and Philippine Trench. On a global scale, the island arcs can be seen to continue the great folded mountain chains on land such as the Himalayas and Western Cordilleras. The arcs usually include lines of active volcanoes; moreover, they are associated with many of the world's most severe earthquakes and with very pronounced gravity anomalies. The theory of PLATE TECTONICS provides an explanation for the distribution of these features, together with their associated volcanic and seismic activity: they mark the boundary between two of the plates that make up the Earth's surface. At this margin ocean floor is being destroyed as one plate overrides the other. The friction between the plates accounts for the series of earthquakes whose foci increase in depth away from the ocean side of the arc. The volcanic activity is the result of the increased thermal energy present as a result of the friction, which leads to partial melting of one or both plates. The trench on the ocean-ward side marks the surface expression of the junction between the plates.

isobar A line on a chart joining points of constant barometric pressure. At the ground surface, observing stations are often at different altitudes and corrections have to be made to pressure readings to provide a common level, normally sea level. In some countries, e.g. South Africa, much of the land surface is well above sea level and surface pressure readings are corrected to differences from the 850 mb pressure level. The patterns produced by joining points of equal surface pressure represent synoptic systems, such as anticyclones, depressions, troughs, ridges, and cols, from which interpretations of the weather can be made.

isochemical Describing metamorphic processes that involve little or no change in chemical composition of a whole rock. However, the addition or subtraction of water and carbon dioxide usually occurs during metamorphism.

isoclinal folding Generally upright tight folds whose axial planes and limbs are parallel. (See diagram at FOLD.)

isogon (isogonic line) **1.** A line joining points of constant wind direction.
2. A line joining points of equal magnetic variation.

isograd *See* zone (def. 1).

isohaline A line drawn on certain oceanographic charts and diagrams joining points of equal salinity.

isohyet A line joining points experiencing the same amount of rainfall. This can be used for any time interval from a single period of rain to mean annual totals.

isoline A line on a map that joints points having the same value of some specified quantity, such as an ISOBAR, ISOGON, ISOHALINE, ISOHYET, ISOPACH, and ISOPYCNIC.

isomeric line A line that joins points having the same average monthly rainfall expressed as a percentage of the mean annual rainfall. It is helpful in displaying similarities in rainfall by removing the effects of large totals over mountainous areas.

isometric *See* cubic.

isomorphism The existence of crystals that have the same form but different chemical compositions. *See* crystal form.

isopach (isopachyte) A line on a geologic map joining points of equal thickness for a particular rock bed.

isopleth map A map showing quantitative spatial distributions indicated by lines of equal value, such as isotherms.

isopycnic (isopycnal) A line drawn on certain oceanographic and meteorological charts joining points of equal density.

isoseismal line A line on a map joining points on the Earth's surface of equal earthquake intensity.

isostasy A condition of theoretical balance for all large portions of the Earth's crust, which assumes that they are floating on an underlying more dense medium. As a result of erosion or deposition, this balance is put out of equilibrium and has to be compensated for by movements of the Earth's crust. Areas of deposition sink, whereas areas of erosion rise. In this way the roots of old mountain chains formed many kilometers beneath the Earth's surface are brought up to the surface. *See also* Airy's hypothesis of isostasy.

isotach **1.** A line of constant wind speed. **2.** A line joining points of equal distance traveled in a period of time.

isotherm A line joining points of equal temperature. Isotherm maps are the most common methods of showing temperature changes on a climatic scale. As with pressure, the effects of altitude are important and corrections to sea level are normally made.

isothermal layer Any vertical section of the atmosphere in which the temperature remains constant with height, i.e. there is a zero lapse rate, as revealed by an upper-air sounding. Isothermal layers are frequent but are usually transitory. The term formerly referred to the stratosphere.

isotope One of two or more forms of the same element whose atoms differ in their numbers of neutrons (although the numbers of protons and electrons is the same). Isotopes therefore have the same atomic number; they differ in mass but not in chemical properties. Most elements occur naturally as a number of different isotopes.

isotopic ratio The ratio of the amount of one isotope of an element to the amount of another isotope of the same element, in the same sample of rock. It can be useful in studying the origins of rocks.

isotropic Having the same physical properties if measured in different directions. The term usually pertains to optical properties, particularly the speed with which light is transmitted in crystals. Crystals of the cubic system are isotropic. *Compare* anisotropic.

isthmus A narrow strip of land that connects two large land masses. For example, North and South America are connected by the isthmus of Panama.

J

jacupirangite A nepheline-bearing basic plutonic rock. *See* ijolite.

jade A green or white semiprecious stone. Jade is a tough compact variety of either jadeite or nephrite.

jadeite A PYROXENE of composition $NaAlSi_2O_6$, found in high-pressure metamorphic rocks such as glaucophane schists. *See also* jade.

jasper A red opaque variety of chalcedonic quartz. *See* silica minerals.

jet A type of hard COAL that can be carved and polished for decoration.

jet stream A narrow band of winds at high altitudes, with speeds above 60 knots and having strong vertical and lateral shear. Two main regions of jet streams are known in the troposphere, one in the subtropical westerly circulation, which is fairly constant in location for any given season and so appears clearly on mean charts, and a second associated with the polar front. This is highly variable in its location, being dependent on the transitory thermal gradients of mid-latitude depressions. Because surface depressions form where divergence in the upper atmosphere is strong, they are frequently initiated downstream of a trough in the upper westerly circulation, with anticyclones forming in areas of upper air convergence upstream of the trough. This development takes place in zones of large surface temperature gradient, which in turn causes marked temperature contrasts in the upper atmosphere as depressions intensify. The jet streams are the response to this upper atmospheric thermal gradient, and frequently encourage further development of surface pressure features by the marked convergences and divergences that are experienced in different parts of the jet.

Easterly jets have been reported between India and Africa, and low-level jets in East Africa, but these are weaker and less persistent than those of the westerly zone. Above the troposphere, a POLAR NIGHT JET STREAM occurs at high latitudes in winter at about 50 mb level (40 km).

joint A surface crack in a rock, with no displacement of the pieces each side of it. *See* jointing.

jointing The system of cracks in a mass of rock constituting, in many rock types, the only means by which water can infiltrate the rock mass. Consequently, within an area of a single rock type in which there are variations in the extent of jointing, well-jointed areas will become weathered more quickly and to a deeper level than the more massive parts of the outcrop, which may even remain as solid rock.

joule The unit of work or energy in the Système International (SI), equal to the amount of work done by a force of one newton in moving an object one meter in the direction of the force. It is equivalent to 1 watt second or 0.239 calories.

junction The plane of contact between two adjacent and distinct bodies of rock. Such a junction may be conformable, unconformable, or produced by intrusion or faulting.

jungle An imprecise term for a tropical forest or mangrove region that has dense undergrowth. *See* tropical rainforest.

Jurassic The period of the MESOZOIC Era that followed the TRIASSIC and preceded the CRETACEOUS. It lasted for about 65 million years, from 208 to 144 million years ago. It was named for the Jura Mountains in Europe, where rocks of this age were first studied. The Jurassic System is subdivided into eleven stages. The Lower Jurassic consists of the Hettangian, Sinemurian, Pliensbachian, and Toarcian Stages; the Middle Jurassic consists of the Aalenian, Bajocian, Bathonian, and Callovian Stages; the Oxfordian, Kimmeridgian, and Tithonian Stages form the Upper Jurassic. Continental rifting of the supercontinent Pangaea began toward the end of the Triassic; by the Middle Jurassic North America was moving away from Eurasia and Africa and the Atlantic Ocean was opening. Following a marine transgression, the characteristic rocks of the Jurassic System are clays and limestones, some of them oolitic. There were coral reefs, and other important invertebrates included brachiopods, bivalves such as oysters, ammonites (on which much of the system is zoned), and irregular echinoids. On land, dinosaurs continued to flourish and diversify and other reptiles included aerial pterosaurs and marine forms. Fossils of both *Archaeopteryx*, the earliest known bird, and the first mammals come from Jurassic rocks. A marine regression toward the end of the period resulted in nonmarine deposits.

juvenile water Water that exists as deposits within the Earth's magma and first appears at the surface during volcanic activity.

K

kaersutite A monoclinic AMPHIBOLE.

kainite A mineral consisting of a hydrated double salt of magnesium sulfate and potassium chloride, $MgSO_4.KCl.3H_2O$. It crystallizes in the monoclinic system and occurs as underground granular masses, often associated with rock salt deposits. It is used for making other compounds of magnesium and potassium.

Kainozoic *See* Cenozoic.

kaliophilite A rare type of kalsilite. *See* feldspathoid.

kalsilite A potassium aluminosilicate that occurs in nepheline. *See* feldspathoid.

kame An isolated hummock of stratified FLUVIOGLACIAL sediments, mainly sands and gravels, associated with slow-moving or stagnant ice. Kames may be formed in a variety of ways. Debris-laden meltwater streams may emerge from the snout of a decaying glacier, depositing their load into an ice-dammed lake. On melting, that part of the accumulation formerly adjacent to the snout may form a steep ice-contact slope, although a certain amount of slumping will occur. Alternatively, material may be deposited in meltwater pools on the surface of a stagnant glacier. The accumulated sediments will be lowered to ground level on melting, forming a kame.

kame terrace An accumulation of FLUVIOGLACIAL sediments deposited by meltwater flowing between a decaying glacier and its valley sides. As with KAMES, the ice contact slope suffers collapse on melting of the ice. Kame terraces can be continuous or discontinuous, deposition occurring only where meltwater channels could widen, thereby reducing velocities of flow. These features are flat-topped and steep-sided, sloping gently down-valley. Long continued existence of these terraces after deglaciation depends upon the extent of subsequent fluvial and slope processes.

kaolin (china clay) A soft white clay used in medicine, as a filler for paper, and in the manufacture of china. It consists of deposits of minerals of the KAOLINITE group that are produced by the weathering and hydrothermal alteration of feldspars in granitic rocks. *See also* clay minerals.

kaolinite A white or gray CLAY MINERAL consisting of hydrated aluminum silicate, $Al_2Si_2O_5(OH)_4$. It crystallizes in the triclinic system and is derived by alteration of alkali feldspar, feldspathoids, and other silicates. It is the main component of KAOLIN (china clay).

karren Relatively small irregular hollows and elongated grooves cut on limestone surfaces as a result of solution. At the most basic level they may be divided into *rundkarren*, having rounded separations between grooves, and *rillerkarren*, with sharp separations. It is believed that whereas the latter are the result of running water, the former may be at least partly created beneath a cover of snow, soil, or peat. The term is also used for the whole dissected surface and is the German equivalent of LIMESTONE PAVEMENT, the French equivalent being *lapiés*.

karst Denoting any area within which limestone solution landforms such as SINKHOLES, UVALAS, and LIMESTONE CAVERNS have become well developed. The term stems from the regional name for the massive limestone area on the Dalmatian coast of the Adriatic Sea, where practically all the landforms have been produced by solution and almost all the drainage is underground. Prerequisites for a fully developed karst landscape include a considerable thickness of strong soluble well-bedded and well-jointed limestone, moderately heavy rainfall, and sufficient altitude to allow an extensive flow of underground water.

katabatic wind On calm clear nights the ground surface cools by radiational losses and the air in contact with the ground also cools. The density of air increases when the temperature falls, and if the ground surface is sloping, these heavier layers will give rise to a downslope gravitational flow. Where concentrated in a valley, night-time mountain breezes can be quite strong, but the most extreme form of katabatic wind is off the Greenland and Antarctic ice caps. A mean hourly speed of 156 km per hour was reported at Cape Denison (67.9°S, 142.7°E) under such conditions. *See also* anabatic wind.

katafront Any frontal surface at which the warm air is descending relative to the cold. As a result of this subsidence, weather activity at a katafront is normally weak with only a belt of stratiform cloud indicating its presence. Some fronts vary in character along their length and may take the form of an ANAFRONT near the depression center and a katafront farther away.

katophorite A monoclinic AMPHIBOLE.

kelp Certain species of seaweed of considerable length and thickness. These seaweeds, which are usually brownish in color, thrive below low-tide level, and may be luxuriant enough to form *kelp forests.* Of particular importance are the large laminarians, which anchor themselves to rocky or stony bottoms and, depending on their rigidity, either stand fairly erect or move with the tidal currents. They influence the movement of bottom materials, and may, if uprooted by strong tidal flow or wave action, carry their stony attachments shoreward or alongshore. They are also effective at times in damping wave action. Kelp forests are important as fish feeding grounds.

kelvin The unit of thermodynamic temperature in the Système International (SI), equal to 1/273.16 of the temperature of the triple point of water. 1 kelvin is equal to 1°C and a temperature expressed in kelvins is the same as the ABSOLUTE TEMPERATURE.

Kelvin-Helmholtz instability The development of waves at the surface separating two layers of the atmosphere having different temperatures and wind speeds. These waves sometimes increase in amplitude and become unstable, rolling into vortices, which break and give CLEAR AIR TURBULENCE. Very occasionally moisture conditions may be such that clouds form on the waves and make the vortices visible.

Kelvin wave A tidal phenomenon in which a progressive wave traveling along a tidal channel is affected by the Earth's rotation to the extent that the tidal range (in the N hemisphere) is increased along the right-hand side of the channel and decreased on the left-hand side (with the observer facing the direction in which the progressive wave is traveling). Usually, the energy of the progressive wave is much diminished, so preventing a true amphidromic tidal system from operating (*see* amphidromic system). The Kelvin wave phenomenon can be observed in the English Channel, where the tidal range along the English coast is small compared with that along the French coast. During any tidal state, the sea surface slopes from one side of the Channel to the other.

kelyphitic rim *See* reaction rim.

kentallenite A syenogabbro containing olivine and augite together with plagioclase and orthoclase feldspars and large poikilitic crystals of biotite. *See* alkali gabbro.

kenyte A TRACHYTE containing large rhombic orthoclase phenocrysts. Medium-grained varieties may be termed *rhomb-porphyry*.

keratophyre A fine-grained igneous rock associated with basaltic lava. *See* spilite.

Kerguelen-Gaussberg Ridge A large ocean feature extending southward from the East African coast, reaching almost to Antarctica. The ridge branches out, one branch reaching westward to the MID-ATLANTIC RIDGE, and another trending eastward, and then circumscribing the Australian continent and extending toward the East Pacific Rise. Surveys conducted off the continent of Antarctica show that the continental slope south of the Kerguelen-Gaussberg Ridge is in places made up of several tilted plainlike features with gentle undulations. Rocks dredged from the Kerguelen-Gaussberg Ridge yield samples of basalt and continental-type rock.

kernite A soft white or colorless mineral form of hydrated sodium borate, $Na_2B_4O_7.4H_2O$. It crystallizes in the monoclinic system and is an important source of boron compounds, including BORAX.

kernlose winter In Antarctica, temperatures decrease toward winter, as in most parts of the world, but by the end of the long winter night, temperatures are little lower than they were at its commencement. This peculiarity has been called the kernlose (or coreless) winter. It represents a balance between surface radiational cooling and atmospheric counterradiation from above the intense surface inversion. Some theories suggest that advection of warmer maritime air may play a part in the kernlose effect, but this is disputed.

kettle hole A depression in the ground surface in a glaciated area. As a mass of ice stagnates, material is deposited from above the ice surface, and FLUVIOGLACIAL deposits also accumulate. In this way masses of ice can become buried beneath DRIFT. Once buried the ice takes a considerable time to melt, and when this finally happens, collapse occurs and a depression is formed, which may subsequently become filled by a lake or by sediments. Not all kettle holes are associated with stagnant ice, because small lakes and streams may freeze while sediments are accumulating, but those that are will be found in association with KAMES and ESKERS.

key *See* cay.

khamsin A hot dry southerly wind blowing across Egypt from the Sahara. It precedes depressions passing eastward along the North African coast and hence is most common between April and June. According to Arab tradition it blows for 50 days.

kidney ore A form of the iron-containing mineral HEMATITE, which occurs in masses whose shape resembles solidified bubbles or kidneys.

kimberlite (blue ground) A brecciated carbonate-rich phlogopite-bearing PERIDOTITE found in pipes and diatremes piercing very old metamorphic rocks. Many kimberlites contain DIAMONDS.

kinetic energy The energy of a body derived from its movement. Its magnitude is expressed as $\frac{1}{2}mv^2$ where m is the mass of the body, and v is its velocity. The atmosphere possesses large amounts of kinetic energy by virtue of its motion, but this is constantly being lost through surface friction and turbulence. It is replenished by conversion of potential energy derived originally from solar radiation and upward motion of air.

kingdom The largest group in the taxonomic classification of living organisms. Five kingdoms are usually recognized: Animalia includes all animals; Plantae includes all plants; Fungi includes all fungi; Protista (or Protoctista) includes algae, protozoa, and slime molds; and Monera (Bacteria or Prokaryotae) comprises the bacteria. Kingdoms are divided into phyla. *See also* taxonomy.

kink plane A deformation structure in which the orientation of a foliation is changed as a result of slippage or gliding.

klippe Originally, any rock body isolated by erosion. The term is now restricted to the outlier or erosional remnant of a nappe.

knee fold A GRAVITY FOLD, generally zigzag in shape. (See diagram at FOLD.)

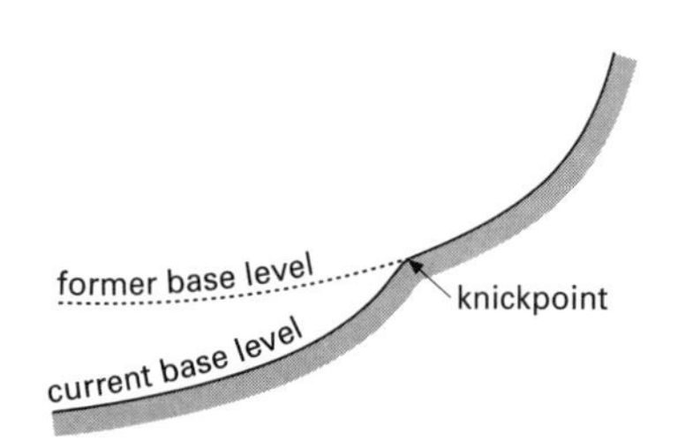

Knickpoint

knickpoint (rejuvenation head) A break of slope in the long profile of a stream, resulting from a fall of base level rejuvenating the seaward portion of the stream. This REJUVENATION leads to renewed downcutting, and the creation of a new lower long profile; the point where this profile meets the established profile is the knickpoint.

Knickpoints are often marked by waterfalls if the break of slope is steep, or rapids if less steep. Once created, the knickpoint tends to migrate upstream and be lowered in height; the rate at which it advances and the degree to which it loses its height depends on the lithology in which it is cut. In unconsolidated material it will rapidly disappear. The increased turbulence associated with break of slope accelerates erosion and rapidly lowers its height.

knoll A small rounded hill. *See also* hummock.

knot A unit of speed, being one NAUTICAL MILE per hour; 1 knot is equal to 0.515 meters per second.

komatiite A series of ancient extrusive igneous rocks that contain some of the oldest PERIDOTITE lavas. They are characterized by large branching crystals, indicative of rapid cooling of the magma from which they derived.

K/T boundary event The time when the Cretaceous period changed into the Tertiary, when (according to fossil evidence) large-scale animal extinctions took place. For example, dinosaurs, ammonites, and belemnites ceased to exist. The cause is thought to be a general and rapid rise in world temperatures. Because of high concentrations of the element iridium in rocks of the time, some geologists think that the climatic change resulted from a major meteorite impact which filled the Earth's atmosphere with debris, cutting off most of the sunlight (iridium is relatively abundant in meteoroids).

Kuroshio Current An ocean current constituting one of the important western BOUNDARY CURRENTS in the world's ocean circulation system. It develops following a gradual deflection of part of the North Equatorial Current in the Pacific Ocean, being at first rather diffuse, but in more northern tracts becoming concentrated between the shelf region off the Ryukyu Islands and a submarine ridge. At about 35°N, the current swings away from the Japanese mainland and flows for some 8000 km before again splitting up into branches. In contrast to the water beneath the Gulf Stream, water that is relatively warm and fairly saline because of its continual replenishment of very saline water from the European Mediterranean Sea, the water beneath the Kuroshio Current is only partly saline and cold, being of sub-Antarctic origin. The Kuroshio Current and the Florida Current have roughly the same flow rates.

kyanite (cyanite) A light blue, white, or gray aluminum silicate mineral, Al_2SiO_5. It crystallizes in the triclinic system and occurs in pressure-metamorphosed rocks, such as amphibolite and some mica schists. *See also* aluminum silicates.

L

Laborde projection A map projection resembling Mercator's transverse, the only difference being that Laborde uses a spheroid and not a sphere. It is generally restricted to areas of east-west extent, for which it is orthomorphic.

Labrador Current One of several major surface currents in the N Atlantic Ocean, bringing cold water southward from the Arctic Ocean region. The Labrador Current flows in on the W flank of Greenland, and is matched on the E flank by the East Greenland Current. These influxes of cold water meet, in the surface layers, with the relatively warm and saline Gulf Stream and North Atlantic Drift water. Both cold currents carry icebergs southward, and these tend to become concentrated in the "Gateway of Icebergs" east of the Grand Banks. This assembly of bergs is most common between mid-March and mid-July. Another climatological feature of the region is the high frequency of fogs, especially off the coast of Newfoundland. Winter cooling of the combined waters from the North Atlantic Drift, the Labrador Current, and the East Greenland Current may increase their density sufficiently to cause the surface waters to sink, perhaps to depths in excess of 1000 m.

labradorite A variety of plagioclase FELDSPAR.

laccolith A concordant intrusive igneous body with a dikelike feeder, usually forming small lenslike features less than five kilometers in diameter. As the magma is injected it has sufficient pressure to arch up the overlying strata. Therefore although it has a flat base, its upper surface is convex.

lacuna The time interval missing between beds above and below an unconformity, including both the nondepositional and erosional components.

lacustrine Describing something that lives in, is related to, or is produced by a lake.

lagoon A stretch of water that is more or less enclosed and often shallow, protected from the open sea by a spit, tombolo, bay-mouth bar, or other kind of barrier. These barriers are not complete, for lagoons invariably possess a free connection with the open sea, however narrow this may be. The barrier provides protection from waves, although wave overwash is common to many lagoonal areas. Tidal current action is usually minimal within the lagoon itself. The water may be fresh, saline, or brackish. Moreover, the salinity level in a lagoon may vary with time. The deposits of lagoons are characteristically fine-grained, with a significant proportion of organic matter.

lahar A mudflow composed mainly of volcanic debris. If the unstable debris accumulating on the sides of a volcano is lubricated by heavy rain, it will flow under gravity. The flow that covered Herculanium in AD 79 may have been of this type.

lake A body of water completely surrounded by land. It lies in a depression in the Earth's surface or in an artificially created location behind a dam. Most lakes contain fresh water, although some very large ones (such as the Caspian Sea and Dead Sea) have saline water. A very small lake is generally referred to as a pond or pool. Some lakes have streams or rivers flowing into them, and some dry up at certain times of the year. In ecological terms, a lake can be considered as a complete ECOSYSTEM.

Lambert's cylindrical equal-area projection A map projection drawn so that the scale along the Equator is correct, but all the other parallels are exaggerated in length. The scale on the meridians decreases away from the Equator, causing the parallels to become closer poleward. The projection is equal-area because the ratio between the meridians and parallels remains constant. It is seldom used outside the equatorial zones because of the degree of distortion of shape north and south of about 40°.

Lambert's zenithal projection An equal-area map projection used mainly in polar areas, where the pole is the center of the projection and the meridians are straight lines radiating from it, making the bearings true. The parallels are concentric circles. Lambert also constructed an oblique (again equal-area, showing countries within 40° with good shape) and an equatorial projection (used for countries astride the Equator, equal-area).

Lamellibranchia *See* Bivalvia.

lamina A thin layer of sedimentary rock, up to 1 mm thick, that differs in some respect from the layers on each side of it. It may, for example, have a particular color, particle size, or composition.

laminar flow The nature of movement of a fluid over a smooth surface at very low velocities, whereby parallel "slabs" of air or water move by sliding over each other, with no mixing. Energy is transmitted from one layer to the next by the viscosity between the layers, while at the boundary (ground surface, stream bed, stream wall, etc.) movement is nil. Laminar flow is rare in natural streams or air flows, except at boundary contacts.

lamprophyre One of a group of medium-grained basic alkali-rich igneous rocks occurring as dikes and sills. These rocks are characterized by the presence of mafic phenocrysts, biotite, amphibole, or augite, and a groundmass containing orthoclase or sodic plagioclase feldspar.

Minette and *vogesite* are varieties containing orthoclase feldspar and in which biotite and hornblende respectively is the predominant mafic mineral. Analogous rocks containing plagioclase are called *kersantite* and spessartite. *Camptonite* contains plagioclase and a brown amphibole called *barkevikite*. *Monchiquite* is a feldspar-free lamprophyre containing phenocrysts of biotite, augite, amphibole, and olivine in an analcime-rich matrix.

Lamprophyres are melanocratic hypabyssal equivalents of diorites and syenites and are associated with such plutonic intrusions.

land breeze The weaker nocturnal equivalent of the SEA BREEZE. It results from the unequal cooling of the different surfaces at night, but because the temperature difference is small, wind speeds are low. It is most evident when the pressure gradient is weak and skies are clear, so allowing radiational cooling over the land to be at a maximum.

The surface wind is part of a larger convection cell, which maintains air continuity. Above the land breeze is a weak airflow from sea to land.

land bridge A strip of land that joins two larger landmasses and may provide a route for the migration or spread of plants and animals. *See also* isthmus.

landform The various surface features of the Earth that contribute to the landscape, such as plains, mountains, and valleys.

Landsat Any of the unmanned orbiting earth resources satellites, launched by NASA from July 1972. They use remote-sensing devices, such as infrared cameras to provide information about land use, vegetation, and pollution.

landscape The geomorphological and

artificial features that make up the Earth's scenery. It thus includes the natural LANDFORM as well as buildings, civil engineering works, and agricultural modifications.

landslide The sudden downslope movement of rock and debris due to failure along a shear plane. The movement is started when the stress in the FREE FACE exceeds the resisting power of the potential shear plane: this loss of resistance may be due to increased weight of the face, e.g. if it becomes wetted; it may be due to undercutting of support at the foot of the slope by some process of BASAL SAPPING; or it may be due to a sudden triggering by an earthquake.

As a result, a convex scar is left upslope, with a mass of jumbled debris at the base in a tonguelike form. Areas of rapid downcutting by rivers, or with variable lithologies, or swelling clays that expand on wetting are susceptible; so too are undercut sea cliffs and river banks, and areas of ice formation. Like all MASS MOVEMENTS, the effect is to render an unstable slope more stable.

landslide surge A large and often destructive wave caused by the falling of a mass of rock, or perhaps parts of a glacier, from cliffs or hill slopes bordering a body of water. Such events have been witnessed on the open coast and in lakes. Because of the generally unpredictable nature of rockfalls and rockslides, detailed studies of landslide surges have rarely been possible. One rockfall in Lituya Bay, Alaska, which resulted from a 1958 earthquake, caused the fiord water to swamp some 500 m up the mountainside on the opposite shore of the fiord. A 15-m high wave was generated along the fiord, traveling at an estimated 160 km per hour. Somewhat similar occurrences have been recorded in Japan, Norway, and elsewhere, but they are comparatively rare.

langley A unit of energy equal to 1 calorie per square centimeter. It is mainly used for radiation, but as the calorie has been largely superseded by the JOULE, the langley is decreasing in use.

lapiés The French word for KARREN.

lapilli Round fragments of solid lava thrown out by an erupting volcano; small volcanic bombs. They are a form of EJECTA up to 65 mm across. *See also* pyroclastic rock.

lapis-lazuli The semiprecious variety of LAZURITE. *See* feldspathoid.

lapse rate The change of temperature in the atmosphere with height. Distinctions are made between the decrease of temperature in the free atmosphere, the ENVIRONMENTAL LAPSE RATE, and the decrease of temperature under adiabatic conditions in thermals, the DRY ADIABATIC LAPSE RATE and SATURATED ADIABATIC LAPSE RATE.

larvikite (laurvikite) A dark blue iridescent type of SYENITE, used as an ornamental stone.

latent heat The energy required or released on a change of phase. In the atmosphere, water can exist as a solid, a liquid, or a vapor, representing increasing energy states respectively. Thus freezing and condensation release latent heat, melting and evaporation require heat. Latent heat is extremely important in atmospheric processes, enabling energy to be transferred within the general circulation, from ground to atmosphere, and as a reservoir of heat in the atmosphere.

Specific latent heat is the latent heat required or released per unit mass.

lateral accretion Material that accumulates at the side of a channel, such as that deposited on the inner sides of bends on a meandering river.

lateral erosion River erosion in a horizontal direction. Within the meandering reach of a river, erosion on the outside of bends leads to downstream migration and constant paring away of the bluffs at the side of the valley; this effectively erodes the valley width laterally. In floods, the elevated water level fills the channel and raises the water table, saturating the banks.

When the floods subside, the support on the weakened (wetted) banks is removed and mass slumping occurs, hence widening the channel laterally. By these processes, some FLOODPLAINS are eroded rather than deposited; they consist of a rock-cut bench with a thin veneer of alluvium, unless they form because of a base-level rise, when a deep purely aggradational floodplain will result.

lateral fault *See* tear fault.

lateral moraine A ridge of material along the side of a glacier (or along the side of a valley formerly occupied by a glacier), primarily composed of angular and unsorted material that has fallen from overhanging rock outcrops as a result of FREEZE-THAW weathering, although it is believed that a certain proportion of the material is derived from within the glacier itself. The ridge may consist extensively of ice, because once a layer of waste about 0.5 m thick has developed, the ice beneath is protected from melting and hence it stands above the surface level of the rest of the glacier.

laterite A layer of deposits formed from the weathering of rocks in humid tropical conditions and consisting mostly of iron and aluminum oxides. It hardens on exposure to the atmosphere and is therefore used for building. It is not a soil, because it is produced by processes operating on a far larger scale than typical pedologic processes. It can occur in two forms, either buried and soft or as a superficial capping (cuirass) of hardened red material on uplifted areas that have had their overburden stripped off. Most laterites developed in the Tertiary period, and are hence relic. In profile they display a strong hardened reddish surface horizon, dominated by ferric oxide, which breaks up with depth into a mottled horizon, which in turn passes down to a depleted pallid zone, which may be 50 m thick and reach to the bedrock. In a pure laterite, the red horizon will be composed of 90–100% iron, aluminum, and titanium oxides. *See also* laterization; limonite.

lateritic soil **1.** A soil developed on part of a relic LATERITE, or with an indurated layer within its profile.
2. A red tropical soil developing into laterite. It undergoes leaching of silica and bases, the concentration of aluminum and iron oxides increasing, as happened in the formation of the relic laterites. As leaching of silica proceeds, the silica:sesquioxide ratio falls; in 1927 Martin and Doyne calculated that while laterites had a ratio of <1.33, lateritic soils had a ratio of 1.33–2.0, showing that they were following the same development trend but had not yet been desilicified as much as true laterite. However, relic laterite probably developed by much larger-scale processes and the term lateritic soil has now been replaced by latosol (ferrallitic soil) or, in Australia, red loam.

laterization The concentration of iron and aluminum oxides in the upper layers of the soil and the removal of silica and bases, either in the formation of relic LATERITES or in current soil-forming processes in the tropics. In the case of relic laterites, laterization is thought to be a result of a fluctuating water table. Opinions vary: the rising and falling groundwater (resulting from seasonal dryness and wetness) either brings up iron and aluminum from the pallid zone and concentrates it at a point of maximum upward extension, or it causes iron and aluminum being washed down in percolating rainwater to be precipitated on meeting the alkaline groundwater, or it causes a combination of the two. It is a large-scale process, operating over geologic timescales. In the case of current tropical soil formation laterization denotes the process of individualization of iron and aluminum oxides, and to a lesser extent oxides of titanium, chromium, and nickel, by intense weathering and leaching, removing bases and silica. This process now tends to be called ferrallitization, leading to the development of FERRALLITIC SOIL (latosol).

laterolog A subsurface logging technique, which measures variations in the conductivity of strata.

latite An extrusive igneous rock consisting of alkali feldspar and plagioclase. *See* syenodiorite.

latitude The angular distance of a point on the Earth's surface north or south of the Equator, measured as the angle subtended at the center of the Earth by an arc along a meridian between the point and the Equator. The Equator has no angle whereas the poles are at 90°. Latitude is measured in degrees, minutes, and seconds. Parallels of latitude form approximate concentric circles, with the poles as the center (see diagram). *See also* longitude.

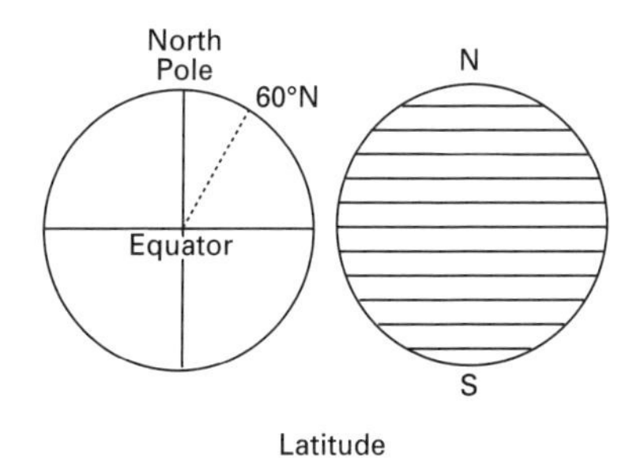

Latitude

latosol *See* ferrallitic soil.

laumonite A hydrated silicate of calcium and aluminum, $Ca(Al_2Si_4O_{12}).4H_2O$. It crystallizes in the monoclinic system, and occurs in veins in schists and cavities in igneous rocks. It is a member of the ZEOLITE group of minerals.

Laurasia The N hemisphere supercontinent believed to have been present before 200 million years ago. This supercontinent subsequently fragmented to form the present Greenland, Europe, Asia (excluding India), and North America. *See also* Pangaea.

lava A molten or partly molten mass of MAGMA extruded from a volcano or volcanic fissure and the rock that forms as a result of cooling and solidification. The accumulation of single extrusions, termed lava flows, produces a volcano or lava pile.

There are numerous textural terms used to describe lavas, such as aphyric, porphyritic, vesicular, glassy, holocrystalline, etc. *See also* aa; block lava; columnar joint; pahoehoe; pillow lava. *See* lava flow.

lava flow A stream of molten LAVA flowing from a fissure or vent of an active volcano, and the solidified rock it produces when it cools. An acidic siliceous lava, with a high viscosity, tends to move slowly in short thick flows whereas basic basaltic lava, with a low viscosity, flows quickly in long thin streams.

lava plateau A large area of elevated land gradually built up by a series of LAVA FLOWS from a volcanic FISSURE. *See also* flood basalt.

lava tunnel A long hollow below the surface of a solidified LAVA FLOW, created when lava that was still molten continued to flow after the surface had become solid. Such hollows occur most frequently in basaltic lava, which has a low viscosity when molten.

lawsonite A white mineral of composition $CaAl_2(OH)_2(Si_2O_7)H_2O$, found in low-temperature regionally metamorphosed rocks, particularly glaucophane schists.

layer cloud A cloud having little vertical development. Such clouds are formed under conditions of zero or weak uplift; precipitation may be extensive from nimbostratus clouds but because uplift is weak, intensity is low. Clouds of this type are cirrostratus, altostratus, stratus, stratocumulus, and nimbostratus.

layered igneous rock Many large intrusions of gabbroic igneous rocks are stratified, containing layers in which different minerals are concentrated, which suggests that the dominant differentiation process (*see* magmatic differentiation; metamorphic differentiation) was the gravitational settling of early-formed crystals.

Often mineralogically similar layers are repeated in a rhythmic fashion. Systematic changes in mineral composition from the bottom to the top of a layered sequence are termed *cryptic layering*. Typically, olivines and pyroxenes become progressively richer in iron (relative to magnesium) and plagioclase trends to more sodic compositions, changes which are compatible with crystallization from residual liquids that are differentiating toward more acid compositions.

lazurite A deep blue to violet complex sodalite mineral, $(Na,Ca)_8(Al,Si)_{12}O_{24}(S,SO_4)$, the chief component of lapis-lazuli. It crystallizes in the monoclinic system, and occurs in pegmatites and metamorphic rocks. *See also* feldspathoid.

leaching The action of water moving down through a soil profile carrying soil materials with it in suspension or solution. It includes the mechanical action of water, e.g. LESSIVAGE; its chemical action, as in the weathering processes of HYDRATION, HYDROLYSIS, and CARBONATION; and its partly biochemical action, as when it contains leaf leachates and moves material by CHELUVIATION. Leaching removes bases, clays, organic matter, and the various sesquioxides; in its varying forms it is responsible for the processes of lessivage, PODZOLIZATION and LATERIZATION. Soils subject to leaching become increasingly acid with strongly developed eluvial and, if the leached material is redeposited, illuvial horizons.

lechatelierite A rare silica glass. *See* silica minerals.

lee The side of a hill or other prominence that is sheltered from the prevailing wind.

lee waves Under certain conditions of atmospheric stability, stationary waves of air may be generated on the leeward side of hills or mountain ranges. Their wavelength is usually between 3 and 30 km. Lee waves are frequently seen when the crests of the wave reach saturation point to give a lens-shaped cloud outlining the upward motion. This cloud is constantly forming on the upwind side and being dissipated in the area of descent while remaining stationary; it is classified as *altocumulus lenticularis*. Lee waves occur when a stable layer of air is sandwiched between layers of lower stability above and below, with a strong and steady wind flow.

lepidolite A pink, purple, or gray lithium-bearing mineral, a hydrated silicate of aluminum, lithium, and potassium, sometimes also containing fluorine and rubidium. It crystallizes in the monoclinic system as platy aggregates occurring in pegmatites, and is a member of the MICA group of minerals. It is a principal source of lithium.

lessivage The mechanical movement of clay down a soil profile under the influence of water and its redeposition lower down. It occurs in soils subject to LEACHING but where acidity is neutral or slight, so that the clay is not broken down into its constituents but moves as a whole. It does nevertheless operate better in less alkaline conditions, and DECALCIFICATION generally precedes lessivage.

Diagnostic signs include an increase in the heaviness of texture with depth, and micromorphological examination can show clay skins on ped surfaces and along channels. Clay skins are evidence of redeposition of clay; increased heaviness alone is only suggestive, as Bt horizons (texturally heavier B horizons) can be due to a lighter horizon above or increased weathering in the B due to increased moisture.

leucite A white or gray potassium aluminum silicate, $KAlSi_2O_6$, which occurs in basic lavas and other igneous rocks. It crystallizes in the tetragonal system and is used in making fertilizers. *See also* feldspathoid.

leucitite A volcanic rock that is composed of LEUCITE and a pyroxene. *See* nephelinite.

leucocratic *See* color index.

leveche A hot dry wind that blows from the south in southern Spain. It generally precedes a depression moving along the Mediterranean Sea. *See also* khamsin; sirocco.

levée A berm or bank elevated above the general level of the floodplain, bordering many streams in their alluvial floodplain sections. Usually made of coarser material than the rest of the floodplain, they originate in times of flooding when the overbank flow decreases in speed and volume away from the stream, depositing the coarse materials first as it rapidly loses COMPETENCE. If the river lacks a coarse fraction, alluvium being mostly silt and clay size, levées may be absent.

levéed channel A submarine channel in the deep ocean flanked by marked levées, not unlike some of the levée features that border many river channels on land. Such submarine channels usually lie at the lower ends of SUBMARINE CANYONS, such as the La Jolla Canyon off the coast of California. Here, at a depth of some 600 m, the true canyon terminates in a channel flanked by natural levées. At its seaward extremity, the channel merges into a submarine fan. Sand layers have been found in the fan and on the levées on each side of the channel. Channels with levées are also strongly developed across many abyssal fans.

level A surveying instrument used in conjunction with a LEVELING STAFF to determine heights. A level is basically a telescope fitted to an adjustable base, which clamps onto a tripod. Before use the instrument must be adjusted until the line of sight is perfectly horizontal, by centering a bubble that moves as the basal footscrews are turned. Once leveled it can be used to sight onto the staff. The telescope incorporates a diaphragm with engraved lines, the reading at the central line being that used for heighting calculations.

leveling A method of surveying used for obtaining heights of points in relation to that of a point of known height. A LEVEL is set up between a point of known height and one for which a height is required; it is then adjusted to give a horizontal line of sight. A sight, known as a *backsight*, is made onto a LEVELING STAFF at the known point and the staff reading noted. A reading is then obtained by making a *foresight* onto the staff at the other point. The difference between the two staff readings gives the difference in height between the two points. The process continues by moving the level between the second point, now of known height, and another unknown one, and so on. A leveling traverse (*see* traversing) returns to the first known point, as a check on accuracy.

leveling staff A staff used in surveying in conjunction with a LEVEL for determining heights. Most staffs consist of three sections, which are either telescopic or simply clip together, to reach a height of 4 m at their full extent. They embody a small bubble to enable the user to hold them vertically. The staff has heights marked clearly in black or red on white or yellow, which can be read to the nearest millimeter. However, when viewed through the telescope of a level, the figures appear upside-down, and care must be taken in reading them.

lherzolite An ultramafic rock consisting largely of olivine together with orthopyroxenes and clinopyroxenes.

lightning The visible discharge of a thunderstorm and the natural mechanism for neutralizing the high electrical fields that build up. Air is a good insulator but as charge production increases the potential gradient in a thunder cloud, and water droplets decrease the insulational properties of air, eventually the insulation is overcome and the lightning flash results. It consists of a line of highly ionized air molecules along which electricity can flow for a very short time.

High-speed film recordings of lightning discharges have clarified the development of a flash. A cloud-to-ground discharge is started by a streamer, which develops downward from the cloud in a series of steps. It is of low luminosity. When this streamer has approached to within about 5 to 30 m of the ground, a highly luminous return stroke comes up to meet it and travels up the ionized channel to the cloud. This is the main visible flash, which travels at speeds of about 0.1 times that of light, passing a current of 10 000 amperes but only lasting for about 100 microseconds. Several more streamers and return strokes may develop. *See also* thunder.

lignite (brown coal) Low-grade brown coal with a high moisture content.

limb *See* fold limb.

limburgite An undersaturated basic igneous rock occurring as flows and dikes and consisting of phenocrysts of augite and olivine embedded in a matrix of brown basaltic glass.

lime An imprecise term with several meanings. To a chemist, lime (also called quicklime) is calcium oxide, CaO, made by heating LIMESTONE. Slaked lime (made by treating lime with water) is calcium hydroxide, $Ca(OH)_2$. To an oceanographer, the term lime is generally applied to calcium carbonate, $CaCO_3$, especially when it is dissolved in seawater or laid down in the shells of mollusks and other aquatic animals. *See also* carbonate minerals.

limestone A rock formed from carbonate minerals, principally CALCITE but including others such as DOLOMITE. Limestones can be classified broadly on a genetic basis as CLASTIC, organic, precipitated, or metasomatic. Organic limestones are formed from the calcareous skeletons of living organisms, and include in situ BIOHERMS and accumulates such as CHALK. Among the various precipitated limestones are those that are part of an EVAPORITE sequence, and OOLITES. Clastic limestones are those derived from preexisting calcareous rocks. Metasomatic limestones are other limestones that have undergone diagenetic change. The most common change involves the alteration of calcite to dolomite. Descriptive classifications employing the relative proportions of ALLOCHEMS, MICRITE, and SPARITE are increasingly used.

limestone cavern A subterranean cave developed in limestone as a result of the enlargement of joints and fissures by solution, larger caverns occurring in less resistant well-jointed limestone. Most cave systems comprise both large open caves and small restricted passages, all of which continue to grow in size as long as unsaturated water flows through them. Some caves are considerably extended horizontally, which may reflect a position corresponding to the present or a former water table. The largest caves are found in hard well-jointed limestones of low porosity; soft limestones, such as chalk, are unsuitable for major cave development.

limestone pavement A flat exposure of bare limestone resembling a rough pavement and consisting of flat or ridged irregularly sized and shaped blocks (*clints*), separated by clefts (*grikes* or *grykes*) formed by the widening of joints by solution. Grikes can measure up to half a meter across and frequently two to three meters deep.

limestone solution Limestone is the only common rock type in which SOLUTION is important in the development of its characteristic landforms. Because it is composed of calcium carbonate, solution is easily achieved by very weak carbonic acid (formed from water and dissolved carbon dioxide from the air and soil). Many limestones possess a well-developed system of joints and bedding planes; the vast majority of the water flowing through the rock passes along these fissures, in which solution takes place and they therefore become enlarged, forming better watercourses. In theory more limestone is dissolved as the temperature or acidity of the solution increases, but it is probably more effective to have a rapid turbulent flow of ample supplies of water over a large surface area of rock.

limnology The scientific study of lakes and other bodies of fresh water. It considers all the living and nonliving factors that affect them.

limonite The yellow to brown amorphous and cryptocrystalline oxidation and hydration products of iron. Goethite and hematite are important constituents together with colloidal silica and clay minerals. Limonite is found as an alteration product in all kinds of rocks, as a precipitate in bog iron ore, and as a major constituent of gossan and laterite.

lineament A large-scale linear topographic feature that is structurally controlled. Lineaments often appear as long linear features on aerial photographs and frequently result from faults, where erosion has worked selectively on the softer material of the fault zone.

line scale A divided scale given on a map to enable distances to be directly measured and read off in terms of distances on the ground.

linkage analysis A statistical procedure used to group individuals into clusters. The measures of similarity used are coefficients of association, coefficients of correlation, or distance measures, depending upon the type of data and the nature of the problem being analyzed

litchfieldite A type of alkaline SYENITE containing albite, nepheline, and potassium feldspar.

lithic Describing anything composed of or related to stone.

lithification The process by which sedimentary deposits develop into sedimentary rocks.

lithographic limestone A dense fine-textured type of limestone, so called because it was formerly used for lithography. It is a kind of MICRITE.

lithophile An element that occurs mainly as silicates and is therefore most concentrated in the Earth's crust. Lithophile elements include lithium, oxygen, silicon, and sodium. *See also* atmophile; chalcophile; siderophile.

lithosol AZONAL SOIL with an AC or (A)C profile (i.e. no illuvial B horizon) developed on fresh and imperfectly weathered rock or rock debris. These soils often form on steep slopes where runoff erodes developing soil and little water percolates down through the profile to promote LEACHING and soil development. The surface horizon is typically dark due to organic matter accumulation, but may be coarse-textured and light colored if slope wash is intense. On eroded land, subject to poor agricultural practices, these soils can be artificial. A special type is the *ranker* developed on acidic rocks. These soils are part of the entisol order of the SEVENTH APPROXIMATION.

lithosphere That part of the Earth including the crust and the upper mantle above the ASTHENOSPHERE.

lithospheric plate (tectonic plate) Any of the large plates of rock that go to make up the Earth's crust (the lithosphere). *See also* continental crust; oceanic crust.

lithostratigraphy The branch of STRATIGRAPHY concerned solely with lithological features and with naming and elucidating the spatial relations of rock units. It is the most purely descriptive of the systems of stratigraphic classification, incorporating the least amount of inference, and does not take account of the evolution of organisms as shown by contained fossils (*see also* biostratigraphy) or of time (*see also* chronostratigraphy). Hence lithostratigraphic units are often diachronous (*see* diachronism). The hierarchy of lithostratigraphic divisions consists of SUPERGROUP, GROUP, FORMATION, MEMBER, and BED (*see also* complex). Formally designated lithostratigraphic units (formations) are given geographical names with capitalized initial letters.

lit-par-lit gneiss A coarsely banded or foliated gneiss consisting of alternating layers of quartz-feldspar and mafic minerals. The felsic material, which may have originated by partial melting, has intruded

LOAD		
Major division	*Subdivision*	*Nature of load*
(1) solution load		chemical content, dissolved in water
(2) solid debris load	(a) suspension load	finest solid particles, supported by water
	(b) saltation load	sand/fine gravel, bouncing along bed
	(c) bed load	coarsest material, rolling and sliding on bed

along the schistosity planes of the metamorphic rock. *See also* migmatite.

litter The accumulation of leaves and twigs on the surface of the soil. It is the raw material for the formation of the soil humus; beneath the fresh litter there is a fermenting layer and below this the humus.

Little Ice Age The period between the middle of the 16th and 19th centuries that was characterized in northern regions by heavy snowfalls, expansion of glaciers, and prolonged winters. The world's climate, however, did not undergo any overall change.

littoral drift *See* longshore drift.

littoral zone The region immediately flanking the coast on its seaward side, strictly the combination of the intertidal and SURF ZONES, i.e. the region lying between highest high water and the outer edge of the surf zone. The width of the littoral zone varies according to the prevailing wave conditions and the slope of the sea floor. The term *eulittoral* refers to the zone that extends between high water and the limit of attached plants, which is usually at a depth between 40 and 60 m. In global terms the littoral zone occupies an area of about 150 000 sq km.

llanos A SAVANNA type of grassland found in northern South America, particularly in the Orinoco River basin.

load The material carried along by a river. Material is incorporated into a river by its erosive action on the valley bed and walls. It can also be supplied by groundwater flow, or direct from valley sides by wash, CREEP, or LANDSLIDES. Once in the river, load can be split up as in the table. The predominance of the different types of load varies between rivers and in the same river over time, e.g. DISSOLVED LOAD is important in rivers receiving a steady groundwater flow and little overland runoff; in floods, transporting ability is increased owing to velocity increases, and debris that at low flow moves as BED LOAD may start to saltate, and saltating load may be carried in suspension. Generally bed load is the least important part: in the Mississippi, only 8% of load moves as bed load, compared with about 65% in suspension and saltation.

loam A type of soil with characteristics between those of sand and clay. It is usually rich in humus. It drains reasonably well and yet retains sufficient moisture to promote good plant growth.

loch The Scottish name for a lake. *Compare* lough.

lode A deposit of minerals occupying a seam or vein. Lodes probably formed when hot gases or liquids were forced through existing rocks before cooling and solidifying.

lodestone MAGNETITE that is strongly magnetic.

lodgment till Unstratified glacial material deposited directly from beneath a glacier or ice sheet. Such deposits tend to be somewhat compacted due to the weight of ice at one time above them. *Compare* ablation till.

loess A fine-grained yellowish sedimentary deposit originating as windblown soil. It is very soft and easily eroded to form gullies. It is porous but can form extremely fertile soil.

Lomonosov Ridge A vast submarine ridge that effectively divides the Arctic Ocean into two separate basins. The ridge was first discovered by Russian scientists while they were conducting their oceanographic surveys from field stations set up on drifting pack ice. The ridge runs directly underneath the North Pole, stretching from Ellesmere Island, north of Canada, across to the Novo Sibirsk Islands off the Siberian mainland. Such data as has been collected suggests that the ridge may be a faulted feature rather than volcanic. Soundings across the ridge have revealed what appears to be a truncated upper surface lying at a depth of some 1400 m.

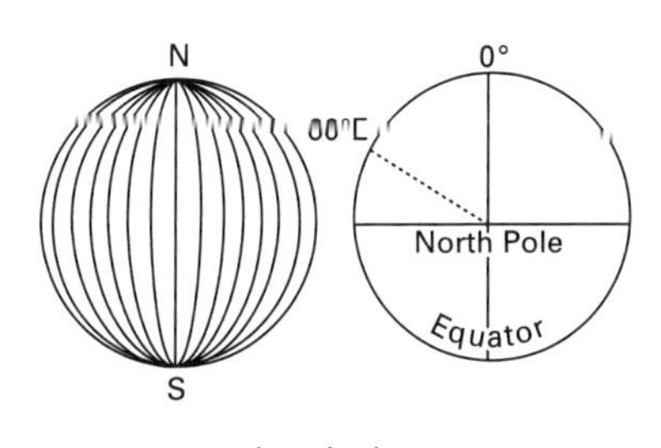

Longitude

longitude The angular distance of a point on the Earth's surface east or west of a central meridian (see diagram), measured by the angle between the plane of the meridian through the point and that of the central meridian. In 1884 it was internationally agreed that the meridian from which the readings would be taken would be the one passing through Greenwich, London, now considered as 0 degrees east, west. In passing through 15 degrees of longitude at the Equator there is a local time difference of one hour. *See also* latitude.

longitudinal wave *See* primary wave.

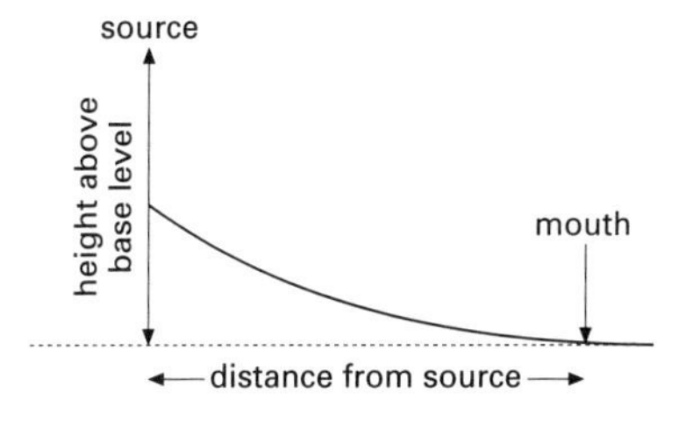

Ideal long profile

long profile A section drawn from source to mouth along a stream or river, showing the changes in gradient throughout its length. The ideal form is an exponential concave curve, with far greater gradients at the head than at the mouth. This profile results from the fact that discharge increases downstream and hence velocity increases, so less gradient is needed to transport the load supplied. Simultaneously the size of debris decreases downstream, most probably owing to abrasion of the debris, but possibly owing to sorting, and so there is a further reduction in energy needed, allowing a further reduction in gradient.

Surveyed profiles in fact rarely appear as the ideal profile; rather they are a series of partial curves, due to past base-level changes, with further irregularities due to lakes, pools, bars, hard rock bands, etc. Insufficient time to grade a smooth profile is a general cause of irregularity, as well as the fact that if a river flows over different lithologies, the size of debris supplied will vary: the coarser the debris the steeper the slope needed, so gradients over sandstone are steeper than those on shale and limestone. *See also* cross profile; river.

longshore bar (ridge, full) A sand BAR lying roughly parallel to the coastline, in the intertidal zone or in the surf zone.

longshore current The current that is generated over the whole width of the SURF ZONE at those times when waves approach and break at an angle to the shore. Longshore currents tend to flow parallel to the shore, and probably attain their maximum rate when the wave crests near breakpoint are angled approximately 30° to the shoreline. Field measurements have revealed that such wave-induced currents are roughly of uniform speed across the whole width of the breaker zone, but that they rapidly weaken seaward of the surf zone. Tidal and other currents may be superimposed upon the longshore currents, tending either to strengthen or to nullify them.

longshore drift (littoral drift) The movement of beach materials, sand or shingle, along the shore parallel with the coastline. This process is basically due to the effects of waves breaking on the beach obliquely. Despite WAVE REFRACTION, waves infrequently break perfectly perpendicularly to a beach and their SWASH tends to move material landward diagonally. However, the returning water, in the form of the BACKWASH, moves directly down the beach and the net effect of swash and backwash is a small longshore movement. Longshore currents aid this process. In most areas one set of waves predominates, resulting in an undirectional longshore movement, but some coasts may be influenced by two sets in opposition, causing the material to move to and fro along the same stretch of beach.

long wave (Rossby wave) The smooth broad wave-shaped contour patterns on a pressure chart of the upper atmosphere. They have wavelengths up to about 2000 km and there are usually either four or five such waves around the westerly wind belt of the N hemisphere. Both heat and momentum are transferred poleward by the long waves and they also have a significant effect on CONVERGENCE and DIVERGENCE at the surface.

lopolith A large concordant igneous intrusion, which is saucer-shaped in cross section. The type example is the Duluth gabbro, Canada, which has a volume of approximately 200 000 km^3.

lough The Irish name for a lake. *Compare* loch.

love wave A surface wave produced by an earthquake in which the particle motion is in a horizontal direction, perpendicular to the direction of wave propagation. *See also* Rayleigh wave.

low *See* depression (def. 1).

low index The state of the westerly circulation when flow is weak and there is appreciable meridional transfer, as in a BLOCKING situation.

low-velocity zone *See* asthenosphere.

loxodrome *See* rhumb line.

luster The appearance of the light reflected from the surface of a mineral, which is a qualitative characteristic often providing an aid to identification. The kinds of luster are:

1. metallic – displayed by galena and pyrites.
2. vitreous – displayed by glass and quartz.
3. resinous – displayed by amber.
4. silky – displayed by gypsum.
5. pearly – displayed by talc and muscovite.
6. adamantine – displayed by diamond.

lutite Any fine-grained sedimentary rock formed of material that was once mud. Examples include MUDSTONE, SHALE, and SILTSTONE.

luxullianite A TOURMALINE-QUARTZ-FELDSPAR rock produced by the boron pneumatolysis of granite.

L wave A seismic wave that travels in the thin uppermost layer of the Earth's crust. It causes multiple reflections between the surface and the top of the layer of rock below. *See also* primary wave; secondary wave.

M

maar A shallow volcanic crater not associated with a single vent or cone. Maars are thought to be formed by many small explosive eruptions, sometimes at the top of a DIATREME. Many fill with water and form lakes.

macroclimate The general climate of a large area. Observations of the macroclimate are normally recorded in Stevenson screens to provide a standard exposure of the instruments and to reduce the effect of the local ground surface. In this way it is easier to obtain more comparable values between observing stations, which are determined by the larger-scale features of the climate rather than local conditions.

macrometeorology The study of the atmospheric processes responsible for the differentiation of the macroclimates on the Earth's surface.

maculose Having a knotted or spotted appearance, usually describing a metamorphic rock. The irregularities are thought to result from the growth of new minerals during heating.

mafic Describing dark-colored minerals. Mafic minerals are approximately equivalent to FERROMAGNESIAN MINERALS. *Compare* felsic.

magma A hot silicate liquid beneath the Earth's surface containing suspended crystals and dissolved gases. IGNEOUS ROCKS are formed from the crystallization and solidification of magma and are said to be magmatic. A lava represents magma extruded at the Earth's surface whereas magma solidifying at depth gives rise to HYPABYSSAL and PLUTONIC ROCKS. During the solidification of magma, the volatile constituents escape and thus compositions of igneous rocks are only approximations of the compositions of the magmas from which they originated.

magma chamber A reservoir within the Earth's lithosphere in which molten rock accumulates before being emptied at the Earth's surface through a volcano. Such a chamber exists beneath most volcanoes, and is replenished from below after being emptied during a volcanic eruption.

magmatic differentiation A number of processes causing the gradual evolution of the composition of an igneous melt and the separation of the MAGMA into two or more fractions of contrasting compositions. Several mechanisms are recognized:

1. *Fractional crystallization.* One of the main processes producing compositional variation in igneous rocks is the gravity settling of early-formed minerals. Such crystals having densities higher than the enclosing magma sink to the bottom of the magma chamber. In some cases, light minerals such as leucite float upward. The composition of the residual liquid differs from that of the original melt because of the removal of the solid crystalline material.
2. *Filter pressing.* During the later stages of crystallization, residual liquid may be squeezed out from a crystal mush and rise upward in the crust.
3. *Flow differentiation.* The movement of a magmatic liquid within a dike or lava flow often concentrates phenocrysts toward the center of the body.
4. *Liquid immiscibility.* During cooling, two liquid fractions of a magma may become immiscible and separate. Sulfides

separate from a silicate magma during the early stages of crystallization. The role of liquid immiscibility in igneous petrogenesis is limited, but this process may account for the segregation of carbonate liquids from mafic alkaline magma.
5. *Gaseous transfer*. The upward streaming of gas bubbles and dissolved materials results in the concentration of volatiles near to the top of a magma body.

magnesite A white, yellow, or gray major mineral form of magnesium carbonate, $MgCO_3$. It crystallizes in the trigonal system and occurs in masses in sedimentary rocks. It is used as a refractory material and as a source of magnesium. *See also* carbonate minerals.

magnetic anomaly A departure from the predicted value of the Earth's magnetic field at a point on the Earth's surface.

magnetic declination *See* angle of declination.

magnetic division A stratigraphical unit (*see* stratigraphy) consisting of the rocks formed during one MAGNETIC INTERVAL (*see also* paleomagnetic correlation). The establishment of magnetic divisions constitutes an important method in the CALIBRATION and CORRELATION of stratigraphic sequences, especially in the Quaternary and late Tertiary.

magnetic elements The seven parameters of the geomagnetic field at a particular locality on the Earth's surface. They include a value for the field with respect to geographic north (X), east (Y), the vertical (Z), the horizontal (B), the total induction (F), its ANGLE OF DECLINATION (D), and inclination (I) (*see* dip).

magnetic equator An imaginary line round the Earth joining all points where the ANGLE OF DECLINATION is zero.

magnetic field A field of force that exists as a result of a circuit carrying an electric current, or the presence of a permanent magnet.

magnetic inclination *See* angle of inclination.

magnetic interval The variable period of time between MAGNETIC REVERSALS in the Earth's polarity. The paleomagnetic scale must be calibrated by some independent method of assessing time, such as radiometric dating. The rocks formed during a magnetic interval are known as a MAGNETIC DIVISION, which can be used in the correlation of stratigraphic successions. *See also* paleomagnetic correlation.

magnetic meridian An imaginary line on the surface of the Earth joining the magnetic poles. It indicates the direction of the horizontal component of the Earth's magnetic field.

magnetic pole Either two of the points on the Earth's surface at which the lines of magnetic force are vertical. These points are slowly changing with time and do not coincide with the geographical poles.

magnetic reversal (polarity reversal) Paleomagnetic studies (*see* paleomagnetic correlation) have shown that the direction of the Earth's magnetic field's induction has periodically changed by 180°, that is north became south and vice versa. Based upon these reversals, which have been recorded in rocks forming at that time, a polarity/time scale has been established for the last 4.5 million years of the Earth's history.

magnetic storm A more or less violent disturbance of the Earth's magnetic field, probably caused by the arrival of charged particles ejected from the Sun in solar flares.

magnetic stripes The roughly parallel bands of ocean floor with alternating directions of induction: a normally magnetized stripe will be enclosed by two reversely magnetized stripes. These patterns can be picked up by shipborne instruments and used to correlate areas of ocean floor offset along transform faults.

magnetite (magnetic iron ore) A shiny black magnetic mineral, an oxide of iron, Fe_3O_4, sometimes containing some magnesium. It crystallizes in the cubic system and occurs as granular or compact masses in various igneous and metamorphic rocks. It is a member of the SPINEL group and an important source of iron.

magnetometer An instrument used for measuring the strength of the Earth's magnetic field.

magnetopause The region where the Earth's MAGNETOSPHERE ends and interplanetary space begins.

magnetosphere The area around the Earth in which the Earth's magnetic field is present. The field is confined to this area as a result of the interaction between the Earth's magnetic field and the solar wind. The magnetosphere extends out away from the Earth's surface for a much greater distance on the side away from the Sun.

malachite A blue-green mineral form of hydrated copper carbonate, $Cu_2CO_3(OH)_2$. It crystallizes in the monoclinic system and occurs in the oxidation zone of copper deposits. It is used for making ornaments and as a semiprecious gemstone, and is a minor source of copper.

malignite An alkaline nepheline-containing igneous rock, a type of SYENITE.

mallee A type of scrubland that occurs in semiarid regions of Australia. The vegetation consist mostly of low-growing kinds of eucalyptus.

Mammalia A class of warm-blooded vertebrates that nourish their young on milk secreted by mammary glands in the female. Mammals have a relatively large brain, morphologically differentiated teeth, and (in most) a body covering of hair or fur. They are skeletally distinguished from reptiles by having only one bone in each lower jaw, which also differs in its articulation with the cranium.

Mammals evolved from mammal-like reptiles in the late Triassic; the earliest forms were small flesh-eating active creatures similar to modern shrews. Fossils of these primitive Mesozoic mammals are rare and consist mainly of teeth. Mammals existed throughout the dominance of the dinosaurs but did not achieve supremacy on land until the Cenozoic, after the extinction of most reptilian groups, when their main radiation took place. The two main mammal groups, the Marsupialia and Eutheria, separated from the ancestral stock by the late Cretaceous. They both give birth to living young (as opposed to laying eggs). The most primitive group is the Monotremata, which comprises egg-laying mammals whose modern representatives are the duck-billed platypus and echidnas (spiny anteaters).

man Modern man, *Homo sapiens*, belongs to the family Hominidae of the mammalian order PRIMATES. Primates that are sometimes classified as hominids are known from the Miocene and perhaps even earlier. In the Pliocene and early Pleistocene of Africa the remains of a variety of manlike creatures have been found; they are usually placed in two genera, *Australopithecus* and *Homo*. Most authorities believe that *Homo* evolved from one of the species of *Australopithecus*. *Homo erectus*, a hominid of robust construction, was present in Africa, Asia, and possibly Europe during the mid-Pleistocene. In the later Pleistocene two species of *Homo* are recognized: modern man, *H. sapiens*, and Neanderthal man, *H. neanderthalensis*. Primitive stone tools have been found from rocks dated as Pliocene in age.

manganese nodule Somewhat similar in form to PHOSPHORITE nodules, manganese nodules (one of the AUTHIGENIC minerals) are the only potential deep-sea mineral deposit that could be economically recovered from the deep ocean floor utilizing present-day technology. They seem to be largely confined to deep-sea clays, but have also been found in association with oozes. The nodules are surprisingly heavy, averaging some 24% manganese, 14%

iron, 1% nickel, 0.5% cobalt, and 0.5% copper. The various minerals frequently form roughly concentric layerings on stones and other hard objects, such as sharks' teeth, present on the sea floor. These layerings build up gradually on account of the slow precipitation from sea water of manganese oxides and other mineral salts. Surveys have revealed that the nodules are widely distributed on the deep ocean floor and are sometimes surprisingly abundant. For example, in certain parts of the Pacific Ocean, concentrations of 20 000 tonnes per sq km have been located.

manganite A black or gray semiopaque lustrous mineral form of hydrated manganese oxide, MnO(OH). It crystallizes in the monoclinic system and occurs in hydrothermal veins, usually associated with BARYTES or CALCITE, and in deep-sea MANGANESE NODULES. It is a minor source of manganese.

mantle That part of the Earth's interior beneath the base of the CRUST and above the Earth's OUTER CORE, roughly between 30–2900 km beneath the Earth's surface. It is thought to consist of ferromagnesian silicate minerals such as olivine and pyroxene. Essentially it behaves as a solid, although within the asthenosphere it may be partly molten.

mantled dome (gneiss dome) A structure present in Precambrian shield areas, consisting of a central core of granite rock surrounded by regionally metamorphosed rocks, mainly gneisses, that has been exposed by erosion and covered by a series of younger sediments. The area was then subjected to orogenesis, during which the granitic core was remobilized and rose up, causing the overlying strata to be arched upward.

mantle nodule A small fragment or block of rock from the Earth's MANTLE that has been brought to the surface by a volcanic explosion.

mantle plume A mechanism that may account for the formation of volcanoes and constructive plate boundaries. Mantle rock begins to melt, thereby becoming less dense. As a result this material rises toward the surface as a plume. In so doing it causes the overlying crust to be domed up. *See also* hot spot.

map A graphic representation on a plane surface of the Earth's surface or part of it, showing its geographical features. These are positioned according to pre-established geodetic control, grids, projections, and scales.

map projection Any method of representing the curved surface of the Earth on a flat plane surface such as a piece of paper. A grid printed on the map represents lines of latitude and longitude, which are drawn according to different mathematical formulas, depending on the area to be mapped. Some projections give the property of equal area (homolographic), others true bearings (azimuthal), and others achieve a compromise.

map scale The ratio of the distance measured on the map to the distance on the ground between the same points. On a map with a scale of 1:50 000, for example, one unit on the map is equal to 50 000 of the same unit (e.g. centimeters) on the ground. Most maps now are metric and therefore have ratios such as 1:50 000, 1:100 000. The larger the ratio, the smaller the scale of the map.

The scale used is largely determined by the purpose of the map: for index maps and wall charts of the world a small-scale map would normally be used; for town plans the larger the scale, the more accurate and detailed the plan can be.

map series A collection of maps, usually at the same scale, using the same specifications and generally identified by a series number. Map series normally cover individual countries but can cover the entire Earth's surface, e.g. the International Map of the World 1:1 000 000.

maquis A type of scrubland that occurs in the Mediterranean districts of France. The vegetation consists mainly of deep-rooted low-growing evergreen bushes and trees, whose leaves are able to survive the hot dry summers.

marble A rock composed largely of calcite or dolomite, produced by the regional or contact metamorphism of limestones. During the metamorphism of a pure limestone, calcite recrystallizes to produce coarser interlocking grains, often with a preferred orientation imparting a weak schistosity. A variety of minerals may form when impure limestones are metamorphosed, including quartz, diopside, forsterite, grossular, tremolite, micas, and epidote.

marcasite A pale bronze-yellow iron sulfide mineral, FeS_2, polymorphous with pyrite but having orthorhombic symmetry. Marcasite is found mainly as a replacement mineral in sedimentary rocks and as radiating nodules, especially in chalk. It is used as a semiprecious gemstone.

margarite A pearly pink mineral form of calcium aluminum silicate, $(Ca,Al)Si_2O_{10}(OH)_2$. It crystallizes in the monoclinic system and is a member of the MICA group of minerals.

margin *See* plate boundary.

marginal sea A category of ADJACENT SEA (*compare* inland sea) consisting of a sea area that is significantly open to the adjacent ocean. Marginal seas may be of the deep type or the shallow (shelf sea) type. The deep marginal seas usually lie between the continental blocks and offshore submarine ridges, which may or may not have islands, and often display depths well in excess of 200 m. The shallow marginal seas occupy the continental shelf areas and are thus generally less than 200 m deep. There is an obvious contrast between the open ocean and those sea areas juxtaposed between mainland and groups of offshore islands.

marigram The graphic record obtained from the use of an instrument that monitors the rise and fall of water level throughout a tide cycle. From this, an average water-level curve, known as a tide curve, can be plotted. Such curves or records enable oceanographers to identify the types of tidal motion affecting particular areas.

marine ecosystem The littoral and sublittoral zones are characterized by an enormous diversity of plant and animal life which, taken with their physical environment, constitute the marine ecosystem. The system is highly dynamic. Most sublittoral plant life occurs as plankton but there are also a number of bottom-living seaweeds. The range of species of sublittoral animals is enormous. The food chain of sublittoral life begins with plant plankton (phytoplankton), which consume nutrient salts present in the sea water and also carbon dioxide. Animal plankton (zooplankton) feed actively on the phytoplankton and are themselves eaten by the smaller species of fish. The small fish are consumed by larger fish, and so on. Decomposition of dead fauna and flora refurbishes the supply of nutrients. Thus, the individual organisms in the marine ecosystem are interdependent. The marine ecosystem is highly susceptible to damage from marine pollution and other forms of human intervention.

marine geology The scientific study of the floor of the ocean and its various physical features, rocks, and sediments.

maritime air mass A large mass of air that moves from over the sea to over the land. Such air masses have significant effects on weather, although the effects of, say, a tropical air mass and a polar air mass will be very different.

maritime climate The generalized characteristic climate of the temperate parts of the Earth's surface that are affected by the sea. The sea acts as a vast storehouse of the Sun's energy and changes its temperature only slightly from winter to summer. Hence areas experiencing a maritime climate have low annual ranges of tempera-

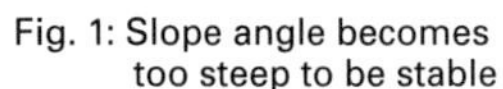

Fig. 1: Slope angle becomes too steep to be stable

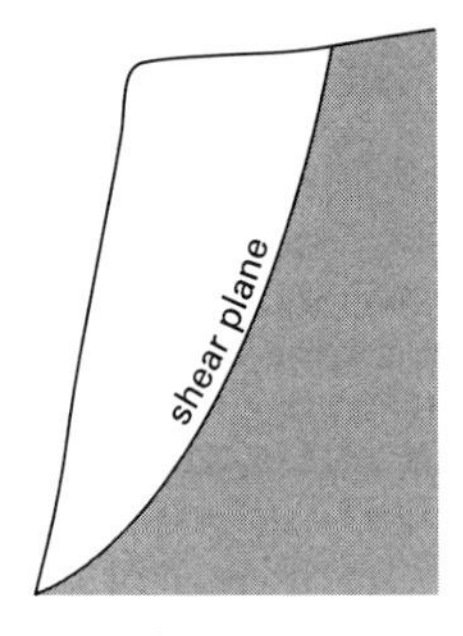

Fig. 2: Shear plane develops

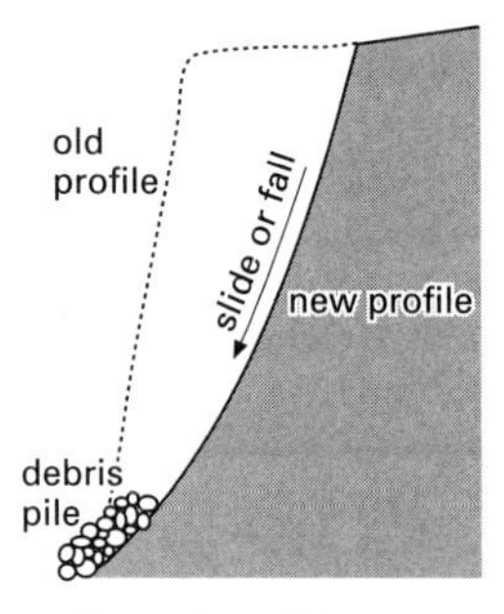

Fig. 3: Slide/fall lowers slope angle

Mass movement

ture (cool summers and mild winters) and an even distribution of precipitation throughout the year with no dry season. Some coastal areas may not experience a maritime climate if the prevailing winds are offshore.

marker bed (marker horizon) A bed of rock having some distinctive characteristic of lithology or contained fossils that permits its easy recognition wherever it occurs. Such a bed can often be inferred to represent a very short period of time or, in the case of beds resulting from certain kinds of volcanic eruption, to be virtually instantaneous in its formation. Marker beds thus have great use in STRATIGRAPHY and in the CORRELATION of strata.

markfieldite A porphyritic microdiorite. *See* diorite.

marl An ARGILLACEOUS sedimentary rock in the form of MUDSTONE that has a high proportion of calcareous material in its composition.

marsh An area of soft wet land that has poor drainage and frequently becomes waterlogged. The dominant vegetation consists of grasses, reeds, and sedges. Marshes occur in depressions left by retreating glaciers and in estuaries, especially where there is a DELTA. On flat low-lying land they generally overlie impermeable rock where the water table is near the surface. If an estuarine marsh is periodically covered by the tide, it can become a *salt marsh*. The draining of marshes to create arable land completely changes the ECOSYSTEM. *See also* bog; swamp.

Marsupialia Mammals whose offspring are born at an immature stage and continue their development within a special pouch (*marsupium*) on the belly of the mother. This and other (skeletal) differences distinguish them from the placental mammals (*see* Eutheria). Fossils show that the marsupials diverged from the placentals in the Cretaceous, when they appear to have been much more abundant than they are today. They are known from the Tertiary of Europe and America, but as the placentals became the more successful group, the marsupials gradually became confined to isolated areas such as Australasia and South America. In these regions marsupial species evolved to resemble placental species elsewhere (*see* parallel evolution).

massif An area of high ground or mountain landscape with no level areas and features that distinguish it from the surrounding terrain. The constituent rocks are generally stiffer and older than those around the massif.

mass movement (mass wasting) On unstable slopes, that is slopes whose angle is greater than the natural ANGLE OF REST of the constituent material and whose SHEAR strength is not capable of maintaining this angle, gravity along with the help of other agents such as wind and water may act to produce a mass movement of some kind, whereby slope angle becomes reduced to a more stable value. Mass movements constitute flows, slides, and falls of rock material from a slope; they may be sudden and short-lived, as in the case of ROCKFALLS, or very slow and long-lived, as in the case of CREEP. Slides and falls act along a shear plane within the rock. Flows, on the other hand, involve internal deformation and movement by overturning of the material. Commonly mass movements are compound: flows on gentler slopes often start on steeper slopes above as a slide or fall. *See also* basal sapping; earthflow; earthslide; landslide.

mass transport current The slow drift of water beneath waves, in the direction of wave propagation, that arises from the nature of ORBITAL MOTION beneath waves. Because, in the case of deepwater waves, the orbits of water particles are almost circular but not quite closed 1oops, each particle of water in motion not only orbits but also moves slowly forward in the direction of wave propagation. This asymmetry or orbital motion in deep water becomes even more pronounced in shallow water, for here the orbits become increasingly elliptical with decreasing depth. Theoretically a cork placed on the sea surface in deep water should both bob up and down and shift slowly in the direction of wave travel.

mass wasting *See* mass movement.

matrix The fine-grained component of a rock into which the coarser components are set. In sedimentary rocks, the matrix is often the material that cements the larger grains or pebbles together. In igneous rocks the GROUNDMASS is equivalent to the matrix.

maximum temperature The highest value recorded by a thermometer in a specified time. The normal period between observations is the day, but the mean monthly maximum and mean annual maximum are often extracted from the daily data. The maximum temperature is recorded by a mercury thermometer, which has a constriction near its bulb. As the temperature rises, the force of expansion is strong and pushes the thread of mercury up the column. If the temperature then falls, the contraction force is weaker and the constriction prevents the mercury returning to the bulb. It remains at the position of the highest temperature. On a world scale, the highest maximum air temperature recorded, 57.8°C (136°F), was reported from Azizia, Libya, on September 13 1922.

meadow soil *See* gley soil.

mean The arithmetic mean or average is a statistical method of representing the magnitude of a data set. It is obtained by summing the individual values and dividing by the number of individuals. It gives meaningful results only when the data have a normal distribution. For example, the mean rainfall in desert areas is often calculated on the basis of many years without rain interspersed with exceptionally heavy falls and may indicate an annual total that has never been experienced.

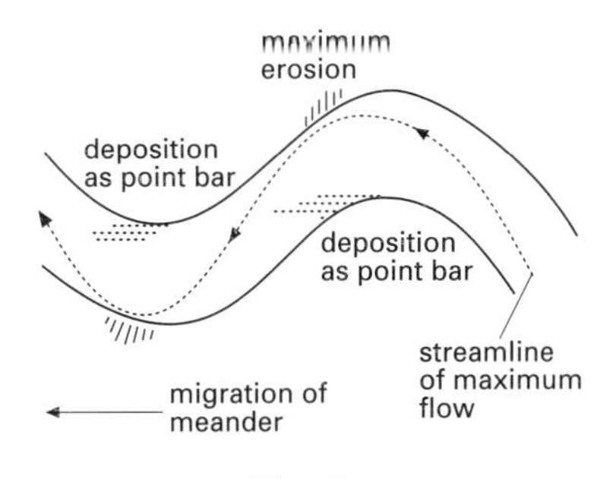

Meander

meander A marked curve in a river channel. W. M. Davis stated that meanders originated when a river used up excess energy by side-cutting when it could no longer cut down, but since Davis other causes have been put forward:

1. Local erosion and deposition patterns will lead to the development of one mean-

der, which will then influence the streamflow pattern, propagating the system of bends down river. The initiation of local erosion and depositions is due to a chance abrasion in the channel, e.g. a fallen tree or stream-bed shoals.
2. Long profiles of rivers characteristically develop POOLS AND RIFFLES. Shallowness over the riffles leads to increased velocity and bed roughness, and hence greater energy loss over the riffles than the pools. Since rivers try to distribute energy loss as evenly as possible throughout their lengths, it may be that meander development in the pools would be one way of equalizing energy loss. Others have said the riffles are the site of initiation.
3. Irrespective of the cause, the divergence of primary downstream flow will set up a secondary HELICAL FLOW concentrating erosion on the outside of bends with deposition on the inside, and therefore continuing meander development downstream. Individual meanders will tend to migrate downstream by this process of erosion of the outside of meander bends, and deposition on the insides. If the erosion and deposition exactly complement each other, the meander train retains its form but migrates down valley. However, erosion rates typically exceed deposition rates in some meanders, leading to modifications.

The distribution of velocity of stream flow in a meandering reach is not even: because of the pattern of helical flow, velocity is highest on the outside of meander bends, least on the inside. There is therefore active erosion in the former areas, deposition in the latter. As the outsides erode and move downstream, so the insides of the bends sediment and move downstream as well. Thus the whole reach migrates down valley.

mean sea level The level of the surface of the sea averaged over all tide states, being actually determined from the average of hourly heights. The periods over which readings are recorded vary; computations have been based on a month, a year, and even a 19-year period. All levels, including those of the sea surface, coastal land elevations, and the levels of coastal structures are referred to a chosen datum plane known as the *tidal datum*. *See also* sea level.

mechanical weathering (physical weathering) The breakdown of solid rock either through the development of cracks produced by strain as a result of internal or external stresses, or by the movement away from each other of the individual grains within a rock. The cause of stresses within rocks are many and mechanical weathering can operate at a number of different scales. The major types of mechanical weathering are: FREEZE-THAW weathering, EXFOLIATION, GRANULAR DISINTEGRATION, UNLOADING phenomena, WETTING-AND-DRYING WEATHERING, FLAKING, and INSOLATION WEATHERING.

medial moraine A ridge or spread of angular material extending lengthwise along the center of a glacier. Moraines may be formed as a result of the joining of two LATERAL MORAINES as two glaciers converge, or of material originating at a spur between two confluent glaciers. Beaded forms occur, reflecting a greater availability of rock waste in summer periods. On the melting of the glacier the medial moraine may be left as a ridge running down the middle of the valley.

median ridge *See* mid-ocean ridge.

median valley (median rift, axial rift zone) A deep cleft that follows the axis of a MID-OCEAN RIDGE, marking the site of a constructive PLATE BOUNDARY.

Mediterranean climate In Köppen's climatic classification, the Mediterranean basin was shown to have a distinctive climate of hot summers and mild wet winters, which he designated a Cs climate. Similar climatic regimes are found in California, Chile, S Africa, and W and S Australia.

meerschaum (sepiolite) A mineral form of hydrated magnesium silicate, $Mg_2Si_3O_6(OH)_4$. It is yellowish, pink, or white and claylike, and is used as a building stone and for making tobacco pipes.

megabreccia A type of very coarse BRECCIA, with the largest fragments more than 1 km across, often the result of rockfalls or landslides. *Compare* microbreccia.

megaripple *See* sand wave.

melange A sedimentary rock consisting of a jumbled mass of various rock fragments including schists, limestones, cherts, quartzites, graywackes, and a wide range of other rock types. They are thought to have been deposited as a result of the slumping of large masses of unstable rock debris, associated with destructive plate boundaries.

melanite A black or dark brown member of the GARNET group of minerals, containing significant amounts of titanium.

melanocratic *See* color index.

melilite A member of a group of minerals with composition $(Ca,Na)_2(Mg,Fe^{2+},Al,Si)_3O_7$ found mostly in thermally metamorphosed limestones and some strongly undersaturated basic igneous rocks such as NEPHELINITES.

melilitite A member of the NEPHELINITE group of minerals.

melteigite *See* ijolite.

meltwater Melted ice from a glacier, which may travel over the surface of the ice, within it, or beneath it, carrying material with it and achieving a certain amount of sorting and rounding of rock fragments in the process. At ice margins meltwater is responsible for the formation of distinctive landforms, such as KAMES and ESKERS, while the presence of meltwater at the ice-rock interface is essential for successful glacial erosion in the form of GLACIAL PLUCKING. *See also* meltwater channel.

meltwater channel A channel cut by glacial meltwater in either solid rock or drift deposits, the most favorable locations being beneath the ice, where water restricted to tunnels can move at high velocity, carrying an erosively powerful load, and marginal to the ice, where large volumes of water derived from the ice surface can accumulate and flow. After the disappearance of the ice, these channels bear little similarity to normal fluvially-developed ones. They may start abruptly, frequently in unlikely locations, crossing present-day drainage divides; they may be unusually straight and steep-sided and may contain only a small stream compared with the channel size, or sometimes no stream at all.

member A division in the lithostratigraphic classification of bodies of rock (*see* lithostratigraphy; stratigraphy). It is part of a FORMATION, distinguished by some particular lithological characteristic.

Mercalli scale A scale devised in 1931 by Giuseppe Mercalli for measuring the intensity of earthquakes in terms of the damage they cause. It ranges from I (no damage or visible effects) to XII (total destruction). It has been superseded by the RICHTER SCALE.

Mercator projection An orthomorphic cylindrical map projection most commonly used for navigation charts (see diagram overleaf). Although there is great exaggeration of area in the higher latitudes, the straight-line bearings are correct. It is therefore used for portraying information of a directional nature, e.g. ocean currents. Similar to the Mercator is the TRANSVERSE MERCATOR PROJECTION. *See also* orthomorphic projection.

meridian A line of longitude passing through any given point and the North and South Poles. Meridians are semicircles of equal length and cut the Equator and all other parallels at right angles. The Greenwich meridian is taken to be the *prime meridian* from which all others are measured. Distance between two adjacent meridians decreases with distance from the Equator. *See also* magnetic meridian.

meridional circulation Any large-scale air movement from south to north or vice versa. *Compare* zonal circulation.

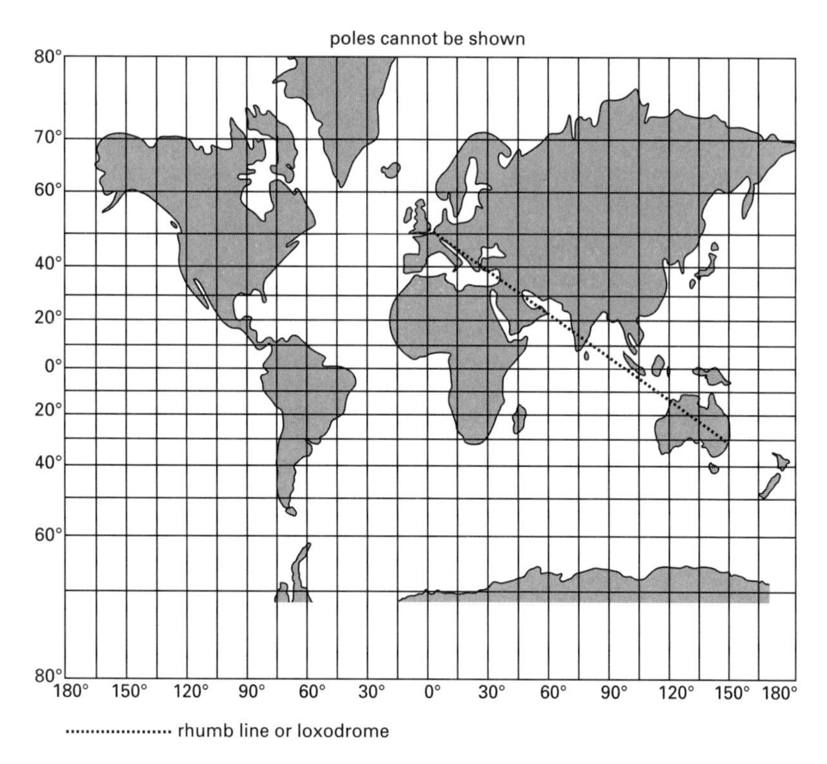

Mercator map projection

mesa An isolated flat-topped hill with steep sides most frequently found in old-established landscapes based on horizontally bedded strata. This often exhibits alternate layers of comparatively hard and soft rocks, which on dissection produce steep and flat slopes respectively. Mesas often have almost vertical upper slopes, developed on the resistant cap rock, followed by much flatter slopes extending down to a general plain. The cap rock may be a lava flow or the same effect can be produced where a resistant soil horizon has been created.

meseta The high plateau region in central Spain that tilts to the west.

mesocratic *See* color index.

mesolite A white fibrous calcium-containing ZEOLITE mineral, whose composition is intermediate between those of natrolite and scolecite.

mesometeorology The study of the atmospheric processes that give rise to weather phenomena having scales that are intermediate in size between those of MACROMETEOROLOGY and MICROMETEOROLOGY. The actual size of systems in mesometeorology is a matter of some dispute but it is somewhere from approximately 10 to 150 km. Mesosystems can be subdivided into those topographically induced, free-atmosphere convective, and free-atmosphere nonconvective, depending upon their mode of origin. A denser network of observing stations is required to obtain adequate data for this scale of investigation than would be needed in macrometeorology.

mesopause The upper limit of the mesosphere at a height of about 80 km. It represents a temperature minimum (–80°C) between the mesosphere and the ionosphere.

mesosphere **1.** The layer of the upper atmosphere between 50 km and 80 km above the Earth's surface. Through the layer temperatures decrease from values of about 0°C at the stratopause, its lower boundary, to about –80°C at the mesopause, its upper boundary. It is not thought to affect weather conditions in the troposphere.
2. That part of the Earth's interior lying beneath the ASTHENOSPHERE.

mesostasis Glassy or late-stage crystallization minerals, such as quartz, alkali feldspar, zeolites, or calcite, which infill INTERSTICES in igneous rocks.

mesothermal *See* hydrothermal process.

Mesozoic The era of PHANEROZOIC time that followed the PALEOZOIC and preceded the CENOZOIC, comprising three periods: the TRIASSIC, JURASSIC, and CRETACEOUS. It lasted about 179 million years, from 245 million years ago to the beginning of the Paleocene, 66.4 million years ago. It is often known as the age of reptiles, for it was during this time that the Reptilia underwent their major evolutionary radiation, producing such diverse forms as the dinosaurs, pterosaurs, and ichthyosaurs. Most reptiles became extinct before the end of the era, and the mammals, birds, and flowering plants first appeared. Invertebrate groups important during the Mesozoic include the Ammonoidea and Belemnoidea. The Alpine episode of orogenic activity began at the end of this era. *See also* geologic timescale.

metabasite Any metamorphosed basic rock.

metamorphic differentiation In medium- to high-grade metamorphic rocks some constituents, especially quartz and feldspar, migrate by a diffusion process into preferred zones, leaving behind micas and amphiboles. Layers of contrasting mineralogical and chemical compositions result and constitute a FOLIATION parallel to the SCHISTOSITY.

metamorphic facies A method of classifying METAMORPHIC ROCKS in terms of the chemical composition of their constituent minerals, which is thought to depend on the temperatures and pressures at which they were formed.

metamorphic rock One of the three main types of rock, formed from one of the other types (sedimentary or igneous) by METAMORPHISM. Well-known metamorphic rocks include amphibolite, gneiss, hornfels, phyllite, marble, migmatite, quartzite, schist, shale, and slate.

metamorphism The process by which mineralogical and chemical changes take place in the solid state in preexisting sedimentary or igneous rocks as a response to the imposition of new physical or chemical conditions. Metamorphic changes take place at temperatures ranging from 100°C to those in excess of 600°C, when rocks begin to melt. The mineral assemblages produced when rocks have attained chemical equilibrium are dependent upon the chemical compositions of the original rocks and the temperature and pressure conditions prevailing, provided that the metamorphism was isochemical and involved no METASOMATISM.

Different types of metamorphism are recognized:
1. *Contact metamorphism*, i.e. the thermal metamorphism developed in rocks that are intruded by hot magma.
2. *Regional or dynamothermal metamorphism*, developed over large areas that have suffered intense deformation and the regional emplacement of granitic bodies. The temperatures and pressures associated with regional metamorphism are those incurred at depth within the Earth's crust.

3. *Dislocation metamorphism*, developed as a result of the mechanical deformation or CATACLASIS of preexisting rocks and localized along shear belts, fault planes, and thrust planes.

Prograde metamorphic assemblages are formed in response to an increase in intensity or grade of metamorphism. Retrograde mineral assemblages are formed in metamorphic rocks that have become subjected to a lower grade as the intensity of the metamorphism wanes. *See also* facies (def. 1), ZONE (def. 1).

metasomatism A metamorphic process by which mineralogical and chemical changes occur in rocks as a result of interaction with migrating fluids introduced from an external source. Metasomatic changes should be distinguished from those produced by pneumatolysis. *See* fenitization; granitization.

Metazoa A group of animals whose bodies are formed of many cells. The Metazoa includes all animals except the sponges (PORIFERA).

meteoric dust Extraterrestrial fine dust that reaches the Earth's surface through entering the Earth's gravitational field. It contributes to a small degree to marine sediments. Some deep-sea cores have been found to contain meteoric dust, often in the form of magnetic spherules, perhaps only 0.1 to 0.5 mm in diameter. Meteoric dust may be transported in the atmosphere for considerable distances across the oceans, and, if present in significant amounts, may later be identifiable in certain seabed samples. One estimate suggests that some five million tonnes of meteoric dust fall onto the Earth's surface annually. Meteoric dust is probably a small though important part of some deep-sea RED CLAY deposits, and meteoric spherules have been found to be plentiful in some slowly deposited deep-sea lutites.

meteorite A solid body that has fallen to the Earth's surface from an extraterrestrial source. Meteorites are composed of a nickel-iron alloy (typically 90% iron and 10% nickel) and silicate minerals, mainly olivine and orthopyroxene. Three main types are distinguished:
1. *Stony meteorites* containing mainly silicate minerals plus some nickel-iron alloy. They are divided into chondrites and achondrites depending upon the presence or absence of small rounded bodies consisting of olivine or orthopyroxene, termed chondrules.
2. *Stony-iron meteorites* contain about equal proportions of silicate minerals and nickel-iron alloy.
3. *Iron meteorites* consist largely of nickel-iron with only accessory silicates.
Certain nonvolcanic glasses, termed *tektites*, are also thought to be of extraterrestrial origin. These are small bodies composed of acid glass and containing lechatelierite. No tektite fall has been witnessed and their origin remains controversial. *See also* SNC meteorite.

meteorology The science of the atmosphere. It is concerned with the physics, chemistry, and movement of the atmosphere and its interactions with the ground surface. Some meteorological interest has also developed in studying the atmospheres of other planets by applying the principles obtained from the Earth's atmosphere to different environmental conditions. Because the main emphasis of meteorology has been on explaining and forecasting surface weather, meteorology is primarily concerned with the troposphere and stratosphere. Higher levels of the atmosphere, where different techniques of data collections are needed, tend to be regarded as part of geophysics. *See also* climatology; macrometeorology; mesometeorology; micrometeorology; synoptic meteorology.

miarolitic cavity An irregular cavity within a plutonic rock into which large euhedral crystals project. *Compare* druse; vugh.

mica A mineral that has a layered structure in which cations are sandwiched between sheets of $(Si,Al)O_4$ tetrahedra and hydroxyl ions. The general formula is

$X_2Y_{4-6},Z_8O_{20}(OH,F)_4$, where X = K,Na,-Ca; Y = Al,Mg,Fe,Li, and Z = Si,Al. The main mica minerals are:
muscovite $K_2Al_4(Si_6Al_2O_{20})(OH,F)_4$
paragonite $Na_2Al_4(Si_6Al_2O_{20})(OH)_4$
margarite $Ca_2Al_4(Si_4Al_4O_{20})(OH)_4$
phlogopite
$K_2(Mg,Fe^{2+})_6(Si_6Al_2O_{20})(OH,F)_4$
biotite $K_2(Mg,Fe^{2+})_{6-4}(Fe^{3+},Al,Ti)_{0-2}(Si_{6-5}Al_{2-3}O_{20})(OH,F)_4$
zinnwaldite
$K_2(Fe^{2+},Li,Al)_6(Si_{6-7}Al_{2-1}O_{20})(OH,F)_4$
lepidolite
$K_2(Li,Al)_{5-6}(Si_{6-7}Al_{2-1}O_{20})(OH,F)_4$
Clintonite and *xanthophyllite* are related to phlogopite by the substitution of calcium for potassium.

All micas have a perfect basal cleavage, which reflects their layered structure. "Books" of mica, found in pegmatites, are so called because they have the appearance of piles of pages. The cleavage flakes of micas are flexible and elastic. Muscovite and paragonite are colorless whereas biotite and phlogopite are usually dark shades of red, brown, and black. Muscovite and biotite are common in schists, gneisses, and granitic rocks. Biotite is also found in basic and intermediate igneous rocks. Phlogopite occurs in some peridotites and metamorphosed limestones. The lithium micas are found almost exclusively in pegmatites.

micrite Chemically precipitated CALCITE having microscopic grains of less than 0.01 mm in diameter, present in some LIMESTONES; the terminology used in the petrographic description and classification of limestones is based on the presence of micrite, SPARITE and ALLOCHEMS.

microbiome A COMMUNITY of living organisms that occupies a comparatively small area. Examples include a hedgerow, pond, or wood. *See also* biome; habitat.

microbreccia A type of rock consisting of very small angular fragments in a finer matrix. Typical particle size is less than 0.2 mm. *Compare* megabreccia.

microclimate The climate within a few meters of the ground surface, resulting from the interaction between soil, atmosphere, and vegetation.

microcline A variety of alkali FELDSPAR.

microfossil A fossil so small that a microscope is required for its study. Microfossils include bacteria, diatoms, protozoa, some Crustacea, the larvae of certain animals, and the isolated skeletal parts or fragments of organisms. Apart from their taxonomic and evolutionary interest, microfossils are important in the correlation of rocks of which only small samples are available, such as the cores from boreholes. The study of microfossils is known as *micropaleontology.*

microgabbro *See* diabase.

microlite A small incipient crystal that is found in glassy rocks. Unlike CRYSTALLITES, microlites are sufficiently large for their nature to be determined under the microscope.

micrometeorology The study of small-scale atmospheric processes, usually operating near the ground surface. Experiments and research in this subject require large numbers of accurate instruments, which result in most studies being limited to small areas and short duration under ideal conditions. The enormous variety of surface conditions results in a complex interaction between the atmosphere and the ground to give multitudinous microclimates; in fact each instrumentation site could be classed as unique.

micropaleontology The study of MICROFOSSILS.

microseism A small irregular Earth tremor and quavering, which makes up the background pattern recorded on seismometers. Such tremors can result from natural sources such as waves breaking on a beach, or from the passing of heavy trucks.

Mid-Atlantic Ridge A very large relief feature forming part of the MID-OCEAN

RIDGE system, which trends north-south through the Atlantic Ocean, connected by Kerguelen Island and St. Paul Island with the Indian Ocean Ridge. The relief of the elevation above the adjacent ocean-basin floor is 1 to 3 km, and the width in most places is more than 1000 km. It is essentially a wide fractured arch taking up the center third of the Atlantic Ocean. Along the crest region are the rift mountains and high fractured plateaus, jointly making up a strip 80 to 300 km wide. Some of the higher parts of the ridge rise above sea level to form islands.

mid-ocean ridge (median ridge) A massive submarine elevation that extends, in global terms, over a distance of about 60 000 km. It traverses all of the major oceans. Its general form is that of a broad arch that stands, on average, 1 to 3 km above the adjacent deep-sea floor, and which has experienced fracturing along its crest region. Along much of its length, the ridge is more than 1000 km wide. Parts of the ridge crest region are gashed by a steep-sided rift feature (*see* median valley), whereas other fracture zones (some of them known as TRANSFORM FAULTS) tend to be perpendicular to the main axis of the ridge. The rift is coincident with a belt of shallow earthquake epicenters. The floor beneath the ridge is generally hot and the basaltic rocks of the ridge-crest and ridge-flank region appear to be quite young in geologic terms.

In terms of the PLATE TECTONICS theory, such mid-ocean ridges mark the sites of constructive plate boundaries. It is at this point that basaltic lava from the Earth's mantle reaches the surface to be added to the oceanic crust in the form of pillow lavas and dikes. Because these eruptions are relatively frequent by geologic scales the area does not have time to adjust immediately to regain isostatic balance but sinks as the sea floor spreads away from the ridge.

migmatite A coarse heterogeneous gneissose rock consisting of bands and patches of quartz-feldspar granitic material and mafic material consisting mainly of biotite or hornblende. Migmatites have the appearance of mixed rocks, comprising portions that resemble granitic igneous material and mafic portions that are metamorphic and possess a SCHISTOSITY. They are associated with metamorphic rocks that have formed at the highest grades. Under such conditions, partial melting takes place producing mobile granitic material, which invades and reacts with the metamorphic rocks.

Milankovitch radiation curves The Earth does not revolve in a circular orbit round the Sun at a constant velocity, but over a period of many years changes take place affecting the amount and distribution of solar radiation received by the Earth. First, the tilt of the Earth's axis of rotation relative to the plane of its orbit varies between 21.8° and 24.4° over a period of 40 000 years. This affects the seasonality or thermal range between summer and winter. Secondly, the ellipticity or eccentricity of the Earth's orbit varies over a period of about 100 000 years. This results in a greater seasonal range of radiation receipt. When the orbit is at its most eccentric, there will be a 30% difference between aphelion and perihelion, compared with 7% at present and none when the orbit is circular. Finally, the season when the Earth is nearest the Sun changes over a period of about 21 000 years. At present it is nearest in the N hemisphere winter (January 7) and farthest in N hemisphere summer (July 7). The effects of these radiation changes vary from latitude to latitude. The variations likely to be produced were calculated by Milankovitch in 1940. It is believed that these cyclic changes may have great importance for the commencement of ice ages, but the association is not as close as might be expected.

millibar (mb) One thousandth of a bar, a cgs unit of pressure. Because 1 bar is equal to a pressure of 10^6 dynes per sq cm, one millibar is equal to 1000 dynes/cm^2 or 100 newtons/m^2 (pascals). The millibar is still in use in meteorology.

milligal *See* gal.

Millstone Grit The middle of the three lithological divisions of the CARBONIFEROUS System in Britain. It corresponds to the lower half of the Upper Carboniferous (or Namurian) Stage.

mineral Any naturally occurring substance having a definite chemical composition. When a mineral of commercial value is present in some rock body in economically sufficient quantity to merit extraction, the body constitutes a mineral deposit.

mineralization The process by which minerals are formed within a rock, resulting in a vein (lode) or other deposit. There are various mechanisms by which this may take place, including replacement (of one mineral by another), impregnation (of rock by mineral-laden gases or liquids), and the filling of preexisting fissures.

mineralogy The scientific study of minerals. It includes their classification and deals with their composition, formation, and physical and chemical properties.

minette A type of LAMPROPHYRE consisting of biotite and orthoclase. It occurs in DIKES and other intrusions.

minimum temperature The lowest temperature recorded during a given period either in a Stevenson screen (minimum air temperature) or on the ground (grass minimum temperature). Both temperatures are normally measured by an alcohol thermometer in which the meniscus drags a metal index to the lowest temperature then leaves it as the column rises when temperatures increase. Global minimum temperatures are found on the Antarctic ice caps.

Miocene An epoch of the TERTIARY Period extending from the end of the OLIGOCENE about 23.7 million years ago, for 18.4 million years to the beginning of the PLIOCENE. Grasses evolved and spread and, by their effects on the environment, may have contributed to changes in the mammals, which became more modern in appearance. Miocene mammals included pigs, deer, horses, rhinoceroses, elephants, monkeys, apes, and possibly hominids. Alpine orogenic movements took place during this epoch.

miogeosyncline A geosyncline in which there are no volcanic products or processes associated with its sedimentation.

mirabilite (Glauber's salt) A yellow mineral form of hydrated sodium sulfate, $Na_2SO_4.10H_2O$. It crystallizes in the monoclinic system, and occurs in salt lakes and around hot springs. It is used as a source of sodium sulfate.

mirage An optical phenomenon resulting from the refraction of light through layers of air having very large temperature gradients. The rapid change of temperature vertically produces varying refractive indices of the air and so light appears to travel in curved paths. The most common is the inferior mirage above a heated ground surface. Light is strongly refracted upward near the surface so that the apparent pools of water on the ground are, in reality, refractions from the clear sky. A superior mirage may occur above a flat surface of much lower temperature than the air above, as over ice or a cold sea in summer. Light in this case is bent downward from the object toward the viewer.

misfit river A river that is apparently too small for its valley. It is characterized by meander bends in its channel of smaller amplitude and greater intricacy than the meanders in the valley itself. The length of the valley meanders is ten times that of the current meanders. There is an established relation between meander length and discharge in rivers, and on that basis it has been calculated that the rivers that eroded the valley meanders must have had a bankfull discharge 80–100 times the current streams. This probably occurred at the end of the Ice Age, when the glaciers thawed; the periglacial climate of that period would have reduced vegetation and soils to a minimum, allowing all the meltwater to flow off the land to the rivers, with none lost by infiltration into the soil. The resulting torrents cut the current valley.

Misfits could also evolve through the capture of a river's headwaters, reducing its volume, and leaving it a misfit in its valley. *See* river capture.

mispickel An old name for ARSENOPYRITE.

Mississippian A period within the American classification of geologic time that extends from the end of the DEVONIAN, about 360 million years ago, for about 40 million years to the beginning of the PENNSYLVANIAN. The Mississippian System corresponds approximately to the Lower CARBONIFEROUS elsewhere. It was named for the Mississippi River Valley where the rocks associated with this period are well exposed.

mist The state of atmospheric obscurity intermediate between fog and haze. It is composed of suspended microscopic water droplets with an atmospheric relative humidity of at least 95%. The term is also used for dense but shallow condensation phenomena at ground level when the ground surface may be totally obscured, but hedges or animals can be seen quite clearly.

mistral A strong cold dry northerly wind that is funneled down the Rhône Valley in S France and then blows across the delta into the Mediterranean Sea. Its frequency and strength have necessitated agricultural adaptations to prevent crop damage. These take the form of hedges or screens oriented east-west to protect plants. The winds are linked to intensification of depressions in the Gulf of Genoa.

mixed tide Any tide that is not clearly of the diurnal or semidiurnal type. Such tides might be referred to as intermediate tides. The tidal wave is characterized by a significant inequality in either the low-water or the high-water levels and a double low water and a double high water occur during each tidal day.

mixing ratio The ratio of the mass of an atmospheric gas to the mass of air with which the gas is mixed. It is most frequently used for water vapor, the *humidity mixing ratio*, and is expressed in g kg^{-1}.

mobile belt (orogenic belt, fold belt) A long linear area of the Earth's crust undergoing intense deformation, often accompanied by seismic and volcanic activity. These structures originate as deep trenches, which fill up with a thick wedge of sediment. As two continents later collide this wedge between them is compressed and deformed to form a mountain chain.

mobile dune A coastal sand dune at the stage when its vegetation (mainly marram grass) is open, leaving sometimes 50% or more bare sand subject to wind movement and hence mobile. Their appearance is of a confused mass of sandy depressions and grassy knolls superimposed on a profile that shows a gentle windward slope and a steep lee slope.

mock sun (parhelion) An optical phenomenon seen in the sky as a result of the refraction of sunlight by hexagonal ice crystals with vertical axes. This is most frequent with cirrostratus clouds. Usually two mock suns are seen, equal distances on either side of the Sun, and appear as brighter areas of the cirrostratus clouds.

mode The percentage of each of the component minerals contained in an igneous or metamorphic rock. *Compare* norm.

moder humus Humus intermediate between mull and mor.

mofette A type of SOLFATARA that also produces large amounts of carbon dioxide. *See also* fumarole.

mogote A large limestone hill that rises on a tropical or subtropical KARST landscape.

Mohole project An abandoned project originally proposed in order to obtain samples of the rocks of the upper mantle. It was hoped to drill down through the Earth's crust to the Mohorovičić discontinuity.

Mohorovičić discontinuity (Moho) The seismic boundary that marks the junction between the base of the Earth's crust and the top of the mantle. This boundary varies in depth below the Earth's surface from 5–10 km beneath the ocean and between 55–70 km beneath the continents. The boundary is marked by an increase in the P wave velocity from 6.9 to 8.2 km/sec.

Mohs' scale A scale of hardness devised to aid the identification of minerals. Ten reference minerals were chosen, the softest, talc, being assigned a hardness of 1 and the hardest, diamond, 10. Each mineral on the scale can be scratched by those of higher numbers: 1. talc. 2. gypsum. 3. calcite. 4. fluorite. 5. apatite. 6. orthoclase. 7. quartz. 8. topaz. 9. corundum. 10. diamond.

Other common substances are useful in testing hardness: finger nail (2.5), copper coin (3), glass (5.5), and a penknife blade (6.5). The hardness of many minerals varies with direction. For example, the hardness of kyanite varies between 5.5 and 7.

molasse A term originally used by Alpine geologists for a particular sequence of Miocene sediments. The term has since been used to denote sequences of sediments deposited under continental or freshwater conditions after a period of mountain-building. *Compare* flysch.

mold A fossil in which the original skeletal parts have been dissolved away, leaving a space that preserves their shape. In some cases a mold may preserve features of only the outer or inner surfaces of such structures as shells. These are known respectively as *external* and *internal molds*. The infilling of a mold with secondary material produces a CAST.

mollisol One of the ten soil orders of the SEVENTH APPROXIMATION classification. Mollisols are soft calcimorphic soils, developed in subhumid to semiarid conditions, characterized by a dark strongly structured surface horizon, often almost one meter deep and more than 50% saturated with bases (mainly calcium), which is termed a *mollic epipedon.* It includes the CHERNOZEM, CHESTNUT SOIL, PRAIRIE SOIL, RENDZINA, BROWN EARTH, BROWN CALCAREOUS SOIL, and associated types such as humic gley, and solonetzic soil.

Mollusca The phylum of invertebrate animals that includes snails, mussels, and octopuses. Mollusks are unsegmented creatures having a head, a visceral region, a muscular ventral organ for locomotion known as the *foot*, and a *mantle*, which generally secretes a hard calcareous shell. The geologically important classes of this phylum are the GASTROPODA, the BIVALVIA, and the CEPHALOPODA, which includes important extinct forms such as the AMMONOIDEA and BELEMNOIDEA. Ancient mollusks seem to have been marine animals inhabiting shallow waters and their shells are known as fossils from the Cambrian Period onward. Some, such as the Bivalvia in the Carboniferous, have been used in stratigraphic CORRELATION.

Mollweide's projection A map projection showing the entire Earth's surface on an ellipsoidal base. The major axis (Equator) is twice the length of the minor axis (central meridian). The geographic parallels are straight lines, as is the central meridian; the other meridians are curved, the curvature increasing toward the outside limits of the projection. The distortion of shapes and directions is therefore considerable, although the projection is homolographic and consequently used for distributions. This projection is sometimes interrupted (see diagram overleaf): each continental area has its own central meridian (C.M.), with different ones north and south of the Equator; the interruptions are in the oceans.

molybdenite A silver-gray molybdenum sulfide mineral, MoS_2, found as an accessory in acid igneous rocks and in hydrothermal veins.

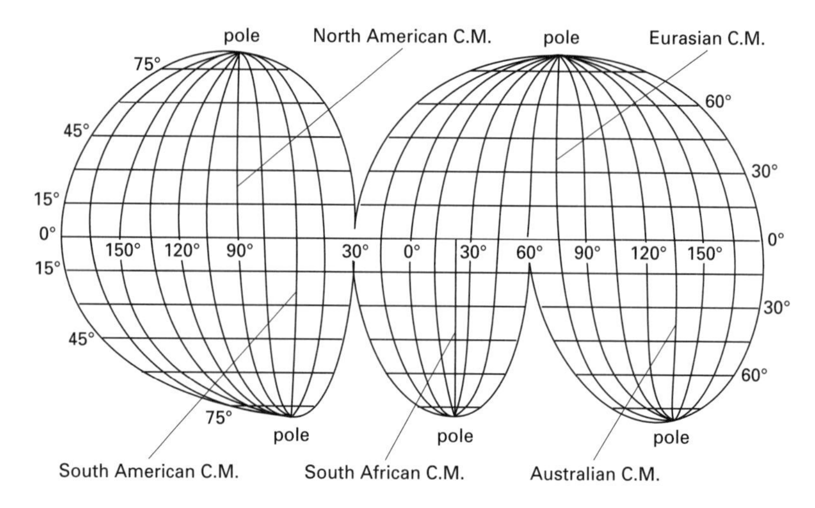

Mollweide's interrupted projection

monadnock Any hill standing up above the surface of a flat or gently undulating plain. These hills may be composed of rocks that are more resistant to weathering than those around them; alternatively they may be located in former drainage divide positions. In the latter case they will have been subjected to active fluvial erosion for a shorter period of time than the rest of the area and it is for this reason that they remain as upstanding masses.

monazite A rare-earth mineral of composition $(Ce,La,Y,Th)PO_4$ found as an accessory in acid igneous rocks and as a detrital mineral in certain beach sands.

monchiquite An alkaline type of LAMPROPHYRE.

monocline A bending of rock strata produced in sedimentary sequences that have deformed under conditions favoring the development of a normal fault. Two areas of horizontally bedded sediments are left at different elevations but still connected by a steeply inclined series of the same beds. See diagram at FOLD.

monoclinic *See* crystal system.

monogeosyncline *See* trench.

monomineralic Describing a rock consisting almost wholly of one mineral, e.g. dunite, anorthosite.

monsoon A large-scale reversal of winds in the tropics (from the Arabic word for *season*) due to differential heating of continent and ocean, but because the circulation can be affected by many other factors, the monsoon varies in character from one area to another. It is best developed in India, SE Asia, and China but N Australia and E and W Africa all show similar wind reversals. The summer season is normally the rainy period of the monsoon with southwesterly winds in the N hemisphere, the dry season having winds from the N sector.

Monsoon Current In the N part of the Indian Ocean, the surface currents reflect closely the seasonal changes in the wind direction as the monsoon season develops. The North Equatorial Current is clearly apparent during February and March, at the time when the northwest monsoon is blowing. An Equatorial Countercurrent is also strongly developed, with its axis aligned roughly along latitude 7°S. However, when the southwest monsoon commences in August–September, the North Equatorial Current is replaced by the Monsoon Current and flows eastward. These currents are generally located within the uppermost layers of water. *See also* equatorial current.

montane Describing a mountain or mountainous area, or something associated with it, such as a montane tropical forest.

monticellite A calcium-bearing variety of OLIVINE.

montmorillonite A type of CLAY MINERAL consisting of a hydrated aluminum silicate with the unusual property of swelling when it absorbs water. It also undergoes ion exchange of its calcium, potassium, and sodium ions. It is the main component of bentonite and fuller's earth.

monzonite *See* syenodiorite.

moonstone A bluish or silvery iridescent type of orthoclase (an alkali FELDSPAR), used as a semiprecious gemstone. *See also* sunstone.

moraine A depositional feature composed of glacial TILL, which may be in association with active ice or deposited by former glaciers and ice sheets. This material may be transported on the surface of the ice, within the ice (*englacial*), or beneath the ice (*subglacial*). *See* ground moraine; lateral moraine; medial moraine; push moraine; terminal moraine.

mor humus Acid humus, which forms in wet lime-deficient environments, e.g. podzolic conditions. Most of the organic breakdown is achieved by the fungi in the soil but it is a slow process with the result that the L, F, and H layers (*see* profile) are clearly differentiated.

morphogenetic zone (in climatic geomorphology) An area of the Earth in which climate and relief are distinctive, through the control of climate on geomorphic process and the control of process on landform. These zones run roughly parallel to major climates, which conform to latitude. Early attempts to divide the Earth into such zones used very simple climatic data (mean annual rainfall and mean annual temperature) but modern attempts use much more sophisticated parameters. Tricart and Cailleux identified four major world zones, running from a cold zone in the north, through a forested zone and subarid zone to the equatorial tropical zone; within each, climates and landforms are in sympathy through the link of process.

moss agate A type of CHALCEDONY that has fernlike inclusions of black dendritic PYROLUSITE (manganese dioxide).

moss animal *See* Bryozoa.

mother-of-pearl *See* nacre.

mountain wind A form of KATABATIC WIND in mountainous areas where, after night-time radiational cooling, strong breezes are concentrated in the valleys as the cold air has subsided downslope.

mud Sediment having a preponderance of particles measuring less than 0.06 mm in diameter. Under natural conditions, muds generally form very plastic masses comprising sediment particles mixed with water. The term applies to both silty and claylike particles. Mud may be deposited, according to prevailing conditions, in both shallow water and deep-sea areas. Sheltered environments are particularly susceptible to mud deposition, as for example in sheltered lagoons and estuaries, and along low-energy beaches and tidal flats. Once partly consolidated into mud banks, the material tends to be remarkably cohesive and therefore resistant to scour; many mud banks are eroded primarily through side-scour because of the gradual migration of tidal channels.

mud crack *See* desiccation crack.

mudflat A low area of fine silt that lies along the shore of an estuary or on the lee side of an island. It supports no vegetation and is generally covered and uncovered by the tide. Various types of arthropods, mollusks, and worms live in the mud, and as a result mudflats are important feeding grounds for several species of wading birds.

mudflow *See* earthflow.

mudrock *See* mudstone.

mudstone (mudrock) An ARGILLACEOUS sedimentary rock that is less fissile along bedding planes than SHALE.

mud volcano *See* sand volcano.

mugearite An alkaline basalt. *See* trachybasalt.

mull humus Alkaline humus, which commonly forms in a mild, moist, and base-rich environment, e.g. chernozem conditions. It is rapidly incorporated into the soil by the abundant soil fauna, particularly earthworms.

mullion A tectonically produced structure in metamorphic and sedimentary rocks that resembles a rod or column.

mullite A colorless mineral form of ALUMINUM SILICATE, approximate formula $Al_6Si_2O_{13}$. It crystallizes in the orthorhombic system and is used as a refractory.

multiple intrusion An intrusion of igneous rock (*see* intrusive rock) consisting of several masses of similar composition. Most DIKES are multiple intrusions.

multiple reflection On a seismic profile, a seismic wave may be recorded more than once as a result of it having been reflected more than once.

multispectral satellite imagery *See* satellite.

muscovite One of the major types of MICA, $K_2Al_4(Si_6Al_2O_{20})(OH,F)_4$. It crystallizes in the monoclinic system as colorless, green, or pale brown plates and occurs in various igneous rocks, mainly pegmatites. It is used as an electrical insulator and lubricant.

muskeg A type of BOG that occurs in TUNDRA areas. It forms when the surface permafrost melts in summer, and generally supports lichens and mosses, such as sphagnum moss. The best-known muskegs are found in N Canada.

mylonite A rock formed during dislocation metamorphism by extreme granulation and shearing. Mylonites are banded or streaky rocks and may contain augen of undestroyed crystals or rock fragments in a fine-grained matrix. They are commonly found at thrust planes along which extensive movements have taken place. *See* cataclasis; pseudotachylite.

myrmekite Wormlike intergrowths of quartz in plagioclase feldspars produced as a result of the replacement of potassium feldspar by sodium feldspar, during which excess silica is liberated.

N

nacre (mother-of-pearl) A lustrous iridescent coating on the inside of the shells of various mollusks, such as oysters. It consists mainly of calcium carbonate, $CaCO_3$, and is the substance that PEARLS are made of.

nappe A large tectonic feature that owes its origin to a combination of folding and thrusting. Nappes are basically large horizontal recumbent folds that have traveled for many tens of kilometers along thrust planes. They are well developed in the Alps, where they were first described.

native Denoting the state of occurrence of certain elements or minerals when they are uncombined with other elements or minerals.

natrolite A colorless, white, or yellow mineral form of hydrated sodium aluminum silicate, $Na_2(Al_2Si_3O_{10}).2H_2O$. It crystallizes in the orthorhombic system as needlelike crystals, usually in cavities in basaltic rocks. It is a member of the ZEOLITE group of minerals.

natural arch An archway on an exposed headland that results from erosion by wave action. Generally, a cave in chalk or limestone rock becomes enlarged so that it cuts through to the other side of the headland.

natural bridge A bridge of rock formed in limestone by erosion. It generally results when the roofs of two underground caverns situated close to each other collapse, leaving a bridge of rock between the resulting holes in the ground.

natural gas A mixture of flammable hydrocarbon gases – mainly ethane and methane – that occurs in underground deposits (often below the sea bed), either alone or in association with oil. It is thought to have formed from the remains of microscopic organisms that became buried and compressed in underground sediments. It is an important FOSSIL FUEL.

natural selection The process proposed by Darwin in 1859 to explain the EVOLUTION of organisms from one species to another. It attempts to account for the origin and diversity of all organisms. Plant and animal species produce relatively large numbers of offspring that differ in various ways. If an organism possesses certain variations that make it better suited to its environment, this organism is more likely to survive and reproduce. If these variations are inheritable the likelihood of its offspring also possessing them is increased. In this way, and over many generations, radical changes may arise.

nautical mile A distance equal to 1 minute of longitude along a GREAT CIRCLE, equivalent to about 1853 m. A speed of 1 nautical mile an hour is a knot.

Nautiloidea An almost extinct subclass of mollusks of the class CEPHALOPODA. The external shell of the nautiloids is similar to that of the related AMMONOIDEA, except that in fossil forms it is straight or only slightly coiled and in all types the suture lines are simple curves, showing none of the complex convolutions characteristic of the Ammonoidea. The oldest fossils come from upper Cambrian rocks and their shells sometimes reached a length of up to 4.5 m. These straight-shelled forms were extinct by the end of the Triassic Period.

The group as a whole declined from the Devonian Period onward; very few genera existed during the Mesozoic and only one, *Nautilus*, has persisted to the present day.

neap tide A TIDE of relatively small range that occurs near the time of quadrature of the Moon (i.e. near the time of the Moon's quarters). The tidal range falls below the average range by about 10 to 30%. Because of this low range, the tidal current speeds during a neap tide are significantly less than those attained during a SPRING TIDE.

nearshore circulation system The system of currents, notably those induced directly or indirectly through wave activity, that operate within and closely adjacent to the BREAKER zone. These currents include the shoreward-directed MASS TRANSPORT CURRENTS beneath waves, the wave-induced LONGSHORE CURRENTS, the powerful seaward-flowing RIP CURRENTS, and other currents associated with the expanding rip-head zones. Superimposed on these currents there is invariably a reversing tidal current, and, on those occasions when local winds are blowing, wind-induced currents. In the vicinity of rivers and estuaries, there may also be density-induced currents. The relative importance of the various types of current varies both spatially and temporally.

nebular hypothesis An idea proposed by the Marquis de Laplace in 1796 in which he suggested that the material now forming the Sun and planets originated as a disk-shaped nebula or gas cloud, which contracted into discrete bodies.

neck (plug) An erosional remnant representing the former conduit that fed a volcano. This conduit was filled with lava and pyroclastic material that was more resistant to erosion than that of the enclosing volcanic cone, which has since been eroded away.

needle ice Needles of ice, often several centimeters long, growing beneath the surface layer of debris and able to lift frost-shattered rock fragments and soil particles perpendicularly from the ground surface. On melting of the needles (often called *pipkrakes*), the fragments are deposited a little way downslope. Material can progress considerable distances and, as the process tends to be more effective in coarser material, it also achieves a certain amount of sorting. Needle ice occurs in periglacial environments, most effectively where there is diurnal freezing and thawing.

negative gravity anomaly *See* gravity anomaly.

negative movement of sea level A rise of the land relative to the sea due either to an actual rise of the land through tectonic movements, or a fall in sea level through the locking up of water in the form of ice, as happened in the Pleistocene period. Whatever the cause, more land is created, and the relative relief of the land, from sea level to mountain tops, is increased. The major result is the extension of river systems out across former sea-bed areas and their REJUVENATION. Raised beaches, marine beaches, knickpoints, river terraces, and incised meanders are all evidence of past negative movements. *See also* positive movement of sea level.

nekton (nektonic organisms) Pelagic organisms that swim and move actively, such as fish. *Compare* plankton.

nematath A series of volcanoes that form as one of the Earth's LITHOSPHERIC PLATES (tectonic plates) moves slowly across a thermal center (*see* hot spot). The farther a volcano is from the thermal center, the older it is. The youngest volcanoes are generally still active, whereas the older ones may be dormant.

Neogene The upper part of the TERTIARY Period, comprising the MIOCENE and PLIOCENE epochs. It lasted for about 22 million years, from the end of the OLIGOCENE Period, 23.7 million years ago, to the beginning of the PLEISTOCENE. Some authorities suggest that the Neogene should be considered as a period, with the Tertiary as a subera.

nephanalysis The analysis of cloud patterns on weather satellite photographs. The photographs cover large areas of the Earth's atmosphere and surface and new techniques were required to describe and analyze the larger features of cloud patterns compared with the apparently random distribution seen by the surface observer. On a nephanalysis the following information can be portrayed: cloud elements, cloud masses, cloud patterns, cloud systems, cloud bands, cloud types (stratiform, cumuliform, cirriform, cumulonimbus, or stratocumuliform), percentage cloud cover, cloud boundaries, and synoptic interpretations of these features.

nepheline A major member of the FELDSPATHOID group of minerals.

nephelinite A strongly undersaturated basic to ultrabasic feldspar-free volcanic rock, having a plutonic equivalent within the IJOLITE-MELTEIGITE series. Nephelinites consist of NEPHELINE which may occur both as euhedral phenocrysts and as anhedral groundmass grains, together with a variety of mafic minerals, typically titanaugite, biotite, olivine, and magnetite.

Assemblages of LEUCITE plus mafic minerals constitute the *leucitites*. Ugandite is an augite-rich olivine-leucitite. *Melilitites* are composed of melilite plus mafics and often contain small amounts of nepheline and calcite in addition. *Alnoite* is a dike rock largely composed of melilite and biotite together with pyroxene, calcite, and olivine. Mafurite is an olivine-kalsilite rock.

Nephelinites and leucitites are compositionally related to phonolites and leucitophyres. These volcanics are restricted to stable continental areas and occur in association with ijolites and carbonatites.

nephrite A tremolitic AMPHIBOLE. *See also* jade.

neptunian dike A roughly vertical body of sediment, usually sandstone, that has infilled a fissure in a preexisting rock body exposed at the Earth's surface or on the sea floor.

neritic Describing the shallow-water marine environment lying between low-tide level and a depth of some 200 m. The environment rarely covers a distance from the shore in excess of several hundred kilometers; on a global scale, it covers roughly 30 million sq km or about 10% of the total ocean area. The environment is characterized by very diverse sediments (of which fine to coarse terrestrially-derived sediments mixed with what are mostly calcareous organic remains tend to predominate) and BENTHOS, and a wide array of environmental conditions, as for example in the degree of light penetration. The whole of the sea floor in the neritic zone probably experiences some measure of wave disturbance and almost certainly tidal current action.

neutron logging A subsurface logging technique used in uncased boreholes to record the presence of fluids. When used in conjunction with a gamma-ray log it can be used to calculate the porosities of the formations encountered. It measures the intensity of radiation in the borehole after it has been exposed to a radioactive source.

nevé 1. *See* firn.
2. The area in which firn exists; firn field.

New American Comprehensive Soil Classification *See* Seventh Approximation.

New Red Sandstone A succession of red sandstones forming the continental facies of the PERMIAN and TRIASSIC Systems in Britain.

niccolite (nickeline) A copper-colored mineral form of nickel arsenide, NiAs. It crystallizes in the hexagonal system, and occurs associated with ores of copper and silver. It is used as a source of nickel.

niche (in ecology) The place or status of an organism in its COMMUNITY or ECOSYS-

TEM. The niche can be defined in terms of its moisture and temperature and, from the organism's point of view, the availability of food and the presence of competitors, and how well adapted it is to live there. No two species can occupy exactly the same niche, or they would be in direct competition all the time.

nickeline *See* niccolite.

nimbostratus cloud The main rain-bearing cloud of temperate latitudes. It consists of a gray cloud layer of appreciable depth beneath which are frequently low ragged clouds, called FRACTOSTRATUS CLOUDS or scud. They are most often associated with low-pressure areas and precipitation may be prolonged, although it is not usually heavy.

niter (saltpeter) A colorless or white mineral form of potassium nitrate, KNO_3. It is found as a surface deposit in caves and in dry regions. It has many uses: in curing meat; in making glass, pyrotechnics, and explosives; and as a fertilizer. *See also* Chile saltpeter.

nitrification The conversion of nitrogen by bacteria to a form usable by higher plants. When material decays, ammonia is often liberated and this is toxic to many plants. Some bacteria, namely *Nitrosomonas*, can oxidize ammonia to nitrite and in turn it may further be oxidized, by *Nitrobacter*, to nitrate, which can be used by plants. *Denitrification*, a conversion of nitrates back to nitrogen gas, most of which is unusable, may occur when soils become poorly aerated and acid. *See* nitrogen cycle.

nitrogen The most abundant gas in the atmosphere and a critical constituent in the soil, which can only be used directly by a few specialized bacteria. To be of widespread value it has to be converted into the nitrate form (*see* nitrification). In nature, nitrogen is involved in cyclic changes termed the NITROGEN CYCLE.

nitrogen cycle The series of chemical reactions by means of which nitrogen circulates through the global ECOSYSTEM. Nitrates and other inorganic nitrogen compounds in the soil are taken up by plants and turned into proteins and other organic nitrogen compounds. These are eaten by herbivores, which are in turn eaten by carnivores. The excreta and dead bodies of all these animals returns nitrogen to the soil, where decomposers (such as fungi and bacteria) reconvert it to an inorganic form. Some bacteria that live in the soil cycle nitrogen between the atmosphere and plants. *See* nitrification; nitrogen fixation.

nitrogen fixation The absorption of atmospheric nitrogen by some heterotrophic bacteria. There are two groups: the non-symbiotic, e.g. *Azotobacter*, which make the nitrogen part of their tissue so that when they die it is made available by nitrification; and the symbiotic bacteria, e.g. *Rhizobium*, which form nodules on the roots of leguminous plants and pass nitrogenous compounds on to their host.

nivation (snow patch erosion) Any of a number of processes associated with a patchy snow cover over unconsolidated rocks, such as glacial deposits. The most important erosional process is FREEZE-THAW, the role of the snow being to supply meltwater to assist in the process. The thickness of snow is an important factor, because if it is too thick it will tend to protect surface materials from the effects of atmospheric freeze-thaw cycles. The most active nivation takes place around the edges of snow patches. The presence of PERMAFROST greatly increases efficiency since it stops the meltwater soaking away. In order to achieve any erosion, weathered material must be removed, and consequently the process is most efficient on slopes, where SOLIFLUCTION and meltwater runoff are the principal transportation agents. The characteristic form produced by nivation is the NIVATION HOLLOW.

nivation hollow A hemispherical hollow of variable size produced by NIVATION processes. These hollows can be described as transverse, longitudinal, or circular, depending on the shape of the associated snow patch and its relationship with the slope. The downslope part of the hollow is generally flat or gently inclined, while the upslope part is steeper. It is considered that this backslope recedes gradually upslope, increasing in height, and that in this way the hollows may develop into cirques, as more and more snow can be accommodated, and FIRN begins to form.

noctilucent clouds Clouds observed in the high atmosphere when it is almost dark at ground level. Sometimes they show brilliant colors, although they more usually have a bluish-white or yellow appearance. Almost all other clouds are limited to the troposphere, but the height of noctilucent clouds has been shown to be about 80 km, sometimes reaching speeds of 300 knots. Their true nature and origins are not completely understood, but they are believed to consist of ice particles, saturation being reached through orographic wave development resonated from the surface.

nodule 1. A rounded concretion in sedimentary rocks.
2. *See* inclusion.
3. A small lump containing nitrogen-fixing bacteria on the roots of certain plants, particularly legumes. *See* nitrogen fixation.

nonconformity (heterolithic unconformity) An unconformable contact between overlying younger sedimentary rocks, and underlying older igneous or metamorphic rocks.

nondipole field That small part of the Earth's magnetic field that is superimposed upon the main DIPOLE FIELD. This small additional component is responsible for the irregularities of the Earth's magnetic field. It is irregular in its distribution and is different in each hemisphere; as a result the Earth's magnetic dip poles are not antipodal, the north magnetic dip pole being at 75°N, 101°W, while the south magnetic dip pole is at 67°S, 143°E.

nonsequence A short break in sedimentation, causing a type of unconformity detectable only by paleontological techniques.

nordmarkite A member of the SYENITE group of minerals.

norite A coarse-grained basic igneous rock consisting mainly of orthopyroxene and plagioclase. *See* gabbro.

norm A chemical analysis of an igneous or metamorphic rock expressed in terms of standard minerals (the *normative minerals*), the proportions of which form a basis for comparison and classification. The norm is distinct from the MODE, which is the actual mineral composition of a rock. A quantitative classification for igneous rocks was devised in 1903, based upon a normative calculation by Cross, Iddings, Pirsson, and Washington, and termed the CIPW classification.

normal 1. (in climatology) Denoting the average value of any climatological element over a specified time period. The usual time period is 30 years, which is believed to be sufficiently long to average out year-to-year variations but not long enough to contain trends in the value of the elements, such as temperature or precipitation.
2. (in statistics) Denoting a distribution of data that corresponds to an accepted frequency distribution about a population mean.

normal fault (gravity fault) A type of fault in which the predominant displacement is downward in the same direction as the inclination of the fault plane. This results from a stress configuration in which the principal maximum stress is vertical with the other two principal stresses (minimum and intermediate) being horizontal. *See* fault.

North Atlantic Drift The portion of the major surface current in the North Atlantic that is an extension of the Gulf Stream current. The flow of the North Atlantic Drift is generally in a northeasterly direction. It tends to be more diffuse than the Gulf Stream, being characterized by several shifting bands of water. Each band displays large meander flows that appear, disappear, and reappear in a constantly altering pattern of flow patterns. Large eddies are often thrown off the main stream. In its more easterly portion, the North Atlantic Drift bifurcates, one stream moving north-northeastward toward the Polar Seas, the other stream (the Canary Current) moving east-southeastward. Being a relatively warm current, especially on account of being partly replenished at depth by fairly warm water from the Mediterranean Sea, it has an important ameliorating influence on the climate of northwest Europe's ocean flank.

North Equatorial Current *See* equatorial current.

northing Any of the east-west grid lines on a map, quoted after the EASTING when coordinates are being given, showing distance north from the origin of the grid.

nosean A member of the sodalite group of FELDSPATHOIDS.

nucleation (accretion) A process by which it has been suggested that continents grow, younger fold belts being added to their margins. This is consistent with the present theory of plate tectonics because these fold belts originate in the deep trenches bordering continents that are overriding the adjacent ocean floor.

nucleus (in meteorology) Any minute solid particle suspended in the atmosphere. Nuclei have been classified on the basis of size into Aitken nuclei (smaller than 2×10^{-5} cm), large nuclei with radii between 0.2 and 1×10^{-4} cm, and giant nuclei with radii greater than 1×10^{-4} cm. The Aitken nuclei are most abundant in the atmosphere, reaching about 10 000 per cm^3, but there are far more in industrial areas than over the oceans. Identification of the nuclei is very difficult, even with electron microscopes, unless they possess a distinct crystalline form.

Some nuclei have an attraction for water and are known as hygroscopic; others are nonhygroscopic. This is very important in enabling nuclei to act as favorable points for condensation as part of the precipitation process. There are undoubtedly sufficient such nuclei in the atmosphere because condensation always takes place as soon as saturation is reached. *Compare* freezing nucleus.

nuée ardente An incandescent cloud of gas and volcanic fragments emitted during certain types of volcanic eruptions. *See* pyroclastic rock.

numerical weather forecasting A system of computing expected values of selected isobaric surfaces. In operational terms this is used for up to 48 hours ahead, as beyond that time differences between observed and computed values become unacceptable. This is because future pressure fields will be influenced by many factors, not all of which are included within the modeling system. Interpretation of the predicted chart has to be made to obtain the weather forecast for that time. For 24-hour periods, numerical methods are as good as if not better than forecasts based on years of experience by professional forecasters.

nummulites Benthonic marine protozoa forming a family of the FORAMINIFERA. Their skeleton, up to 80 mm in diameter in some species, consists of a smooth flattened disk-shaped spiral divided into many separate chambers. Nummulites are known from the Cretaceous Period but were most abundant in the Eocene and Oligocene Epochs; they are used locally in stratigraphic correlation.

nunatak A rock peak sticking out above the surface of an ice sheet. In many instances these peaks were formerly covered with ice and only subsequent reduction in the extent of ice cover has brought about their emergence.

nutation The small oscillation superimposed upon the precessional motion of the Earth. This results from the gravitational attraction on the Earth's equatorial bulge from the Sun and Moon. This attraction varies continuously as the Sun and Moon change their positions with respect to the Earth.

nutrient cycle (in ecology) The transfer of nutrients from one stage in an ECOSYSTEM to another stage. For example, within a forest leaves that fall to the ground rot and are broken down by bacteria to provide nutrients in the soil, which are needed for the growth of plants. Green plants use the energy of sunlight (in PHOTOSYNTHESIS) to make more tissue, which provides food for herbivores (primary consumers). The herbivores, in turn, provide food for carnivores (secondary consumers).

O

oasis A place in a dry landscape (such as a desert) that has water and so can support vegetation. The water may come from a river that flows across the desert on its way to the sea, or there may be a well for raising groundwater to the surface. Natural oases have long been sites of human settlement.

obduction In plate tectonics, the process by which OCEANIC CRUST is forced over the edge of and incorporated into CONTINENTAL CRUST at a destructive PLATE BOUNDARY. *See also* subduction zone.

objective analysis (in meteorology) A method of interpreting a two-dimensional scalar field of some meteorological element, usually pressure, so that the results from a particular data set are independent of human subjectivity. The usual way is by fitting a two-dimensional surface to the data by means of a least-squares method, which is a statistical process in which the sum of the squares of the difference between actual and predicted values is reduced to a minimum.

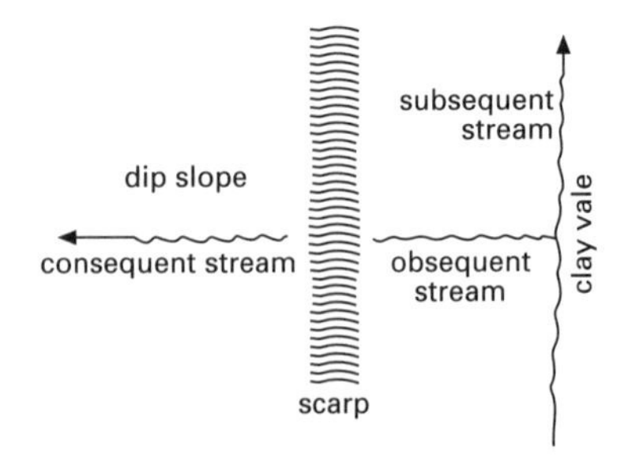

Obsequent, subsequent, and consequent streams

obsequent stream A stream that flows in the opposite direction to a CONSEQUENT STREAM, often against the direction of dip. In scarpland areas, the obsequent streams develop after the consequent and subsequent pattern has dissected the strata to produce the characteristic sequence of limestone and sandstone scarps, with clay vales; obsequent streams will flow down the scarp faces, tributary to the subsequent master streams.

obsidian A black, brown, or red lustrous type of volcanic glass with a conchoidal fracture, derived from rapidly cooled rhyolitic lava (*see* rhyolite). It is sometimes used as a semiprecious gemstone.

occluded front A combination of a cold front and a warm front that occurs in a DEPRESSION. The cold front moves under the warm front as it catches up to it. This type of front is associated with prolonged heavy rainfall. *See also* occlusion.

occlusion A front that develops during the later states of the evolution of a DEPRESSION when the air of the warm sector is no longer at the ground surface. As the cold front normally travels more quickly than the warm front, it slowly reduces the area of the warm sector until it merges with the preceding front to complete the occlusion process. The occlusion is therefore a compound zone with warm and cold front characteristics. If the air behind the original cold front was colder than that ahead of the original warm front it is known as a *cold occlusion*, and if warmer, as a *warm occlusion*. The frontal structure of the occluded system is shown in the diagram. In reality individual occlusions are more com-

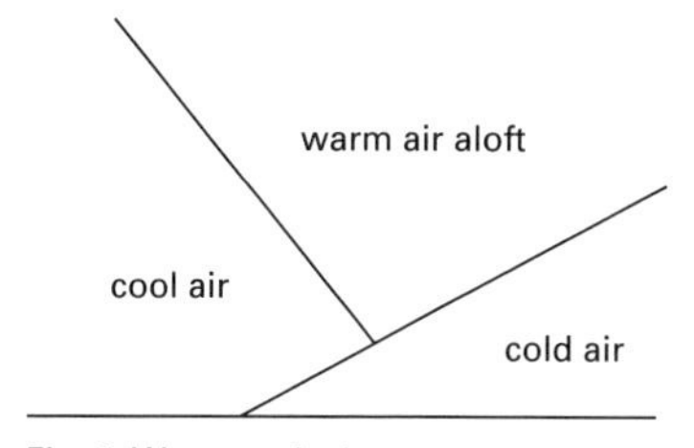

Fig. 1: Warm occlusion

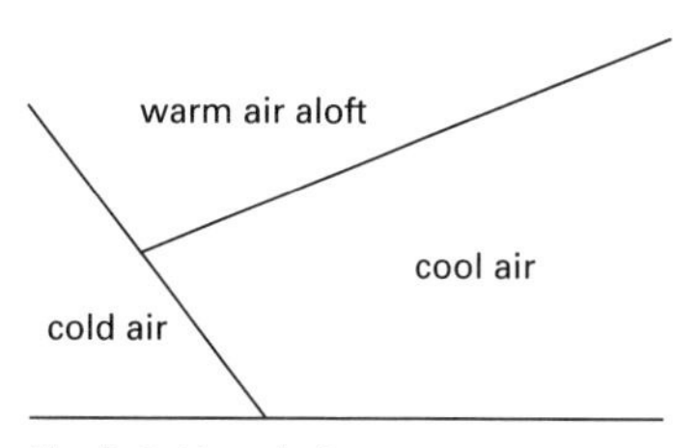

Fig. 2: Cold occlusion

Occlusions

plex. Sometimes, as the depression moves eastward along the occlusion, the line of airmass contrast to the north of the depression swings counterclockwise giving a further period of rain. This is known as a *back-bent occlusion.*

ocean The water mass (excluding lakes and seas completely surrounded by land) that covers part of the surface of the Earth. The water mass as a whole constitutes the HYDROSPHERE. The interface between the hydrosphere and the atmosphere is thus the sea surface, while the interface between the hydrosphere and the lithosphere is the sea bottom. The DEEP OCEAN lies beyond the shelf edge or shelf break.

ocean basin floor The ocean basin floor, the CONTINENTAL MARGIN on the continental flank, and the MID-OCEAN RIDGE on the oceanic flank constitute the three major morphological divisions of the ocean floor. The ocean basin floor has been further subdivided into a number of physiographic provinces, two of which (the elevated oceanic rises and the very deep abyssal floor zones), are of particular importance. The ocean basin floor occupies very extensive areas: approximately a third of the Atlantic Ocean floor, a third of the Indian Ocean floor, and three quarters of the Pacific Ocean floor. The ocean basin floor is also highly important for its SEAMOUNTS (either occurring individually or in large groups), and its linear-type archipelagos.

ocean current A distinct flow of water in the ocean. Almost all the water in the ocean is in a constant state of movement, true stagnation occurring only at the bottom of certain deep basins hemmed in by rock sills. Generally, currents in the oceans occur at fairly low speeds, especially within the deep water masses, but there are also well-defined currents (such as part of the Gulf Stream) that may flow as rapidly and strongly as some of the largest rivers on land. Neglecting those currents that flow within the surf zone, ocean currents result from three particular sets of circumstances: wind stresses acting on the sea surface, tidal motion, and differences in the density of seawater. The density differences arise because of differential heating and cooling, differential salinity (perhaps because of evaporation, ice melting, river discharge, and so on), and changes in turbidity levels. The wind-induced ocean currents may be directly or indirectly driven.

ocean floor The whole of the seabed seaward of lowest low-water mark, including the continental shelf, continental slope, and deep-sea floor. It is characterized by diverse relief forms, some of them of immense size. Significant tracts of the deep-sea floor are remarkably flat and smooth. In spite of considerable depths of water in the ABYSSAL ZONE, most of the ocean floor experiences some current action. In global terms, the ocean floor is vast: some 360 million sq km out of a total of some 510 million sq km for the Earth's surface as a whole. Ocean floor occupies about 61% of the N hemisphere and 81% of the S hemisphere. In general, the ocean floor arches upward rather than downward. Truly concave basins are rare.

oceanic crust That part of the Earth above the Mohorovičić discontinuity forming the floor of the ocean basins. It is usually up to 15 km thick and consists of three layers: layer one composed of young and poorly consolidated sediments, layer two composed of basalt pillow lavas and dolerite dikes, and layer three composed of dolerite dikes and gabbro. The ocean crust is generated at the constructive plate boundaries and destroyed at the destructive plate boundaries. *See also* continental crust.

oceanicity The degree of oceanic influence on the climate of an area. This is indicated by low annual and diurnal ranges of temperature, and abundant precipitation throughout the year.

oceanite A basaltic rock containing a high proportion of olivine phenocrysts. *Compare* ankaramite.

oceanography The study of the oceans, with particular reference to their overall form, the nature of the sea floor and associated sediments, the characteristics of the ocean waters, and the types of fauna and flora living within the oceans. This field has widened its scope to embrace aspects of human intervention in the ocean environment since the oceans have assumed such importance economically and strategically. It is a multidisciplinary subject, embracing physics, mathematics, biology, and several others.

ocean trench A long narrow steep-sided depression in the ocean floor. The deepest trenches – as much as 10 000 m deep – occur along SUBDUCTION ZONES and are associated with seismic activity. Most are located around the edge of the Pacific Ocean.

ocean wave Ocean waves of some kind are operating all the time, some affecting the surface of the sea, others the internal water masses. Some waves are generated by wind blowing over the sea's surface, often traveling far beyond the generation area as SWELL. Large waves with long periods may spread from storm centers and travel thousands of kilometers before affecting distant coasts. Some waves or wave energy may be reflected, or generated, in a seaward direction, as with SURF BEATS or the occasional catastrophic LANDSLIDE SURGES. TSUNAMI waves are relatively rare and result from submarine earthquakes or explosive submarine volcanic eruptions. Like landslide surges and very big storm waves, they may cause damage and loss of life along coasts. Oceans also experience tidal waves: these waves operate as standing oscillations that are modified by the Earth's rotation and move around nodal centers known as amphidromic points (*see* amphidromic system). Within the water masses of the oceans, there occur interfacial waves known as INTERNAL WAVES. *See also* wave.

ocellar Describing a texture of some volcanic rocks in which PHENOCRYSTS are enveloped by small radially or tangentially arranged crystals of another mineral.

ocelli Small spherical or dropletlike bodies found in certain alkaline igneous rocks. Ocelli contain mineral assemblages supposed to have crystallized from a liquid that was immiscible with the host magma.

ocher A yellow, brown, or red type of mineral iron oxide – or clay strongly colored by iron oxide – which is powdered and used as a pigment. Red ocher usually contains HEMATITE, whereas the yellow and brown forms are generally based on LIMONITE.

offlap The disposition of sediments laid down as a sea regresses, the progressively younger sediments being deposited seaward of the shoreline that marked the maximum extent of the former marine transgression. In this way, instead of older rocks being covered by younger deposits, they are exposed. The beds that mark the maximum extent of the former transgression have the largest areal development. *See also* overlap.

offset *See* chaining.

offshore bar A bar that is partly or completely submerged beyond the BREAKER zone. Considerable littoral transport of sediment, especially sand, may occur along the crest of nearshore bars and offshore bars that lie in fairly shallow water because of wave and tidal action. *See also* barrier beach.

offshore zone The zone extending from the outer limit of the BREAKER zone to the depth at which waves cease to influence the seabed. But because experimental evidence suggests that such wave influence occurs almost everywhere over continental shelves (with the exception, possibly, of shelves that have been abnormally deepened), the offshore zone effectively extends out to the SHELF-EDGE zone.

ogive A cumulative frequency curve. It constitutes one of the most effective ways of representing a frequency series, because uneven class intervals do not distort the curve and interpolation is easy. Calculation and plotting are also easy. The accumulated values for each class interval are determined as a percentage of the grand total, or simply plotted as frequency against the element concerned.

oil sand Any porous rock that is impregnated with liquid hydrocarbons or contains deposits of oil (petroleum). *Compare* gas sand; tar sand.

oil shale A type of fine-grained shale that can be strongly heated to produce oil (petroleum). The resulting shale oil is rich in unsaturated hydrocarbons.

oil trap A geologic structure that allows oil and/or natural gas to become trapped and accumulate. This occurs where an upward convex stratum of porous and permeable rock is sealed from above by an impermeable cap rock. A distinction is made between a *structural trap*, formed by tectonic activity (such as folding and faulting), and a *stratigraphic trap*, where, for example, a layer of fine sediments may act as an impermeable cap above coarser sedimentary layers.

okta (in meteorology) A measure of cloud cover equal to one-eighth cover. Cover of 8 oktas corresponds to a completely overcast sky.

Old Red Sandstone A succession of conglomerates, red shales, and sandstones forming the continental facies of the DEVONIAN System in NW Europe. These rocks contain abundant remains of ostracoderm and gnathostome (jawed) fish.

Oligocene The epoch of the TERTIARY Period that followed the EOCENE and preceded the MIOCENE. It began about 36.6 million years ago and lasted for some 13 million years. Many mammalian groups common in the Eocene became extinct but others continued to flourish, including camels, rhinoceroses, and the first pigs. The main phase of Alpine folding began at the end of the Oligocene.

oligoclase A sodic plagioclase FELDSPAR.

oligotrophic Describing a lake or other body of water that is deficient in plant nutrients and so has very clear water (because of the scarcity of plankton). The lowest cold levels of water are rich in dissolved oxygen, but the deposits on the bottom contain little organic material. *See also* eutrophic.

olivenite A rare green mineral form of hydrated basic copper arsenate, $Cu_2(AsO_4)(OH)$. It crystallizes in the orthorhombic system and occurs in deposits of copper.

olivine A member of a group of orthorhombic rock-forming minerals consisting of (SiO_4) tetrahedra linked by divalent metal cations. The general formula is R_2SiO_4, where R = Mg, Fe^{2+}, Mn, or Ca (in part). A complete gradation in chemical and physical properties exists between the two end-members, *forsterite* (Mg_2SiO_4) and *fayalite* (Fe_2SiO_4).

Olivines are green, brown-green, and yellow-green and show little or no cleavage but possess a chonchoidal fracture. Peridot is a pale green semiprecious gemstone variety. Magnesium-rich olivine is common in basic and ultrabasic rocks whereas fayalite occurs in acid igneous rocks. Forsterite and monticellite ($CaMgSiO_4$) are found in metamorphosed limestones.

omnivore An animal that feeds on both plants and other animals. Typical omnivores include cockroaches, bears, chimpanzees, hogs, and human beings. *See also* Carnivora; herbivore.

omphacite A member of the PYROXENE group of minerals.

oncolite A rounded, chalky body that originated as algae. Most oncolites are up to 2 cm across, although some are five times that size. They are thought to have been formed when sediments became trapped round an algal growth that was then rolled around by currents in the shallow sea.

onion weathering Small-scale expansive EXFOLIATION in which the splitting away of layers of rock resembles that of concentric onion skins.

onlap *See* overlap.

onyx A type of CHALCEDONY that has alternating straight bands of various colors (typically whitish with black or brown). It is used for making ornaments and cameos. *See also* banded agate.

oolite An allochemical LIMESTONE that is formed predominantly from *ooliths*. These are more or less spherical bodies formed by the precipitation of carbonate in concentric layers around a nucleus. They are usually small, up to about 2 mm in diameter. Ooliths larger than this are often known as *pisoliths*. Oolites are common in the JURASSIC system of Britain. Some noncalcareous oolites are known, such as those formed from iron minerals.

ooze *See* pelagic ooze.

opal A cryptocrystalline or colloidal variety of hydrous silica, $SiO_2.nH_2O$ (*see* silica minerals). It usually contains between 3 and 9% water, and occurs in various colors, from transparent to multicolored; the reddish type is called fire opal. Found in igneous rocks and near hot springs, it is highly valued as a gemstone.

open fold A fold in which the interlimb angle is greater than 70°. See diagram at FOLD.

ophiolite The sequential association of ultramafic rocks, gabbros, dolerite dikes, pillow lavas, and cherts occurring within eugeosynclinal environments. Ophiolites are thought to represent slices of the basaltic ocean crust that have been tectonically emplaced onto continental margins. *See also* spilite; Steinmann trinity.

ophitic Describing a texture of common occurrence in basalts and dolerites in which large plates of clinopyroxene completely envelop earlier-formed euhedral plagioclase plates. The term *subophitic* may be used when the pyroxenes only partly enclose the plagioclases. Ophitic texture is a specific example of POIKILITIC texture.

orbicular Describing certain plutonic rocks that contain large ovoid bodies made up of concentric shells of alternately light and dark minerals. The ORBICULES are the result of rhythmic crystallization around XENOLITHS, which have acted as nuclei. *Compare* rapakivi structure.

orbicule A large ovoid or spherical rocky mass consisting of concentric shells of minerals. Most orbicules are up to 15 cm in diameter, although some may be as large as 3 m across. *See* orbicular.

orbital motion Waves on the sea's surface experience orbital motion of the water particles beneath each wave; concurrently, the wave form is propagated along the sea's surface, in the direction of wave advance, at speeds that are dependent on the length of the waves. Beneath the waves, the orbital motion of water particles takes place at much lower speeds. The orbits of the water particles are nearly circular and have their greatest diameter just below the sea's surface, but the diameters decrease rapidly with increasing depth. The orbital paths are referred to as open, because the circular orbits are not completely closed loops; with each wave period, a water particle moves slightly in the direction of wave propagation. When the waves enter shallow water, orbital motion becomes distorted, the orbital paths becoming increasingly elliptic as waves run into increasingly shallow water. Beneath a given shallow water wave, the long diameter of the ellipses decreases with increasing depth, while at the bed the water particles no longer orbit but merely oscillate to and fro along a straight line, although again with a slight asymmetry shoreward. *See also* mass transport current.

order A group in the taxonomic classification of organisms. Several related orders constitute a CLASS, and an order itself is composed of one or more FAMILIES. For example, the Primates and Carnivora are two orders of the class Mammalia. *See* taxonomy.

Ordovician The period in the PALEOZOIC Era that followed the CAMBRIAN and preceded the SILURIAN. It began about 505 million years ago and lasted for about 67 million years. The name was proposed for rocks in the Arenig Mountains of N Wales, part of an area inhabited by a Celtic tribe, the Ordovices. Sedimentary rocks containing fossils of the Ordovician occur in all present-day continents; the graptolites, an abundant and rapidly evolving group, are used as the base for the Ordovician geologic timescale. The Early Ordovician consists of the Tremardoc and Arenig series, the Middle Ordovician the Llanvirn and Llandeilo series, and the Late Ordovician the Caradoc and Ashgill series. Fossil evidence suggests that the landmasses were in very different locations to those of today. During the period the major landmass Gondwanaland extended from the South Pole north to the tropics; the landmasses of Laurentia and Baltica began to move toward each other narrowing the Iapetus Ocean between them. There is evidence that tectonic and volcanic activity was extensive and intense, with the beginning of mountain building (the Caledonian orogeny) along the subduction zone of the E margin of what is now North America. Sedimentary rocks of the period are deep-water graptolitic shales and mudstones and calcareous sandstones, mudstones, and limestones. Common fossils include brachiopods (articulate forms were dominant), trilobites, and gastropods. The first corals, echinoids, crinoids, and Bryozoa appeared and the straight-shelled nautiloids reached their peak. The first fossils of ostracoderm vertebrates are found in Ordovician rocks.

ore A naturally occurring usually rocky material from which a useful product (such as a metal or one of its minerals) can be extracted. The commercial worth of a metallic ore depends on the value of the metal and the percentage of it in the ore, bearing in mind how difficult it is to extract the ore and, subsequently, the metal itself. *See also* mineral.

organic acid (in soil science) When organic matter is attacked by the soil microorganisms, certain simple products such as oxalic acid, $(COOH)_2.2H_2O$, acetic acid, CH_3COOH, and lactic acid, $CH_3CH(OH)COOH$, are released. These acids greatly increase the power of leaching in soils. *See* podzolization.

organic horizons The L, F, and H layers in a soil PROFILE, which lie above the mineral soil. However, humified organic matter is also usually found intermixed with the mineral matter in the A horizon and may also be present lower down the soil profile in the B horizon.

organic matter (in soil science) Organic matter rarely forms more than 5% by weight of the soil yet it is highly significant in promoting a productive soil because it improves the structure and provides a source of important mineral nutrients. It forms as a result of the incorporation of plant and animal remains in the soil and their breakdown by the microorganisms present. Some of this material is broken down completely into soluble substances but the more resistant material tends to remain, and with the new resynthesized compounds this forms the soil HUMUS.

organic reef *See* reef.

organic rock Any sedimentary rock that consists essentially of the remains of plants or animals.

organic soil (bog soil) A soil with an accumulation of peat more than about 70 cm in depth. Such soils form in water-saturated environments where the decomposition is slow. In upland regions with high rainfall, rapid leaching of bases results in the formation of an acid peat called *blanket bog*. Similar acid soils, called *raised bogs*, may occur in lowland sites where rainfall is sufficient. The most characteristic lowland organic soil, however, is *fen peat*, which is neutral to alkaline in reaction as a result of saturation by base-rich waters. When this black peat is drained, rich agricultural soils are formed.

organic weathering The breakdown and decomposition of rocks caused by plants and animals. Both mechanical and chemical breakdown are involved. Burrowing animals and plant roots can physically separate rock particles and can mix or transfer material elsewhere. Through the respiration of vegetation, the carbon dioxide content of the soil may be substantially increased, having a marked effect on weathering, while soil bacteria can also be very important, especially in chemically reducing conditions. Lichens can exist on bare rock surfaces and are frequently responsible for the initial decomposition of rocks by their removal of nutrients. Larger plants can affect the surface atmosphere through shading while even in areas of only shallow soils, weathering has been seen surrounding roots at great depths, owing to their localized ability to alter the constituent minerals chemically.

Ornithischia An order of dinosaurs characterized by a pelvis similar to that of birds, although they were not bird ancestors nor very closely related to them. They were all herbivorous and had correspondingly modified dentitions; some ornithischians had a horny beak. For avoiding saurischian predators some developed rapid bipedal locomotion, while the more slow and bulky forms evolved horns and other types of armor. Ornithischians became extinct at the end of the Cretaceous period (*see* K/T boundary event). *Compare* Saurischia.

orocline An orogenic belt that has a marked change in trend. This is thought to have resulted from horizontal forces operating within the Earth's crust.

orogenesis The process by which mountains are formed, i.e. the deformational processes such as thrusting, folding, and faulting, which result from the collision of two continental plates. The sediments between the continents are then strongly deformed, causing them to spill out onto the older stable continents and be compressed into long linear mountain chains. There is frequently associated igneous activity. *Compare* epeirogenesis.

orogenic belt *See* mobile belt.

orogeny A mountain-building period. Major orogenic phases since the Precambrian include the Caledonian, Variscan, and Alpine orogenies.

orographic cloud A cloud that owes its formation to rising air resulting from airflow over mountains. Because air is forced to rise over high ground there will always be some upward component in the wind field. If the air is already moist, or vertical motion extensive, the condensation level may be reached and orographic cloud formed. If there are stable layers within the airstream, LEE-WAVE clouds may be seen. These are the most easily recognizable type of orographic clouds.

orographic rainfall Rainfall resulting from the vertical motion of an airstream, caused by the presence of uplands. However, it is fairly rare for orographic rain to fall without either convective or cyclonic processes also acting. The orographic component is normally weak and merely acts as an enhancement to the other two mechanisms.

orpiment A yellow to orange mineral form of arsenic sulfide, As_2S_3. It crystallizes in the monoclinic system, and occurs in association with CINNABAR and REALGAR, often in lodes containing gold, lead, or sulfur. It may also be found in deposits around hot springs. Despite its highly toxic nature, it has long been used as a pigment; it is also used as a source of arsenic.

orthite (allanite) One of the EPIDOTE group of minerals.

orthoclase A potassic alkali FELDSPAR.

orthoferrosilite One of the PYROXENE group of minerals.

orthogeosyncline A geosyncline that is associated with both continental and oceanic environments, and contains both volcanic and nonvolcanic areas in association with the accumulating sediments.

orthogonal (wave ray) A line that is perpendicular to a series of wave crests. This is usually done on the basis of a wave refraction diagram, which shows a series of wave crests in the vicinity of the coast or a harbor. In theory, the wave energy between any two orthogonals remains roughly the same as they are traced shoreward into decreasing depths. For this reason, where orthogonals bunch together at the coast, or conversely when they tend to spread out, one can delimit zones of wave energy concentration (zones of CONVERGENCE), and zones of energy dissipation (zones of DIVERGENCE) respectively.

orthomagmatic Describing the stage in the crystallization of a magma during which the early-formed primary minerals having high crystallization temperatures are formed. The orthomagmatic stage is followed by periods of pegmatitic, pneumatolytic, and hydrothermal crystallization.

orthomorphic projection (conformal projection) Any map projection on which the shape of an area of the map is the same as the corresponding area of the Earth's surface.

orthopyroxene One of the PYROXENE group of minerals.

orthoquartzite *See* quartzarenite.

orthorhombic *See* crystal system.

orthotectonics (alpinotype tectonics) Mountain belt structures having the characteristics of the Alps, i.e. their deep-seated areas have deformed plastically, with associated igneous activity and metamorphism, while their shallower regions have deformed brittly and are characterized by faults, thrusts, and several small folds.

oscillation (in meteorology) Any cyclic variation of a climatic element above and below its mean value or position. Its most frequent use is as the *Southern oscillation* where an oscillation with a period of 2.33 years was found in the seasonal distribution of pressure and to a lesser extent temperature and rainfall over the oceanic areas of tropical latitudes.

oscillatory current The type of current that waves induce at the seabed arising from the ORBITAL MOTION of the water particles beneath waves. The larger the waves, the deeper such currents operate. As waves, especially long waves, enter shallow water, the orbital motion of the water particles tends to be elliptic, while at the seabed the particles oscillate to and fro about a mean position along a straight line. This gives rise to an oscillatory current at the bed. Beneath the wave crest there occurs a short sharp acceleration shoreward, while beneath the flattish trough of the wave there is a more protracted seaward movement of water. Such an oscillating type of flow beneath waves often has a profound influence on the movement of bed materials (although far less an influence on materials moving continually in suspension), and may cause pebbles to move rapidly shoreward.

oscillatory wave A wave, often of the progressive oscillatory type, in which only the waveform on the sea's surface advances in the direction of wave propagation, while the water particles beneath the wave orbit in almost closed loops. Whereas the waveform may be propagated across the sea's surface quite rapidly, at a speed that is largely dependent on the length of the wave, the gradual shift of the water particles in the direction of wave travel may be quite slow.

Osteichthyes The class constituting the bony fish. It includes the ACTINOPTERYGII (teleosts and more primitive ray-finned fish) as well as the fleshy-finned CROSSOPTERYGII. Some early groups had a body armor of closely fitting scales of complex construction; these were reduced and simplified in later forms. The class is characterized by the development of some kind of lung, which may be converted into a *swim bladder* in advanced forms for maintaining hydrostatic equilibrium. Fossil fish first appeared in the late Silurian, since when the class has continued to expand in diversity and abundance. *Compare* Chondrichthyes.

Ostracoda A group of aquatic arthropod animals that belong to the class CRUSTACEA. The body is completely enclosed within two calcareous valves forming a carapace. Unlike the shells of the BIVALVIA, these have no growth lines; they are periodically completely replaced as the animal increases in size. They are usually very small, less than 1 mm in length, but some species grow up to 20 mm across. Varied ornamentation of the shells is of use in taxonomy. Ostracods range from the Cambrian Period to the Recent and because of their small size, distinctive forms, and widespread distribution, they are valuable in the calibration and correlation of the rocks in which they occur, especially in the cores from boreholes.

Ostracodermi Fossil jawless fish (*see* Agnatha), having an outer protective covering of bony plates and scales. Usually not more than 30 cm long, they were common inhabitants of Silurian and Devonian ponds and rivers and a few fossil fragments are known from the Ordovician. This group probably gave rise to the jawed fishes.

outcrop The area on the Earth's surface where a particular rock type or body is present. This includes both exposed areas and others covered by drift.

outer core The layer of the Earth between 2900 km and 5000 km beneath its surface. It is bounded at 2900 km by a seismic discontinuity (*see* Gutenberg discontinuity) that separates it from the overlying mantle. On geophysical evidence it is thought to have a composition of iron alloy and nickel and to behave essentially as a liquid, unlike the INNER CORE, which is solid.

outgassing The action of heat in removing occluded gases from rocks. The outgassing of water vapor and other gases from molten rocks in the primeval Earth is believed to be the source of the atmosphere.

outlier An area of exposed younger sediments completely surrounded by older rocks, usually as a result of folding, faulting, and erosion. *Compare* inlier.

outwash plain (sandur) A widespread area of fluvioglacial deposits produced by deposition from meltwater streams emerging from the margins of an ice mass. The material decreases in size with distance from the ice and is frequently well bedded. Some unstratified till may be found among the gravels, indicating a period of readvance of the ice margin. These outwash plains may exhibit irregular surfaces, resulting from the presence of kettle holes, and are frequently found in association with other fluvioglacial landforms, such as KAMES, KAME TERRACES, and ESKERS.

overflow channel A valley carved out by water that overflows from a lake. Often meltwater from a glacier or ice cap overflows from the lakes that form when the ice melts. Some overflow channels become deep gorges.

overfold (overturned fold) A type of FOLD in which very strong compression has caused the rock strata to be pushed right over.

overland flow (wash, rain wash, sheet erosion) The movement of water as a shallow unchanneled sheet over the soil surface. Overland flow, which is intermediate between flow of water through the soil and surface channeled flow, occurs once the infiltration capacity, or ability to hold and store water, of the soil is exceeded. Water then collects in depressions on the surface and begins to flow as a broad shallow layer, often giving a glistening sheen to the surface. The flow is not even but turbulent, moving as a series of waves, although LAMINAR FLOW in sheets 8 cm deep has been recorded. Speeds and depth of flow are limited but soil particles can be entrained and moved. The thickness of the sheet determines the size of material moved. Overland flow is most devastating on bare soils of limited permeability, especially on long steep slopes. The depth of flow increases downslope from nil at the hill crest until at the downslope end the sheet becomes channeled into rills and ceases to exist. It tends to erode a hydraulic concave depression, and as such is important in PEDIMENT evolution. Most authorities attribute to it a major capacity to shape the landscape.

overlap (onlap) The disposition of beds lying above an unconformity, where they were deposited by a transgressing sea. Each successively younger bed extends farther onto the previous land surface than its predecessor. *See also* offlap.

oversaturated rock An igneous rock that contains free quartz. Examples include granites and rhyolites. *See* silica saturation.

overstep The disposition of strata associated with an unconformity where younger transgressive marine beds encroach onto progressively older beds of the underlying sequence.

oxbow lake (cutoff) A crescent-shaped lake that was formerly part of a river meander. Oxbow lakes are typical features of meandering rivers, and result from the river's ability to erode laterally being concentrated on the outsides of meander bends. Once created the oxbow rapidly silts up.

An alternative origin of cutoffs has been suggested by laboratory experiments, which showed that cutoffs develop when stream gradient falls below a critical limit, because the stream needs a minimum gradient to flow, and meanders effectively decrease gradient by increasing length over a

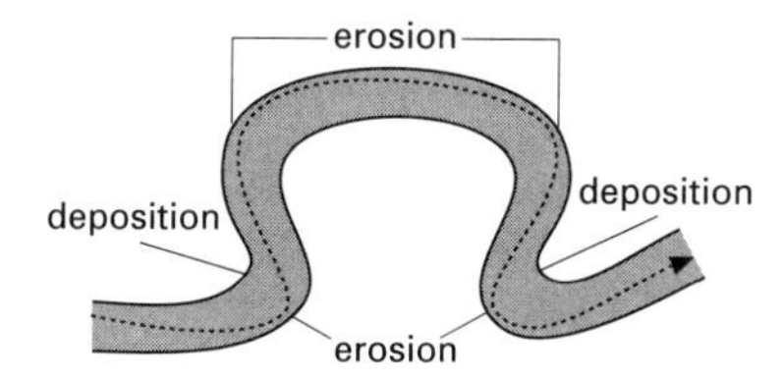

Fig. 1: Pattern of erosion and deposition in a meandering reach

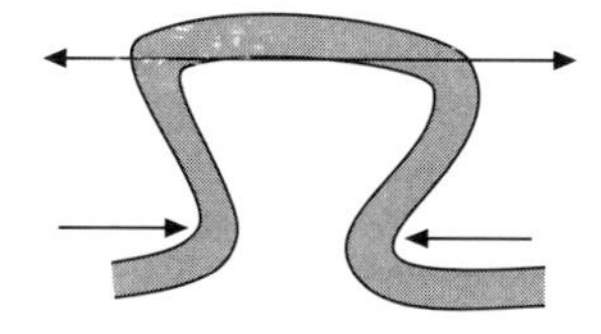

Fig. 2: As a result of the pattern of erosion and deposition, meander migrates as shown

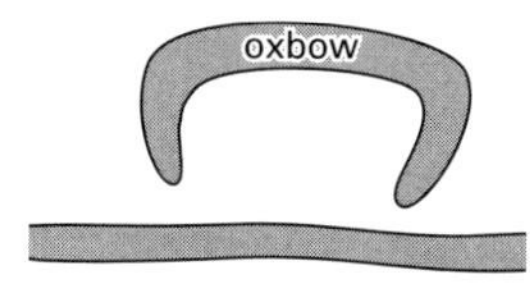

Fig. 3: In a storm, the narrow neck is breached leaving the oxbow

Oxbow lake formation

given fall. If the meanders develop to such an extent that they rob the stream of its minimum necessary gradient, cutting off that meander will restore the gradient. needed. *See also* meander.

oxidation (in geology) The chemical weathering process involving the reaction between rocks and atmospheric oxygen, the oxygen usually being dissolved in water. The products of oxidation are oxides and hydroxides, iron being the mineral most frequently affected and its oxidation products give many weathered rocks their reddish or yellowish color. A certain amount of oxidation may also result from bacteriological activity. *Compare* reduction.

oxide A chemical compound of oxygen and one or more other elements. In mineralogy, the non-oxygen component is a metal or metals; metal oxides make up a large percentage of all minerals.

oxisol One of the ten soil orders of the SEVENTH APPROXIMATION classification characterized by the presence of free iron and aluminum oxides (Fe_2O_3 and Al_2O_3) and possibly other oxides of titanium and chromium, as well as a high proportion of simple clays (especially kaolin) and some quartz. This includes the currently developing ferrallitic soils and the relic laterites of tropical areas. Oxisols are highly weathered and leached, generally lacking in fertility, and in the case of the relic laterites may well include deposits of a more geologic than pedological origin.

oxygen One of the main gases of the atmosphere (about 20% by volume). From the meteorological point of view, oxygen is most important as an absorber of radiation in the ultraviolet wavelengths (0.13–0.17 micrometers). It dissociates at high levels of the atmosphere into monatomic oxygen (O), which combines with oxygen molecules (O_2) to give OZONE (O_3).

oyster reef A REEF largely formed from the consolidation of oyster shells. The maximum development of oysters tends to be in the brackish water areas skirting coasts, as within certain estuaries, or in bays with fairly narrow inlets. A reasonable supply of fresh water from rivers, to reduce salinity levels, is also an important factor. On the other hand, oysters do also occur in the open sea, although in lesser numbers. The reefs they give rise to are common to many lagoonal areas in the humid zone. The reefs may attain a significant thickness, for instance, commonly 10 m or more along parts of the coast of Texas.

ozone Triatomic oxygen (O_3), which forms in the upper atmosphere especially at about 20–25 km level (sometimes called the *ozonosphere*). It results from the dissociation of molecular oxygen into single atoms of oxygen, which then combine with other oxygen molecules to give ozone. Below this level of formation, it gradually disperses toward the ground surface, but being a chemically reactive gas it is quickly reduced to oxygen. Ozone absorbs strongly in the 0.23–0.32 micrometer waveband, preventing potentially damaging ultraviolet radiation from reaching the ground surface. It also absorbs at other wavelengths, but these are of less importance.

The distribution of ozone would be expected to correspond to its rate of production under conditions of prolonged exposure to ultraviolet radiation. However, it seems to reflect vertical and horizontal movements in the stratosphere rather better. High latitudes have the maximum levels of ozone in spring with a minimum in the fall.

In recent years scientists have detected holes in the ozone layer, particularly near the South Pole. They are thought to be caused by chlorofluorocarbons (CFCs), used as aerosol propellents, which enter the atmosphere and break down the ozone. As a result, more damaging ultraviolet radiation can reach the Earth's surface, possibly also leading to climate change.

P

packing The arrangement in space of the atoms or ions in a CRYSTAL or the grains in a rock. The individual particles can be thought of as tiny spheres, and they can be packed in various ways. The closest is hexagonal close packing, in which in any plane each sphere is centered on a corner of a regular hexagon. In cubic packing they are located at the corners of a square. In a rock, the packing of the grains determines its porosity.

Pacific ring of fire The circum-Pacific system of earthquakes and volcanoes that surrounds the Pacific basin's oceanic plates where they are subducting below other plates. It extends around the Philippines, Japan, and the W coast of North and South America. *See also* plate tectonics.

pahoehoe A lava having a surface that resembles coils of twisted rope. *Compare* aa; block lava.

palagonite Hydrated basaltic glass, usually yellow or orange, formed when basalts are erupted beneath water or ice. Palagonite breccias or tuffs produced by the fragmentation of the glassy material are termed *hyaloclastic rocks.*

paleobotany The branch of PALEONTOLOGY that deals with the nature and evolution of plants through geologic time, as shown by fossil remains.

Paleocene The first epoch of geologic time in the Tertiary. Beginning about 66 million years ago and lasting for 8.6 million years, it followed the CRETACEOUS Period and was succeeded by the EOCENE Epoch. Occasionally the time represented by this epoch is included within the Eocene. Paleocene rocks are of both marine and nonmarine facies. Following the extinction of the large reptiles at the end of the Cretaceous (*see* K/T boundary event), mammals became abundant and diversified into a variety of primitive and now mainly extinct groups. Insectivores were common and by the end of the epoch primates and rodents had evolved.

paleoclimatology The study of the climates of earlier geologic periods. The evidence for the former climate is obtained from sediments and fossils. Complications arise from the fact that the continents have been changing their latitudinal positions through time, so the time period and location of deposition have to be determined before conclusions can be deduced.

paleoecology The study of the ecology of fossil organisms, i.e. the relationships of organisms in past geologic ages with their nonliving environment and with other animals and plants. Paleoecology involves the investigation of the rocks in which the organisms are found to determine paleoenvironmental information and details of TAPHONOMY. TRACE FOSSILS often provide information regarding the behavior of an animal.

Paleogene The lower part of the TERTIARY Period, comprising the PALEOCENE, EOCENE, and OLIGOCENE Epochs. It lasted for about 42.5 million years from approximately 66.5 to 24 million years ago, when it was followed by the MIOCENE Epoch of the NEOGENE. Some authorities suggest that the Paleocene should be considered as a period, with the Tertiary as a subera.

paleomagnetic correlation The correlation of the residual magnetism in rocks. The direction of polarization of the magnetic field of the Earth changes periodically. It is possible to measure the residual magnetization of suitable rocks and to discover the direction of polarity at the time of their formation. By this means a sequence can be established showing normal and reversed periods in the Earth's magnetic field. It may be calibrated by some means, such as radiometric dating (*see* calibration), to provide a valuable tool for the CORRELATION of stratigraphic sequences of rock. *See also* magnetic division; magnetic interval.

paleontology The study of all aspects of ancient organisms, including their TAXONOMY, anatomy, ECOLOGY, and EVOLUTION. The evidence for this comes from FOSSILS preserved in rocks, which are either the remains of the organisms themselves or TRACE FOSSILS. Much paleontological work is applied to the elucidation of the stratigraphical relationships between bodies of rock by means of BIOSTRATIGRAPHY. The branch of the subject that deals with microscopic organisms is known as *micropaleontology*; fossils of such organisms are MICROFOSSILS.

Paleozoic The first era into which PHANEROZOIC time is divided. It followed the PRECAMBRIAN and consists of the following periods: the CAMBRIAN, ORDOVICIAN, SILURIAN, DEVONIAN, CARBONIFEROUS, and PERMIAN. Systems of the first three periods constitute the Lower Paleozoic, and the second three the Upper Paleozoic. It lasted for about 325 million years from 570 to 245 million years ago, when it was followed by the MESOZOIC Era. The fauna of the Paleozoic is characterized by an abundance of invertebrate types, many of which, such as the rugose corals (*see* Anthozoa), trilobites, graptolites, and productid brachiopods, became extinct at the end of the era. In the course of the Paleozoic fish evolved and the first amphibians appeared, flourished, and declined. Two episodes of major orogenic activity occurred, giving rise to mountain ranges and associated tectonic activity. The CALEDONIAN OROGENY was the earliest, occurring at the end of the Silurian and the beginning of the Devonian; the VARISCAN orogeny took place toward the end of the era. *See also* geologic timescale.

paleozoology The branch of PALEONTOLOGY concerned with the nature and evolution of animals through geologic time by the investigation of their fossil remains.

pallasite A stony iron METEORITE that has grains of OLIVINE in a nickel-iron matrix.

paludal Describing something that lives in, is related to, or is produced by a marsh.

palustrine Describing plants that grow in or deposits that accumulate in a marsh.

palygorskite Any of a number of claylike minerals, hydrated magnesium aluminum silicates, $(Mg,Al)_2Si_4O_{10}(OH).4H_2O$. They form thin fibrous crystals that resemble cardboard and occur in desert soils.

palynology The study of fossil spores, especially POLLEN. Because spores are usually adapted to resist destruction and to be dispersed over large distances, they are valuable for use in the CORRELATION of the rocks in which they occur. They are also important environmental indicators and have been used in monitoring climatic change during the Quaternary Period. Palynology forms part of micropaleontology.

pampas A region of temperate grassland in South America, found in Argentina and Uruguay near the Plate River estuary. Used originally as pasture for cattle and sheep, today much of it has been plowed for growing crops such as alfalfa, corn, and wheat.

Pangaea A hypothetical supercontinent that is thought to have existed in the very distant geologic past. This single continent was composed of all the present-day continents, which have since been derived from it as a result of repeated episodes of SEA-FLOOR SPREADING. This supercontinent is thought to have fragmented initially into two segments, Laurasia to the north and Gondwanaland to the south.

paragonite One of the major types of MICA, a hydrated sodium aluminum silicate, $Na_2Al_4(Si_6Al_2O_{20})(OH)_4$. It forms colorless, yellowish, or green crystals and occurs mainly in various metamorphic rocks.

parallel A line of latitude, running parallel to the Equator round the Earth through any given point. The plane of each parallel is at right angles to the Earth's axis. Parallels range in length from a point at each pole to the circumference of the Earth at the Equator, which is the only parallel that is a GREAT CIRCLE. Parallels are uniformly spaced along the meridians.

parallel drainage A type of drainage pattern that develops where there is strong structural control in one direction, streams being strictly guided in that direction. A uniform regional slope, rigid alternation of hard and soft lithologies, or parallel folds will produce parallel streams; recently glaciated areas may also have parallel drainage nets if the surface is covered in fluted ground moraine or drumlin fields.

parallel evolution The CONVERGENT EVOLUTION of closely related and anatomically similar organisms, which undergo nearly identical evolutionary changes through time.

parallel retreat As opposed to SLOPE DECLINE, parallel retreat involves a slope eating into a rock but maintaining its original angle as it does so, hence remaining parallel in new positions to its previous positions. W. Penck and later L. King used this principle as the basis for their scheme of the CYCLE OF EROSION. As a method of SLOPE EVOLUTION it is strongly related to geomorphic processes, in that a slope will remain parallel only if the debris being eroded in its retreat is removed at an equal rate from its base. If the debris accumulates, the slope declines.

The idea was first formulated in relation to the tropical scheme, with the original work being in Africa; since then, Dury claims to have identified it as a viable process in the humid temperate lands, and it can be seen in areas of BASAL SAPPING, e.g. active sea cliffs. S. Schumm has suggested it operates preferentially on certain lithologies through their control on process: on sandstones, dominated by overland flow, parallel retreat operates, whereas on clays, dominated by creep, the slope declines. It seems that a number of situations can favor parallel retreat, not merely certain climates; lithology and processes, in various permutations, are also vital.

paramagnetic Describing a material that has a magnetic permeability slightly greater than 1. In a magnetic field, a paramagnetic mineral is magnetized in proportion to the field strength. Platinum is a paramagnetic metal.

parameterization (in numerical weather prediction) The inclusion of the effects of meteorological systems that are smaller than the grid on which the model is based. It involves approximating the statistical effects of these small-scale processes in terms of the large-scale factors.

parasite An organism that lives on or inside another organism (the host) on which it is totally dependent for food and energy. This may or may not harm the host. External parasites are called *ectoparasites*; internal ones are *endoparasites*.

parasitic cone A small cone on the flank of a major volcanic cone. These cones develop frequently and indicate that the magma trying to reach the surface has found an easier route than via the main vent. They are usually several tens of meters beneath the main vent.

parataxitic *See* eutaxitic.

paratectonics The tectonics of stable areas, such as old fold belts and cratons.

paraunconformity *See* disconformity.

parautochthonous Describing rock bodies that are midway between ALLOCHTHONOUS and AUTOCHTHONOUS, i.e. that have been displaced only short distances.

parent material The material from which the soil forms, which may not be the same as the underlying bedrock. Before the importance of climate was recognized, parent material was thought to be the dominant soil-forming factor. Certain features of the parent material, such as texture, mineral composition, and porosity, have a considerable influence on the soil-forming processes, e.g. a limestone soil rich in calcium will delay the process of acidification in a humid climate. Parent material is likely to have a greater influence in temperate regions where soils have been forming for a relatively short time, e.g. in Scotland brown earths have been recorded on basic igneous rocks and podzols on acid igneous rocks.

parhelion *See* mock sun.

partial melting The geologic process in which a rock is subjected to extremely high temperature and pressure so that it partly melts and the liquid flows away. The liquid then solidifies to form of a different type of rock.

partial pressure In a mixture of gases, such as the atmosphere, the partial pressure of one constituent is the pressure it would exert on its own while occupying the same volume as the whole mixture. The sum of all the individual partial pressures of the components is equal to the pressure of the whole mixture.

particle size The effective diameter of a particle. Particle size analysis is the technique of determining the proportion of the different particle sizes present (each particle size is termed a *separate*) in a sample. There are several classifications in use; the main classes in one of them are shown in the table. There are three main techniques of analyzing particle size: by sieving in a stack of sieves of decreasing mesh size, by allowing particles to settle in a tube of water, the rate of settling being proportional to size, and by micrometric methods. It is a technique commonly used in soil science and geomorphology, e.g. to describe texture classes in soil, or to infer the origin of a sedimentary deposit in geomorphology.

CLASSIFICATION OF PARTICLE SIZES

Class	*Size range (mm)*
boulder	>200
cobble	60–200
gravel	2–60
coarse sand	0.6–2
medium sand	0.2–0.6
fine sand	0.06–0.2
silt	0.002–0.06
clay	<0.002

passive glacier A GLACIER that accumulates little material and loses little material; technically, its rates of alimentation and ablation are about equal. This occurs when there is not much winter snow and a cool summer with little melting of the ice. Such glaciers move very slowly, causing little erosion and transporting few materials.

patina A film or skin on the surface of a boulder, produced by chemical weathering. Patinas are distinctive in color and physical properties, their thicknesses often reflecting age.

patterned ground A distinctive geomorphological feature mostly found in periglacial areas. Where the ground surface is flat, STONE POLYGONS or circles may develop, either in groups or in isolation, which may grade into STONE STRIPES even on very gentle slopes. In nonperiglacial climatic areas the most likely cause of patterning is cracking due to heat shrinkage (*see* desiccation crack).

pays A low plateau of land in northern France.

peacock ore Another name for the copper-containing mineral BORNITE.

pearl An accumulation of white or cream-colored, usually spherical, NACRE inside the shell of a mollusk, such as a clam or oyster. It forms when layers of nacre gradually build up on a foreign body, such as a grain of sand. Pearls are valued as precious gemstones; most prized are black pearls.

peat A mass of fibrous plant debris, which is only partly decomposed, and is often dark brown in color. This is a result of wet anaerobic conditions, which retard microorganism activity. Certain soils such as peaty gleys and peaty gleyed podzols have peat as one of their main horizons; other soils may have a thicker accumulation of peat. (*See* organic soil.)

peaty gley podzol A hydromorphic variant of PODZOL, characterized by a blue-gray waterlogged area above the impermeable iron pan, capped by a thick peat horizon. Above and below the gleyed area the color is orange-brown because of the free drainage and oxidation, as contrasted with the reducing conditions in the gleyed area. Such soils are most probably polygenetic, being originally BROWN EARTHS that became degraded to podzols with forest clearance, and subsequently became gleyed when the iron pan developed to the point where it became impermeable. Once gleying had set in, microbiological activity was inhibited and so litter was not broken down but accumulated as peat. It is characteristic of many gently sloping uplands.

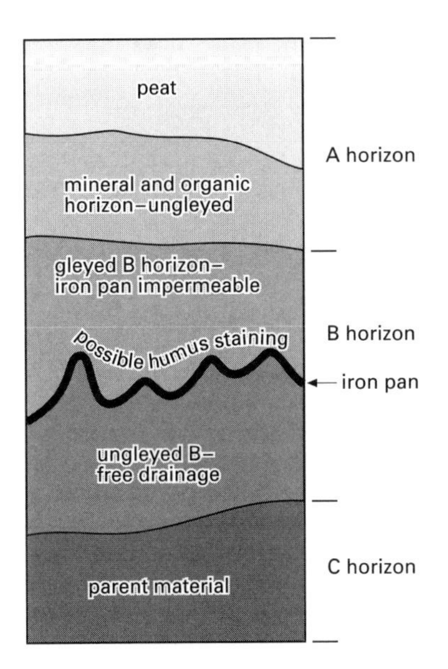

Peaty gley podzol

pebble A rounded rock fragment from 5 mm to 65 mm across. It is thus larger than a granule but smaller than a cobble.

ped An aggregate of soil particles, forming a crumb of soil.

pedalfer A leached soil in which iron and aluminum increase relative to silica during pedogenesis. Pedalfers are soils of humid climates, including PODZOLS, BROWN EARTHS, LATERITES, and PRAIRIE SOILS. *See also* soil formation.

pedestal rock A rock that has been shaped like a mushroom by the erosive action of windblown sand (corrasion). In desert regions, the corrasive action of sand is strongest about a meter above the ground and most pedestal rocks have "stems" of this height.

pediment A surface of low relief that slopes away from residual uplands at angles of between 9° and 0.5° or less, being bounded on their upslope ends by a sharp break of slope, the PIEDMONT ANGLE, at the point where they meet rock residuals. Such features are said to be typical of arid and semiarid regions, although some would attribute this type of surface with a much wider occurrence.

Pediments have a concave long profile, with the angle of slope decreasing away from the piedmont angle. They are cut in bedrock, often with a capping of gravel or alluvium, and display great regularity of form over wide areas. Pediments have been identified at various scales, ranging from valley-side types to the continental-scale forms that L. King has identified in Africa.

Pediments evolve by PARALLEL RETREAT of the upstanding residuals, leaving a broad area of low relief as they retreat. L. King and W. Penck, who see this as the usual tendency for landscape evolution, therefore see pediments as the end products of denudation. The processes responsible are not certain, but much discussion has centered on erosion by running water. Because the pediments have a concave profile, analogies have been drawn between their evolution and that of the long profiles of rivers: both are hydraulic curves. The form of this water erosion is disputed: some authorities consider that the scarp retreats by headward erosion of streams emerging from the upland, leaving the low-level surface as it retreats; others consider that pediments are due to lateral planation of streams, while a major school of thought has stressed the importance of sheet wash, which has been recorded up to a meter deep on pediments. L. King states that pediments begin on weak rocks and then extend into harder rocks as the scarp retreats by stream erosion at the base. Several different processes may be responsible, with different emphases in different areas.

pediplain The coalescence of a number of pediments toward the end of the cycle of erosion produces a pediplain, which is a broad landscape of low relief broken by isolated residual uplands, which meet the pediplain in a sharp PIEDMONT ANGLE. According to L. King and W. Penck a pediplain is the logical final product of subaerial denudation in an area of PARALLEL RETREAT (compared with a PENEPLAIN in areas of SLOPE DECLINE). Pediments tend to be smooth with gentle concave profiles, broken by few stream lines. It is difficult to be categorical as to whether a surface of low relief constitutes a true pediment, owing to the fact that landscapes have not evolved under constant climates and hence a single cycle, but are POLYCYCLIC in origin.

pedocal Unleached soil in which lime accumulates during pedogenesis, e.g. CHERNOZEM, CHESTNUT SOIL, and SIEROZEM.

pedogenesis *See* soil formation.

pedology The scientific study of soil, including its classification, origins, and composition. *See* soil; soil formation.

pedon A three-dimensional body of soil, including the full SOLUM and the PARENT MATERIAL, with a circular or hexagonal cross section, ranging in size from 1–10 square meters. Introduced by the SEVENTH APPROXIMATION method of classification, it replaces the earlier use of the two-dimensional profile as the lowest unit of classification. It includes all the properties formerly noted from the profile, plus notes on interfingering of horizons. Similar contiguous pedons are grouped into *polypedons*, which can then be amalgamated into soil series, and hence mapping of soil is dovetailed into classification.

pegmatite A very coarse-grained igneous rock occurring as segregations, veins, and dikes within and emanating from granite bodies. Pegmatites consist largely of alkali feldspar and quartz, often graphically intergrown, and contain abundant accessory minerals such as micas, tourmaline, topaz, spodumene, beryl, cassiterite, and wolframite. The largest crystals occur in pegmatites, occasionally measuring meters in length, and because of the concentration of rare elements, pegmatites may be of economic importance. Coarse-grained facies of igneous rocks other than granite may be termed pegmatitic. Pegmatites represent a stage in the crystallization of a magma during which a residual silicate melt and an aqueous gas phase coexist over the temperature range 500–700°C. The growth of large crystals may be ascribed to conditions of slow cooling (and therefore slow crystallization), rapid diffusion, and low viscosity due to the high concentration of volatiles. *See also* aplites.

pelagic **1.** Denoting the environment of the open ocean. A classification of ocean sediments, based on distance from the coast, might include beach, shelf, hemipelagic, and pelagic sediments. The

latter have been precipitated from the open ocean or have accumulated from the organic remains of planktonic marine animals. The main types of pelagic sediment are the PELAGIC OOZES and RED CLAY. The hemipelagic-abyssal environment is taken to lie within several hundred kilometers of the coast, at depths of over 1000 m. The pelagic-abyssal environment also occurs in depths greater than 1000 m but is farther from the coast. Similar sediments occur in the hemipelagic environment but in this case they are often mixed with land-derived material.
2. Denoting communities of marine organisms (free-swimming nekton and floating plankton) that survive within the ocean waters quite independent of shore or seabed environments. *Compare* benthos.

pelagic ooze Organic deposits made up of shells and other hard parts of marine organisms, which cover large tracts of the deep-sea floor. The nature of the skeletons of the various organisms determines the kind of deposit ultimately formed, the deposits being named after the most prominent organisms present. They include the calcareous GLOBIGERINA and PTEROPOD OOZES, and the siliceous RADIOLARIAN and DIATOM OOZES. Nearly all the oozes are soft and are easily disturbed by crawling or burrowing organisms and deep-sea currents, a fact confirmed by deep-sea photographs. The radiolarian and pteropod oozes occupy very limited areas of deep-sea floor; diatom ooze, a cold-water deposit, occupies some 9% of the total ocean floor, while globigerina ooze, the most widespread of the oozes, globally occupies some 130 million sq km.

Pelecypoda *See* Bivalvia.

Pele's hair, Pele's tears *See* pyroclastic rock.

pelite A metamorphosed ARGILLACEOUS rock. *Compare* psammite; psephitic rock.

pellet An ALLOCHEM that is well rounded but has no internal structure. Pellets may be produced by algae or they may be the feces of invertebrates.

pendant *See* roof pendant.

penecontemporaneous Describing geologic events that appear to have occurred at almost the same time. For example, the deposition and erosion of a bed of sediment are described as penecontemporaneous if the bed underwent erosion during, or just after, its deposition.

peneplain An extensive area of low relief dominated by broad floodplains and gentle interfluves, with isolated residuals (monadnocks) left upstanding in areas of resistant rocks.

According to W. M. Davis's original CYCLE OF EROSION scheme, the peneplain is the end product of subaerial denudation. His scheme being formulated mostly in the humid temperate climate of New England, he regarded this as the normal cycle, with the PEDIMENT and INSELBERG landscapes of arid and semiarid areas being deviations due to climatic controls.

There are no large peneplains on the Earth, because no landscape has evolved under just one cycle. All have polycyclic origins, interrupted by base-level movements and process variations. The peneplain concept is developed from the theory that slopes evolve by decline, such that as the cycle nears its end the remaining interfluves gradually become inconspicuous as weathering and rill wash continue to erode them, with the slope foot zones in the valleys being areas of little activity, mostly accretional in nature. Relative relief diminishes, therefore leaving an almost flat plain. This contrasts with the ideas of King and Penck, who stated that slopes evolve by PARALLEL RETREAT, leaving broad PEDIPLAINS and small upstanding residuals as the final products.

Pennsylvanian A period within the American classification of geologic time that followed the MISSISSIPPIAN, and preceded the Permian. It began about 320 million years ago and lasted for about 34 million years. The Pennsylvanian System corresponds approximately to the Upper CARBONIFEROUS elsewhere. It was named for the State of Pennsylvania, where rocks from this period are widespread and have produced much coal.

pentad A period of five successive days. It is often used in climatological work because the period divides exactly into the number of days in a year, leap years excepted.

pentlandite A yellow or orange mineral, an iron-nickel sulfide, $(FeNi)_9S_8$. It crystallizes in the cubic system, and occurs as masses, generally associated with PYRRHOTITE. It is used as a source of nickel.

percentile The percentage division of a ranked data series. For example, the upper five percentile indicates the value that is exceeded by 5% of the data, and the lower five percentile is the value above which 95% of the values occur.

percolation The process by which water moves downward, under the influence of gravity, through pores in soil or cracks and joints in rock. The water may carry dissolved chemicals, or may dissolve minerals from the soil (*see* leaching).

percoline A line just below the surface that denotes the extent of water seepage through soil. It generally follows the slope of the surface, but may be affected by plant roots and the activities of burrowing animals.

percussion mark A circular chip out of a pebble, caused by a sharp blow against another pebble when they were moving rapidly, probably transported by water.

perennial stream A stream or river that flows permanently throughout the year. *Compare* intermittent stream.

perfect elasticity The property of a material that can regain its original form after an applied force has been removed.

perfect plasticity The property of a material that retains a new form after an applied force is removed.

periclase A pale-colored cubic magnesium oxide mineral of composition MgO, found in metamorphosed dolomites and limestones.

pericline A fold structure, such as a basin or dome, in which the beds dip around a central point. In a basin they dip toward the center (*centroclinal*); in a dome they dip away from the center (*quaquaversal*).

peridot An olive-green gem variety of OLIVINE.

peridotite An ultramafic rock consisting wholly or largely of olivine, together with other ferromagnesian minerals, and devoid of feldspar.

periglacial Describing the climate, physical processes (*see* altiplanation; congelifraction; congeliturbation; freeze-thaw; frost-shattering; needle ice; solifluction), and resultant landforms (*see* involution; patterned ground; permafrost; pingo; stone polygons; stone stripes) characteristic of an area bordering an ice sheet. Although ice sheets and glaciers do not occur in the periglacial zone, it is still extremely cold, temperatures remaining below freezing throughout the winter and rising slightly above only during the summer. Periglacial zones are restricted to high latitudes today but were far more widespread during the PLEISTOCENE glacial periods.

perihelion The point on the Earth's orbit that is nearest the Sun. It is also applied to the nearest point to the focus of any orbiting body. *Compare* aphelion.

period An interval of geologic time in the Chronomeric Standard scale of chronostratigraphic classification (*see* chronostratigraphy). The equivalent Stratomeric Standard term, indicating the body of rock formed during this time, is the SYSTEM. Period and system are usually known by the same name; for example, rocks constituting the Jurassic System were formed during the Jurassic Period. Periods are formed of a number of EPOCHS grouped together and are themselves compounded to form ERAS. Thus the Triassic, Jurassic, and Cretaceous Periods together constitute the Mesozoic Era.

periodicity An oscillation in a time series of a climatic element that is found to recur at approximately equal time intervals. There have been numerous searches for periodicities in climatic data because they would be of great assistance in long-range forecasting. Unfortunately, although periodicities can be found for short time periods, they are rarely persistent.

Perissodactyla The order consisting of hoofed mammals having an odd number of toes, such as the horse and rhinoceros. Horses evolved from hyracotherium, a small dog-sized five-toed animal of the Eocene, by increasing in size and, as an adaptation to fast locomotion of the plains, by gradually losing all but one toe on each foot. They acquired high-crowned teeth with a complicated enamel pattern, well suited for chewing a diet of coarse grass. Horses originated in North America and spread to Europe, Asia, and Africa. Other, now extinct, types of perissodactyl lived in the early Tertiary, some reaching a large size, but the order has not achieved the success of the Artiodactyla.

perlitic Describing glassy rocks that contain concentric onion-like cracks caused by contraction during cooling.

permafrost Permanently frozen soil occurring in PERIGLACIAL areas, where the winter temperatures rarely exceed freezing point and the soil is frozen to considerable depths. In summer only the top meter or so of soil thaws, that beneath remaining frozen and acting as an impermeable barrier to percolating water. This top layer, which is subject to freezing and thawing, is known as the *active layer* and when melting occurs it rapidly becomes saturated and hence mobile (*see* muskeg; solifluction). Pedogenic and cryoturbation processes are confined to this layer.

permeability The ability of a rock to allow pore fluids and gases to pass through it. This is governed by the extent to which the pore spaces within the rock are connected.

Permian The final period of the PALEOZOIC Era. Beginning about 286 million years ago and lasting some 40 million years, it followed the CARBONIFEROUS and was succeeded by the TRIASSIC (which marks the start of the Mesozoic Era). It is named for the Perm region of the Russian Urals. The lower part of the Permian System consists of three stages: the Sakmarian, Artinskian, and Kungurian. The Kazanian and Tatarian Stages compose the upper division.

Large-scale movements of the crustal plates continued into the Permian; the continent of Laurasia in the N hemisphere became linked to Gondwanaland in the S hemisphere and, as a result, by the middle of the period the vast supercontinent, Pangaea, was created. The continuation of the Allegheny orogeny in E North America saw the culmination of the formation of the Appalachian fold belt; in Europe and Asia the corresponding Variscan orogeny of the Hercynian Mountains was also diminishing in intensity by the end of the period. Characteristic rocks of the Permian are red continental sediments and evaporites. They are widespread in North America with great thicknesses of deposits occurring in Texas, New Mexico, Nevada, and Utah.

The period saw the extinction of many animal groups, including the tabulate and rugose corals (*see* Anthozoa), productid brachiopods, trilobites, and blastoids, and a great reduction in others. Reptiles became abundant and pteridophyte plants were superseded by gymnosperms. *See also* Permo-Triassic.

Permo-Triassic The rocks of the PERMIAN and TRIASSIC periods in Britain, when these are considered as a single unit. Because continental conditions prevailed for much of the Permian and Triassic Periods in Britain recognition of any boundary between the two systems is difficult.

perovskite A black or brown cubic mineral form of calcium titanate, $CaTiO_3$, but usually with some substitution of rare earth elements for calcium and niobium for titanium. It is found as an accessory mineral in some highly undersaturated alkaline igneous rocks and in thermally metamorphosed limestones.

persistence The continuance of any synoptic event or climatic element beyond its expected duration; the term is frequently used for sea-surface temperatures or pressure anomalies. In a time series of climatic data, persistence also indicates a nonrandom tendency for relatively high or low values to occur in succession.

perthite An intergrowth of orthoclase and albite feldspar in which the albite occurs as patches or lenticles in the orthoclase host. Perthite may be produced by the reaction of potassium feldspar with sodic fluids or by the unmixing of an originally homogeneous alkali feldspar. *See* antiperthite; exsolution.

perthosite A member of the SYENITE group of minerals.

Peru Current *See* Humboldt Current.

pervious Describing a rock through which water can pass along cracks and fissures. The water can in this way percolate as far as the water table and form part of the supply of groundwater.

petrogenesis The origin or mode of formation of rocks.

petrography The study of the mineralogical and textural relationships within rocks revealed by observation of thin sections and hand specimens.

petroleum A naturally occurring complex mixture of flammable hydrocarbons, known also as oil or crude oil. It is a green to black liquid, often of high viscosity, and occurs seeping out of the ground or in underground deposits (from which it is extracted by drilling). The crude extract is distilled to produce various fuels and starting materials for the petrochemical industry.

petrology The scientific study of rocks, which deals with their classification, origins, distribution, and composition. Petrology involves mineralogy, chemistry, petrography, and petrogenesis.

pH A measure of the acidity or alkalinity of a solution. It is the negative logarithm of the hydrogen ion concentration. A neutral solution has a pH of 7; a pH of more than 7 is acidic, less than 7 is alkaline.

phacolith A minor concordant igneous intrusion, usually lensoid (the concave surface facing downward) and present in folded strata.

phanerocrystalline (phaneritic) Describing an igneous rock in which the constituent crystals can be seen with the naked eye. *Compare* aphanitic.

Phanerozoic The geologic time that has elapsed since the end of the PRECAMBRIAN, comprising the PALEOZOIC, MESOZOIC, and CENOZOIC Eras. It has lasted for some 570 million years, from the CAMBRIAN Period to the present. The name, meaning visible life, refers to the fact that clearly recognizable fossils are found in rocks laid down during these periods. In Precambrian rocks fossils are extremely rare and often of obscure biological affinities.

phase boundary (boundary line) (in geophysics) The line along which two constituents meet where only two are present. In a more complex system it is the line along which any two liquid phases meet.

phenakite (phenacite) A white or colorless quartzlike mineral form of beryllium silicate, Be_2SiO_4. It crystallizes in the hexagonal system, and occurs in veins in granite.

When cut it has a brilliant luster and it is used as a semiprecious gemstone.

phenoclast A large fragment or clast set in a finer-grained matrix of a sedimentary rock.

phenocryst A large generally euhedral crystal contained in many igneous rocks and set in a fine-grained matrix or groundmass. Such rocks are said to be porphyritic. The phenocrysts are the first crystals to form as a melt cools and are thus minerals having the highest crystallization temperatures.

phillipsite A fibrous mineral form of a hydrated aluminosilicate of calcium, sodium, and potassium, $(Ca,Na,K)_3(Al_3Si_5O_{16}).6H_2O$. It crystallizes in the orthorhombic system, and is one of the ZEOLITE group of minerals.

phlogopite A brownish mineral form of potassium magnesium aluminum silicate, with some iron, $K_2(Mg,Fe)_6(Si_6Al_2O_{20}(OH,F)_4$. It crystallizes in the monoclinic system, and occurs mainly in ultrabasic igneous rocks and metamorphosed limestones. It is a member of the MICA group of minerals.

phonolite A strongly undersaturated lava, the volcanic equivalent of nepheline-syenite. Phonolites are leucocratic rocks containing a high proportion of sanidine or anorthoclase feldspar and nepheline. Mafic minerals include soda pyriboles and biotite. Most phonolites are porphyritic and possess phenocrysts of feldspar, nepheline, and aegirine-augite. The typically aphanitic groundmass has a flinty appearance and the rocks ring when struck with a hammer and fracture subconchoidally. Trachytes containing accessory nepheline up to 10% are termed *phonolitic trachytes*. In many phonolites, analcite, sodalite, or leucite accompany nepheline. With an increase in the content of leucite and the disappearance of nepheline, leucite-phonolites pass into leucitophyres.

Tinguaite is an obsolete term for phonolite when it occurs as a medium-grained dike rock.

Phonolites often occur with strongly undersaturated basic lavas, nephelinites, and in association with nepheline-syenites in oceanic islands and continental rift environments.

phosphate A salt of phosphoric acid, containing the ion PO_4^{3-}. Many minerals are composed of phosphates, although few are common, except for APATITE.

phosphatic nodule A rounded mass that occurs in sedimentary rocks that have been laid down on the sea floor. Up to 30 cm across, they consist of pieces of corals, mica flakes, sand grains, and shell fragments.

phosphorite (phosphate rock) A calcic phosphate deposit that occurs as a marine deposit in roughly nodular form and (according to sampling and photographs) sometimes in quite large concentrations. It is mainly used for the manufacture of fertilizers. It is found on some shelf areas, and over parts of the continental slope. It is currently dredged at a number of offshore sites, for example, off the E coast of the USA.

photocontour map A topographic map produced by aerial photography. All the detail normally apparent on topographic maps is shown, the information being extracted from the photographs using stereoscopic equipment. This technique is now extensively used for 1:50 000 series mapping of level underdeveloped areas of the world, e.g. part of Arabia and Botswana, especially if a fairly rapid result is required.

photogrammetry The technique of producing maps and charts using stereoscopic equipment to obtain reliable measurements from aerial photography. *See* photomap.

photomap A map prepared by adding grid information, names, boundaries, and other map data to a reproduction of photographs or photomosaics. It is a quick method of producing maps, and is often used in areas of little relief, e.g. deserts, or for towns. Contours can be added to these maps for use in areas of relief difference, although this greatly lengthens the time taken to produce the map.

photoperiodism The response of plants to variations in day length. Some plants flower only when the day length (*photoperiod*) exceeds a specific time; others flower only in photoperiods of less than ten hours. This is obviously important in the natural distribution of plants.

photosynthesis The chemical processes by which green plants, algae, and other chlorophyll-containing organisms use the energy of sunlight to make complex organic compounds from carbon dioxide (from the atmosphere) and water. The main reaction is catalyzed by chlorophyll, and oxygen is released into the atmosphere as a by-product. It is a vital process because nearly all plants and animals rely either directly or indirectly on it for their existence.

phototroph A living organism that obtains energy from sunlight, such as all green plants. *See* photosynthesis.

phreatic activity The violent reaction that results from hot lava coming into contact with cold water. The surface of the lava chills and forms a glassy skin. The pressure of gasses such as water vapor present within the lava then cause this surface to fracture. Small fragments are hurled into the air, the reaction continuing until a tuff ring has built up, which separates the lava and water.

phreatic eruption A volcanic eruption caused by escaping steam generated when a lava flow comes into contact with groundwater.

phreatic water GROUNDWATER, especially that occurring below the water table, i.e. where all fissures are filled with groundwater. Phreatic water may supply springs and wells. *Compare* vadose water.

phyllite A metamorphosed rock resembling slate but of a coarser grain size. The cleavage or schistosity surfaces have a lustrous sheen caused by muscovite and chlorite. Phyllites may possess incipient banding due to the segregation of quartz and feldspar into layers parallel to the cleavage. The characteristics of phyllites are essentially intermediate between those of slates and schists.

phylum (*pl.* phyla) A major group in the taxonomic classification of organisms. Animal phyla include the Mollusca, Cnidaria, and Chordata. Each phylum is composed of one or more CLASSES. In traditional plant classification systems phyla are known as *divisions*. *See* taxonomy.

physical geography The branch of geography that includes aspects of hydrology, climatology, meteorology, oceanography, and pedology.

physical weathering *See* mechanical weathering.

phytoplankton Small, often microscopic, plants and algae that float passively in the sea or other bodies of water (*see* plankton).

picotite A dark brown variety of SPINEL that is rich in chromium.

picrite An ultramafic rock consisting of FERROMAGNESIAN MINERALS and accessory calcic plagioclase feldspar.

piedmont A gentle slope leading from the foot of a mountain range down to comparatively flat land. In arid and semiarid areas, where these features most frequently occur, they consist of an eroded upper segment (known as a PEDIMENT), which makes an abrupt angle with the mountain front, followed by an accumulation form or BAJADA, consisting of transported debris from the mountains. This merges into a flat INLAND BASIN or playa.

piedmont angle The sharp inflexion between lowlands and uplands in pediment and inselberg landscapes. Characteristically, the scarp of the upland slopes at 25–35°, passing to the slope of the pediment at 9° or less over a very short distance, leaving an abrupt angle. This angle can be very sharply defined, but usually it is a concavity, with the break of slope spread over a considerable horizontal distance.

Its origin is closely linked with the processes of PARALLEL RETREAT of the upland and beveling of the pediment. Some geologists have found that the angle is maintained by intense weathering in the scarp foot zone, caused by runoff from the pediment and through flow emerging at the angle; in other cases it coincides with a tectonic or geologic boundary. Others consider it to be a product of contrasted processes, e.g. unconfined wash on the residual with channeled water on the pediment (Bryan), or turbulent wash on the residual with laminar wash on the pediment (King); still others have explained it by slope process, the increased volume of water in the scarp foot area undercutting the slope of the residual, leading to mass movements which keep the angle sharp. It is almost definitely produced by different mechanisms in different areas. *See also* pediment.

piemontite A member of the EPIDOTE group of minerals.

pigeonite A calcium-poor monoclinic PYROXENE.

pillow lava A lava extruded under water and having the appearance of pillows piled one upon another. The outer skin of the lava is chilled on extrusion and a bubble of lava grows, flattening under its own weight and producing the characteristic pillow shape.

pilotaxitic Describing a close-packed felted arrangement of acicular MICROLITES. In many trachytes and andesites, feldspar laths exhibit parallelism due to flow and are deflected around phenocrysts in the direction of flow, this texture being called *trachytic.*

pinch and swell A deformation feature developed when competent rocks are squeezed. The more competent beds extend and thin toward their margins and eventually break. The less competent beds deform plastically to fill any space. *See* boudinage.

pingo A dome-shaped hill found in PERMAFROST areas. Pingos vary in size but seldom exceed 60 m high and 300 m in diameter. The top may be broken, so that it resembles a crater, and often contains a lake. Internally they frequently consist of outward dipping beds of stratified sand or silt. Many have a visible or supposed core of ice. They are believed to have been formed as a result of bulging, produced by subsurface pressure build-up within isolated groundwaters before the complete extension of permafrost over the area.

pipkrake *See* needle ice.

pisolith A pea-sized accretion that occurs in some sedimentary rocks (such as pisolite). It consists of concentric layers of calcium carbonate, possibly resulting from the biochemical encrustation of algae. *See also* oolite.

pitch The direction of dip of the axis of a fold, measured by the angle between the axis and the horizontal on the axial plane.

pitchblende A massive form of URANINITE.

pitchstone A member of the RHYOLITE group of minerals.

pivot fault *See* hinge fault.

placentals Placental mammals (*see* Eutheria).

placer A surface deposit of sand or gravel that contains significant quantities of valuable minerals, such as chromite, diamonds, gold, platinum, or tin. Small amounts of the minerals can be removed by panning; large-scale extraction usually involves dredging and concentration of the mineral by various processes.

Placodermi A class of extinct jawed fish comprising a number of primitive but varied groups common in the Devonian. They had paired fins and, in the arthrodires and antiarchs, an armor of protective bony plates. Some placoderms were of great size, reaching a length of up to 10 m.

plagioclase feldspar Any member of a series of minerals with compositions varying between two end-members, albite ($NaAlSi_3O_8$) and anorthite ($CaAl_2Si_2O_8$). *See* feldspar.

plain An extensive region of low-lying land, which is generally flat or gently undulating. Most plains are formed by deposition of eroded sediments; others are created by the wearing away of higher land (DENUDATION), forming a peneplain.

planation surface A surface of low relief, the end product of a cycle of subaerial or marine erosion. Characteristically such surfaces bevel indiscriminately across structures; they may be PEDIPLAINS, PENEPLAINS, or surfaces of marine erosion. In order to develop, a planation surface needs a long period of base-level and climatic stability: such conditions have not existed since the end of the Tertiary period, because of the oscillations in the base level and climate of the Pleistocene and post-Pleistocene periods. Those planation surfaces that do exist therefore are relic and much dissected. They have been reconstructed by analysis of relief, which shows in general concordance of summit levels at the level of former planation surfaces.

plane tabling A rapid surveying method for fixing detail in fairly open areas. The plane table is a flat board with an attached sheet of paper onto which a baseline is drawn to scale between two known points. The whole stands on a tripod, which is set up over one end of the baseline, with the drawn line aligned with the actual line. Points of detail are then sighted using an ALIDADE and rays to them drawn directly onto the paper. Once enough points have been covered, the plane table is set up at the other end of the baseline, aligned, and rays drawn to the same detail points. Their actual positions on the map occur at the points of ray intersection on the paper. More accurate fixing can be achieved by drawing rays from three known points.

planetary wind Any of the world's major winds, affecting large areas of the globe. They include the TRADE WINDS of the tropics, the south-westerlies of northern temperate regions, and the north-westerlies or ROARING FORTIES in the S hemisphere.

planimetric map A map on which, unlike topographic maps, no vertical information such as contouring is shown.

plankton Aquatic and usually microscopic organisms that float and drift passively in the water (*compare* nekton). Plankton consists mainly of animal larvae, PROTOZOA, and algae, such as DIATOMS. *Phytoplankton* refers to plant and algal forms, on which all other marine organisms depend, directly or indirectly, for their survival; *zooplankton* refers to the immense variety of animal forms. The distribution of both phytoplankton and zooplankton throughout the oceans is very patchy; plankton-rich areas form highly fertile zones in the oceans, for example where UPWELLING currents occur.

Plankton is of considerable geologic importance as it contributes to rock-building; for example, much of the chalk (*see* Cretaceous) consists of COCCOLITHS, the skeletal remains of past planktonic organisms. Fossil plankton is studied in micropaleontology (*see* microfossil), which has important stratigraphical applications.

planosol An INTRAZONAL SOIL developed in flat areas, principally under humid continental warm summer climates. Planosols are characterized by a whitened A_2 horizon passing abruptly to a B horizon of high clay content, either due to in-situ weathering or LESSIVAGE from the A horizon. Periodic waterlogging results from the poor drainage due to the high clay content in the B horizon and lack of relief. They intergrade with a number of soils, including GRAY-BROWN PODZOLIC SOILS, PRAIRIE SOILS, and surface-water GLEY SOILS. In the SEVENTH APPROXIMATION classification they are split between a number of orders, including mollisols, alfisols, and ultisols.

plant cover The geomorphological significance of the plant cover is as a balance between the soil and weathered rock beneath and the processes acting on the surface from above. Its loss greatly accelerates rates of erosion, e.g. in areas of cultivation, overgrazing, human trampling, or destruction of vegetation by burning or pollution. The beneficial effects of vegetation include the binding action of the roots, the diverting action of its mass on water flowing on the surface, shelter from wind action, and the increase in organic matter on the soil surface, which improves aggregation and hence the ability of the soil to hold water, thereby reducing runoff. Vegetation can hold soil on a slope that would otherwise be too steep for stability and hence be eroded; it also intercepts rainfall, preventing RAINSPLASH erosion. In ecological terms, plant cover provides HABITATS for various organisms.

plastic deformation of ice A process operating in glacier flow. Glaciers do not move down their valleys as rigid masses simply sliding across the bedrock. Ice is a crystalline solid and as such changes shape (or *deforms*) at temperatures near its melting point. Movements between adjacent crystals are limited but individual crystals can deform internally, although little change in shape appears to occur over a period of time: this indicates that recrystallization accompanies the deformation. Individual crystal deformation is achieved by relative movements of layers, one above another, parallel to the basal plane. Field evidence has shown that ice is molded as it moves across its bed, but that resultant features are maintained within the ice beyond the location of molding. This suggests that the ice acts plastically under pressure, but that it remains rigid on removal of the pressure.

plastic relief map A three-dimensional map produced by printing a topographic map on plastic and then molding the plastic to fit the relief. It is generally used only for demonstration purposes because there are still certain inaccuracies involved in this method.

plastic shading *See* hill shading.

plate One of a series of large blocks of continental or oceanic material of which the Earth's crust is composed. These plates, whose base is marked by the asthenosphere, move across the surface of the Earth as a result of SEA-FLOOR SPREADING. The six major tectonic plates are named Eurasian, Indian, Pacific, American, African, and Antarctic.

plateau A fairly flat elevated area of land. A *dissected plateau* is broken up by river valleys or canyons, an *intermontane plateau* is surrounded by mountains, and some plateaus may incorporate the mountains themselves.

plate boundary (plate margin) The edge of a LITHOSPHERIC PLATE (tectonic plate). There are three types of boundary:

1. *Constructive*, where new sea floor is added to the plates on each side of the boundary, e.g. at the Mid-Atlantic Ridge.
2. *Destructive*, where sea floor is destroyed through subduction. This is usually marked by a deep ocean TRENCH and results from one plate overriding another. *See also* subduction zone.
3. *Conservative*, where plates slip passively past each other without destroying or adding sea floor. This occurs along TRANSFORM FAULTS.

plate tectonics A theory arising from a series of ideas developed in the early 1960s proposing that the surface of the Earth is composed of a number of relatively thin PLATES of rigid material. These lithospheric (or tectonic) plates extend down to the low-velocity zone of the upper mantle. They are all in motion relative to one another and it is through these movements and the consequent collisions between the plates that the present distribution of almost all volcanic, seismic, and orogenic activity is controlled. The plates are bounded by PLATE BOUNDARIES. The plates themselves are composed of either oceanic or continental crustal types, or a combination of both. Only the oceanic parts of the plates grow or are destroyed; the continents ride along on these plates and grow only slowly by the addition of volcanic material and sediment deposition along their margins. These sediments are compressed and folded when two continental regions collide by orogenic processes, which results in the formation of fold mountain chains. These have a central nucleus of older rocks, frequently cratons.

playa A flat-bottomed enclosed basin in a desert, sometimes occupied by an ephemeral lake. There may be deposits of EVAPORITES just below the surface, and similar deposits on the surface when any lake dries up. *See also* inland basin.

Pleistocene The earlier epoch of the QUATERNARY, extending from the end of the PLIOCENE about 1.6 million years ago, until the beginning of the HOLOCENE. During this period, often referred to as the ICE AGE, the world experienced great fluctuations in temperature, resulting in cold periods (*glacials*), separated by warmer periods (*interglacials*). The Earth's climate cooled by about 5–10°C during the glacials. In North America ice sheets developed in Canada and advanced S into the N USA at least four times before retreating. In the Alps, four main glacial episodes can be recognized: the Gunz, Mindel, Riss, and Würm.

pleochroism A property of some crystals that display different colors when viewed from different directions under transmitted plane-polarized light. It is caused by selective absorption of some wavelengths along the different crystal axes.

Plesiosauria A group of Mesozoic reptiles that were secondarily adapted to an aquatic life but, unlike the ICHTHYOSAURIA, were not fishlike. They had a rigid flattened trunk and well-developed powerful paddles for propulsion. A long neck provided the necessary rapid flexibility for catching the fish on which they fed. Some plesiosaurs attained a length of 17 m. The group became extinct by the beginning of the Cenozoic.

Pliocene The final epoch of the TERTIARY Period, preceded by the MIOCENE and followed by the PLEISTOCENE. It began about 5.3 million years ago and lasted some 3.7 million years. Mammals similar to modern forms existed, and there were definite hominids, including species of *Australopithecus* and *Homo* (*see* man).

plucking *See* glacial plucking.

plug **1.** Any roughly cylindrical vertical body of intrusive igneous rock, usually relatively small.
2. *See* neck.

plumbago *See* graphite.

plunge The tilt of the axis of a fold from the horizontal, the value of plunge being the angle between the axis and the horizontal lying in the same vertical plane. See diagram at FOLD.

pluton A deep-seated major intrusive body of coarse-grained igneous rock, generally of granitic composition.

plutonic rock An igneous rock that crystallizes at depth and cools slowly, resulting in a coarse grain size. Granite is an example. Igneous rocks of the same composition may show striking textural differences depending on their mode of occurrence. Those that are extruded on the Earth's surface as lava flows and cool rapidly, producing a fine grain size, are called VOLCANIC ROCKS. *Hypabyssal rocks* are those that have crystallized at levels intermediate between those of plutonic and volcanic conditions. They are medium-grained and occur mostly as sills and dikes. Gabbro and dolerite are, respectively, the plutonic and hypabyssal equivalents of basalt.

pluvial period A period of time experiencing greater rainfall than preceding or succeeding periods, usually on a geologic timescale. Many of the semidesert areas of the tropics experienced pluvials during full glaciation in polar regions. This was a response to the southward movement of the circulation belts and depression tracks. Evidence for these wetter periods comes from greater water erosion, increased plant growth, and animal remains found in areas in which they would not be found under present conditions. However, this has been disputed because lower temperatures could have reduced evaporation, making the same rainfall amounts more effective.

pneumatolysis A process occurring during the final stages of the crystallization of acid igneous rocks when residual borofluoric gas escapes along joints and fissures and brings about mineralogical changes in the crystalline parent rock (*compare* metasomatism). Three kinds of pneumatolysis are recognized:

1. *Greisening.* Alteration at the margins of granites to assemblages of muscovite, quartz, and topaz (greisen) is the result of the action of fluorine-bearing vapors. Original feldspars are pseudomorphed by aggregates of mica, often varieties rich in lithium such as zinnwaldite. The end product of the process is a quartz-topaz rock known as topazfels.
2. *Tourmalinization.* The boron pneumatolysis of granitic rocks results in the growth of TOURMALINE at the expense of feldspar and biotite. The rock, luxullianite, contains radiating aggregates of black tourmaline but some feldspar survives. The final stage is a quartz-tourmaline assemblage known as *schorl rock*.
3. *Kaolinization.* Feldspars in granitic rocks may be altered to aggregates of kaolinite and sericite under the action of high-temperature aqueous fluids. The end product of this pneumatolytic/hydrothermal process is KAOLIN.

podzol (podsol) A soil characterized by an ashen-colored acid eluviated A horizon and a B horizon illuviated with iron or humus, possibly in the form of a compact pan. The A horizon is depleted of bases and sesquioxides by LEACHING processes, especially CHELUVIATION, while the B horizon is divided into an upper part (Bh) dominated by the redeposited humus and a lower part (Bs or Bfe) dominated by the redeposited iron, and to a lesser extent, aluminum. If both the Bh and Bs horizons are present, the soil is an iron-humus podzol; if the Bh only is present, a humus podzol; if Bs only, an iron podzol.

Podzols are zonally developed soils in the TAIGA zones of Russia and North America, where they form a mosaic with regosols, gleys, and peats, and develop from initiation directly on the parent material. Elsewhere podzol is intrazonal, e.g. on coarse deposits such as sands and gravels or human-induced by forest clearance. As the profile matures, the Bh-s horizons may become impermeable and lead to GLEYING producing hydromorphic variants such as PEATY GLEY PODZOLS.

podzolic soil (podsolic soil) A soil with certain features of morphology and genesis akin to true podzols. There is a large number of varieties. In North America there are gray-brown podzolics in a belt below the true podzols, which develop slowly by increased leaching, destruction of the Bt horizon, and graying of the A_2 horizon to brown podzolics. In the south, in the subtropical areas, are red-yellow podzolics, which have free aluminum and ferric oxides like other tropical soils but also clay skins in the B horizon.

podzolization The movement of clays and sesquioxides down the soil profile and their deposition in the B horizon. The dominant process is CHELUVIATION by organic acids, but opinions vary as to the other processes involved. Some pedologists believe that podzolization operates best in acidic and anaerobic conditions, so that the cheluviation is preceded by DECALCIFICATION, LESSIVAGE, and clay destruction by weathering. Others have suggested that acidic conditions are not necessary, and decalcification is therefore not a necessary preliminary; it has also been argued that anaerobic conditions (i.e. waterlogging) are not necessary because the organic acids can reduce the sesquioxides themselves, and do not rely on the anaerobic reduction of iron from its insoluble ferric to its soluble ferrous state. Podzolization is a zonal process in areas of natural LEACHING and coniferous vegetation, which together favor intense cheluviation, the chelates of pine litter being among the most powerful. Intrazonally, it is the dominant process on coarse-textured parent materials, or where forest clearance has increased leaching.

poikilitic Describing igneous rocks in which large crystals of a mineral completely enclose crystals of earlier-formed minerals. *Poikiloblastic* denotes a similar texture in metamorphic rocks. *See also* ophitic.

poikiloblastic *See* poikilitic.

point bar A depositional feature that develops on the inside of meanders, complementing the erosion that occurs on the outside of meander bends. Point bars extend downstream from the point of maximum curvature of the meander bend, leaving a trough between themselves and the bank, which eventually becomes filled up by the sedimentation of fine material. *See* meander.

Poisson's ratio The ratio between the fractional longitudinal strain and the fractional lateral strain in a deformed material. It is equal to the ratio of change in diameter divided by change in length.

polar Describing the climate that is characteristic of the regions within the Arctic and Antarctic Circles (around the North and South Poles). They have permanently low temperatures, a short growing season, and no trees.

polar air depression (polar low) A small depression that forms within cold north or northwesterly airstreams between Iceland and Scotland. It was formerly thought to be the result of thermal contrasts between warm sea and cold air, but radar investigations have shown that these depressions are very similar to the normal baroclinic depression, although on a smaller scale. The small scale prevented the frontal positions being identified in the sparse network of observing stations over sea areas.

polar easterlies The easterly wind belt between the weak polar anticyclone and the westerly depression tracks. Because of the variability of the intensity and tracks of the mid-latitude depressions and the weakness of the polar anticyclone, the polar easterlies are rarely strong or persistent.

polar front The major frontal system, situated in the N Pacific and N Atlantic, separating tropical and polar air masses. Depressions are often initiated on this pronounced thermal gradient and Bjerknes based his theory of frontal evolution on it. In summer the thermal contrast across the front is less and its position is much more variable than in winter.

polar glacier *See* cold glacier.

polarity reversal *See* magnetic reversal.

polarization colors (interference colors) *See* birefringence.

polarized light In ordinary light the electromagnetic vibrations take place in all directions in a plane at right angles to the direction of propagation of the ray of light. When vibrations are confined to a single direction in this plane, the light is said to be plane-polarized. Certain crystals constrain ordinary light to vibrate in only two directions at right angles to one another; i.e. they polarize it. *See* anisotropic; birefringence.

polar night jet stream A very steep thermal gradient in winter giving correspondingly very strong westerly winds, situated around the stratospheric cold pole at levels of 20-30 km. During summer the stratosphere at this level warms up because of absorption of ultraviolet radiation, and a reversal of the wind system takes place to give stratospheric easterlies.

polar wandering curve A theoretical line produced for a particular point on the Earth's surface by joining successive paleomagnetic pole positions through time.

polar wind Cold air that blows from the high-pressure regions near the North and South Poles. As they become warmer, polar winds pick up moisture from the oceans and may bring heavy snowfalls in winter.

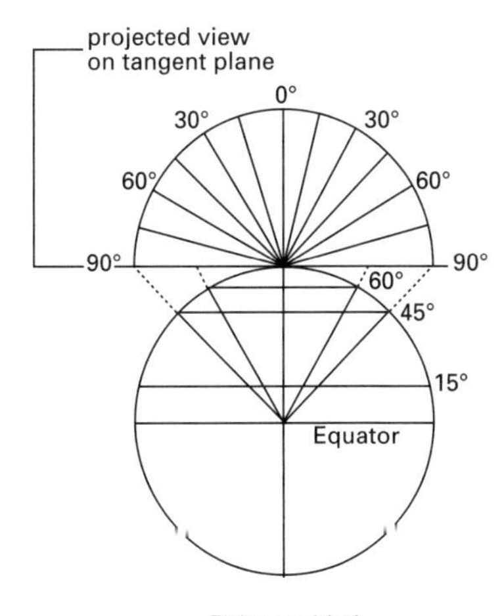

Polar zenithal

polar zenithal gnomonic projection A map projection in which the center of the projection is the pole, from which the straight lines representing the great circles radiate (see diagram). The parallels are portrayed as concentric circles. It is commonly used to show the polar areas, the exaggeration becoming too great in areas farther away from the center.

pole of rotation A pole about which a lithospheric plate rotates during sea-floor spreading and continental drift.

polje The largest type of solution depression found in limestone areas, extending up to 250 sq km. The floors support a certain amount of surface drainage, are alluvium-covered, and often uneven, reflecting the collapse of former cave systems. Poljes are frequently surrounded by steep marginal walls, up to 100 m in height, while considerable numbers show signs of being tectonically bounded basins.

pollen Microscopic spores of the higher plants (gymnosperms and angiosperms). Because pollen is very resistant to destruction and may be carried for large distances by the wind, fossilized pollen provides a valuable means of correlating the rocks in which it occurs. Pollen may also be used as a sensitive environmental indicator and has proved particularly useful in monitoring climatic change during the Quaternary Period. The study of pollen and other spores is known as *palynology*.

pollen analysis A technique of both relative dating and environmental reconstruction, consisting of the identification and counting of pollen types preserved in peats and lake beds. It is based on the assumption that the vegetation in an environment is in equilibrium with it, and therefore reflects its conditions; dispersal of pollen leads to some of it being preserved in reducing environments, such as peat beds and lake bottoms. Subsequent careful close sampling and analysis allows the structure of the vegetation at each particular horizon in the deposit to be worked out. Reference to the present-day vegetation in different conditions allows deduction, by analogy, of the environment that the vegetation lived in. This method has shown well-defined sequences of changes for each of the interglacials and the postglacial period, which allowed relative correlation between different sites, and hence dating.

polyconic map projection A modified CONICAL PROJECTION in which each parallel is treated as a standard parallel on a normal conical projection. All the parallels and the central meridian are truly divided, the meridians being constructed by joining the divisions on the parallels. This projection is neither an ORTHOMORPHIC PROJECTION nor a HOMOLOGRAPHIC PROJECTION and is therefore restricted to showing small areas.

polycyclic landscape A landscape that has evolved under a number of geomorphic cycles (*see* cycle of erosion) or part cycles of different types. The independent variables of geomorphology are climate, geology, and base level: a change in any one starts a new cycle. Regional studies of landscape evolution emphasize that most if not all landscapes have experienced changes in these factors and have hence evolved under several different cycles. This realization has led to the abandonment of the youthful, mature, and senile labels for landscapes, because they assume that each landscape is a product of only the current cycle, variation being due to the STAGE of that cycle. Instead each landscape can be seen as composed of elements from several different cycles, only some of which are related to current conditions.

polygenetic (in geomorphology) Describing a soil or landscape that has evolved under a number of differing conditions that occurred in successive phases of its history. Each of these conditions has played a part in influencing its current appearance. With landscapes, a change in base level, climate, or geology will institute a new cycle and hence superimposition of varying conditions in its evolution; with soil, a change in climate, vegetation, topography, or drainage will institute a new set of soil-forming conditions and hence change the direction of the soil's evolution. Some authorities consider this term to be synonymous with POLYCYCLIC when applied to landscapes; others consider that polygenetic should be used to describe only the influence of minor changes in conditions, leaving polycyclic for the major changes.

polymorphic transition A change in minerals that involves a change in their atomic structure but not their chemical composition, e.g. aragonite ($CaCO_3$) to calcite (also $CaCO_3$).

polymorphism The ability of some substances to exist in two or more structurally distinct forms, each having its own characteristic properties but having identical chemical compositions. Calcium carbonate, $CaCO_3$ occurs as aragonite and calcite. Andalusite, kyanite, and sillimanite are polymorphs of the aluminum silicate, Al_2SiO_5. Graphite and diamond are mineral polymorphs of the element carbon.

polyphase deformation Folding and faulting occurring in several periods during a single orogeny.

pools and riffles In detail, the long profiles of many streams have alternating gravel bars (riffles) and pools: these features are regularly spaced, usually such that the distance between successive pools is 5–7 times the channel width. In meandering streams, lengths of meanders are about twice the distance between pools, and there may be a relationship between the pool and riffle formations and MEANDER creation.

Porifera The sponges: a phylum of simple sessile aquatic (mostly marine) multicellular animals with a saclike body, often containing small calcareous or siliceous skeletal elements known as *spicules*. Water is drawn into the body through small holes and expelled from an opening, the *oscula*, during which food particles are filtered off. The spicules may become fossilized. Fossil sponges are useful environmental indicators and have contributed to the growth of bioherms and formation of rocks. They are known to have extended from the Cambrian Period to the present day, and structures found in Precambrian rocks have been attributed to this group.

porosity The extent to which a body of soil, rock, or sediment is permeated with cavities between grains, usually expressed as a percentage of the volume. These pores are filled by air and water (air is mainly in the larger macropores and water in the micropores), which impede air movement and allow water to move by CAPILLARITY only. In sands and compact soils, where the particles lie close together, porosity is low (25–30% in compacted soils); conversely, where organic matter content is high, promoting good aggregation, porosity is high, reaching possibly 60%. Cropping, which removes organic matter, eventually lowers porosity, and hence aeration.

porphyritic Describing an igneous rock that possesses large crystals called *phenocrysts* set in a finer-grained groundmass.

porphyroblastic Describing a metamorphosed rock containing large crystals set in a fine-grained matrix. The large crystals are called *porphyroblasts* by analogy with the corresponding igneous term, PORPHYRITIC.

porphyry A medium- or fine-grained igneous rock containing numerous PHENOCRYSTS. The porphyritic mineral may be indicated by the prefix of a mineral name, e.g. quartz-porphyry. A rhomb-porphyry is so called on account of the distinctive shape of the feldspar phenocrysts.

positive area A large area of the Earth's crust that has remained above sea level for a long period of geologic time.

positive gravity anomaly *See* gravity anomaly.

positive movement of sea level A fall of the land relative to the sea, usually due to sea-level rise, as has happened since the end of the last Ice Age. The results are basically twofold: flooding, and deposition of material creating broad low plains.

Flooding produces estuaries or rias from inundated river valleys; estuaries characterize lower-lying areas (e.g. the Chesapeake and Delaware bays on the Atlantic Coast of the USA), rias the more rugged areas (e.g. in SW Britain, SW Ireland). If the valleys inundated are glacial in origin, fiords result (e.g. in Norway). Low-lying parts of the land may be flooded wholesale, creating new seas and severing islands from the mainland (the British Isles were severed from the rest of Europe by the postglacial sea-level rise).

The effect on rivers is that CAPACITY and COMPETENCE are reduced, leading to AGGRADATION in the lower reaches, which fills up the newly created estuaries and creates broad alluvial floodplains. The lengths of drainage systems are reduced, leaving buried portions beneath the elevated sea level. *See also* negative movement of sea level.

postkinematic Describing minerals developed after a period of deformation (usually metamorphic minerals).

potassium feldspar A type of alkali FELDSPAR that contains potassium aluminosilicate, $KAlSi_3O_8$. The two principal crystal forms are the monoclinic ORTHOCLASE and triclinic MICROCLINE. Adularia and sandine are also alkaline feldspars.

potential energy The energy possessed by a body as a result of its position. It is measured by the amount of work required to move that body from a position of zero potential energy, usually sea level, to its new position. Stationary air at high levels of the troposphere therefore has a high potential energy.

potential evapotranspiration (PE) The evaporation from an extended surface of a short green crop, actively growing, completely shading the ground, of uniform height, and possessing an adequate supply of soil moisture. It is essentially a function of climate rather than the nature of the vegetation, and as such has been used by Thornthwaite as a method of climatic classification. It can be estimated or measured more easily and accurately than actual evapotranspiration and so has achieved more universal acceptability in studies of evaporation.

potential temperature The temperature of a parcel of air if it is uplifted or subsides at the DRY ADIABATIC LAPSE RATE to a standard pressure, usually 1000 mb. Use is also made of the wet-bulb potential temperature by changing the temperature of the parcel of air at the SATURATED ADIABATIC LAPSE RATE from the wet-bulb temperature of the original level to 1000 mb. This is useful because the wet-bulb potential temperature is a conservative property in such processes as evaporation and condensation, and both dry and saturated adiabatic temperature changes.

pothole 1. A circular hole worn in rock in an eddy of a stream or river, often at the foot of a waterfall, by moving pebbles and gravel.
2. *See* sinkhole.

prairie A region of flat treeless temperate grassland of Canada and the USA, extending from Alberta to Texas. The prairies have summer rain and fairly dry winters. Once home to large herds of bison, cattle, and sheep, much of the eastern part has been plowed for growing cereal crops, particularly wheat. Repeated plowing in the drier western prairies has resulted in erosion and the formation of a DUST BOWL.

prairie soil (brunizem) Dark soil of subhumid grasslands in which the profile has been decarbonated but leaching has not produced any movement of sesquioxides. The carbonate appears as flecks in the C horizon. These soils develop under tall grass vegetation and have a brown or gray-brown A horizon grading to a yellow-brown B horizon, with blocky structures and a light parent material at around one meter. Their exact status is not certain. They may be degraded CHERNOZEMS or more related to the podzolic group, with which they intergrade at their boundary, or their distinctive profile may be caused by a water table standing in the profile producing GLEYING. In the SEVENTH APPROXIMATION classification they fall in the order of MOLLISOLS.

Pratt's hypothesis A proposed mechanism of hydrostatic support for the Earth's crust. This hypothesis relies upon crustal density being greater under mountain chains than under oceans. If this were true at a datum level, called the COMPENSATION LEVEL by Pratt, rock columns of equal diameter would have equal mass.

Precambrian The geologic time prior to the Cambrian Period or the rocks stratigraphically below the Cambrian System. Because the age of the Earth's crust has been estimated at about 4600 million years, and the beginning of the Cambrian is dated around 570 million years ago, the Precambrian represents most of geologic time. Precambrian rocks contain rare FOSSIL evidence and, because of their extreme age, have often been subjected to a great deal of subsequent alteration. The original stratigraphical relationships are therefore frequently obscure. The SHIELD areas represent the largest areas of exposed Precambrian rocks.

The Precambrian is often now divided into three eons. The earliest time span is the *Hadean* ("beneath the Earth"), from which time virtually no rocks or fossils remain. During the *Archean* eon (about 3.8 to 2.5 billion years ago) evidence for the earliest known life forms (early bacteria) appears in rocks (*see* stromatolite). Rocks of the *Proterozoic* eon contain fossil evidence of the first multicelled organisms, including jellyfish and soft marine worms. Calibration and correlation of isolated successions is achieved mainly by radiometric dating.

precession A form of motion of a rotating body that results when a couple, having its axis at right angles to the axis of rotation, is applied to the body so that it then turns about a third mutually perpendicular axis. As a result of the gravitational attraction of the Sun and Moon, the Earth's axis precesses and traces out a conical figure.

precipitable water The amount of water that could be obtained if all the vapor in a standard column of air was condensed onto a horizontal surface of unit area. It is a useful index of the moisture content of air above a specified point, although precipitation processes are never so efficient that all the water would be precipitated and it neglects the effects of vapor advection. It is calculated by the formula:

$$M_w = (1/g)\int_{p_2}^{p_1} r\,dp$$

where M_w is the precipitable water, g is the acceleration of free fall, p_1 and p_2 are the pressures (mb) at the top and bottom of the layer, r is the mixing ratio, and dp is the depth of the individual layer of mixing ratio.

precipitation The deposition of water in solid or liquid form from the atmosphere. It therefore covers a wide range of particles including rain, drizzle, sleet, snow, hail, and dew. Precipitation is initiated within clouds by the Bergeron-Findeisen process (*see* Bergeron-Findeisen theory), by the coalescence process, or by a combination of both operating together. For precipitation to reach the ground a number of other conditions must be satisfied. The droplet or ice crystal produced by one of the above mechanisms must be sufficiently heavy to overcome upward vertical motion and large enough to withstand evaporation beneath the cloud base. Thus, some clouds can be seen to be giving precipitation that does not reach the ground.

precipitation effectiveness A measure of the usefulness of the annual rainfall total for agricultural purposes or for hydrology. For example, 600 mm of rain in temperate latitudes with low evaporation maintains humid vegetation and surplus runoff for river flow. In the tropics, such an annual total would give semidesert. Precipitation effectiveness has been used by Thornthwaite as a basis for climatic classification. However, the sparsity of data of both measured and calculated POTENTIAL EVAPOTRANSPIRATION, from which precipitation effectiveness is obtained, has prevented its use on a global scale.

precipitation variability The annual variation in rainfall totals, although shorter time periods can be used. It is assessed by many statistical methods but the most common is the *coefficient of variation*, which is the standard deviation of the annual totals divided by the mean annual value, multiplied by 100 to be expressed as a percentage. In Britain, values range from below 10% in the NW to 20% in S England, but these are low compared with many other parts of the world. Malden Island in the central Pacific is believed to have the highest rainfall variability with annual rainfall totals varying between 30 mm and 1422 mm in a thirty-year period.

predator An animal that kills other animals (the prey) for food. It is a secondary (and sometimes tertiary) CONSUMER in a FOOD CHAIN.

prehnite An orthorhombic mineral of composition $Ca_2Al_2Si_3O_{10}(OH)_2$ found chiefly in basic lavas as a secondary mineral associated with ZEOLITES. It is also found with pumpellyite in low-grade regionally metamorphosed rocks.

pre-kinematic Describing minerals that are formed before deformation.

pressure (in meteorology) The weight of air vertically above a unit area centered on a point. It is measured by balancing the force exerted by the atmosphere with a dense liquid, usually mercury. As a result of this, atmospheric pressure was formerly quoted as a number of millimeters of mercury representing the length of the mercury column required to balance air pressure. It is now measured in MILLIBARS. Sea-level values range from extremes of about 890 mb in hurricanes to 1060 mb in strong anticyclones.

pressure gradient force The force that acts on an air molecule by virtue of the spatial variations in pressure at any horizontal level. It is the primary motivating force of air movement in the atmosphere.

pressure release *See* unloading.

pressure system A pattern of isobars that exhibits coherent weather characteristics. It normally denotes low-pressure areas or anticyclones.

pressure tendency The local rate of change of surface pressure. It is recorded at most observing stations with a 10 point code to distinguish various trends of the pressure trace in the preceding time period, usually 3 hours. It is very useful for indicating future movements of pressure systems.

prevailing wind The most frequently occurring wind at any site. In most parts of the globe this will correspond with the climatic zone. Occasionally local factors, such as wind funneling in a valley, become dominant. Also for shorter time periods, the prevailing wind may be different from its annual value, as in monsoon areas.

primary consumer (in ecology) An organism in the second level of a FOOD CHAIN, usually a HERBIVORE. It provides food for secondary consumers (carnivores).

primary magma A magma whose composition has remained unchanged since it was first formed.

primary mineral A mineral that was formed at the same time as the rock containing it and which has remained unchanged in composition. *Compare* secondary mineral.

primary producer (in ecology) An organism at the lowest level of a FOOD CHAIN. It "feeds" on inorganic substances. Green plants and photosynthetic algae are examples; they produce carbohydrates from carbon dioxide and water using the energy of sunlight in PHOTOSYNTHESIS. They can then become the food for consumers in the next level of the food chain.

primary wave (P wave) A type of seismic wave, the first to be recorded on a seismogram of an earthquake. These longitudinal waves travel by a series of compressions and rarefactions (pushes and pulls) in the direction of propagation, i.e. each particle vibrates backward and forward along the direction in which the wave is traveling. P waves can pass through solids, liquids, and gases. *See also* secondary wave.

Primates The order of mammals that includes man, apes, monkeys, and lemurs. They are characterized by adaptations for an arboreal habitat, such as flexible limbs and hands, anteriorly directed eyes providing stereoscopic vision, often a long tail for balance, a relatively large brain, and an unspecialized dentition. Fossil primates are known from the beginning of the TERTIARY onward; they probably evolved from the INSECTIVORA. Monkeys appeared in the Oligocene and the remains of humanoid forms have been found in Pliocene strata (*see* man).

prime meridian *See* meridian.

primitive equations The physical equations governing momentum in the atmosphere. They have been used most extensively in models simulating the general circulation of the atmosphere.

principal shock The suite of waves produced by the main movement in an earthquake. *See also* aftershocks; foreshock.

prismatic Describing the habit of some crystals that are elongated and show well-developed prism faces. *See* crystal habit.

prismatic compass A magnetic compass within which the needle, or the zero of a 0° to 360° scale, always points toward magnetic north. Also incorporated is a sighting arrangement, consisting of a vertical hairline at the far side and a prism with a sight at the observer's side, which allows the object and the graduated scale to be viewed simultaneously. Readings taken with such a compass are true bearings from magnetic north.

Proboscidea The order of mammals that includes the modern elephants and the extinct mammoths. Fossil proboscideans are known from the Oligocene. They were small creatures with long jaws, both upper and lower bearing tusks, and short trunks. Evolutionary modifications associated with the subsequent increase in size have occurred: solid limbs and flat feet support the great weight; the trunk has evolved into an efficient food-gathering organ; the teeth have lengthened, and the jaws have shortened, in modern forms accommodating only one large grinding tooth at a time, which is replaced through life. Formerly a widespread and diverse group, the Proboscidea are today confined to Africa and S Asia.

producer *See* primary producer.

profile **1.** The arrangement of soil HORIZONS found between the ground surface and the parent material. The normal depth of a profile in temperate latitudes is about a meter. Pedologists, notably Dokuchayev, have assigned letters to certain horizons. The surface horizon is known by the letter A and is usually rich in organic matter and plant nutrients. Because material is often leached from this horizon down the profile to the subsoil B horizon, it is known as the *eluvial horizon*. The B horizon (*illuvial horizon*) contains less organic matter and more closely resembles the parent material owing to the lesser effect of the soil-forming processes. This horizon usually merges into the C horizon, which is the weathered parent material. Well-developed soils consist of more than these three horizons and subdivision occurs. Characteristics of certain horizons are symbolized by a suffix or prefix, e.g. Bh = humus-rich B horizon; Ap = plowed A horizon. Superficial organic horizons are denoted by the letters L (litter), F (fermenting), and H (humus). Groups of soils having a similar suite of horizons are linked in broad soil zones.
2. *See* river.

progressive wave A wave that can be recognized by the progressive movement of the waveform at the sea surface. For example, such a wave would be propagated along a channel of infinite length, having been generated at one or other end of it by a wave paddle. In progressive waves, the speed of propagation of the waveform will largely depend upon the depth of water, one example being a tide wave that may be very long in relation to the water depth. As with wind-generated waves, the speed of wave propagation far exceeds the speed at which the water particles themselves advance. The wave-induced currents associated with progressive waves are at their maximum at the highest point of the wave crest and the lowest point of the wave trough, although the direction of the current as between crest and trough is reversed.

projection (in cartography) *See* map projection.

promontory A small peninsula or headland that projects into the sea. Wave action can cause erosion on both sides, which are likely to have cliffs and possibly caves.

Proterozoic *See* Precambrian.

Proto Atlantic Ocean An ocean that existed between the late Precambrian and early Devonian times. Within this ocean, which reached its maximum extent during the Ordovician, a series of sediments were deposited on an ocean floor of pillow lavas. When the two continents on each side of the ocean closed, these sediments were subjected to orogenic processes that resulted in the formation of the Caledonian mountains of Norway and Scotland, and the Appalachian mountains of North America. Evidence for the closure of this ocean can be found in Anglesey, where typical rocks associated with a destructive plate boundary are present in fairly close association with pillow lavas.

protozoa Small, usually microscopic, unicellular organisms, formerly classified as animals but now usually included with certain other organisms in the kingdom Protista (or Protoctista). Protozoa of the orders FORAMINIFERA and RADIOLARIA possess skeletons, either secreted or of agglutinated material, which may be fossilized. They are present in large numbers in marine plankton, and some Foraminifera are benthonic. Fossils are known from the Cambrian Period onward, perhaps even the Precambrian; they are important in micropaleontology, and in stratigraphic correlation, especially of rocks from borehole cores. Under certain conditions their remains accumulate in sufficient numbers to contribute to rock formation.

provenance The source area from which the particles composing sediments are derived.

province (faunal province) A large region characterized by a particular assemblage of animal species, which differs from contemporaneous assemblages in similar environments elsewhere. The differences between faunal provinces have arisen because the communities have been geographically isolated from each other over long periods of time and have evolved in different ways. The detection of ancient faunal provinces is valuable in reconstructing paleogeography.

psammite A metamorphosed ARENACEOUS rock (*see* sandstone). *Compare* pelite; psephitic rock.

psephitic rock A metamorphosed RUDACEOUS rock. *Compare* pelite; psammite.

pseudo-bedding *See* unloading.

pseudo-karren KARREN (grooves) found on the surfaces of rock types not normally associated with the process of solution.

pseudomorph **1.** A mineral that has assumed the external form of another earlier mineral. Pseudomorphs may be produced as a result of:
(a) Replacement or alteration of one mineral by another. In certain cases the pseudomorph may be a polymorph of the original mineral, e.g. calcite after aragonite (*see* polymorphism). Other common examples of pseudomorphism include gypsum after anhydrite, iddingsite after olivine, and kaolinite after feldspar.
(b) Encrustation or investment when a mineral is deposited as a crust on crystals of another, e.g. quartz on fluorite. Sometimes the first mineral is removed, leaving a negative pseudomorph.
(c) Infiltration by a different mineral or substance of a cavity or mold previously occupied by a soluble crystal, e.g. clay pseudomorphs after halite.
2. A fossil in which the original skeletal substance of the organism has been replaced by a secondary material in the course of fossilization, which preserves its shape. *See also* cast; mold.

pseudotachylite A glassy material produced by the fusion of crushed rock by frictional heating, and found as narrow veinlets and streaks in mylonite zones.

pteridophytes Plants that have a vascular system, leaves, stems, and roots but reproduce by spores (rather than seeds). They are predominantly terrestrial and include the ferns and horsetails. Pteridophytes were especially abundant in the late Paleozoic but declined during the Mesozoic. These plants formed much of the coal-measure forests in the Carboniferous Period, when many grew to the size of modern trees.

pterodactyl A colloquial term for one of the PTEROSAURIA, taken from the name of one of the genera of this order (*Pterodactylus*).

pteropod ooze A calcareous deep-sea ooze deposit (*see* pelagic ooze) containing more than 30% organisms. It comprises the shells of pelagic mollusks and may also include the dead bodies of tiny swimming snails and marine butterflies, especially in the vicinity of the Equator. It is limited to a depth range of 1500 to 3000 m, i.e. it lies in generally shallower water than GLOBIGERINA OOZE, being particularly common in the vicinity of coral islands and on submarine elevations that lie well offshore. Viewed on a global scale, it is very limited in extent, occupying a fairly small north-south strip in the center of the S Atlantic and several deep-water sites off Brazil and in the N Atlantic. Collectively, these areas amount to only 1% of the total ocean floor.

Pterosauria An order of extinct flying reptiles (subclass ARCHOSAURIA), prominent in the Mesozoic. The body was small in relation to the size of the wings, which were formed of a thin membrane of skin supported by an elongation of the fourth finger. This arrangement differs from that found in the birds (*see* Aves). The early Jurassic species were suited to flapping flight, but by the late Cretaceous Period pterosaurs were characteristically adapted for gliding, and were among the most efficient gliders known. They ranged from sparrow-sized creatures to the enormous *Pteranodon*, which had a wing span of up to 9 m. Some had teeth; others were toothless.

ptygmatic structure Any one of a series of highly contorted parallel folds, as commonly shown by veins of PEGMATITE in MIGMATITE or other high-grade metamorphic rock.

puddingstone A type of CONGLOMERATE consisting of small rock fragments or pebbles in a sandy matrix, so called because of its supposed resemblance to a fruit pudding.

pulaskite A member of the SYENITE group of minerals.

pumice Highly vesicular, usually acid, volcanic rock. *See* pyroclastic rock.

pumpellyite A hydrous calcium-bearing mineral found in low-grade regionally metamorphosed rocks.

puna A region of high, flat, and bleak land in the South American Andes, up to 4000 m above sea level. Because of the altitude, the air is thin and temperatures are low, even in summer; they fall below freezing point at night. In the Bolivian and Peruvian puna there are some rich deposits of minerals.

push moraine A ridge of material accumulated by the bulldozing action of an advancing glacier or ice sheet. These moraines tend to be more convex in profile than TERMINAL MORAINES and can easily be differentiated by considering the internal structures, which show signs of faulting and thrusting. If really distinct faults occur in the material, then it was almost certainly frozen when incorporated into the push moraine.

puy A plug of volcanic rock, sometimes left standing when the surrounding rock has been eroded away.

P wave *See* primary wave.

pycnometer (density bottle) An instrument for measuring specific gravity. It consists of a small glass bottle of known volume which is filled and weighed (to find the specific gravity of the liquid). Alternatively, it is filled with a dense liquid and a weighed sample of mineral grains introduced, which displaces some liquid. The specific gravity of the mineral can then be found.

pyralspite A chemical series of GARNET minerals.

pyramidal peak (horn, horn peak) An individual steep-sided mountain peak formed in a similar way to an ARÊTE but in this case involving three or more converging CIRQUE headwalls, thereby isolating a single residual rock mass rather than a linear divide.

pyranometer Any instrument for measuring scattered and global radiation on a horizontal surface. Instruments that measure global (direct and diffuse) radiation are also known as *solarimeters*.

pyrargyrite A dark red to black mineral form of silver antimony sulfide, Ag_3SbS_3. It crystallizes in the trigonal system and commonly occurs in association with other silver-bearing minerals. It is an important source of silver.

pyrheliometer A radiation instrument for measuring direct solar radiation only. Diffuse radiation is excluded by having a 5° aperture continuously facing the Sun.

pyriboles A group of minerals consisting of PYROXENES and AMPHIBOLES.

pyrite (iron pyrites) A pale brass-yellow mineral form of iron sulfide, FeS_2, commonly occurring as cubes and octahedra. It is the most widespread and abundant sulfide mineral, found as an accessory in igneous rocks, in hydrothermal and replacement deposits, in contact metamorphic rocks, and in sediments laid down under reducing conditions. It is used as a source of sulfur.

pyroclast Any material, from fine dust and ash to large blocks of rock, ejected from the vent of an erupting volcano.

pyroclastic flow A mixture of hot gases and PYROCLASTS that moves under gravity along the surface from the vent of an erupting volcano. Such flows can move quickly and be extremely dangerous, destroying anything in their path. *See also* pyroclastic rock.

pyroclastic rock A rock formed by the accumulation of fragmental materials thrown out by volcanic explosions (literally, fire-broken). Such material is known as *tephra* or *ejectamenta* and may be expelled as solid fragments or in the molten state, chilling in the air and producing vitreous material. All volcanic eruptions are the result of the release of gas that has been confined under pressure. This gas may be derived from the magma or from the transformation to steam of water from the sea or a crater lake coming into contact with magma. Phreatic eruptions are caused by steam produced when lava flows come into contact with groundwater.

The main kinds of pyroclastic materials and rocks are as follows:

Bombs and *blocks*. Bombs are large fragments with a rounded to subangular shape and erupted in a plastic condition. During flight, aerodynamically modified shapes are produced, the most common being a spindle form. Blocks are large angular fragments ejected in a solid condition. Accumulations of blocks and bombs, known as *agglomerate*, occur near volcanic vents. Showers of hot blebs of magma that flatten and weld on impact with the ground are called *spatter* or *agglutinate*.

Lapilli. Round to angular fragments of diameter between 64 and 2 mm. Most of the irregular vesicular fragments known as cinders or *scoriae* are of lapilli size. A special form of lapilli, known as *Pele's tears*, are droplets of lava that solidified in the air to pear-shaped pieces of glass. The droplets trail behind them threads of liquid that solidify to glass filaments, known as *Pele's hair*.

Ash. Tephra fragments less than 2 mm in diameter. Consolidated ash is known as *tuff*. The most common variety, vitric ash, is formed by the disruption of magma by expanding gas. The gas causes a frothing of the magma, which becomes torn apart. Some highly vesicular masses survive as pumice. Most is completely disintegrated to *shards*. These small glass septa, which separated individual bubbles, have characteristically curved and forked shapes. Ash may be carried a great distance from the volcanic source by the prevailing winds and is deposited in layers, often graded, draping over the underlying topography. Consolidated ash with a high proportion of larger fragments is sometimes called *lapilli tuff*. *Crystal tuff* contains a high proportion of broken crystals, which represent the PHENOCRYSTS present in the magma before eruption.

Ignimbrites (*welded tuffs*). Fragmental flows having some of the characteristics of lava and some of air-fall pyroclastic material. Ignimbrites typically have a streaky or banded appearance known as eutaxitic structure. The glossy streaks, usually darker in color than the surrounding matrix, are laterally discontinuous, unlike flow banding. Individual streaks called *fiamme* are arcuate or elliptical in plan and are thought to represent pumice fragments or lava blebs flattened by the weight of the ignimbrite and welded together as the flow deflates. The shards that compose the matrix are also flattened and drawn out.

The flows producing these deposits are called *ash flows* and the rocks produced are ignimbrites, ash-flow tuffs, or welded tuffs. Not all ash flows become welded and in many flows welded eutaxitic material passes up into unwelded material. Ash flows are mostly of acid composition and are the result of deposition from NUÉES ARDENTES, incandescent clouds of gas and ash produced by the explosive vesiculation of magma.

pyrolusite A soft powdery or fibrous black mineral form of manganese dioxide, MnO_2. It crystallizes in the tetragonal system, and occurs in association with other manganese minerals. It is used a source of manganese, and as a decolorizer and oxidizing agent.

pyrope A deep yellow-red to black member of the GARNET group of minerals, $Mg_3Al_2Si_3O_{12}$. It crystallizes in the cubic system, and occurs in kimberlite and other ultrabasic igneous rocks. Transparent examples are valued as semiprecious stones.

pyrophyllite A soft cream-colored, gray, or green mineral form of hydrated aluminum silicate, $AlSi_2O_5(OH)$, which resembles talc. It crystallizes in the monoclinic system, and occurs in metamorphic rocks such as schist as thin flakes.

pyroxene A member of a group of ferromagnesian rock-forming minerals. Pyroxenes are orthorhombic or monoclinic and have a continuous chain structure of SiO_4 tetrahedra linked by sharing two of the four corners. Cations link the chains laterally. The general formula of pyroxenes is $X_{1-p}Y_{1+p}Z_2O_6$, where X = Ca,Na; Y = Mg,Fe^{2+},Mn,Li,Al,Fe^{3+},Ti, and Z = Si,Al. In the *orthopyroxenes* (orthorhombic pyroxenes), p = 1 and the content of trivalent ions is small. In *clinopyroxenes* (monoclinic pyroxenes), p varies from 0 as in diopside ($CaMgSi_2O_6$) to 1 as in spodumene ($LiAlSi_2O_6$).

Orthorhombic pyroxenes

The orthopyroxenes have a composition $(Mg,Fe^{2+})_2Si_2O_6$ and form a series from *enstatite* ($MgSiO_3$) to *orthoferrosilite* ($FeSiO_3$) that is produced by the replacement $Mg \leftrightarrow Fe^{2+}$. Intermediate compositions in the series include *bronzite* and *hypersthene*. Orthopyroxenes are found in basic and ultrabasic igneous rocks and high-grade metamorphic rocks.

Monoclinic pyroxenes

Most clinopyroxenes can be considered to be members of the four-component system, $CaMgSi_2O_6$–$CaFeSi_2O_6$–$MgSiO_3$–$FeSiO_3$. Within the system, three series are recognized:

1. diopside–salite–hedenbergite $Ca(Mg,Fe)Si_2O_6$
2. augite–ferroaugite $(Ca,Na,Mg,Fe^{2+},MnFe^{3+}Al,Ti)_2(Si,Al)_2O_6$
3. pigeonite $(Mg,Fe^{2+},Ca)(Mg,Fe^{2+})Si_2O_6$.

The pigeonite series are calcium-poor clinopyroxenes. The diopside series have higher Ca/Mg + Fe ratios than the augites. Clinopyroxenes rich in titanium are called *titanaugites*. Pigeonite is found in basic lavas but under the slow cooling conditions associated with plutonic crystallization, pigeonite expels calcium-rich ions as lamellae of augite and inverts to an orthorhombic structure. Diopside and hedenbergite are found in calcium-rich and iron-rich thermally metamorphosed sediments respectively. Hedenbergite also occurs in intermediate and acid igneous rocks. Pyroxenes of the augite-ferroaugite series are found in basic igneous rocks. *Omphacite*, a high-pressure pyroxene resembling augite but with some CaMg replaced by NaAl, is found in ECLOGITES.

Sodic pyroxenes may be considered to be members of the system $CaMgSi_2O_6$–$CaFeSi_2O_6$–$NaFe^{3+}Si_2O_6$, in which the replacements $Ca(Mg,Fe^{2+}) \leftrightarrow NaFe^{3+}$ occur. The sodic end-member is aegirine (or acmite), $NaFe^{3+}Si_2O_6$, and as the name implies, aegirine-augite, $(Na,Ca)(Fe^{3+},Fe^{2+},Mg)Si_2O_6$ is of intermediate composition. Sodic pyroxenes are found in intermediate and acid alkaline igneous rocks and glaucophane schists.

Other pyroxenes include *spodumene* ($LiAlSi_2O_6$), found in lithium-rich pegmatites, and jadeite ($NaAlSi_2O_6$), found in high-pressure metamorphic rocks. *Compare* amphibole.

pyroxenite An ultramafic rock consisting wholly of PYROXENES. Pyroxenites may be monomineralic or bimineralic.

pyrrhotite A yellow-brown magnetic mineral form of iron sulfide with a variable amount of sulfur, average formula Fe_7S_8. It crystallizes in the hexagonal system, and often occurs associated with nickel sulfide, when it is a major source of nickel.

Q

quaquaversal A type of PERICLINE.

quartile If a data series is arranged in order of magnitude, a quartile indicates a quarter of the values. It is more frequently quoted as upper and lower quartile, the former separating the highest quarter of the values and the latter the lowest quarter. It can therefore be used as a measure of the dispersion or range of the data.

quartz One of the most important rock-forming minerals, consisting of silicon dioxide, SiO_2 (*see* silica minerals). It crystallizes in the trigonal system, and occurs as sand and as deposits of highly transparent colorless crystals (rock crystal), sometimes tinted by impurities. These give rise to the varieties known as rose quartz and smoky quartz, as well as to amethyst and citrine. Quartz is also the basis of cryptocrystalline minerals such as agate, chalcedony, jasper, and opal. Many of these are used as gemstones. Quartz is a component of granite and other acid igneous rocks; it also occurs in metamorphic rocks (quartzite) and some sedimentary rocks (sandstone). It has many uses, mainly in the manufacture of abrasives and glass.

quartzarenite (orthoquartzite) An ARENACEOUS rock having a composition including more than 95% QUARTZ.

quartz diorite A coarse-grained igneous rock consisting mainly of PLAGIOCLASE FELDSPAR, sometimes also with some ORTHOCLASE. The minerals biotite and hornblende occur as the dark constituents, with up to 30% quartz as the light component.

quartzite A tough massive rock consisting almost wholly of QUARTZ and usually having a pale color. A quartzite is the product of the metamorphism of a pure sandstone, during which process the quartz grains recrystallize and become interlocking. *See also* psammite.

quasi-geostrophic motion Atmospheric motion that approximates to the GEOSTROPHIC WIND. It is one of the assumptions used in the preparation of numerical weather forecasts. Because it is not quite geostrophic flow, it cannot be used for deriving other quantities based on the geostrophic wind.

Quaternary The period of geologic time following the TERTIARY Period (recent recommendations suggest it be considered as a subera and the term be restricted to informal usage). In either scheme it is formed of two epochs, the PLEISTOCENE and HOLOCENE and extends from 1.6 million years ago until the present. During this time parts of Europe were subjected to four major advances of the ice sheets, which were separated by warmer interglacial episodes, and most British Pleistocene rocks consist of glacial and associated fluvial deposits. Fossil animals and plants were essentially modern. The land fauna alternated between forms adapted to cold conditions, such as the mammoth and woolly rhinoceros, and species now restricted to the tropics. Pollen and the remains of beetles have proved valuable in monitoring climatic change. The Quaternary is the time when humans became the dominant terrestrial species.

quicksand An area of mud and sand containing a large amount of water and thus almost liquid in composition. It can occur on the shore or near a river, where the water table is near the surface and drainage is poor.

R

radar meteorology The investigation or determination of aspects and properties of the atmosphere by radar. Using suitable wavelengths, radar offers great possibilities of investigating the areal distribution of many meteorological elements, which otherwise would be unobserved. By scanning in the horizontal field (*Plan Position Indicator*), radar can measure the distribution and intensity of surface precipitation and distinguish particularly intense storms such as tornadoes, hurricanes, thunderstorms, and hailstorms. Scanning about the vertical (*Range-Height Indicator*) enables the height and structure of precipitating clouds to be obtained as well as the altitude of the melting level, if present. Operation over a period of time clarifies the evolution of many of these atmospheric systems and it is at the subsynoptic scale that most advances in radar meteorology have occurred.

radial dike A dike that radiates from a central-vent volcano. This radial fracture pattern is the result of the volcano's superstructure swelling prior to an eruption as pressure builds up.

radial drainage A drainage pattern that develops when structural control is in the form of a volcanic dome or cone, or some other sort of dome. Slope will cause streams to radiate out from their common center at the crest of the dome like the spokes of a wheel, e.g. in the English Lake District.

radiation 1. (in meteorology) The portion of the ELECTROMAGNETIC RADIATION spectrum that is emitted by the Earth and the Sun. In terms of wavelength, this is radiation between 0.1 micrometers and 70 micrometers, encompassing part of the ultraviolet, all the visible, and part of the infrared spectrum. This type of wave differs from most other waveforms in that it does not require an intervening medium for its propagation; it can travel through a vacuum.

The behavior of electromagnetic radiation is described by certain physical laws. All substances above the absolute zero of temperature (–273°C) emit radiation in amounts and wavelengths dependent on their temperature. Some bodies emit and absorb radiation in certain wavelengths only; this is particularly true of gases (*see* atmospheric window). The ideal radiating body is a black body (*see* black-body radiation), which emits the maximum amount of radiation for its temperature, the amount being proportional to the fourth power of its temperature on the Kelvin scale. Absorption and emission take place at the same wavelengths for any substance. The hotter a body, the shorter will be the wavelengths at which the maximum amount of emission will take place. These laws determine the way in which bodies react to radiation and enable the utilization of solar energy by the Earth and atmosphere to be explained.

The Sun has an emission temperature of 6000 K, giving an energy maximum in the visible light wavelength. This reaches the top of our atmosphere at the rate of about 2 calories per sq cm per minute or 1.35 kW m^{-2}. On its passage through the atmosphere, some of this short-wave radiation is reflected back to space by clouds and dust, some is scattered by gas molecules and dust particles to give diffuse radiation, and some (about 18%) is absorbed by water vapor, carbon dioxide, and dust. The remainder reaches the ground surface where some is reflected and the rest is absorbed. On a global average, this amounts to approximately 47% of the radiation reaching the top of the atmosphere.

The Earth's surface has a mean emission temperature of about 290 K as a result of solar heating. Thus terrestrial radiation is in the longer wavelengths with a maximum about 10 micrometers. The gases of the atmosphere have a very different response to this radiation and much of it is absorbed (*see* absorption). Certain wavelengths are unaffected, especially those between 8 and 12 micrometers (*radiation windows*). The atmosphere warms as a result of this absorption and atmospheric counterradiation is returned to the surface and helps to maintain higher temperatures than would otherwise be expected.

2. A type of PLANE TABLING in which points of detail are fixed by ascertaining direction using an ALIDADE and distance by measuring on the ground with a tape. The method may be used over short distances, dispensing with the necessity to set up the plane table at two stations, as in INTERSECTION methods.

radiation balance The net effect of the difference between incoming and outgoing radiation at any point. These two factors rarely balance. Normally there is a surplus of radiation on the ground surface during the day, which helps to warm the surface and atmosphere, and a deficit at night when cooling takes place. The atmosphere has a negative radiation balance at all times. Taking the Earth and atmosphere together, the areas equatorward of 38° have a radiation surplus and poleward there is a radiation deficit.

radiation fog Nocturnal cooling resulting from terrestrial radiation losses can lead to the air near the ground surface reaching the saturation point so that condensation takes place, causing radiation fog. It is most likely to occur during long, clear, and calm nights with a moist atmosphere. In industrial areas, the abundance of condensation and hygroscopic nuclei facilitates fog formation and it may even occur with relative humidities less than 100%. However, urban warmth frequently counteracts this factor and so city centers may be almost clear of fog while it is still quite dense in the cooler suburbs.

radioactive decay The spontaneous change of one atomic nucleus into another through the emission of a photon (gamma ray) or a particle (alpha ray or beta ray), or by electron capture. It is the basis of RADIOMETRIC DATING. *See also* half-life; isotope.

radiocarbon A radioactive ISOTOPE of carbon, commonly one of mass 14. Its detection is the basis of radiocarbon dating of organic remains. *See also* half-life.

Radiolaria An order of PROTOZOA whose members have an internal skeleton composed of silica. They are one of the constituents of marine plankton. In certain conditions their remains accumulate to form RADIOLARIAN OOZE, and analogous ancient radiolarian cherts are known. Fossils from rocks as old as the Precambrian have been attributed to this group.

radiolarian ooze A siliceous ooze (*see* pelagic ooze) deposit containing more than 30% organisms. It is reddish or brownish in color and results from the deposition of minute siliceous skeletal remains of radiolarians. The deposits are very limited in extent when viewed on a global scale, being almost confined to a narrow tract of deep-sea floor in the Pacific Ocean, located just north of the Equator.

radiolarite A hard fine-grained siliceous rock formed predominantly from the skeletal material of RADIOLARIA.

radiometric dating A method of dating rocks by determining the relative proportions present of parent and daughter isotopes of a radioactive element. Because these decay reactions progress at a known rate, the age of the rock is calculated from the ratio present.

radiosonde An instrument for measuring pressure, winds, temperature, and humidity in the upper atmosphere. It consists of a small radio transmitter attached to a helium-filled meteorological balloon, which lifts the radiosonde to altitudes of about 30 000 m. Values of temperature and humidity are obtained by electrical means from a bimetallic strip and a goldbeater's skin hygrometer, respectively, and then transmitted to the receiver. The balloon is located by radar and from its position wind velocities can be calculated.

rain PRECIPITATION in the form of liquid water drops ranging in size from about 0.5 mm to 5 mm in diameter.

rainbow An optical phenomenon consisting of an arc of light across the sky broken up into the spectral colors. It results from the refraction and internal reflection of sunlight through falling water drops. The intensity of the rainbow depends upon the size of the drops, larger sizes producing brighter colors. Some of the light may be reflected twice within the raindrop to produce a double rainbow effect.

rain day A climatological DAY from 9 am GMT within which 0.2 mm or more of rainfall is recorded. *See also* wet day.

raindrop size spectrum The range of raindrop sizes within or beneath a precipitating cloud. The range of sizes varies with the rate of rainfall. At low rates (0.1 mm hr^{-1}) most drops will have a diameter of less than 0.75 mm and there will be very few above 1.25 mm. With heavier rain, the upper range of drop sizes increases and the density of drops rises markedly.

rainfall The water equivalent of all forms of precipitation from the atmosphere, received in a rain gauge. This includes dew, hoar frost, and rime on the collecting area of the gauge, because on melting it is indistinguishable from the rest of the catch. It is now recorded in millimeters in most countries of the world. The value quoted represents the amount of water that has fallen on a horizontal surface assuming no runoff, percolation, or evaporation.

Rainfall has been classified into three types depending on its mode of formation: convective, cyclonic, or orographic.

rainforest A type of dense forest that grows in regions that have heavy rainfall all the year round. Most rainforests occur in the tropics, although there are some in warm temperate regions. They are among the most complex ECOSYSTEMS in the world, with a great diversity of plants and animals. Their plants are also an important source of atmospheric oxygen (from PHOTOSYNTHESIS). *See also* deforestation.

rain gauge An instrument for measuring rainfall. In its simplest form it is essentially a funnel and a collecting vessel placed vertically into the ground. Different types of rain gauge are used throughout the world but most are set about 30 cm above the ground surface. This is to avoid splashing during heavy rain, to retain snow, and to reduce the effects of wind eddying around the gauge. It is widely accepted that a rain gauge does not catch the precise amount of rain falling on a surface, largely because of the wind eddying effect, but as long as all sites are standardized there is comparability between records. The rainfall is measured by emptying the contents of the collecting vessel into a special cylindrical flask, which is graduated in relation to the diameter of the gauge. The rainfall can be read off directly from the water level in the flask.

rain pit A sedimentary structure formed in subaerially exposed soft fine-grained sediments as a result of the impact of raindrops upon the sediment's surface.

rain shadow An area on the leeward side of a range of hills where rainfall totals are less than would be expected from their latitudinal position. It is the result of a drying-out of the winds through precipitation on the upwind side of the hills.

rainsplash The impact of raindrops on the soil. It influences slope evolution in two ways: firstly, as the drop hits the ground and rebounds, it brings up with it particles of soil, which tend to fall downslope; secondly, this process will, if continued, compact the soil surface, often breaking up soil structures, and reduce the infiltration capacity (ability to hold and store water) of the soil. This promotes surface flow, either as an OVERLAND FLOW or as RILLS, which then erode the surface.

rain wash *See* overland flow.

raised beach An inland terrace of deposits that marks the location of a former shoreline, but is now well above the present sea level. There may be cliffs inland of the raised beach. Most of these features result from uplift of the land following glaciation, although some were produced by the movement of lithospheric (tectonic) plates.

raised bog *See* organic soil.

range 1. A line of mountains that forms a continuous barrier, perhaps with some gaps or passes.
2. An open area of pasture, as in the western USA.
3. The difference between the highest and lowest values of a variable, such as temperatures or the heights of the tides.

rapakivi structure An ORBICULAR structure exhibited by some granites in which large oval pink orthoclase crystals are mantled by white oligoclase.

rapids A stretch of fast-flowing turbulent broken water at a break in the long profile of a river that is not a vertical drop. Rapids often mark KNICKPOINTS, sometimes in the form of the remains of an eroded waterfall. If the fall in base level causing rejuvenation results in a retreat of the shoreline some distance offshore, because the coastal area is gently sloping, the knickpoint may be spread over a considerable horizontal distance and rapids will be created rather than a waterfall. Rapids can also occur where a river crosses a hard rock band, not yet graded to fit a smooth long profile. Rapids and waterfalls tend to be fairly quickly eliminated owing to the increased turbulence and erosion they themselves create.

rarefaction wave After the impact of a meteorite with the ground, a wave of decompression that follows the initial compressional shock wave through the rocks at the impact site. It is the rarefaction wave that causes the ejection of debris from the crater (*see* ejecta).

ravine A small narrow deep-sided valley carved by the erosive action of a river or produced by faulting.

raw soil Soil with an (A)C profile, produced by incipient soil-forming processes on fresh rock.

Rayleigh wave One of two types of surface wave produced either naturally by earthquakes or volcanic eruptions, or artifically by an explosion. In Rayleigh waves the motion of surface particles is in the vertical plane containing the direction of wave propagation. *See also* love wave.

reaction rim A CORONA STRUCTURE consisting of a concentric shell (or shells) of a mineral often arranged in a radial or fibrous manner around a primary mineral. Such rims, in igneous rocks, may be the result of a reaction between an early-formed mafic mineral and the liquid with which it is no longer in equilibrium. Thus a later member of the REACTION SERIES is produced, e.g. olivine may be mantled by pyroxene or amphibole. *Kelyphitic rims* are secondary coronas formed by late-stage or metasomatic fluids reacting with primary crystals, e.g. iddingsite rimming olivine.

reaction series During magmatic crystallization, in order to maintain equilibrium between crystals and liquid, minerals interact chemically with the liquid. The reaction may be progressive so that a continuous series of homogeneous solid solutions is produced. In the plagioclase feldspars, the first-formed crystals are richest in calcium and as the temperature falls these react with the liquid to become progressively more sodic. The reaction series in-

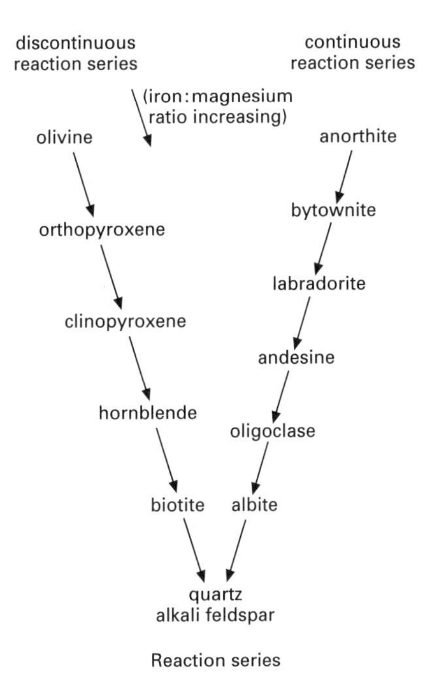

Reaction series

volving the ferromagnesian minerals is discontinuous, each reaction taking place only over discrete temperature intervals and corresponding to the transformation of one mineral to another of different crystal structure. At high temperatures, magnesium-rich olivine crystals are the first to appear in a basic magma but as cooling proceeds, olivine is no longer stable and magnesium-rich orthopyroxene crystallizes. Likewise orthopyroxene is stable over a limited temperature range and reacts to produce clinopyroxene. Ferromagnesian minerals have both high- and low-temperature variants, those rich in magnesium preceding those rich in iron. The common rock-forming minerals have been arranged by N. L. Bowen in the order in which the reactions take place and constitute two series.

The early high-temperature members of both series generally crystallize together; hence gabbros contain magnesium-rich olivine and pyroxenes together with calcic plagioclases. The low-temperature minerals, alkali feldspar, mica, and quartz, are associated in granitic rocks. When the reaction between crystals and the liquid is unable to go to completion because of too rapid cooling, early-formed members of the reaction series persist as relics in the final rock and zoned crystals (*see* crystal zoning) and REACTION RIMS are often observed.

realgar A rare orange-red mineral form of arsenic sulfide, As_2S_2. It crystallizes in the monoclinic system as compact aggregates in hydrothermal veins and hot springs; it may be found in association with ORPIMENT. It is used as a pigment and as a source of arsenic.

Recent *See* Holocene.

recompilation Once a map has been compiled, at a particular scale and according to certain specifications, any updating of the information can be by either revision or recompilation. The latter is necessary if the basic detail, i.e. relief, drainage, and planimetry, needs to be redrawn.

recrystallization The formation of new mineral grains in a rock while it is still in the solid state. The composition of the new material may be the same as or different from the original (primary) grains, and they may be significantly larger.

rectangular drainage A drainage pattern that is characterized by streams flowing in two directions at right angles to each other. Neither direction is subordinate, both being equally developed. The most

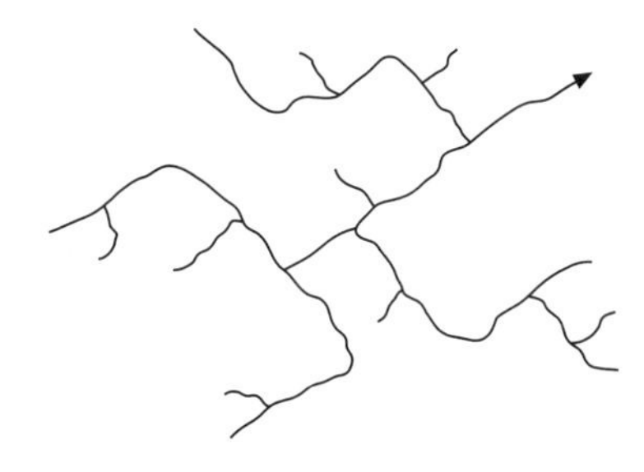
Rectangular drainage

usual control is right-angled faulting in an area of little other structural guidance.

rectilinear slope A slope or individual facet of a slope that is straight, i.e. has a single angle of slope.

recumbent fold A fold type in which the axial plane is almost horizontal. See diagram at FOLD.

red beds Layers of sedimentary rocks consisting mainly of sandstone, siltstone, and shale colored red by iron oxide (HEMATITE). An example is OLD RED SANDSTONE.

red clay (brown clay) A pelagic deposit that covers large parts of the abyssal floor, actually about a quarter of the Atlantic and Indian Oceans respectively, and something approaching a half of the Pacific Ocean, or collectively some 28% (approximately 130 million sq km) of the total ocean floor. The deposit is characterized by a low silica and carbonate content. The clay particles may be mixed in places with manganese nodules, whale bones, sharks' teeth, pumice, and other assorted materials.

red earth *See* ferrallitic soil.

red rain Rain colored red by dust particles, usually fine sand. A phenomenon of mid-latitudes, an example is the red rain that sometimes falls on S Europe containing dust picked up in the Sahara.

reduction A chemical process that occurs in rocks and in the gleyed and leached parts of soil, whereby oxygen is removed from the weathering material. The continuous presence of water renders oxygen scarce, leading to its displacement, for example from ferric iron oxide (Fe_2O_3) to give ferrous iron oxide (FeO), in which state it is rendered more soluble and hence potentially more mobile. The characteristic red and yellow colors of OXIDATION are absent, being replaced by greens and grays. Bacteria are important in this process.

reef A ridge, island chain, or area of rocks that projects above the surface of the sea for all or part of the tide or it may be permanently submerged. Reefs constitute distinct navigational hazards. A reef may be formed of solid rock or pebbles, but the term is more commonly applied to organic reefs (BIOHERMS). They are sometimes attached to the coast and act as groins by collecting debris against them (*fringing reefs*), sometimes parallel to the coast and separated from it by a lagoon too deep for coral growth (*see* barrier reef), sometimes in tabular sheetlike masses (*apron reefs*), and sometimes in circular form (*see* atoll). Coral, a hard calcareous material, dead or alive, may be predominant in all of these (corals are at present the commonest reef-building organisms). CORAL REEFS are common in the tropics. Other organic reefs include OYSTER REEFS and serpulid reefs (which result from the cementing action brought about by certain types of marine worms). The presence of a freshwater outflow from reefs locally inhibits organic growth, causing gaps through the reefs.

reentrant A small valley or area of lowland that stretches into higher ground. It generally results from erosion by water, but may be caused by FREEZE-THAW weathering.

reflection profile A seismic profile produced and recorded by equipment designed to reflect energy from layered rock bodies.

reflux The process by means of which dense concentrated salt solutions move downward through the EVAPORITE on the floor of a PLAYA or other basin. It is probably one of the mechanisms that leads to the increasing magnesium content in some sedimentary rocks.

refolded fold A fold that has been subjected to more than one period of folding. The folds produced in the first period of folding have further fold trends overprinted on them by subsequent periods of folding.

reforestation The planned replanting of trees in an area where they have been cut down or destroyed (*see* deforestation). It is particularly important as a method of preventing soil erosion and the formation of infertile land.

refractory inclusion A type of irregular mineral grain found, along with carbonaceous chondrules, in some chondrite METEORITES. Up to 2 mm across, the inclusions consist of high temperature-resistant oxides of aluminum, calcium, magnesium, and titanium.

reg A type of arid desert plain consisting of gravel, especially in the Sahara. *See also* erg; hammada.

regelation A process of thawing and refreezing within a GLACIER, which contributes toward the down-valley movement of ice. It is believed that pressure within the ice of a glacier causes the melting of some ice crystals. The resultant meltwater will move to locations at which pressure is less, i.e. normally downslope, and then refreeze. The process has been classed as the primary cause of ice motion but many authorities on the subject consider it to be only a secondary factor arising from flow.

regeneration (in ecology) The natural regrowth of plants that follows destruction (such as of grassland after fires or of trees after forest fires or felling). It is an important factor in the long-term stability of many ECOSYSTEMS.

regime The total economy or habit of a natural system, e.g. fluvial regime, estuarine regime, coastal regime. *See also* equilibrium regime.

regolith (waste mantle) Unconsolidated weathered material between the ground surface and the bedrock, which may reach thicknesses of 60 m in the tropics. It is formed either in situ, by the weathering of the underlying bedrock, or it is transported into the area by water, wind, or ice. The upper biochemically weathered portion of the regolith is the soil.

regosol A thin azonal soil that may develop where one soil-forming factor delays the soil-forming processes. Such soils are young in development and lack a B horizon, resulting in an AC profile with the A horizon, because of organic matter, being darker than the C. Regosols are typically developed in unconsolidated deposits such as loess, recent till, or sand dunes, and the parent material dominates the characteristics of the profile. They are classified in the ENTISOL order of the Seventh Approximation.

regression The retreat of the sea from a land area.

rejuvenation The revitalizing of streams by an increase in their erosive capacities resulting from a fall in BASE LEVEL. This causes oversteeping of the lower reaches, and increased erosion cuts out a new long profile which will intersect the original long profile at a KNICKPOINT. Tributary valleys meeting the main stream seaward of the knickpoint will also be rejuvenated and begin to cut down; if they do not keep pace with the incision of the master stream, there will be a marked break of slope at the junction, leaving them "perched." Below the knickpoint, the former valley floor will be left hanging above the stream as it cuts down to a new level, becoming a RIVER TERRACE. The knickpoint will advance upstream at a rate dependent on the lithology of the bed material and the nature of the flow over it; if it meets a hard rock band it may be halted, and so the effects of the rejuvenation will not be felt upstream.

Valley-side slopes will also be rejuvenated, because the incision of streams will oversteepen their lower portions, and temporarily create a valley-side facet adjacent to the stream.

rejuvenation head *See* knickpoint.

relative dating The ordering of rocks or fossils in terms of the GEOLOGIC TIMESCALE, without taking into account their absolute ages.

relative humidity The most frequently used index of atmospheric HUMIDITY. It is the actual vapor pressure of the air expressed as a percentage of the saturation vapor pressure at the same temperature. It can be measured directly by changes in length of a hair or a skin in a hygrometer, or indirectly using wet- and dry-bulb temperatures. Because the value of relative humidity changes inversely with temperature, even for the same moisture content of the atmosphere, it is not a very precise unit.

relative relief The degree of dissection of a landscape, shown by the difference in height between the tops of drainage divides and the bottoms of adjacent valley floors. Cartographically it is approached by calculating the difference between the highest and lowest points in each grid square of the relief map, and then expressing the results as a series of isolines joining areas of equal relative relief. According to the normal CYCLE OF EROSION, relative relief increases to a maximum in maturity and thereafter decreases. One geomorphologist has based a scheme for landscape attractiveness on relative relief, high relative relief producing high scenic appeal.

relic sediment (in oceanography) A marine deposit, usually present on shelf areas, that is incompatible with contemporary marine environments. For example, much relic sediment on the continental shelf surrounding Britain was initially deposited there under glacial or interglacial conditions. Much of the material lying on the shelf off S California is relic, especially the sands that contain extinct shallow-water foraminifera, and which currently lie on outer parts of the shelf. Some sands off Long Beach, California, have been found to include Pleistocene fauna. Many relic shelf sediments are now undergoing sorting and transport, under the action of waves and tidal flow, gradually shifting to more compatible sedimentary environments. *Modern sediments* are those being derived under contemporaneous conditions.

relief The variation in elevation or physical outline of a landscape, shown on maps by the use of contours, spot heights, hypsometric tinting, and hill-shading. Relief is also used synonymously with RELATIVE RELIEF. Positive relief indicates land rising above the general level, i.e. hills.

relief map A map that uses contours, coloring, or shading to indicate the different heights of various features. *See* contour; hachures.

remnant magnetism The magnetism "frozen" into a mineral or rock as it cools through its Curie point.

remote sensing The use of orbiting satellites and spacecraft to study the surface of the Earth and the weather, and transmit the information down to a ground station. The most commonly employed techniques are infrared photography and radar, and usually computers are employed in interpreting the data. *See also* Landsat.

rendzina Soil developed on soft calcareous parent material with an AC profile (no illuvial B horizon). The A is usually dark brown or black, rich in carbonates and humus, with a good crumb or granular structure, and a pH value above neutral. There are no free sesquioxides as in the related TERRA ROSSA soils, which also develop on calcareous material. It may be that the rendzinas are less mature than the terra rossa soils, or that the terra rossas develop on hard limestone, whereas rendzinas develop on soft. Rendzinas occur in humid to semiarid climates, under grass or grass-and-tree vegetation, and are perhaps best known as the soils of chalk downland.

reniform Denoting the massive kidney-shaped form in which some minerals occur. Reniform hematite is called kidney iron ore. *Compare* botryoidal.

replacement (in geology) The natural substitution of one mineral for another, generally by the action of gases or solutions. *See* mineralization; reflux.

representative fraction (R.F.) The ratio between distance on a map and distance on the ground, expressed as a fraction.

Reptilia A class of cold-blooded terrestrial vertebrates whose young complete their development within a tough-shelled egg; they are therefore protected from desiccation and totally independent of an aquatic environment. Reptiles evolved from the AMPHIBIA in the Carboniferous Period and their main radiation began at the end of the Permian. During the Jurassic and Cretaceous a great variety of reptiles came to dominate life on land. These ranged from the giant DINOSAURS to the flying PTEROSAURIA and the secondarily aquatic ICHTHYOSAURIA and PLESIOSAURIA. All these groups became extinct at the beginning of the Cenozoic, when they were replaced by the mammals (*see* K/T boundary event). Modern reptiles include crocodiles, lizards, and snakes.

reservoir (in geology) A rock that has a high porosity and good permeability and is able to store and transmit fluids or gases.

residual clay A type of clay that is formed where it lies by the weathering of rock. It may result from the removal of nonclay minerals from the rock, or from chemical changes to FELDSPAR.

residual deposit Rock fragments that are left behind after weathering and erosion of the preexisting rock. SCREE, for example, consists of such rock fragments left on a hillside. Rockfalls or transport by ice or water may move some residual deposits from their original site. *See also* residual clay.

residual hill An isolated small hill, all that is left of a larger mass of high ground that has suffered erosion or land movement. *See* monadnock; puy.

resistant rock Any rock that has resisted weathering and erosion better than surrounding rocks because it is relatively harder. Such rocks may form higher ground (*see* residual hill).

resorption The remelting of crystals due to reaction with the magma from which they originally crystallized. Early-formed phenocrysts frequently fail to maintain equilibrium with the melt and are partly resorbed. Markedly anhedral crystal forms are produced and a REACTION RIM of a new mineral may result.

resultant wind A wind velocity may be divided into zonal (east) and meridional (north) components. To determine the resultant wind of a series of wind velocities, the individual components of the actual wind are summed, squared, added together, and the square root taken. That is:

$$\text{resultant wind} = \sqrt{(\Sigma V_N)^2 + (\Sigma V_E)^2}$$

resurgence The emergence of an underground stream, usually where it encounters impermeable rock after having flowed through permeable strata.

reticulated Describing something, such as a lode or vein, that has a netlike appearance or structure.

reverse fault A type of fault in which the movement along the inclined fault plane has been up-dip. This results from a principal stress configuration in which the maximum principal stress is vertical, whereas the intermediate and minimum principal stress directions are horizontal. (See diagram at FAULT.)

Reynolds number (Re) The extent to which viscosity modifies a fluid flow pattern depends upon the speed of flow (u), the width of an obstacle placed across the flow or of the flow passage (D), and a value for the kinematic viscosity (ν). The relationship, uD/ν gives the Reynolds number.

It is an important number in laboratory simulations of prototype flow conditions, because ideally the Reynolds number for scale hydraulic model flow should be the same as for the prototype, i.e. to maintain scale relationships, the small model size has to be compensated by increasing the flow rates in the model.

rheidity Deformation of a substance by plastic flow as a result of an applied stress over a long period.

rheomorphism The process in which a rock changes composition while it flows. It may or may not result from intense heat or pressure.

rhodochrosite A pink, gray, or brown mineral form of manganese carbonate, $MnCO_3$. It crystallizes in the hexagonal system, and occurs in masses in hydrothermal veins. It is used as a source of manganese and as a semiprecious gemstone.

rhodonite A pink or brownish translucent mineral form of manganese calcium silicate, $(Mn,Ca)SiO_3$. It crystallizes in the triclinic system, and occurs in metamorphic rocks as granular masses containing veins of black manganese dioxide. It is sometimes used for making ornaments.

rhombochasm A parallel-sided break in the Earth's continental crust, the break being filled by oceanic crust. The fracture is thought to be the result of SEA-FLOOR SPREADING.

rhumb line (loxodrome) A line of constant bearing, used in navigation, cutting all meridians at the same angle. *Compare* great circle.

rhyodacite *See* rhyolite.

rhyolite Rhyolites, rhyodacites, and dacites are the volcanic equivalents of GRANITES, adamellites, and granodiorites respectively. Rhyolites usually contain sanidine, the high-temperature form of potassium feldspar, both as phenocrysts and microlites in the groundmass. Phenocrysts of sodic plagioclase and quartz are also common, the latter often markedly anhedral owing to resorption. The dominant mafic mineral is biotite, although hornblende and clinopyroxene also occur.

In dacites, sodic plagioclase is dominant over sanidine and, together with quartz, occurs as phenocrysts. With a decrease in the amount of quartz, rhyodacites and dacites pass into LATITES and ANDESITES but the relative proportions of minerals are difficult to assess in fine-grained volcanic rocks and classification is often based upon chemical analyses. Rhyolitic and andesitic lavas constitute the calc-alkaline volcanic suite, which is characteristic of island arcs and orogenic regions.

Strongly alkaline rhyolites, rich in sodium and poor in aluminum, are found in continental areas, particularly those subjected to rift movements, and are the volcanic equivalents of alkali granites. These lavas contain anorthoclase feldspar and soda PYRIBOLES. An arbitrary division at 12.5% femic constituents separates pantellerites from the more leucocratic comendites. With a decrease in the amount of quartz, alkali rhyolites pass via quartz-trachytes into TRACHYTES.

Many lavas of rhyolitic composition occur in the glassy state, as do *obsidian* and *pitchstone*. Obsidian is a black natural glass with a conchoidal fracture but pitchstone has a dull appearance. Whereas obsidian contains very little water, pitchstone may contain up to 10% and owes its appearance to secondary hydration and devitrification. Most volcanic glasses contain crystallites; many contain spherulites and exhibit perlitic cracking.

Rhyolitic lavas are often flow-banded and it is sometimes difficult to distinguish such rocks from ignimbrites with parataxitic structure. *See* pyroclastic rock.

rhythmite A sedimentary deposit that has alternate layers of coarse-grained light and fine-textured dark material. Most were formed on the beds of coldwater lakes. *See* varve.

ria A DROWNED VALLEY eroded by subaerial (nonglacial) processes at a time when the sea level was lower than it now is. Like river valleys of the present, rias have a V-shaped cross section, which deepens seaward and narrows inland, possibly bifurcating. Former MONADNOCKS may remain as small islands. Rias are produced wherever a dissected area of hills and valleys, perpendicular to the coastline, has been submerged as a result of a postglacial rise in sea level. Erosion of the projecting interfluves eventually reduces the indentation of such coasts.

richterite A monoclinic AMPHIBOLE.

Richter scale A logarithmic scale that was devised by C. F. Richter in 1935 for comparing the magnitude of earthquakes. Although originally devised for Californian earthquakes, it has since been adopted for worldwide use. The upper limit of the scale at 9 is a function of the strength of crustal rocks.

ridge *See* mid-ocean ridge.

ridge of high pressure A long narrow region of high atmospheric pressure, wider than a WEDGE, that extends from an ANTICYCLONE. It is associated with short-lived anticyclonic weather, with generally short dry sunny spells.

riebeckite A monoclinic AMPHIBOLE.

rift valley (taphrogeosyncline) A structural and topographical feature formed within the Earth's crust consisting of a steep-sided flat-bottomed valley, the rocks of the valley floor having subsided between two parallel faults or two parallel series of step faults. There are conflicting views as to the origin of rift valleys, the main ones favoring tension in the Earth's crust, compression, or the cracking of a crustal dome along the crest. There is often associated volcanic activity along the sides of the valley floor. Famous rift valleys include the East African system, extending more than 4000 km from Syria to the Zambezi, and the Rhine rift valley between Mainz and Basle. *See also* graben.

rigidity modulus The ratio of stress to strain when the stress is a shear. This is calculated by dividing the tangential force per unit area by the angular deformation.

rill A minute ephemeral channel at the head of drainage systems, forming at the point where unconfined sheet wash becomes concentrated into definite channels. This usually occurs on the lower concave portions of slopes, because there the volume of sheet flow has built up to a sufficient amount for channeling to occur. Rills are said to be responsible for shaping the concavity of the basal part of the slope. Downslope they run into more permanent gullies, which constitute the smallest tributaries of drainage systems. Rills carry water only in storms. *See also* gully erosion; overland flow.

rill mark A small channel found on beaches, aligned more or less up and down the beach. The channels develop because of the seaward flow of water that has percolated into the upper parts of the foreshore during the period around mid- to high-water or just after, or during the action of STORM waves with powerful swash action between low- to mid-water. As the tide retreats, or as the waves subside, the infiltrated water may flow out on to the beach face and cut the rills. Once developed, the rills tend to channel this flow seaward and possibly enlarge themselves. The seawater may be reinforced by freshwater seepages from landward, for instance where springs issue from a rock platform beneath the beach.

rime A deposit of white ice crystals formed by the freezing of supercooled water droplets onto surfaces below 0°C. It grows on the windward side of the surface, especially on sharp edges, by impaction as the droplets drift past in the wind.

ring complex A circular igneous intrusive body of rock including both concentric dikes and cone sheets.

ring dike *See* concentric dike.

rip A turbulent area of water where tidal currents meet head-on, where waves meet a current flowing in the opposite direction, where a tidal current suddenly enters shallow water, or in a RIP CURRENT.

rip current A strong well-defined current that flows away from the shore. Such currents are wave-induced, providing the mechanism whereby water piled up at the coast because of mass transport, and possibly wind, is returned seaward. The current speeds are usually fairly high. They are usually manifested by a fairly narrow zone of agitated water and by an accentuated amount of sandy material in suspension. They consist of *feeder currents* that flow inshore of the breakers and more or less parallel to the shore as longshore currents, the *neck*, where the current is concentrated into a RIP and flows seaward through the surf zone, and the expanding *rip-head* where the rip loses impetus within or beyond the surf zone. They account for many drownings.

ripple bedding Small-scale cross-bedding resulting from the rapid deposition of sand in the form of ripples.

ripple mark The most common minor beach morphological form, consisting of fairly regular and generally small ridges formed in sediment (usually sand, but possibly shell and fine gravel) on a river bed, in the inter-tidal zone, or on the seabed seaward of low-water mark. Fossil ripples or ripples formed by the wind may occur above high-water mark. Ripples are caused by water or wind flow, and are aligned more or less perpendicularly to the flow direction. They can be used, therefore, to indicate the existence of and something of the nature of wind and water currents. Ripples tend to be larger in more exposed areas and in deeper water offshore. Very large ripples are usually termed megaripples or SAND WAVES. Dune areas above high-water mark may display numerous ripple features.

rise A broad elevated area of the sea floor, similar to a MID-OCEAN RIDGE but lacking a median rift valley.

river A channeled flow of water running downhill under the force of gravity. Most rivers flow into the sea, although some enter lakes and a few flow into desert areas, where they dry up. The course of a river from start to finish is called its *profile*. The source is usually on high ground, at a spring or marsh, or where the water table reaches the surface. It then generally flows through mountains, typically through rocky areas as a turbulent flow along narrow, steep-sided valleys. There may be waterfalls or cascades. The river then broadens along its middle course through an upland area. As the valley widens, meanders may form; there may also be bluffs. Finally the lower course of the river crosses a wide floodplain, where large meanders may become cut off as oxbow lakes. The widest part of a large river as it enters the sea is an estuary, which may include a delta.

river bar A build-up of waterborne silt (alluvium) in the channel, along the banks or at the mouth of a river. It is generally exposed at low tide. *See also* braided stream.

river basin The area from which water is collected by a river. *See* catchment area; drainage basin.

river capture The diversion of the headstreams of a river into an adjacent river with more powerful headward erosion. It cuts back its valley into that of the weaker river, thus enlarging its drainage basin at the expense of the other. Rivers whose headstreams have been captured often flow as MISFIT RIVERS in valleys that are too large to have been eroded by the present capacity of the river. The ELBOW OF CAPTURE is the bend at which the captured headstreams are diverted.

river cliff *See* bluff.

river profile A section of a river along its length from source to mouth. *See* river.

river regime The variations in the volume of water in a river that take place with the changing seasons, which can be plotted as a graph. In temperate regions, the greatest flow is usually in the winter (corresponding to the greatest precipitation). In monsoon and savanna regions, it is usually in the summer (the rainy season); a river may even dry up in the winter dry season.

river terrace A benchlike landform bordering many rivers, elevated above the current streams and ending in scarps suspended on the valley side. These terraces constitute all that remains today of past abandoned valley floors formed when the river was flowing at a higher level than today. Most commonly they are paired on each side of the valley, the product of REJUVENATION of the river or climatic changes (via their influence on stream discharge, base level, and sediment load). They consist of a rock-cut bench, with or without a thin veneer of alluvium. Some British rivers, such as the Thames, Avon, Severn, and Trent, have staircases of terraces rising above the current stream, produced in the climatic fluctuations of the Pleistocene.

Roaring Forties The region between 40°S and 50°S where the westerlies blow with great regularity across the open ocean. The frequent depressions and strong winds produce a stormy climate unfavorable to shipping.

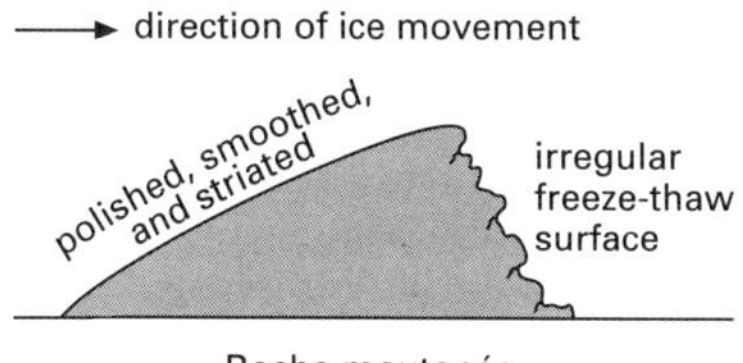

Roche moutonée

roche moutonnée A glacial erosion form found in areas of low relief and consisting of a resistant rock mass, which has been eroded to comprise a gently sloping up-valley face and a steep down-valley face. The up-valley face is smoothed, polished, and striated, the effects of glacial abrasion, while the down-valley face has a broken shattered appearance as a result of freeze-thaw processes, which are assisted by increased pressure-melting induced by the presence of the rock mass.

The term is extremely old and is widely used for similar landforms of greatly contrasting size. As a more suitable alternative, the term *stoss and lee topography* has been suggested.

rock An assemblage of MINERALS, generally cemented or consolidated together. It may contain only one type of mineral, or many. The surface of the Earth is made up of rocks. They are classified into three major groups: IGNEOUS ROCK, SEDIMENTARY ROCK, and METAMORPHIC ROCK.

rock crystal A colorless transparent variety of QUARTZ. *See* silica minerals.

rock cycle *See* geochemical cycle.

rockfall A type of landslide involving purely dry materials. Basal sapping or some other type of trigger mechanism will institute an instantaneous downslope movement, leaving a pock-marked slope at the source and a confused pile of talus at the foot. The nature of this talus or colluvium depends on the height of fall and the nature of the weathering, lower falls and less rigorous weathering leaving larger blocks. *See also* mass movement.

rock flour Fine powdery rock produced by the abrading action of a glacier, which grinds particles off the rock it flows past. Most of the rock flour is transported by subglacial streams, which have a milky or inky appearance as a result.

rock pedestal *See* pedestal rock.

rock salt The mineral halite (sodium chloride, NaCl) when it occurs as granular or massive aggregates.

roller A large wave rolling in onto an exposed coast, often following a storm, increasing its height before breaking in a rather destructive manner. Such waves make up a series of long-crested forms. They characterize such coasts as the South Atlantic, the South Indian Ocean, and the West Indian islands. A hydrographic chart of, for example, the E coast of North Island, New Zealand, would indicate the presence of rollers on its more exposed beaches.

roof pendant A small downward projection of country rock into an underlying igneous body. *See also* cupola.

ropy lava Volcanic lava that has solidified in long strands. A thin crust forms on the molten lava while it is still moving; the liquid center continues to move and form the linear structure. *See also* pahoehoe.

rose quartz A pink QUARTZ sometimes used as a semiprecious gemstone.

Rossby wave *See* long wave.

rotational slip The semicircular movement of a mass of loose material, such as ice, rock, or soil, down a concave slope.

rotation of the Earth The turning of the Earth on its axis, which it completes every 24 hours. As different parts of the Earth face the Sun, day and night occur, whose lengths vary because of the 2.5° tilt of the Earth's axis. At the EQUINOXES, there are 12 hours daylight and 12 hours darkness everywhere.

roughness length A measure used in micrometeorological studies as an indicator of degree of roughness of a surface to airflow. It is determined by extrapolating the observed relationship between wind speed and height to the point where the wind speed becomes zero. Its value is about one tenth that of the true height of the roughness elements, i.e. for a lawn of thickness 1 cm it is about 0.1 cm.

rubellite A pink or red form of TOURMALINE, used as a semiprecious gemstone.

ruby A deep red transparent gem variety of CORUNDUM (the color is caused by chromium impurities). It crystallizes in the trigonal system, and has been synthesized.

rudaceous Describing a CLASTIC sedimentary deposit or rock in which the constituent fragments are of relatively large size. It is formed of GRAVEL, i.e. the clasts are greater than 2 mm in diameter.

rudite A type of sedimentary rock that consists of coarse grains (more than 2 mm across). The category includes BRECCIA and CONGLOMERATE.

runnel A depression extending along a beach perpendicular to the direction of wave approach. Runnels separate and lie parallel with BEACH RIDGES.

running mean A statistical method of smoothing the variations in a time series. It is calculated by determining the mean of a set of numbers (frequently five or ten year periods), then omitting the first value and taking the next succeeding value, the mean is again calculated. A series of running means is thus obtained for a specified time period, and helps illustrate the overall trend by reducing the influence of individual extreme years.

runoff That part of total precipitation left to flow into rivers after evaporation and transpiration by plants have taken place. It has several components: rain falling into the channel of the river, surface runoff, rainfall that soaks into the soil moving laterally toward the river to reach it as "interflow", and water that percolates through the soil to the water table, feeding steadily all the year round to the stream as groundwater. Over the world as a whole, around one fifth of rainfall becomes runoff, but regional variations are very great, dependent on climate, the nature of vegetation and soil, whether the rain comes in severe storms or widespread gentle showers, etc. In any one region it varies over time according to climatic cycles and the seasons.

rutile A brown, violet, or black tetragonal mineral. It is the commonest polymorph of titanium dioxide, TiO_2, and is widespread as an accessory mineral in igneous and metamorphic rocks. It also occurs in high-temperature veins with quartz and apatite and as an alteration product of sphene and ilmenite. It is an important source of titanium.

S

sabkha A salt flat that has a surface encrusted with HALITE; there are also calcium and magnesium salts in the deposit. It runs along the coast just inland and is periodically flooded by the sea.

saccharoidal Describing a fine- to medium-grained granular rock resembling sugar.

saddle A broad shallow depresion in a mountain ridge forming a pass.

saddle reef A deposit of minerals that occurs at the top of an anticline fold. In some places in Australia and Canada, saddle reefs in folded slate beds contain gold.

saeter An area of pasture in the Norwegian mountains, above the treeline, used for the summer grazing of cattle.

Sahel A large region to the south of the Sahara, extending from Mauritania and Senegal in the west to Ethiopia and Sudan in the east. Overgrazing has removed much of the vegetation, with consequent soil erosion. DESERTIFICATION (from the Sahara) is occurring in the north.

salic Denoting the silicon- and aluminum-rich minerals of the CIPW normative classification. *Compare* femic.

salina A saltpan, named after those that occur in semiarid areas of Spain. *See* playa.

saline Describing any solution that contains a salt, particularly common salt, sodium chloride, NaCl. The salinity of seawater, for example, is 35 000 parts per million (of which 85% is sodium chloride).

salinity The extent to which water contains dissolved salts. Together with temperature and oxygen content, it is one of the fundamental properties of sea water. A wide range of salinity occurs in the various seas of the world: normal sea water has a salinity of approximately 34.33 grams per thousand grams (3.433%), although it may be as high as 4.0% (in the Red Sea and Persian Gulf) or less than 3.3% (at high latitudes). The differences arise largely as a result of variations in the rate of evaporation. Increased salinity raises the relative density, lowers the freezing point, and greatly influences marine ecosystems to the extent that fauna and flora have to adapt to saline conditions. A knowledge of salinity is also important in understanding the circulation of ocean water.

salinization The accumulation in water or soil of salts of magnesium, potassium, and sodium. It often occurs in arid and semiarid regions, where the rate of evaporation is greater than the rate of precipitation. Plants cannot grow in the salty soil and in this way salinization contributes to DESERTIFICATION.

salite A monoclinic PYROXENE.

salt 1. Chemically, a compound formed when an acid is neutralized by a base. Most salts are crystalline ionic compounds, and most minerals are composed of salts.
2. In everyday language, the term salt usually means common salt, which is sodium chloride, NaCl. *See* halite.

saltation A mechanism of sediment transport, whereby individual grains move by bouncing off the surface of the land or

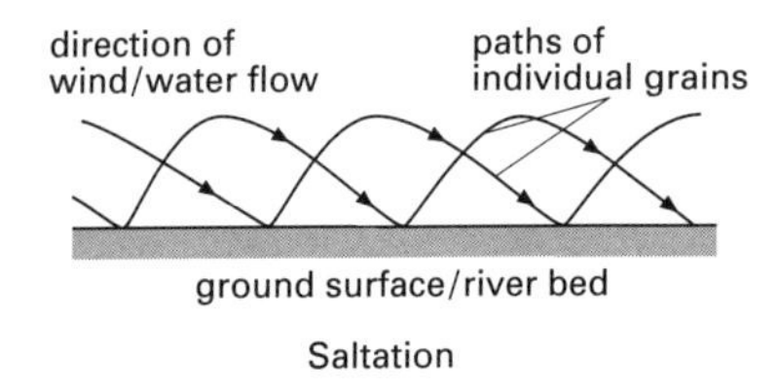

Saltation

bed. It is the most common form of transport by wind. It is much less significant in rivers because of the viscosity of water as compared with air; the "stickiness" of water slows down the rise of the grain and cushions its fall, so the height of saltation is but a few grain diameters as compared with several thousand grain diameters in air. The initial rise of the grain is near vertical, followed by a gentle downward flight in the direction of wind or stream flow, terminated by impact again with the surface at an angle of 10–16°. In air, this impact may cause the grain itself to rebound, or dislodge another grain on the surface, which will then saltate. In water, the lifting force is eddy turbulence, which temporarily overcomes the weight of the grain, imparting an upward movement.

salt dome A circular domelike structure formed by the upward movement of a column of salt (halite), generally below strata of sedimentary rock. It may be capped by a layer of anhydrite or gypsum (calcium sulfate) or calcite (calcium carbonate), possibly containing some sulfur. The rocks below the dome are deformed, and may hold deposits of oil and natural gas.

salt-earth podzol *See* solod.

salt flat A level salt surface constituting the bed of a former salt lake.

salt lake A highly saline lake formed in an INLAND BASIN.

salt marsh *See* marsh.

saltpeter *See* niter.

San Andreas fault A 1125-km fault that runs through California. It is on the boundary between two crustal plates that intermittently move laterally, producing friction that causes earthquakes.

sand Particles of rock with a size range of 0.06–2.00 mm in diameter (i.e. between silt and gravel):

0.6–2.0 mm coarse sand.
0.2–0.6 mm medium sand.
0.06–0.2 mm fine sand.

Rocks formed from sediment in this range are known as sandstones.

According to Bagnold (1941), sand is that material lying above a lower limit of material capable of carriage by suspension (mostly dust) and below an upper limit of material incapable of movement either by direct wind force or by the impact of falling grains. Windblown sands are said to be mostly 0.3 to 0.15 mm.

Most sands are predominantly made of QUARTZ, other material being too easily eroded to survive for long. The quartz is derived from the weathering of quartz-bearing rocks, subsequently reduced in size by water or ice abrasion. Desert sand grains are more rounded than fluvial or marine types.

sand bank A bank of sand and other sediments that occurs in estuaries and along the open coast where the tidal streams tend to flow in a rectilinear fashion and where transportable sediments are fairly abundant. The form of the bank and the pattern of a group of banks often reflect the nature of tidal flow. The surrounding or included channels may be of the ebb or the flood type, depending on the relative dominance of one or other tidal stream. The banks tend either to be linear or parabolic in shape.

sand dune *See* dune.

sand ribbon A submarine morphological form of linear shape that, like a sand wave, occurs where tidal streams are suitable for its formation and where transportable sediment is sufficiently abundant. The ribbons occur within certain estuaries and along parts of the open coast, seaward of low-water mark. They are often only a few centimeters thick, but may be up to 100 m wide and extend for several kilometers. The general trend of sand ribbons off the S coasts of Britain is parallel to both the coast and the directions of the strongest tidal streams. Where surface streams are less than 1 knot, the boundaries of the ribbons tend to be somewhat obscure; in contrast, where surface tidal streams attain more than 1 knot, the boundaries become more strongly defined. Such morphological information is clear from high-precision echo sounding.

sand shadow *See* aeolian form.

sand sheet *See* aeolian form.

sandstone A lithified ARENACEOUS deposit. It is a CLASTIC sedimentary rock composed of grains between 1/16 mm and 2 mm in diameter. The particles may be bound by a secondary cement such as CALCITE or various iron minerals, or be welded together by pressure. A great variety of sandstones are known, and they can be classified by both mineralogical and textural characteristics.

sand stream A broad tract of moving sediment (largely of sand grade) seaward of low-water mark, possibly tens of kilometers in width. Sand streams tend to extend over considerable distances, roughly parallel to the coast. The streams are evident from various field data, including morphological forms such as SAND RIBBONS, SAND BANKS, and SAND WAVES, information relating to sediment grain-size distribution, and a knowledge of waves and tidal streams. Sand streams may occur anywhere on shelf areas, but are especially developed near to coasts when the tidal currents flow in rectilinear fashion. One such stream moves northward out of the southern bight of the North Sea, toward the Netherlands and Germany.

sandur *See* outwash plain.

sand volcano (mud volcano) A sedimentary structure produced in poorly consolidated sands and clays that have a high water content. As a result of shaking, a lower layer of sediment will be injected up and through the overlying bed. When this process is completed the lower bed builds up as a small cone above the upper bed. Where the process does not reach completion or the overlying bed is very thick, FLAME STRUCTURES are produced.

sand wave (megaripple) A morphological form that resembles a desert dune and has been called an underwater dune. The sediment is usually of sand grade. The structures are created by tidal flow, and, if large, may have many smaller sand waves superimposed upon them. They often occur on the surface of SAND RIBBONS. They tend to be orientated perpendicular to the main tidal flow direction. Most display a measure of asymmetry, which enables the net direction of sediment transport to be determined, the steepest face indicating the direction of travel. A small difference, perhaps only 0.1 knot, in the flood and ebb peak velocities, is sufficient to cause asymmetry of form and to make the sand waves migrate in one or other direction. The crests of sand waves may be straight or sinuous; their height, as in the case of the southern North Sea, may be as much as 15 m. In contrast, off Florida, many of the sand waves are less than 1 m in height. Some sand waves are formed in the relatively deep water near the shelf edge, for example off SW England, where their heights vary between 8 and 12 m, and their average length is some 850 m.

sanidine A potassic alkali FELDSPAR.

Sanson–Flamsteed projection A modified CONICAL PROJECTION and a special case of Bonne's projection. Both the central meridian and the Equator (standard parallel) are straight lines that are truly divided and at right angles to each other. All other parallels are horizontal straight lines, equally divided and equal distances apart. This is a sinusoidal map projection because the meridians are sine curves drawn through predetermined points on the parallels. Although the shapes are distorted away from the center of the projection, it is an equal-area projection and therefore commonly used in atlases for areas astride the Equator, e.g. Africa.

saponite A magnesium-rich claylike mineral related to montmorillonite. *See* clay minerals.

sapphire A colored transparent variety of CORUNDUM (the color is caused by traces of impurities such as oxides of chromium, cobalt, and titanium), valued as a precious gemstone. It may be deep blue (the most prized), pale blue, green, or yellow. It crystallizes in the trigonal system, and has been synthesized.

saprolite The part of the weathering profile that, although weathered, remains in situ and retains such original structures of the parent rock as jointing, banding, and veining.

sapropel A loose deposit made up mainly of the remains of algae, with some mineral fragments. Sapropels form in anaerobic conditions at the bottoms of lakes or shallow seas, and when compacted change into shale containing coal, bitumen, or oil (*see* oil shale).

saprotroph (saprobe, saprophyte) An organism that feeds on dead or decaying organic matter (plant or animal). The best-known saprotrophs are fungi.

sardonyx A type of CHALCEDONY that has alternate parallel bands of reddish-brown and white. It is used as a semiprecious gemstone.

satellite Since 1957, numerous satellites have been placed in Earth orbit. Some of these have had specific meteorological purposes and tremendous amounts of new information have appeared as a result of these instruments. Continuous records are now received of cloud patterns, and infrared photographs and experiments have shown that it is possible to obtain vertical temperature profiles. Geostationary satellites, which are orbiting at the same rate as the Earth and are therefore providing a continuous record of the same portion of the atmosphere and surface, can determine the wind field at cloud level from the varying cloud patterns, in addition to the other facilities offered by the normal satellite. The movements of tropical cyclones (hurricanes, typhoons) can be tracked from their formation enabling advisories and warnings to be issued to lands in their predicted path. *See* Landsat.

Developments in satellite imagery and data have contributed greatly to the observation of features on, above, and below the Earth's surface and the information thus obtained can be incorporated into GEOGRAPHICAL INFORMATION SYSTEMS (GIS). Originally restricted to simple photographs, satellite imagery can now provide widely diverse imaging at a variety of resolutions. *High-resolution satellite imagery*, in which images with 1-meter resolutions are now available in the public domain, has widespread applications. For example, in cartography it is used in updating and creating maps and navigation charts; other uses include locating potential sites in mineral exploration, showing areas of deforestation, and locating pollution and its sources. *Multispectral satellite imagery* can be used to reveal subtle topographic and vegetation features. *Infrared satellite imagery*, which depicts the intensity of infrared radiation emitted from Earth, can be applied in the study of volcanoes. Satellite imagery has been used to study variations in ocean characteristics over wide areas; for example, data from AVHRR instruments (high-resolution visible and infrared light radiometers) on such satellites as the NOAA Polar Orbiting Environmental Satellite series enable sea surface temperature and cloud patterns to be mapped. In 1999 the first of a series of TERRA satellites to monitor effects of human activity on the global environment was launched; measurements recorded are to include that of pollution in the lower atmosphere. *See also* Global Positioning System.

satin spar A type of GYPSUM (calcium sulfate) that has translucent white fibrous crystals. The same name is also sometimes given to a type of calcite (calcium carbonate) of similar appearance.

saturated adiabatic lapse rate The rate of cooling of a parcel of saturated air as it rises. Because the amount of latent heat released depends upon air temperature, the lapse rate is not a constant. At high temperatures the rate is slow because much latent heat is released on condensation, whereas when it is very cold little heat is released and its value approaches the DRY ADIABATIC LAPSE RATE. *See also* environmental lapse rate.

saturated rock An igneous rock that contains neither quartz nor feldspathoids. *See* silica saturation.

saturation **1.** The condition in which a sample of moist air is in equilibrium with an open water surface at the same temperature and pressure. In this state there is a balance between the number of molecules leaving the water surface and the number returning from air to water. The point of saturation can be reached by adding moisture to a system or cooling it. The capacity of air to hold water is largely determined by temperature as shown in the diagram of the saturation vapor pressure curve.
2. *See* silica saturation.

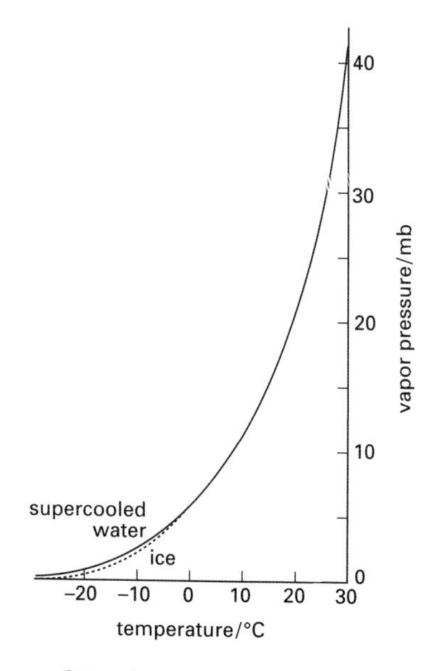

Saturation vapor pressure curve

Saurischia An order of dinosaurs distinguished from the ornithischians by having a pelvis constructed on the typical reptilian plan. These dinosaurs were primarily carnivorous and some grew to a very large size: *Tyrannosaurus*, nearly 7 m high, was the largest carnivorous land animal known and the herbivorous species were even larger, *Diplodocus*, for example, reaching a length of nearly 28 m. Like other dinosaurs the group became extinct at the end of the Cretaceous period. *See* K/T boundary event. *Compare* Ornithischia.

saussuritization The alteration of plagioclase to a fine-grained assemblage of albite, epidote, calcite, and sericite during low-grade metamorphism and hydrothermal processes.

savanna A region of tropical grassland, occurring mostly in a band between the tropical forests and the deserts (between 5° and 20° north and south of the Equator). There are few trees, except for solitary drought-resistant acacias, baobabs, and eucalyptuses. *See also* savanna climate.

savanna climate A tropical climate with a seasonal rainfall regime. It has similar properties to the monsoon climate, the rain being associated with the INTERTROPICAL CONVERGENCE ZONE, although the seasonal reversal of winds is not necessarily as well developed. The name is derived from savanna vegetation characterized by tall grasses.

scablands A type of highly eroded terrain that was formed by sudden extensive flooding after the melting of glaciers and ice sheets. There is little or no soil on the bare rock surface, and so very few plants grow. A well-known example occurs in the northwestern USA.

scapolite A member of a range of minerals that vary in composition between $Na_4(Al_3Si_9O_{24})Cl$ and $Ca_4(Al_6Si_6O_{24})CO_3$, mainly by the substitution NaSi↔CaAl. They are found in a wide variety of metamorphic rocks.

scar A short steep slope of generally bare craggy rock, usually in a limestone region.

scarp (escarpment) A steep clifflike slope often of considerable size, which rises above the surrounding land surface. Such structures result from faulting (*see* fault scarp; fault-line scarp) or the differential erosion of gently inclined strata (*see* cuesta).

scattering The electromagnetic radiation from the Sun is affected by air, dust, and water vapor molecules in the atmosphere, which break up or scatter its unidirectional beam to give diffuse radiation. This scattering is selective in that blue wavelengths are scattered more easily than the longer red wavelengths and so diffuse radiation from the sky appears blue. The reduction of blue light from the direct beam results in the Sun's disk appearing red or yellow.

Larger dust particles scatter light without any wavelength dependence. Hazy conditions with high dust concentrations therefore give a whitish hue to the sky.

scheelite A white, yellow, green, or brown mineral form of calcium tungstate, $CaWO_4$. It crystallizes in the tetragonal system, and occurs in quartz veins and as deposits in contact metamorphic rocks. It is used as a source of tungsten.

schiller A striking play of colors exhibited by some minerals, particularly feldspars and pyroxenes, in hand specimen. This is due to very small inclusions of iron ore, orientated along certain cleavages or partings, which reflect the light when the crystal is rotated. *See* diallage.

schist A strongly foliated coarse-grained rock in which mica minerals are abundant and their subparallel orientation produces a marked SCHISTOSITY. The schistosity is accentuated by the segregation of minerals into thin layers alternately rich in micaceous minerals and quartz/feldspar. Schists in general are the product of regional metamorphism.

schistosity A variety of slaty CLEAVAGE that is well developed in SCHISTS and to a lesser extent in GNEISSES. Platy minerals such as micas, amphiboles, and chlorites recrystallize or crystallize perpendicular to the direction of maximum stress. All traces of the original bedding are destroyed and the rock readily splits along schistosity planes. *See also* foliation.

schlieren Streaked-out patches found in igneous rocks, representing softened xenolithic material drawn out along the direction of flow.

schorl A black variety of TOURMALINE.

scirocco *See* sirocco.

scolecite A white mineral form of hydrated calcium aluminum silicate, $Ca(Al_2Si_3O_{10}).3H_2O$. It crystallizes in the monoclinic system as groups of radiating crystals. It is a member of the ZEOLITE group of minerals.

scoriae *See* pyroclastic rock.

Scotch mist A mixture of thick cloud and heavy drizzle, which derives its name from frequent occurrences in the hillier parts of Scotland. It is formed in maritime tropical air when uplift over the hills produces large amounts of condensation and cloud droplets, some of which are large enough to give drizzle. It can occur anywhere that the conditions of weak uplift of almost saturated but stable air are fulfilled.

scour Localized erosion, for example tidal scour in estuaries, removing sediment periodically and depositing it again at another stage; in rivers periodic bed scour occurs during periods of high flow, compensated at low flow by infill of depressions, a process sometimes called *scour-and-fill*. In connection with glacial erosion, scour refers to the etching and polishing of solid rocks by rock material incorporated in the ice.

scree An accumulation of loose rock fragments on a hillside or at the base of a weathered cliff. Scree that slips downhill (when it is wet) becomes slightly sorted, with the larger pieces at the bottom.

screw dislocation A type of defect in a crystal lattice that results from the twisting of one part of the lattice relative to the other.

scrub A type of vegetation consisting of low shrubs and small trees that grows in regions that have unreliable rainfall, such as the margins of deserts. Examples include the chaco of Paraguay, CHAPARRAL of North America, the MALLEE of Australia, and the MAQUIS of the Mediterranean area.

scud *See* fractostratus cloud.

Scyphozoa The class of the phylum CNIDARIA that includes the jellyfish. Because they possess no hard parts they are rare as fossils, although remains attributable to them have been found in rocks as old as the Precambrian.

sea breeze A local wind resulting from differential surface heating between land and sea. It blows only during the day when land areas heat up relative to the sea to produce a weak thermal pressure gradient from sea to land. Initially the wind blows down the pressure gradient force at right angles to the coast, but as the system continues the Coriolis effect comes into play, causing a veering of the wind until eventually the breeze may be almost parallel to the coast. It can extend up to 80 km or even more inland and is separated from the warmer air by a sea-breeze front, rather like a cold front. Above the surface sea breeze is a returning circulation from land to sea maintaining continuity. *See also* land breeze.

sea-floor spreading A concept formulated in the 1960s by which a satisfactory mechanism for CONTINENTAL DRIFT was found. This theory showed that the ocean floor is one of the youngest and most active parts of the Earth's surface. Magma rising from the Earth's mantle reaches the surface along the MID-OCEAN RIDGES. The magma cools and becomes part of the Earth's crust. Repeated rifting of the area accompanied by the addition of further magma causes the older material to be displaced sideways. Because the Earth is not expanding in diameter, the generation of new sea floor is compensated by its destruction and reabsorption into the mantle at destructive plate boundaries. As the oceans spread the continents are moved across the surface of the Earth as part of a large PLATE. The ocean continues to expand until the movement of other plates causes its rate of destruction to exceed its rate of construction. At that time the ocean begins to close. Examples of spreading oceans are the Red Sea and Atlantic Ocean; an example of a closing ocean is the Mediterranean Sea. Spreading rates are quite low, being in the order of 4 cm per year but varying from place to place along a mid-ocean ridge, and from ocean to ocean.

sea fog Any fog over the sea, whatever its origin. It is most frequent over cold sea surfaces where contact cooling can lead to saturation of the overlying moist air. An exception to this is ARCTIC SEA SMOKE, which is an evaporational fog.

sea level The average level of the sea surface at a given time and in relation to a chosen datum (*see* mean sea level). Because of the numerous irregularities and slopes present on the sea surface, an accurate measure of sea level is rather difficult. Apart from day-to-day variations, sea level has changed with time, sometimes showing a net upward trend, at other times showing the reverse. This has happened for many different reasons; among these are isostatic readjustment, geosynclinal sinking, local tectonic instability, geodetic changes, and glacioeustatism. Hence, superimposed on the short-term fluctuations of sea level due to tides, winds, etc., are longer-term fluctuations. *See also* base level.

seamount A submerged and isolated elevation on the deep-sea floor. Seamounts stand over 1000 m above the surrounding sea floor. Their slopes are comparatively steep and they display relatively small circular or elliptic summit areas. On a global scale, seamounts are very numerous and widespread, being especially numerous in the Pacific Ocean. Their summits lie in depths of between 200 and 2500 m, although the majority of them lie between 1000 and 2000 m. Many stand on the CONTINENTAL RISE. Although some seamounts are found clustered in roughly linear groups, many others rise from the sea floor quite independently of their immediate neighbors, the floor between them often being relatively flat. Three or more seamounts in a line comprise a *seamount chain*, three or more not in a line make up a *seamount group*, and three or more sited on a rise or ridge comprise a *seamount range*.

season A subdivision of the year consisting of a period of supposedly uniform or similar climatic characteristics. The length and properties of the seasons vary across the globe. In tropical areas, temperatures are fairly uniform throughout the year and distinction is made between the wet and dry seasons only. In temperate latitudes, the seasons are based on the equinoxes and solstices, but for climatological purposes the divisions are on a monthly basis: (in the N hemisphere) spring – March, April, May; summer – June, July, August; fall – September, October, November; winter – December, January, February. These periods do have some average significance, but individual days may have weather conditions appropriate to any season. In polar regions, the changeover from winter to summer and vice versa is very sudden and there is thus a two-season year as in the tropics.

sea surge A large movement of water in the oceans, which generally causes large waves and high tides. Sea surges may be caused by strong winds associated with deep DEPRESSIONS, and can cause extensive flooding of low-lying areas, especially if they travel up an estuary.

sea valley An elongated depression cut in the sea floor and having a valley-like form. Such features display a variety of forms: SUBMARINE CANYONS are deeply incised and possess steep walls, whereas many others have a broad cross-sectional valley form, with relatively gently sloping sides that contrast markedly with the steep canyon walls. Some valleys have a trough-like form, like that off the Mississippi Delta or that off the Ganges Delta. Their floors are broad and flat. Some resemble land valleys by having levée features along one or both flanks (*see* levéed channel). Some are merely short gullies, sometimes arranged in a subparallel series. Eroded submarine valleys have been grouped into shelf channels and submarine canyons. The former group may be further subdivided into glacial troughs, tidal scour channels, and drowned river valleys.

sea wave A wave in the generating area, i.e. the area in which waves are being actively formed. As the waves travel beyond this area they become longer-crested and more regular, being known as SWELL. Because there may be present at any one time waves of quite different lengths (perhaps ranging from a few centimeters to several hundred meters) and different heights, superimposed one upon the other, the sea's surface may present a very confused picture. The size of sea waves depends largely on three variables: the speed of the wind, wind duration, and available fetch. *See also* wave.

secondary depression A depression that forms along a front within the general circulation of a primary depression. Initially it is less intense than the parent low, but on many occasions it intensifies and eventually incorporates the original depression.

secondary enrichment The concentration of valuable metals in the lower part of a vein as a result of the weathering of the upper part. Dissolved metals seep downward and react with the lower-grade compounds, enriching them with metal content.

secondary mineral A mineral present in a rock as a result of the chemical alteration or breakdown of the original mineral.

secondary reflection A signal received at a receiver such as a geophone (hydrophone) from a particular reflecting rock horizon after it has been reversed in its direction of propagation at least twice by other reflecting horizons, and therefore arrives considerably later than a primary reflection from that horizon which has had a direct path of propagation.

secondary vegetation Plants that grow in an area that has been cleared of its primary vegetation, for example following a fire or human activity.

secondary wave (S wave, shear wave) A type of seismic wave that travels through the solid body of the Earth. It is propagated by the oscillation of the particles at right angles to its direction of propagation. This type of wave cannot be transmitted by liquids. *See also* primary wave.

secular trend A climatic variation that can be observed during the period of instrumental records and indicates a persistent tendency for the mean value to increase or decrease. It is most easily determined by smoothing the data series by RUNNING MEANS.

secular variation Changes in the properties of the Earth's magnetic field over a long period of time.

sediment Particulate material that has been deposited in a fluid medium. The fluid concerned is mostly water, but aeolian sediments are not uncommon.

sedimentary rock A rock formed by lithification (consolidation and compression) of sediments laid down by wind or water. The sediments may consist of fine particles of preexisting rock, which may be igneous, sedimentary, or metamorphic, or be derived from plant or animal remains. The resulting rock strata may subsequently be uplifted, tilted, folded or faulted. Sedimentary rocks include CLASTIC rocks formed of land sediments, including clay, silt, sand, and gravel; consolidated types are sandstone and shale. Organic sedimentary rocks, usually lacking land sediment, include LIMESTONE and DIATOMITE (composed of the fragmentary remains of sea creatures), COAL (derived from plants), and coralline (derived from coral). Chemical sedimentary rocks are those formed by chemical processes, and include the EVAPORITES and some ironstones.

sedimentary structure A structure formed within sedimentary rocks during their deposition (*primary structure*) or after deposition (*secondary structure*). Such structures include CROSS-BEDDING, FLUTE casts, BIOTURBATION and RAIN PITS.

sedimentation The process of sediment deposition. Sedimentation of any particular particle size takes place when the velocity of flow in the transporting medium falls below its terminal, or settling, velocity.

sedimentology The study of sediments and sedimentary rocks.

sediment transport The movement of mineral grains by wind, water, or ice. The force of gravity also moves sediments in rockfalls and on scree slopes. *See* aeolian transport; saltation; suspended load.

sediment trap A natural or artificial device that serves to trap part of the sediments moving across or above the sea floor. Artificial traps have been successfully used for experimental purposes in certain limited situations, for example in some Alpine lakes, and in scale hydraulic model experiments. In nature, certain types of seaweed, mangrove, spartina, marram grass, and so on perform similar functions. Soft algae, which in temperate seas may attain lengths of up to 10 m, not only break the force of waves but may reduce the orbital velocity beneath waves to the extent that sediments being carried in suspension may settle out on the seabed. Off S France, poseidonia beds in depths of between 5 and 50 m have led to deposition rates of up to 1 m per century. Synthetic seaweed, which is made of a special plastic material, has been used in experiments to test its effectiveness in building up beaches. Spartina traps marsh muds, marram grass traps dune sands, and mangroves collect estuarine and tidal lagoon muds.

seiche A fluctuation in water level sometimes occurring in the open sea, in semi-enclosed bays, harbors, lakes, and other water bodies. The period of oscillation generally ranges from a few minutes to several hours. Small seiches may be a permanent characteristic of many harbor basins. The period of oscillation is determined by the resonance characteristics of the water body in question, usually being a function of the actual dimensions and shape of the water body. The more severe seiches often result from rapid changes in wind fields and barometric pressure. Harbor basin seiches may result from wave penetration.

seif dune (longitudinal dune) A sand dune occurring as a long chain running parallel to the prevailing wind direction. Such dunes tend to form in areas where the wind regime is two-directional, with a gentle prevailing wind that supplies the sand and short-term cross winds, often of greater strength, that help build the dune, producing curved faces. They can originate ab initio, or from BARCHANS when the wind regime changes. Their characteristic form is of lines of ridges, running parallel over long distances, rising to summits periodically. The dunes grow downwind of the prevailing wind, with little lateral movement. Individual chains reach 100 km in length, but several distinct chains can continue one line for much greater distances. Heights up to 210 m have been recorded, while the distance between successive lines varies from a maximum of 500 m to a minimum of 20 m.

seismic array A series of seismometers laid out in an L- or T-shaped pattern.

seismic discontinuity A boundary between rocks at which seismic waves suddenly change speed, because of the different densities of the rocks on each side of the boundary. Major examples include the GUTENBERG DISCONTINUITY and the MOHOROVIČIĆ DISCONTINUITY.

seismic event A short-lived event that acts as a source of energy for the formation of seismic waves, e.g. an earthquake or explosion.

seismic noise *See* microseism.

seismic reflection The technique of determining the structure of a rock body by measuring the time taken for a pulse from a source at the surface to travel to that body, be reflected from a seismic discontinuity within that rock body, and return to the surface.

seismic refraction The technique of determining the velocity and attitude of a subsurface rock body by measuring the shortest-time travel path from a source to a set of receivers distributed around the rock body.

seismic shooting A technique used in geologic surveys in which explosives are detonated underground and the resulting seismic waves detected at various locations (using seismometers).

seismic wave A wave generated by an explosion or earthquake within the Earth or on its surface. There are four main types of seismic wave: PRIMARY WAVE, SECONDARY WAVE, RAYLEIGH WAVE, and LOVE WAVE.

seismic zone (earthquake zone) A narrow well-defined belt in which the majority of earthquakes occur. These zones usually coincide with the junctions between lithospheric plates, especially in association with island arcs, mid-oceanic ridges, major fracture zones, and young orogenic belts.

There are two main continental seismic zones, one bordering the Pacific and the other corresponding to the Alpine-Himalayan chain. There are larger areas which experience virtually no seismicity.

seismogram A record of the frequency and magnitude of the oscillations produced during an earthquake, recorded on a time-correlating device.

seismograph An instrument that detects and records the seismic waves generated by earthquakes or other tectonic movements and explosions. The epicenter of the earthquake can be calculated from the recordings of a number of seismograph stations.

seismology The scientific study of earthquakes, including their origins and manifestations. One branch of the science concentrates on seeking methods for predicting earthquakes.

seismometer An electronic or mechanical device that detects, amplifies, filters, and records the motions of the Earth in a particular direction. A seismic set is a series of three seismometers orientated at right angles to each other.

selenite A colorless transparent mineral form of calcium sulfate, $CaSO_4$, a type of GYPSUM. It crystallizes in the monoclinic system, as separate large crystals or in crystalline masses.

selva A type of dense tropical rainforest, as in the Amazon basin and parts of central Africa. *See* tropical rainforest.

sepiolite *See* meerschaum.

septarium A large rounded nodule of clay ironstone or limestone, with a characteristic network of radiating cracks that contain calcite or other minerals.

séracs An extremely irregular ice surface, usually at the foot of an ICEFALL on a glacier. It is formed when the slowly flowing ice decreases speed, and crevassed ice piles up into impenetrable pillars and pinnacles.

sericite Secondary muscovite occurring as fine-grained flaky aggregates formed from the alteration of FELDSPARS.

series A division of rock in the Stratomeric Standard scheme of stratigraphic nomenclature (*see* chronostratigraphy). It indicates the body of rock that has formed during one EPOCH. A series consists of several STAGES grouped together and several series may be combined to form a SYSTEM. In the past the term has been applied loosely to bodies of rock characterized on the basis of gross lithology and contained fossils, but a continuation of this informal practice is not generally recommended.

serozem *See* sierozem.

serpentine One of a group of minerals having the general composition $Mg_3Si_2O_5(OH)_4$ and a layered structure. They are monoclinic, typically green or white, and occur in two main forms: CHRYSOTILE and ANTIGORITE. Chrysotile is fibrous and is used in the manufacture of asbestos, whereas antigorite has a platy habit. Serpentines are found in altered basic and ultrabasic rocks where they have formed from the breakdown of olivines and pyroxenes.

serpentinite A rock consisting largely of SERPENTINE, formed by the hydrothermal alteration of ultramafic rocks such as dunite and peridotite in which olivine and pyroxene have been converted to serpentine.

Seventh Approximation (New American Comprehensive Soil Classification) The US soil classification system, published in 1960. Its name comes from the fact that it represented the seventh attempt by its authors to find an "ideal" classification. Unlike earlier classifications based on pedogenesis, the emphasis is placed on properties of soils and diagnostic horizons that affect pedogenesis or result from it. As a result the ten orders of soils have some unusual groupings; for example, there is no separate group for gleys, and podzols occur in more than one order. Although the nomenclature is somewhat obtuse, and laboratory analysis to measure certain properties is necessary before definite classification, there is the great advantage that the major levels are as tightly defined as the field mapping units, so that there is no awkward gap between mapping and broad classification, as so typically happened with many older systems. To some extent, the Seventh Approximation is now being used at a world level, and the FAO has published a world soil map based on it. *See* alfisol; aridisol; entisol; histosol; inceptisol; mollisol; oxisol; spodosol; ultisol; vertisol.

shade temperature (in meteorology) The temperature of the air measured in the shade, usually by a thermometer in a STEVENSON SCREEN. In this way, direct sunlight, breezes, and variations in shade do not affect the measurement, which is the basis of all official meteorological readings.

shadow zone An area on the Earth's surface whose EPICENTRAL ANGLE is greater than 103° and less than 143° from an earthquake focus. Within this area seismic waves are received only after they have been reflected at the Earth's surface.

shale A well-laminated ARGILLACEOUS sedimentary rock that is fissile, and splits easily along bedding planes (*compare* mudstone). The fissility is related to the disposition of clay minerals within the rock.

shallow inland sea *See* inland sea.

shard *See* pyroclastic rock.

shatter cone A striped conical feature in rock formed by compressive shock waves that follow the impact of a meteorite. The cones, which may be up to several meters across, are aligned with their tops oriented toward the center of impact. They are found in most sedimentary rocks and in granite.

shear (shearing stress) A force tending to deform a rock mass through the movement of one part of it relative to another, as for example at a fault or thrust plane. *See also* differential shear.

shear joint A shear fracture, constituting a site of possible future shearing.

shear wave *See* secondary wave.

sheet erosion *See* overland flow.

sheeting *See* unloading.

shelf edge (shelf break) The break of slope between the CONTINENTAL SHELF and the CONTINENTAL SLOPE, where a marked slope increase occurs. The shelf-edge depth varies considerably from place to place, ranging between about 100 and 350 m, but the norm is taken as roughly 200 m.

shell The hard casing of certain marine organisms. On the death of the organisms the shells are deposited on beaches or on the sea floor. Many remain virtually intact for long periods, especially in environments devoid of marked wave and tidal action. Others rapidly become pulverized into minute particles or broken into several fragments. In tropical areas, beaches often consist of shell sands, perhaps mixed with Bryozoa, Foraminifera, and echinoid fragments. Outside tropical areas, shell beaches are much less common. They are usually of rather coarse material and therefore tend to be steep. Shelly material may figure prominently in reef-building, for instance in the case of OYSTER REEFS.

shield A very large rigid area of the Earth's crust made up of Precambrian rocks, which have been unaffected by later orogenic episodes. Shields represent areas of the Earth's earliest formed continental crust. Examples include the Baltic Shield and the CANADIAN SHIELD.

shield volcano A VOLCANO with gently sloping sides and a wide base. It is built up by successive flows of highly liquid basaltic lava, usually issuing from many fissures. *See also* flood basalt.

shingle Pebbles that make up a beach on a seashore. They have been rounded by rolling back and forth up the sloping beach with the tides (attrition). *See also* gravel.

shock wave A wave of compression created when the speed of something exceeds the speed of sound in the medium in which it is traveling. Within a rock, for example, a shock wave (as from a meteorite impact) may deform, melt, or even vaporize the material, perhaps altering its composition at the same time (termed *shock metamorphism*).

shonkinite A type of alkaline SYENITE containing orthoclase feldspar and various mafic minerals.

shoreface The narrow zone that lies to seaward of low-water mark and which is, therefore, always covered by water but which is shallow enough to permit active movement of sea-floor sediments under the influence of waves. Some authorities use the term to mean that part of the beach that is periodically covered then uncovered by the tide, i.e. the FORESHORE zone.

shoreline The line that is produced by the intersection of a sea or lake surface with the sloping beach-face or shoreface. With regard to US Coast and Geodetic Survey marine charts and surveys, the shoreline approximates to the average line of high tides. However, the term is often used loosely to indicate coastline.

shoshonite An extrusive igneous rock with a groundmass of potassium-rich ORTHOCLASE and augite, olivine, and sometimes biotite mica.

sial The Earth's CONTINENTAL CRUST, which is composed of granitic rock types that are rich in silica (Si) and aluminum (Al).

siallitic soil A soil with a medium to high silica to aluminum ratio. Such soils are typical of the temperate zone, because tropical weathering removes silica and lowers the ratio, producing a soil in which the aluminum content is high.

Siberian anticyclone A large area of high pressure that dominates the mean pressure field over Russia in winter. It is emphasized by extensive radiational cooling of the Eurasian landmass.

siderite (chalybite) A yellow to brownish-black mineral form of iron carbonate, $FeCO_3$. It crystallizes in the hexagonal system, and occurs mainly as veins in sedimentary rocks. It is used as a source of iron. *See also* carbonate minerals.

siderophile An element that has an affinity for iron, such as cobalt or nickel. Siderophile elements are found in metallic meteorites and are probably concentrated in the Earth's core. *See also* atmophile; chalcophile; lithophile.

sienna A yellow-brown earthy pigment containing iron oxide, also called raw sienna. When it is roasted, its color changes to reddish brown (called burnt sienna).

sierozem (serozem, gray desert soil) A soil occurring in semidesert areas in more arid zones than the CHESTNUT SOILS, and characterized by low organic matter contents, a lack of leaching, and accumulations of lime at the top of the C horizon or lower B, typically within 30 cm of the surface, and possibly reaching up to the surface. Vegetation cover is scanty: short grass and scattered brush. These soils are included in the aridisol order of the SEVENTH APPROXIMATION classification.

significant wave A wave that has the average period and height of the highest 33.3% of the waves in the WAVE SPECTRUM. Calculation of significant wave characteristics requires careful wave measurement at sea and subsequent analysis of this information, or the use of a wave analyzer in a scale model that involves wave action.

silcrete A surface deposit, that occurs in semiarid regions, consisting of gravel and sand cemented together by chert, opal, and quartz.

silica Silicon dioxide, SiO_2. It is the chief mineral in chert, diatomite, and sand, and in crystalline form it makes up all the various varieties of QUARTZ and other SILICA MINERALS. Opal consists of hydrated noncrystalline (amorphous) silica.

silica concentration For an igneous rock, the percentage of SILICA it contains. It is a common way of classifying such rocks. Those with more than 66% silica are termed acid, 55–66% silica are intermediate, 45–55% are basic, and less than 45% are ultrabasic.

silica minerals *Quartz*, *tridymite*, and *cristobalite* are the three commonly occurring polymorphs of silica (SiO_2). Tridymite and cristobalite are stable at intermediate and high temperatures, respectively, but neither form persists at high pressures and both are absent from plutonic rocks. The polymorphs have distinct crystal structures built from SiO_4 tetrahedra but the pattern of linking is different in each case. In addition, each polymorph has a low- and high-temperature modification, the transitions taking place rapidly at the particular inversion temperature involved. Changes from one polymorphic form to another are very sluggish and tridymite and cristobalite both exist metastably at ordinary temperatures. *Coesite* and stishovite are high-pressure high-density forms of silica found near meteorite impact craters. *Chalcedony* is the cryptocrystalline form of silica containing minute crystals of quartz with submicroscopic pores. *Opal* is a hydrous cryptocrystalline or amorphous form of silica. *Lechatelierite* is a rare metastable silica glass.

Quartz is stable over a wide range of physical conditions and because silica is the most abundant oxide in the Earth's crust, it is a very common mineral. The different colored varieties are: colorless – *rock crystal*, yellow – *citrine*, gray-brown to black – *smoky quartz*, pink – *rose quartz*, and violet – *amethyst*. Quartz is trigonal and the absence of cleavage is characteristic. On Mohs' scale it is the standard mineral of hardness 7. Quartz is an essential constituent of silica-rich igneous rocks, such as granites, rhyolites, and pegmatites, and is found in oversaturated, intermediate, and basic rocks. Quartz often occurs intergrown with alkali feldspar in graphic granite. Vein-quartz, precipitated from hydrothermal solutions, is often associated with ore minerals. Because of its hardness and resistance to chemical weathering, quartz is the most abundant detrital mineral and is concentrated to give rise to sands and gravels of various types, which, on lithification, constitute the arenaceous rocks. Authigenic quartz is often deposited around detrital grains, cementing them together. During metamorphism, the quartz of sediments and igneous rocks recrystallizes to a coarser grain size. Quartz is also produced by the released silica during the metamorphic breakdown of preexisting minerals. At high grades, segregation and mobilization take place giving rise to quartz veins and pegmatites.

Tridymite and *cristobalite* are typically found in cavities in acid volcanic rocks, such as rhyolites and andesites, and may be produced as a result of the passage of hot gases.

Chalcedony occurs in numerous varieties. *Carnelian* is red. The banded variety, *agate*, is formed by intermittent deposition in cavities. *Jasper* is a red-brown color. *Chert* and *flint* are opaque, dull-colored, or black varieties, the former being the massive or stratified form and the latter being found as nodules in chalk. Chalcedony is found in hydrothermal veins, amygdales, and sediments.

Opal occurs in many color varieties, the precious variety being iridescent. It is deposited at low temperatures from silica-bearing waters and is found around geysers and hot springs. *Diatomite* is a rock made up almost entirely of the accumulated opaline skeletons of diatoms. Radiolarian and diatomaceous deposits may recrystallize to form chert.

silica saturation The extent to which a rock contains silica, this measure constituting a convenient method of classifying igneous rocks into three groups:

1. oversaturated – rocks containing free silica as quartz.
2. saturated – rocks in which all the silica is combined and in sufficient quantity to exclude feldspathoids.
3. undersaturated – rocks containing feldspathoids.

silicates A group of minerals constituting about one third of all minerals and approximately 90% of the Earth's crust, in which feldspar (60%) and quartz (12%) are the most abundant. The basic structural unit common to all silicates is the SiO_4 tetrahedron. Silicates are classified on a structural basis according to how the tetrahedra are linked together. The six classes of silicates are as follows: nesosilicates (olivine), sorosilicates (hemimorphite, melilite), cyclosilicates (beryl, axinite), inosilicates (amphiboles and pyroxenes), phyllosilicates (micas and clay minerals), and tektosilicates (quartz, zeolites, feldspars).

silicification The process by which SILICA (usually in the form of chalcedony or quartz) enters pores or replaces other minerals in a rock.

sill A near-horizontal tabular intrusive body of igneous rock, usually dolerite, of roughly uniform thickness but thin relative to its area. It is concordant with the planar structures of the rock types into which it intrudes. *See also* dike.

sillimanite A pale green, gray, or brown mineral form of aluminum silicate, Al_2SiO_5. It crystallizes in the orthorhombic system, generally as fibrous aggregates (fibrolite) in metamorphic rocks. It is used as a refractory. *See also* aluminum silicates.

silt A fine-grained deposit that ranges in sediment particle size from 0.06 mm to 0.002 mm, and is therefore significantly finer than sand. Like clays, silts may include clay minerals, and also hydroxides and oxides of iron, silica dioxide, and numerous other fine material particles. Clay and silt together form the ARGILLACEOUS division of sediments. Silts result not only from decomposition of certain in-situ rocks and rock particles but also from gradual abrasion processes (impact, grinding, and rubbing). Fine silty material tends to collect in sheltered quiescent marine environments and can be a serious threat to navigation. It is sometimes difficult to dredge, often returns rapidly from spoilgrounds to the dredged areas, and can inundate seaweed beds where fish and other marine organisms normally abound.

siltstone A fine-grained layered sedimentary rock derived from SILT. It resembles shale but contains less clay. *See also* mudstone.

Silurian The last period in the Lower PALEOZOIC beginning about 438 million years ago, at the end of the ORDOVICIAN Period, and lasting about 30 million years, until the start of the DEVONIAN. The Silurian System is divided into four series: the Llandovery, Wenlock, Ludlow, and Pridoli, but there are many alternatives throughout the world. It was named by R. I. Murchison for the Celtic tribe, the Silures, that inhabited much of SE and central Wales where Silurian rocks are well developed. Rocks of the Silurian occur throughout the world; in North America they are exposed in the Appalachians and across the Mid-West, notably around the Great Lakes. The Niagara Falls are on an escarpment of Niagaran limestones from the Silurian. Important earth movements were taking place within the period: the later stages of the Taconian orogeny of North America and the Caledonian orogeny in NW Europe. The Iapetus Ocean, which separated Scotland on the North American continent from the rest of the British isles on the Europe and Scandinavian continent, was continuing to close. During the Upper Silurian extensive evaporite deposits were laid down with the widespread regression of the seas.

Similarity in stratigraphy between the Ordovician and Silurian is reflected in the fauna, which seems to be a continuation of the Ordovician fauna. Trilobites and graptolites became less common; the latter were almost extinct by the end of the period. Brachiopods were very varied and crinoids and corals increased in importance. Ostracoderm fish became more abundant and the first gnathostomes (jawed vertebrates) appeared. The first evidence of land plants also comes from Silurian rocks.

sima The Earth's OCEANIC CRUST, which is composed of basaltic rock types that are rich in both silica (Si) and magnesium (Mg). The term was formerly used for that portion of the continental crust below the CONRAD DISCONTINUITY but this was inaccurate and led to confusion with the oceanic crust.

simatic crust The denser lower regions of the Earth's oceanic crust that consists of sima.

singularity The annual recurrence of particular types of weather at certain times of year as illustrated by the BUCHAN SPELLS. Unfortunately their reliability is low except in a very general sense.

sinistral fault A fault at which movement is to the left relative to the side of the fault plane on which the observer is standing. *Compare* dextral fault.

sinkhole (sink, pothole) A saucer-like hollow, typical of limestone areas, varying from 1 to 1000 m in diameter and from a few centimeters to 300 m in depth, produced by solution (*see* limestone solution) or by rock collapse. The ratio of diameter to depth is usually about 3:1. They are usually formed on flat or gently sloping surfaces, because here water can accumulate and stand, thereby causing solution. Sinkholes may also be simply enlarged vertical joints, but whatever their shapes they often act as channels down which surface water seeps to underground drainage systems. *See also* swallow hole.

sinusoidal projection *See* Sanson–Flamsteed projection.

sirocco (scirocco) A wind of similar origin to the Egyptian KHAMSIN occurring farther west between Algeria and Libya. It is a hot dry southerly wind on the North African coast, but on crossing the Mediterranean it can pick up much moisture and bring extensive low stratus cloud to the Italian coast.

SI units (Système International d'Unités) A system of units used, by international agreement, for all scientific purposes. It is based on the meter-kilogram-second (MKS) system and replaces both the centimeter-gram-second (cgs) and Imperial (fps – foot, pound, second) systems. It consists of seven base units and two dimensionless units (see Table 1). Measurements of all other physical quantities are made in derived units, which consist of combinations of two or more base units. Eighteen of these derived units have special names (see Table 2). Base units and derived units with special names have agreed symbols, which are used without a period.

Decimal multiples of both base and derived units are expressed using a set of standard prefixes with standard abbreviations (see Table 3). Where possible a prefix representing 10 raised to a power that is a multiple of three should be used (e.g. mm is preferred to cm).

skarn A calc-silicate mineral assemblage produced during the CONTACT metamorphism and metasomatism of highly calcareous rocks.

slack tide (slack water) The period or state of the tide in which there is negligible horizontal flow of water or when the tidal currents are at virtually zero velocity. This period occurs, in the case of a rectilinear tidal stream, when the currents are reversing their direction. Some authorities consider the period covers that time when flood and ebb velocities are less than 0.1 knot.

slate A fine-grained rock in which the parallel orientation of platy crystals of mica and chlorite results in the production of a perfect slaty cleavage. The individual

TABLE 1: BASE AND SUPPLEMENTARY SI UNITS

Physical quantity	*Name of SI unit*	*Symbol for SI unit*
length	meter	m
mass	kilogram(me)	kg
time	second	s
electric current	ampere	A
thermodynamic temperature	kelvin	K
luminous intensity	candela	cd
amount of substance	mole	mol
*plane angle	radian	rad
*solid angle	steradian	sr

*supplementary units

TABLE 2: DERIVED SI UNITS WITH SPECIAL NAMES

Physical quantity	*Name of SI unit*	*Symbol for SI unit*
frequency	hertz	Hz
energy	joule	J
force	newton	N
power	watt	W
pressure	pascal	Pa
electric charge	coulomb	C
electric potential difference	volt	V
electric resistance	ohm	Ω
electric conductance	siemens	S
electric capacitance	farad	F
magnetic flux	weber	Wb
inductance	henry	H
magnetic flux density	tesla	T
luminous flux	lumen	lm
illuminance (illumination)	lux	lx
absorbed dose	gray	Gy
activity	becquerel	Bq
dose equivalent	sievert	Sv

TABLE 3: DECIMAL MULTIPLES AND SUBMULTIPLES USED WITH SI UNITS

Submultiple	*Prefix*	*Symbol*	*Multiple*	*Prefix*	*Symbol*
10^{-1}	deci-	d	10^{1}	deca-	da
10^{-2}	centi-	c	10^{2}	hecto-	h
10^{-3}	milli-	m	10^{3}	kilo-	k
10^{-6}	micro-	μ	10^{6}	mega-	M
10^{-9}	nano-	n	10^{9}	giga-	G
10^{-12}	pico-	p	10^{12}	tera-	T
10^{-15}	femto-	f	10^{15}	peta-	P
10^{-18}	atto-	a	10^{18}	exa-	E
10^{-21}	zepto-	z	10^{21}	zetta-	Z
10^{-24}	yocto-	y	10^{24}	yotta-	Y

constituent minerals cannot be resolved by the naked eye but some slates contain porphyroblasts or are spotted, the spots representing embryonic porphyroblasts. Slates are formed during the low-grade regional metamorphism of mudstones, siltstones, and other fine-grained argillaceous sediments.

slaty cleavage A parallel alignment of platy minerals in fine-grained rocks, at right angles to the direction of compression. Slaty cleavage is usually also a form of AXIAL PLANE CLEAVAGE.

sleet Precipitation comprising ice pellets with a diameter of 5 mm or less formed when raindrops or melting snowflakes freeze or refreeze. In Britain the term is used for melting snow or a mixture of rain and snow.

slickensides Small parallel grooves or striations formed on the surface of fault or joint surfaces as a result of the movement of rocks against each other. Small platy minerals are developed on them, from which the relative sense of movement of the rocks can be determined. The surface of the slickenside is smooth in the direction of movement, but rough in the opposite direction.

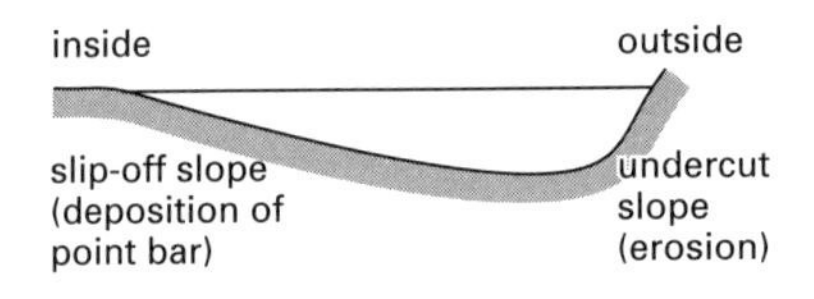

Section through a meander bend

slip-off slope The gentle slope on the inside of a MEANDER bend opposite the undercut slope that occurs on the outside of a meander bend owing to erosion. The deposition of POINT BARS and creation of slip-off slopes is mostly attributable to the HELICAL FLOW in meandering reaches, which constantly transfers sediment from eroding areas on the outside of bends to the depositing areas on the inside.

slip sheet A gravity collapse structure developed in the beds on the limb of an anticline. If the beds at the base of the limb fracture, the overlying beds will slip downward under the influence of gravity and then spill out onto the adjacent strata.

slope An inclined surface whose length is determined by the horizontal distance from crest to foot and whose angle is determined by the inclination of the slope from the horizontal. Usually a slope is said to have one angle in broad terms, but most slopes are composed of a number of different parts, called facets if straight and elements if rounded, each having its own slope angle. *See also* slope concavity; slope convexity; slope decline; slope evolution.

slope concavity The bottom parts of hill slopes are typically concave, except in certain areas where this basal concavity has been buried by aggradation. Analogies have been drawn between the profile of a concave slope and the long profile of a river; this similarity of form has also provoked claims of a similarity of origin, the concavity being essentially a product of running water action, especially rill and gully action. This type of slope form is pronounced on lower slopes because there is a greater catchment area for running water than on upper slopes, there is increased weathering of debris downslope, producing smaller particles that can move on a more gentle slope, and there is increasing impermeability as debris becomes finer downslope, leading to increased running water action. For running water action, an impermeable surface is a prerequisite and consequently concave slopes are well developed on the finer rocks, e.g. clays, shales, and argillites, and less well developed on the permeable rocks, e.g. sandstone, chalk, and limestone. Plentiful rainfall is also important and, other things being equal, the concave parts of slope profiles tend to become more pronounced with increasing age, as weathering of debris advances, making it finer and less permeable, and hence more able to support running water. *See also* slope convexity; slope evolution.

slope convexity The upper parts of hill slopes are characteristically convex in shape, and a good deal of discussion on slope evolution is directed at suggesting causes for this shape. Early workers included Gilbert, who said convexity was a product of CREEP; as the soil mantle moves from upslope, the volume passing a given point increases downslope, so an increasingly steep slope is needed to transmit the debris, hence a convex slope. Today, this idea is still widely accepted, even though creep is now thought to operate as a slide rather than a flow, so that the thickness of the debris does not increase downslope. W. Penck said convex slopes were a product of uplift acting faster than downcutting; Fenneman (1908) thought it might be due to the action of wash; S. Schumm found that creep does produce a convex profile, but because it operates only on clay lithologies, convex profiles tend to form only in clay-like materials. It may also be that because the upper part of a hill is attacked by weathering on all sides, the tendency is to weather out a convex slope. In reality a number of factors may be responsible.

slope correction The reduction of distances measured on a slope in the field to their correct horizontal value for presentation on a map. The corrected figure is obtained by multiplying the measured length by the cosine of the slope angle.

slope decline The original CYCLE OF EROSION, as formulated by W. M. Davis, was based on a progressive lowering of slope angles (slope decline) as contrasted to the later schemes of W. Penck and L. King, emphasizing PARALLEL RETREAT of slopes. Slopes will decline when more material is eroded upslope than is removed from the foot of the slope, leading to a differential rate of evolution, the upper part retreating faster and hence lowering the overall slope. For material to accumulate at the foot of a slope there must be no effective BASAL SAPPING, e.g. in a sea cliff cut off from the sea by the growth of a shingle bank or sand spit, protecting it from wave action, it has been seen that parallel retreat gives way to slope decline. Others have found that lithologies of certain types favor slope decline, e.g. S. Schumm found decline was a feature of slope evolution in clays. Originally, this type of slope evolution was seen as characteristic of the normal cycle in humid temperate areas, i.e. as a climatically controlled process. Today it is known that slopes in the same climate can undergo very different evolutions and slopes in different climates take the same course. Whether or not a slope undergoes decline depends on process and lithology as well as climate.

slope evolution Studies of slope evolution are concerned with the relation between slope form and the processes operating on that slope, the lithology of the rock, and the stage of evolution. It is intimately linked to studies of the general tendencies of landscape form, i.e. whether they tend to retreat parallel to themselves (*see* parallel retreat), leaving INSELBERGS and PEDIMENTS, or whether they tend to decline (*see* slope decline), ultimately producing a PENEPLAIN.

Many different techniques and ideas have resulted from these studies. Some geologists have studied the current evolution of small-scale slopes of different rock types to see the influence of lithology; others have studied the role of climate, through its control of geomorphic process, by comparing slopes in different latitudes, or on the shady and sunny sides of valleys; others have related slope form to tectonic action, or studied the role of base-level changes in producing polycyclic slopes. Early workers inferred a series of sequential stages through which any slope might evolve with time, whereas some modern workers have successfully identified actual time series in the form, for example, of a growing spit at a cliff foot progressively cutting that cliff off from wave action and hence inducing a change of process. Finally, some have employed a statistical approach, measuring the angles of slopes and finding those values around which angles tend to bunch (characteristic angles). Despite all these efforts, slope evolution is still a highly complex topic, although many general models have been produced. In any particular slope, its shape is determined by the balance of losses of rock by erosion and gains by deposition, but because the geomorphic processes operating may have changed through time, there is no reason why current slope forms should have any strong relation to the current processes operating on them.

slump A single mass movement of rock debris and soil downhill. It most often occurs when water-holding permeable rocks overlie impervious ones (such as shale). Rocks may shear away, leaving a scarred rockface. Slumping also occurs underwater at the edges of continental shelves.

slurry A liquid mixture of mud and water, which readily slips downhill. It most commonly occurs in clays and shales.

smectite A member of one of the major groups of CLAY MINERALS.

smithsonite A white, yellow, pink, pale green, or blue mineral form of zinc carbonate, $ZnCO_3$, known also as calamine in the UK. Its color is caused by impurities. It crystallizes in the trigonal system, and occurs in beds and veins in calcareous rocks, often associated with HEMIMORPHITE.

smog The polluted air that occurs above cities and industrial areas. The term is derived from the words *sm*oke and f*og*. The burning of fossil fuels, especially coal, releases sulfur compounds and smoke particles into the atmosphere; *sulfurous smog* forms if these particles are unable to be dispersed because of a persistent INVERSION. The concentration of toxic chemicals may build up to harmful levels. *Photochemical smog*, a hazy brown blanket of smog, is created when sunlight reacts with pollutants in the air. It is most prevalent in urban areas with high concentrations of motor vehicles; heat from burning fossil fuels causes nitrogen and oxygen to chemically combine to form nitrous oxide and hydrocarbon vapors. Under sunlight these undergo photochemical reactions, one of the substances formed being OZONE.

smoker An active volcanic vent on the ocean floor which emits hydrothermal fluids at high pressure. A *black smoker* produces sulfides of copper, iron, and manganese; a *white smoker* releases barytes and silica. Sometimes a vertical *chimney* of deposited minerals forms around the site of the vent.

smoky quartz A yellow to dark grayish-brown type of QUARTZ, which owes its color to the presence of impurities. It is sometimes used as a semiprecious gemstone.

SNC meteorite Any of a few small stony METEORITES (named after Shergotty, Nakhala, and Chassigny, the three main ones) that have significantly different compositions and ages from usual meteorites. Some planetologists believe that they originated from mantle material on Mars up to 1.3 billion years ago.

Snell's law A law stating that the angle of refraction R of a light ray is related to the angle of incidence I by the equation: $\sin I/\sin R = \mu$, where μ is a constant called the refractive index of the material.

snout The down-valley margin of a glacier.

snow Solid precipitation composed of ice crystals or snowflakes. The former occurs when temperatures are well below freezing and the moisture content of the atmosphere is not high. As temperatures approach 0°C, snowflakes are found and these increasingly aggregate into larger flakes at about freezing point. Snow is very difficult to measure accurately. It tends to block rain gauges or be blown out. It is normally recorded by inserting a graduated ruler into a flat surface of undrifted snow. The water equivalent of 30 cm of snow is approximately equal to 25 mm of rainfall.

snowfield A large accumulation of snow in a mountainous region. If it is thick enough, downward pressure gradually converts it to ice, which may form glaciers.

snow line The altitude of the lower limit of permanent snow. In tropical areas the snow line is about 5000 to 7000 m, but gradually decreases poleward, until it reaches sea level in the Arctic Seas, Greenland, and Antarctica. In general terms, its altitude is determined by mean summer temperatures, but local climate conditions such as aspect or shelter from snow-bearing winds can produce variations.

snow patch erosion *See* nivation.

soapstone (steatite) A rock composed largely of TALC.

soda lake An inland area of water containing high concentrations of dissolved sodium salts, particularly the carbonate, chloride, and sulfate. The highly alkaline waters do not support any life.

sodalite A major member of the FELDSPATHOID group of minerals.

soil The natural accumulation of unconsolidated mineral particles (derived from weathered rocks) and organic matter (humus) that covers much of the Earth's surface and forms the supporting medium for plant growth.

The exact classification of soils is difficult because soil occurs as a graded continuum, breaks in this continuum being chosen subjectively for classification. In the past classification was based on the PROFILE, but in the SEVENTH APPROXIMATION classification the unit is the PEDON and rather than being based purely on factors of soil formation (as formerly) it is concerned with morphology, HORIZON arrangement, and other properties. A *soil phase* is the lowest unit in the hierarchy of soil classification, and it reflects the erosional state of the soil, e.g. its depth, stoniness, and degree of erosion. A *soil association* is a collection of soils grouped on the basis of geographic proximity even though they may differ greatly in profile characteristics (this grouping is usually used in the generalization of a detailed map in order to reduce it to a smaller scale). A *soil series* is a collection of soils similar in all respects except for the texture of the surface horizon and its erosional state. Series are the basic soil mapping units. A *soil type* is a subdivision of a soil series on the basis of the texture of the A horizon, these textural variations usually being related to slight changes of slope. A *soil group* is a group of soils sharing a similar type and sequence of horizons, occurring over wide areas with similar temperature and moisture regimes (differences in surface horizon due to agriculture are discounted). A *soil order* is the highest level of generalization in soil classification (the Seventh Approximation has ten orders) and is based on the type and development of horizons present.

Soil structure denotes the mode of binding together of individual soil particles into secondary units (*aggregates* or *peds*), mostly by organic matter. There are seven main structure types (see table) and the grade of structure can be classified into structureless, weak, moderate, or strong, depending on the observable degree of aggregation. Structure is important to soil fertility, because it influences POROSITY (and hence air and water supply), bulk density, heat transfer, etc. Cultivation tends to break down structures by removing organic matter, the major binding agent.

Soil texture denotes the proportion of the various PARTICLE SIZES in a soil. There are four main texture classes (see table) and many intermediate grades. Texture can be determined accurately by laboratory analysis and allocated to a specific class by a *texture triangle*, or more approximately in the field by rubbing a wet sample between the fingers and estimating the proportion of sand (gritty feel), silt (silky feel), and clay (plastic). Loams are generally the best agricultural soils because they contain a good mixture of all particle sizes.

Soil consistency denotes the degree of cohesion and adhesion in a soil, hence its ability to resist deformation or rupture. It is described by the following terms: wet, sticky, plastic, nonplastic, moist, loose, friable, firm, dry, and hard, with a number of intermediate grades.

Soil color as defined by the Munsell Color System is composed of three variables: hue, value, and chroma. Hue is de-

TABLE 1: SOIL STRUCTURE TYPES

Type	*Characteristics*
platy	thin leafy flaky layers
prismatic	prismlike units with flat tops; mainly subsoil
columnar	prismatic units with rounded tops
blocky	rectangular blocks; common in humid zones
subangular blocky	blocky with rounded edges
granular	rounded and porous; mainly surface soil
crumb	soft, rounded, and very porous; mainly surface soil

TABLE 2: SOIL TEXTURE CLASSES

Name	*Composition*
sand	> 85% sand percentage of silt plus 1.5 times the percentage of clay is not more than 15%
loam	7–27% clay 28–50% silt < 52% sand
silt	80% or more silt < 12% clay
clay	< 40% clay with < 45% sand and < 40% silt

termined by the wavelength of the light reflected from it, chroma is the purity or strength of the color, and value is the relative lightness or intensity of the color. Using a color code, a soil sample is allocated to a particular class that is composed of varying degrees of these three variables. color is an important factor in the recognition and description of soils, particularly as it often reflects formation, e.g. blue-gray colors typify gleying, red the concentration of ferric oxide, and gray or grayish brown often shows the accumulation of calcium carbonate.

Soil moisture is fundamental to plant growth and whether it becomes available to plants depends to a large extent on the size and distribution of the soil pores (*see* porosity) within which the water is held. In general the finer the texture of the soil the greater its capacity to store water. Water is held by the soil with varying degrees of tenacity; the less the moisture content the more firmly it is held.

Each soil horizon is determined by the original parent material and subsequent additions, losses, and transformations of materials within it. Especially important in the processes accomplishing horizon differentiation are those that transfer material from one horizon to another, because these differentiate the eluvial A horizon from the illuvial B. *See also* abnormal erosion; cheluviation; decalcification; eluviation; erosion; gleying; humification; illuviation; laterization; leaching; lessivage; podzolization; soil formation.

soil creep The slow movement of soil particles downhill. It occurs particularly in wet areas and is so slow that it cannot normally be seen. Nevertheless, in time large quantities of rock fragments and soil slip to the bottom of the slope.

soil erosion The removal of soil by the action of wind or water. It occurs most often when vegetation has been removed, usually by fire or human activity. *See* badlands; Dust Bowl; erosion.

soil formation (pedogenesis) The early Russians, particularly Dokuchayev, stated that soil forms as a result of five factors: local climate, PARENT MATERIAL, plant and animal organisms, relief and elevation, and the age of the land surface. More recently the American Jenny has tried to formulate the Russian work in a mathematical equation, s = f(cl, o, r, p, t ....), where s = soil, cl = climate, o = organisms (including man), r = relief, p = parent material, and t = time. However, all the factors are interrelated and the equation would be very difficult to solve in the field, but in certain cases it is possible to evaluate the effect of one of the factors against the background of the others where these are constant.

Climatic factor. This is thought to be the most important factor in soil formation. It is very complex because it can considerably influence the effect of the other factors. Parameters of fundamental importance are rainfall (regime and intensity), temperature (maximum and minimum, soil temperatures, and duration of sunshine), and evapotranspiration. The effects of these single components can be measured, e.g. the depth to which calcium carbonate is leached can be broadly correlated with the amount of rainfall. The combined influence of temperature and precipitation is well illustrated in Russia, where the sequence of soil zones changes from north to south in relation to an increase in temperature and a decrease in precipitation. This relationship between soils and climatic zones forms the basis of the zonal classification of soils.

Organic factor. Organisms are responsible for the breakdown of plant tissue resulting in the accumulation of organic matter, the recycling of nutrients, structural stability, and profile mixing in the soil. Vegetation itself, by binding the soil together, prevents erosion. Humans through their farming practices such as cropping, burning, and irrigation are playing an increasing role in soil formation. Organisms are often dependent on climate but they may act as independent variables. For example, certain major soil types are associated with particular vegetation types and when the vegetation type changes in the same climatic region a change in soil type may result.

Time factor. The time a soil takes to develop is measured from when fresh rock is first exposed at the surface, e.g. when the Pleistocene glaciation left the British Isles some 10 000 years ago. Soil develops gradually (the rate is partly dependent on parent rocks, e.g. soils on till will develop in less than 1000 years whereas those on resistant quartzite will take much longer), probably through an (A)C, AC, A(B)C, ABC horizon sequence, and eventually a stage is reached in which the soil is in equilibrium with the environment and can be called mature. However, because the system is dynamic and any environmental change will affect the soil, it is better to use absolute or relative datings than terms such as young, mature, and old.

Topographic factor. Topography affects soil moisture. Drier shallower soils occur at the top of a slope, where drainage is more rapid. Soils at the slope foot tend to be deeper because of an increase in subsurface weathering due to the presence of more moisture and the accumulation of material eroded upslope. As a result of aspect, soils on south-facing slopes (in the N hemisphere) tend to be drier than those facing north. Because vegetation type is closely related to soil moisture, through vegetation topography also has an indirect influence on soil development. *See also* catena.

soil horizon *See* horizon (def. 1).

soil profile *See* profile.

soil temperature The temperature recorded at any level within the soil (standard depths being 30 cm and 100 cm) with a soil thermometer, which consists of a mercury thermometer that has a horizontal section as usual, but is then bent through 90° to a predetermined length where the bulb is located. The diurnal temperature wave decreases very rapidly with depth and below 10 m there is usually little sign of an annual wave.

soil texture The arrangement and sizes of the particles in SOIL. Soils based on clay, for example, have a fine texture but relatively poor drainage, whereas soils containing sand have a coarse texture and good drainage. A mixture of the two types is called LOAM.

solar constant The intensity of RADIATION received from the Sun on a unit area of a horizontal surface at right angles to the solar beam above the atmosphere, at the Earth's mean distance from the Sun. Its value is 1.35 kW m^{-2}. There has been considerable debate as to whether this value is a constant or not. Before the use of satellites it had to be measured from high-altitude observatories where atmospheric interference could be reduced but not eliminated. It now appears that the value is almost constant except in the shortest part of the spectrum where the amount of energy received is small.

sole mark A sedimentary structure formed at the base of sandstone and siltstone beds that overlie softer sediments.

solfatara A FUMAROLE that emits mainly sulfurous gases in addition to steam.

solid geology The geology of rocks underlying glacial and other superficial deposits. *See also* drift.

solifluction A process of mass movement on slopes, usually restricted in its application to areas of periglaciation (*see* periglacial). The major process of solifluction concerns the presence of PERMAFROST. In spring, melting takes place in the top layer of the soil, but the water produced cannot drain away because beneath this active layer the ground remains frozen and hence impermeable. Consequently, the upper soil becomes highly saturated, and its cohesion may be reduced to such an extent that flowage can take place and material be moved downslope, often creating characteristic terrace forms and burying former vegetation layers, although the presence of well-developed vegetation usually hinders the process.

Solifluction sometimes also embraces the effects of NEEDLE ICE and CONGELITURBATION.

solitary wave A water wave consisting of only one elevation which, once formed, may travel practically unchanged as a single hump for a considerable distance. The actual water particles move forward a short distance as the wave passes by, i.e. the passage of a solitary wave causes a net shift of the water particles beneath the solitary wave in the direction of wave propagation. The solitary wave is neither preceded nor followed by other waves, unless these are generated quite independently. TSUNAMI waves, which may be of very limited height at sea, possess some of the characteristics of solitary waves.

sol lessivé A soil that corresponds to the gray-brown podzolic soil of North America. The profile is characterized by a thin surface accumulation of moder humus. Below this the A and Eb horizons have been depleted of clay, which has been eluviated down the profile to form a textural Bt horizon (a process not present in a true brown earth). This is clear from a study of the structure of the soil as the Bt horizon has a prismatic or blocky structure whereas the surface horizon often has a weak crumb/blocky structure. These soils are common in the humid climate of NW Europe on freely drained fine-textured parent materials. They fall into the alfisol order of the SEVENTH APPROXIMATION.

solod (salt-earth podzol) A soil resulting from the leaching of a SOLONETZ soil, forming a white-gray A horizon from the upper B horizon of the solonetz. This bleached A horizon, as in true PODZOLS, is deficient in iron and aluminum, has a sandy texture and an acid reaction. The B horizon is gleyed, with mottlings and a columnar and prismatic structure developed in a compact clay. These soils represent the leached end of the development continuum from SOLONCHAK through SOLONETZ to solod in which salts are progressively removed through leaching, either naturally, sometimes owing to a falling base level, or artificially through leaching by irrigation water.

solonchak (white alkali soil) Saline soil in which sodium chloride and sodium sulfate form more than 0.3% of the total soil. During drought the salts dry out at the surface giving a white color. Saline groundwater exists within two meters of the surface, and this supplies the salts for upward rise by CAPILLARITY to the surface soil; it also imparts gley features to the soil, usually just the lower part of the profile but possibly the whole profile if the water table is near the surface. These soils are characterized by a pH value of more than 8 (more than 15% of exchangeable cations are sodium), a flocculated structure, no illuvial B horizon, and one or two salt layers; they are friable and soft in a moist state, crusty when dry. They are fairly common in arid, semiarid, and subhumid regions, where they may occur naturally in depressions or as a result of secondary salinization of the zonal soils, or they may be artificial due to excessive irrigation or irrigation with water dominated by sodium salts. In the Seventh Approximation they are classified in the ARIDISOL order.

solonetz (black alkali soil) Soil dominated by sodium carbonate. Solonetz develops from SOLONCHAK by the leaching of some of the sodium, due to increased rainfall, improved drainage, or irrigation, which deflocculates the soil mass. As a result the structure is single-grain when wet, but on drying the illuvial horizon develops a massive columnar structure. It is this structure difference that originally divided the solonetz from the solonchak. Whereas the solonchak has an AC profile, increased leaching in the solonetz produces a dark gray or brown B_1 horizon enriched with colloids, and lighter gray B_2. The C horizon is enriched with carbonates and salts, while at the surface alkaline solutions of humic material accumulate in wet periods which on drying produce black crusts. They may be partly zonal soils, being concentrated at climatic transition zones at 15° and 40°N, but traditionally they are considered as intrazonal. In the SEVENTH APPROXIMATION they are classified partly as aridisols, partly as mollisols.

solstice The time of maximum or minimum declination of the Sun. The summer solstice in the N hemisphere occurs on or about June 22 when the Sun is overhead at the Tropic of Cancer, its farthest north. The winter solstice is about December 22 when the sun reaches 23.5°S, the Tropic of Capricorn. At the summer solstice is the year's longest day and at the winter solstice, the shortest.

solum The surface soil and the subsoil, excluding the parent material.

solution (in geology) A basic form of chemical weathering in which solid materials are dissolved by water. Rates of solution vary considerably from one substance to another; silica is fairly insoluble and its removal rate increases with temperature, whereas the most soluble rock minerals are calcium, sodium, magnesium, and potassium. Halite and gypsum are very soluble, subsequent recrystallization sometimes causing mechanical breakdown of rocks (*see* granular disintegration; flaking). The efficiency of the solution process depends largely on the volume of water flowing, but the presence of weak acids within the water may cause increased solution. *See also* limestone solution.

solution breccia A BRECCIA in which the original rock has been fragmented by the removal of soluble material.

sonic log A subsurface logging technique run in a borehole, which measures the time taken for a compressional sound wave to travel through 30 cm of the adjacent formation. In this way a continuous record of the properties of the rocks against depth is recorded.

sonoprobe A continuously recording acoustic reflection instrument, which has made an important contribution to the study of the continental terrace and some deeper areas. Its use provides data on the approximate thickness, distribution, and character of marine sediments and the surface morphology and internal structure of exposed or buried rock areas. Various sound sources are employed to give varying degrees of penetration of the seabed. There are many applications of the sonoprobe technique to offshore mineral exploration and engineering surveys, for example in the laying of submarine pipes and cables and in the siting of oil rigs.

sorting The process by which materials forming a particular sediment are graded according to size by natural processes. If the size range is confined within narrow limits, so that most of the material is silt size (as in aeolian brickearth) or fine sand size (as in many beach sands), the deposit is well sorted. If the deposit has many size ranges (as in a slump deposit, for example) it is poorly sorted.

Sorting is best in sediments subject to a long distance of transport, with repeated reworking, as happens to beach sands being washed along a coast by wave action. In slump deposits, transport is sharp and short with no time for any sorting.

The basis of sorting concerns the settling velocity of a particle; each size and density of grain has a minimum level of energy necessary to overcome its weight and keep it in suspension. If the energy environment falls below that critical level, the grain settles out. Hence on a coast, where wave energy shows a gradient from high on headlands to low in bays, there is a corresponding sorting of sediment, the coarsest fractions in the high-energy areas, the finer fractions in the low-energy areas.

soufrière A volcano that emits sulfurous gases; a large SOLFATARA.

sound **1.** A waterway somewhat wider than a STRAIT linking two bodies of water, or a passage between an island and the mainland.
2. A relatively long sea arm or inlet.

sounding A measure of the depth of water. Before the advent of modern echo-sounding techniques, a sounding line with a lead weight on the end was used, but this often gave very inaccurate readings because of the effect on the sounding line of wave and current action. Almost all soundings are now obtained using echo sounders. These devices are usually set to record at speeds of some 1600 meters a second. Modern sounding techniques largely stem from the use of acoustic equipment in submarine warfare during World War II.

South Equatorial Current *See* equatorial current.

spar A general term for any lustrous translucent or transparent mineral that cleaves easily. Examples include Iceland spar and various other forms of CALCITE.

sparite Coarse CALCITE of a grain size in excess of 0.01 mm in diameter, which may infill cavities within and around ALLOCHEMS in LIMESTONE. The terminology used in the petrographic description and classification of limestones is based on the presence of sparite, MICRITE and allochems.

spatter cone A volcanic structure consisting of a steep-sided mound of lava that has issued from a central vent or along a fissure. The lava takes the form of lumps and cinders, adhering together, that have come from sprays of frothing molten magma.

species The basic group in the taxonomic classification of organisms, which is usually defined as a collection of similar individuals that are capable of interbreeding to produce fertile offspring. In classifying fossil species this criterion is impossible to satisfy, and these species are usually defined on the basis of structural similarities (which in practice are also used in classifying modern species). This type of species, defined solely on morphological characteristics, is known as a *morphological species*. Difficulties in defining fossil species arise because of the absence of soft parts, whose structural characteristics provide an important means of classifying modern species. In addition the boundaries between the successive fossil species of a continuously evolving line must be arbitrary, because the species tend to grade into each other. Some fossil species do not show obvious affinities with any organisms alive today. Examples of such species are those that produced CONODONTS: different skeletal elements belonging to this group are given different specific names, although in life they may have formed part of a single individual.

A species is defined with reference to a TYPE SPECIMEN, which is selected as a standard and formally described. The name of a species is always italicized and preceded by the name of the GENUS to which it belongs. For example, the species that includes modern man is *Homo sapiens*, or *H. sapiens*. A species name is often followed by the name (in abbreviated form) of the person who first formally described the species.

specular iron ore *See* hematite.

speleology The scientific study of caves and caverns, including their exploration, geology, and mineralogy.

speleothem A mineral formation in a cave that results from the action of water. Most consist of CALCITE (calcium carbonate), and the most typical are STALACTITES AND STALAGMITES.

spessartite (spessartine) An orange to deep red member of the GARNET group of minerals, $Mn_3Al_2Si_3O_{12}$. It occurs in igneous and thermally metamorphosed rocks. It is used as a semiprecious gemstone.

sphalerite (zinc blende) A yellow to red-brown lustrous mineral form of zinc sulfide, ZnS, sometimes containing some cadmium and iron. It crystallizes in the cubic system, and occurs in hydrothermal veins and metasomatic replacements. It is the chief source of zinc.

sphene (titanite) A monoclinic mineral of composition $CaTiSiO_5$ but with some replacement of O by OH and F. Sphene is a common ACCESSORY MINERAL in igneous rocks. It is also found in metamorphosed calc-silicate rocks.

sphenochasm A triangular block of oceanic crust between two continental areas.

spheroid A figure resembling a sphere, used as a reference surface for geodetic surveys. Depending on the area of the world in which the survey is being carried out, different spheroids are used as a reference datum, e.g. in the British Isles, Airey's spheroid is used, whereas in much of Africa, Clark's spheroid (1880) is used.

spheroidal weathering A form of EXFOLIATION that takes place on boulders below ground level. Initially angular joint blocks are gradually rounded owing to preferential weathering on the corners, a resultant block consisting of a solid rounded core surrounded by spheroidal shells. The term is sometimes restricted to exfoliation involving no change in volume (any weathering products are removed in groundwater), in contrast to FLAKING. The layering within spheroidal blocks is due to the chemical migration of elements within the rock, creating zones of enrichment and depletion. Any totally rounded blocks found above ground level have probably been exhumed (*see* exhumation).

spherulitic Describing a texture commonly found in glassy or cryptocrystalline acid rocks in which acicular crystals of quartz and alkali feldspar are radially arranged to form a sphere. Spherulitic growth is the result of DEVITRIFICATION of glassy material or rapid crystallization of viscous magma. *See* variolitic.

spilite Any of several basaltic lavas containing albitic plagioclase and in which the primary mafic minerals have been altered to chlorite and epidote. Altered acid rocks associated with spilites are called *keratophyres* and may contain phenocrysts of albite and hornblende in a groundmass of albite, chlorite, epidote, and quartz. Spilitic rocks occur as PILLOW LAVAS interbedded with marine sediments. They appear to be basalts and rhyolites of the ocean floor, which have suffered postmagmatic soda metasomatism due to interaction with sea water. *See also* ophiolite; Steinmann trinity.

spinel One of a group of oxide minerals with cubic symmetry, often occurring as octahedral crystals. The general formula is $R^{2+}R_2^{3+}O_4$ where R^{2+} = Mg, Fe, Zn, Mn, or Ni and R^{3+} = Al, Fe, or Cr. Most spinels fall into one of three series dependent upon the nature of the trivalent cation: *spinel* (Al), *magnetite* (Fe^{3+}), and *chromite* (Cr). *Ulvöspinel* also has a spinel structure.

There is a continuous chemical series between *hercynite* ($Fe^{2+}Al_2O_4$) and true spinel ($MgAl_2O_4$) because of the replacement $Fe^{2+}\leftrightarrow Mg$. Minerals of the spinel series are a wide variety of colors including red, brown, blue, green, and black. Magnetite ($Fe^{2+}Fe_2{}^{3+}O_4$) is black and forms a continuous series with ulvöspinel ($Fe_2{}^{2+}$-TiO_4). Chromite is $Fe^{2+}Cr_2O_4$ but with some replacement $Mg\leftrightarrow Fe^{2+}$. There is a complete chemical series between chromite and hercynite due to the replacement $Cr\leftrightarrow Al$. The brown variety, *picotite*, is a chromium-rich hercynite.

Spinels occur in high-temperature metamorphosed rocks, especially limestones. Hercynite is found in metamorphosed pelites and in basic and ultrabasic rocks. Magnetite is ubiquitous in metamorphic and igneous rocks and is an important iron ore. Chromite is the chief chromium ore and, like magnetite, often occurs in monomineralic bands in layered basic igneous bodies.

SP interval (in seismology) The time that elapses between the arrivals of the first secondary (S) and primary (P) waves following an earthquake. It provides a means of calculating the distance to the earthquake's focus.

spirit level structure *See* geopetal cavity.

spit An elongated accumulation of sand or shingle (or both) attached to the coast at one end and extending out to sea. Spits occur most frequently where the coastline changes direction abruptly. The longshore movement of beach material is an essential factor in the formation of spits. The material is moved along one straight section of the coastline and, rather than rounding a sharp headland, accumulates seaward in the original direction of movement. Spits sometimes form across estuary mouths, or two may approach each other from two facing headlands (a *double spit*). A common feature is a seaward end that curves toward the land. This may be produced by wave refraction or two different sets of waves may be in action, one causing accumulation and the other the rounding of the end.

spodosol One of the ten soil orders of the SEVENTH APPROXIMATION classification including ashen or podzolized soils, characterized by a spodic B horizon containing humus and silicate clays with aluminum or iron that has been moved down the profile from the bleached A horizon above. This includes the true PODZOLS, and the PODZOLIC SOILS and groundwater podzols. Spodosols develop mostly in humid cool temperate climates or on acidic parent materials.

spodumene A lithium-bearing clinopyroxene. *See* pyroxene.

sponges *See* Porifera.

spontaneous potential log (SP log) A subsurface logging technique that continuously records the spontaneous potential present at different depths. This results from electrochemical reactions between the drilling mud and formation fluids, and small electromotive forces resulting from the penetration of drilling fluids into the formation.

spot height A height above the vertical datum, represented on a map by a dot locator with the elevation printed beside the dot. Unlike bench marks or trigonometrical points, spot heights are not identifiable on the ground.

spreading rate The rate at which new ocean floor is added to the oceanic plates at constructive PLATE BOUNDARIES. It is therefore directly related to the rate at which new ocean floor is produced.

spring A place where water flows out of the ground because the water table intersects the surface. Thus the water flows out from above an impervious rock, such as clay, shale, or slate.

spring equinox *See* equinox.

spring line The line where a water table intersects the surface of the ground, between layers of pervious and impervious rocks, and there is a spring or springs. *See* spring.

spring tide A TIDE that has a relatively large range, rises and falls to the greatest extent from mean tide level, and occurs at or near the times of full and new moon. A period of spring tides allows waves to break farther up or down the beach, and at those times when strong onshore winds coincide with the time of high water, this may lead to flooding of the hand behind the beach. A wider tract of beach is intermittently uncovered compared with a period of NEAP TIDES. Also, tidal currents tend to be stronger during spring tides than during neap tides.

spur An area of high ground that projects into lower ground. A spur in the upper parts of a river valley may form INTERLOCKING SPURS, but farther down the valley they are likely to be eroded.

squall A sudden increase of wind speed lasting for several minutes before dying away quickly. To be classed as a squall, the wind should rise by at least 8 m s^{-1} (16 knots) and reach at least 11 m s^{-1} (22 knots). A squall is often, although not necessarily, accompanied by heavy rain or a thunderstorm.

stability The state of the atmosphere where vertical air movement is limited owing to the slow rate of cooling of the environmental lapse rate. If the environmental lapse rate is less than the saturated adiabatic lapse rate then the state is known as *absolute stability*, because no rising thermal can cool at a slower rate and so will quickly become colder than the environment and sink back. These conditions are typical of summer anticyclones.

stabilized dune A coastal dune in which the cutting off of direct wind action by the growth of another dune line in front, and the development of a virtually continuous vegetation cover, has ended sand movement by the wind. These dunes generally have a very subdued topography, unlike a MOBILE DUNE, and a species-rich vegetation.

stable zone A part of the Earth's crust that is not undergoing mountain building (orogeny) or deformation through crustal movements. Such zones are generally near the centers of continents, away from plate boundaries.

stack A small pillarlike island just offshore, generally at the end of a headland. It has been detached from the headland by the erosive action of waves, which may have first formed a cave that was gradually enlarged. *See* natural arch; natural bridge.

stage **1.** The point to which a landscape has evolved in a CYCLE OF EROSION, which determines its appearance. Landscapes used to be labeled youthful, mature, or senile, but this labeling is now being dropped because most landscapes are POLYCYCLIC, so one cannot attribute their appearance purely to the stage of evolution in the current cycle, and because evidence is increasingly being produced that landforms do not adjust steadily with time through equally long stages: they evolve very rapidly at first and then achieve a steady state of little further change.
2. A division of rock in the Standard Stratomeric scheme of stratigraphic classification (*see* chronostratigraphy). It indicates the body of rock that has formed during one AGE. A number of stages grouped together constitute a SERIES and the stages themselves are formed of several CHRONOZONES. Stages are often spoken of in exclusively biostratigraphic terms but they may be calibrated and defined by methods not involving fossils. They are usually named after geographical localities and have the ending *-ian*, as in the *Oxfordian*, a stage of the Jurassic System.

stagnant ice An ice mass that no longer receives an adequate supply of ice in the accumulation zone to maintain movement. The ice melts downward from the surface, meltwater frequently forming lakes marginal to the ice, which may subsequently be evidenced by deltaically-bedded terrace deposits. Debris-covered ice will melt more slowly than that exposed, and in this way large masses can be isolated, resulting in KETTLE HOLE formation at a later stage. The large amounts of meltwater present are responsible for the characteristic ice-stagnation features, namely KAMES and ESKERS.

stalactites and stalagmites Stalactites are tapering pendants projecting from the rocks of limestone caverns. They are formed from drips of water containing dissolved calcium carbonate, which on evaporation deposit a small trace of solid material. Stalactites take many thousands of years to form and are usually associated with stalagmites, which develop on the cave floor in a similar way. Stalagmites tend to be much broader at their bases than stalactites, and in time they extend upward. Eventually stalagmite and stalactite may join to form a complete pillar.

standard atmosphere The idealized atmospheric structure in terms of temperature and pressure for all heights. It is used for the calibration of altimeters, etc. Surface characteristics are taken as 15°C and 1013.25 mb for temperature and pressure with a lapse rate of 6.5°C/km up to 11 km at the tropopause.

standard parallel The line of latitude selected in the construction of a map projection that is projected at its true length.

standing wave (stationary wave) A type of water wave in which there is a surface oscillation that does not travel along the sea's surface. The surface oscillates up and down between fixed points (*nodes*). A crest at one moment becomes a trough at the next, and so on. Water particles have maximum horizontal travel at the nodes and maximum vertical travel at the *antinodes* (or loops). The height of a standing wave is approximately twice that of the initial incident waves. Such a situation is best observed when waves approach a vertical barrier such as a cliff or seawall and are reflected so as to meet the incoming waves. The depth of water must be sufficient to prevent the incoming waves from breaking. Standing waves may also be produced at sea, for instance when similar PROGRESSIVE WAVE trains meet from opposite directions.

staurolite A mineral with an approximate composition $Fe^{2+}Al_4Si_2O_{10}(OH)_2$. It shows orthorhombic symmetry and is brown to yellow. It has a structure of alternating layers of kyanite and a composition $Fe^{2+}(OH)_2$ but with some replacement $Mg \leftrightarrow Fe^{2+}$. Staurolite is found in medium-grade regionally metamorphosed argillaceous rocks, often in association with kyanite and almandine.

steatite (soapstone) A rock composed largely of TALC.

steering The control of the direction and speed of movement of pressure systems by some atmospheric factor. Thermal steering, or movement in the direction of the thermal wind, works quite well for nonintensifying depressions, but in developing lows and anticyclones it is less reliable.

Steinmann trinity The common association within geosynclinal sediments of SPILITES, CHERT, and SERPENTINITES (and other ultramafic rocks). *See also* ophiolite.

steppe An area of flat open temperate grassland in an arid region that will not support trees, particularly that of southeastern Europe and Asia. If the land is plowed, it provides fertile soil for growing cereal crops. In this respect, it is similar to PRAIRIE.

Stevenson screen The standard housing for meteorologial instruments. It consists of a wooden box on legs, the base of the box being 1 m above the ground, in which thermometers are placed. Ventilation is provided by louvered sides, which in normal conditions allow adequate air movement but prevent solar radiation reaching the thermometer bulbs. Access is provided by (in the N hemisphere) the north-facing side of the screen being hinged; this again prevents sunlight affecting the temperature records when records are being taken.

stibnite A lead-gray lustrous mineral form of antimony sulfide, Sb_2S_3. Crystallizing in the orthorhombic system, it occurs in hydrothermal veins and replacement deposits, often with lead, mercury, and silver. It is the principal source of antimony.

stilbite A white, gray, or red-brown mineral form of a hydrated aluminosilicate of calcium, sodium, and potassium, $(Ca,Na_2K_2)(Al_2Si_7O_{18}).7H_2O$. It crystallizes in the monoclinic system as sheaflike aggregates in hydrothermal veins in igneous rocks. It is a member of the ZEOLITE group of minerals.

stishovite A very dense product of QUARTZ, SiO_2, produced at extremely high pressures, possibly following a meteorite impact. *See also* silica minerals.

stock A discordant igneous intrusion, usually of coarse-grained granitic rock, resembling a batholith but having a surface exposure of less than 50 sq km. *Compare* boss.

stone polygons or circles A form of periglacial PATTERNED GROUND found in flat areas. The polygons or circles consist of fine-grained material, which is usually saturated with water and frequently domed upward, surrounded by slightly raised walls of coarser stones, which project into the soil for several centimeters. The diameter of these features, which may occur either in large groups or in total isolation, varies considerably: the larger forms are several meters across and are found in areas of long-continued freezing and extensive summer thaw, whereas smaller examples reflect more frequent freeze-thaw cycles. Their method of formation is still not totally clear, but is probably concerned with pressures set up by repeated freezing and thawing, which move the coarser fragments both upward and outward.

stone stripes A type of periglacial PATTERNED GROUND that forms in a similar way to STONE POLYGONS OR CIRCLES when the processes responsible are acting under the influence of gravity on a slope. The result is a series of alternating coarse and fine parallel downslope stripes of stone. The coarse lines can vary from a few centimeters to about two meters wide, while the finer stripes between them are often at least twice this width. Stone polygons developed on flat ground may become elongated toward a slope and then grade into stripes as the slope angle increases.

stony iron *See* meteorite.

stony meteorite *See* meteorite.

storm A period of strong winds, often accompanied by heavy rain. Major storms are called CYCLONES, HURRICANES, or TYPHOONS, depending on where they occur.

storm beach An accumulation of coarse beach material found at the very highest levels of a beach profile, far above the highest points reached by high spring tides. These boulders and shingle blocks attain such a height as a result of the extreme power of very infrequent storm waves. Although their overall effect on a beach is destructive, the SWASH of these waves can throw some material to great heights and rapid percolation of water prevents its subsequent removal by the BACKWASH.

storm surge (storm tide) The rapid rising of water level during a storm as a result of wind stresses acting on the sea's surface, or a PROGRESSIVE WAVE resulting from certain meteorological circumstances. If this situation occurs at the time of high SPRING TIDES, the tidal levels may significantly exceed predicted values and serious flooding and damage may result. This occurred with disastrous results at Galveston, Texas, in 1850.

stoss The side of a hill that faces the direction from which a glacier is coming. It is exposed to the abrasive action of the moving glacier, and as a result usually has gentle slopes and rounded features.

strain The deformation of a body of rock as a result of stress. It may be a change in shape, volume, or both. *See* heterogeneous strain; homogeneous strain.

strain hardening The increasing strength of a rock as it deforms plastically.

strain rate The rate at which a material deforms. This is usually expressed as a percentage by which the original length is altered per second.

strain-slip cleavage (crenulation cleavage) A type of cleavage superimposed upon slaty cleavage. It is typified by tabular bodies of rock between regularly spaced cleavage planes.

strait A comparatively narrow waterway that links two sea areas or other large bodies of water. Some authorities, however, restrict the term to gaps in isthmian links or island chains.

stratification The condition shown by sedimentary rocks of being disposed in horizontal layers or BEDS, known as *strata.* The term is also occasionally applied to igneous bodies that display analogous parallel textures.

stratigraphic succession The series of the Earth's rock strata arranged in sequence from the earliest to the latest in age. *See also* geologic timescale.

stratigraphy The branch of geology concerned with the description and classification of bodies of rock and their CORRELATION with one another. Various aspects of this are dealt with in LITHOSTRATIGRAPHY, BIOSTRATIGRAPHY, and stratomeric and chronomeric CHRONOSTRATIGRAPHY. This separation of the various concepts involved in stratigraphy is necessary if different levels of inference are to be distinguished.

stratocumulus cloud A layer cloud within which weak convection is taking place. It often forms extensive sheets, which have dark areas separated either by small areas of clear sky or by thinner cloud. It occurs most frequently in winter anticyclones or over cool ocean areas where anticyclones are dominant.

stratomere (in stratigraphy) Any segment of a sequence of rock. Stratomeres are not of standard uniform size nor need they have been formed during equal intervals of geologic time.

stratopause The boundary between the stratosphere and mesosphere. It is found at a height of about 50 km and represents the temperature maximum of about 0°C resulting from energy release in the ozone layer.

stratosphere That layer of the atmosphere above the troposphere extending to a height of about 50 km. The lower stratosphere is almost isothermal but temperatures increase gradually to their secondary maximum of about 0°C at the stratopause. Winds in the stratosphere can be quite strong, especially the POLAR NIGHT JET STREAM.

stratovolcano *See* composite volcano.

stratum *See* bed.

stratus A layer cloud with a fairly uniform base found at lower levels in the atmosphere or actually at the ground surface in the form of hill fog. It may give drizzle or Scotch mist.

streak A colored mark left when a mineral is drawn across a piece of unglazed porcelain. It is characteristic of the mineral and commonly used as an aid to identification.

stream A body of water on the Earth's surface flowing within a natural channel. The term is applied to all such flows, ranging from the smallest creek to large rivers.

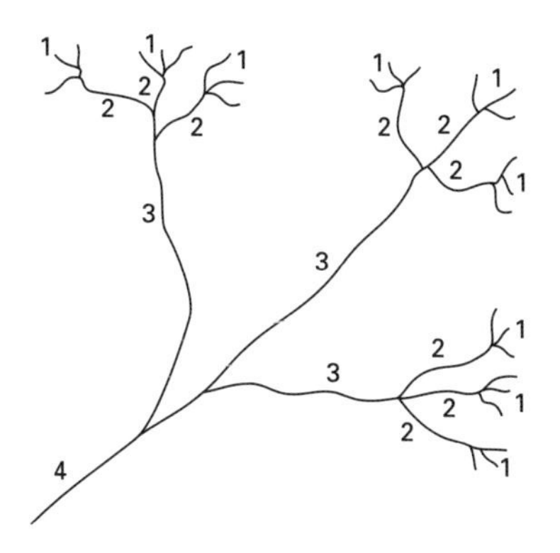

Stream order in a hypothetical drainage pattern

stream frequency *See* drainage density.

streamline A line that is parallel to the instantaneous direction of the wind field at all points along it, showing overall air movement. The term is used particularly in tropical areas where isobars are of little use in indicating the wind field because of the breakdown of the geostrophic relationship.

stream load The material carried along by a stream. *See* load.

stream order The classification of a stream in an integrated drainage pattern that is broken down into stream segments and the segments then ranked in a hierarchy according to their allotted order.

At the head of the basin, the first small tributaries are first order; two or more first-order tributaries unite to give a second-order tributary; two or more second-order tributaries unite to produce a third-order tributary, etc., up to the major stream of the basin, whose order defines the order of the basin. Quantitative comparisons can then be made between basins, and studies made of the relationships between orders. Studies of this kind have produced apparent laws of drainage basin development. *See also* bifurcation ratio.

stress An applied force per unit area, causing rocks to undergo STRAIN. Stress can take the form of tension, compression, or shear.

striation A line that has been cut or scratched into a rock surface over which a glacier has passed. Such lines are usually short, but sometimes exceed one meter, and are best seen on gently sloping faces of fine-grained hard rock, up and over which the ice moved. Ice itself cannot cause such abrasions; the projections of rock debris, held rigidly in the basal ice, act as the cutting tools. Each striation can indicate two diametrically opposed directions of ice movement, and they have often been used to discover movements of former ice masses.

strike The direction along a rock stratum at right angles to the true dip.

strike line (structure contour) A line joining points of equal elevation above or below a selected stratigraphic datum.

stromatolite A laminated concentric structure formed of calcium carbonate and produced by blue-green bacteria (cyanobacteria). Fossilized stromatolites dating back to Precambrian (Archean) times have been found.

stromatoporoid An extinct organism that built laminated reef structures in the Paleozoic. They were particularly common during the Silurian Period. Their taxonomic affinities are obscure but they may belong to the Hydrozoa, a class of the phylum CNIDARIA. They and the tabulate corals (*see* Anthozoa), which were also important reef-builders in this era, seem to have preferred different conditions, for they are rarely found in association.

strontianite A colorless, white, yellow, or green mineral form of strontium carbonate, $SrCO_3$. It crystallizes in the orthorhombic system and occurs in nodules and veins in limestone rocks. It is used as a source of strontium and its compounds.

structural geology The branch of geology that deals with the structural features of rocks, their analysis, and their description, including the forces that create the features.

structural high A geologic structure of positive relief, such as a dome or anticline.

structural low A geologic structure of negative relief, such as a syncline or basin.

structure contour *See* strike line.

subaerial erosion EROSION that takes place at the surface of the Earth.

subcrop The disposition of a bed of rock beneath the surface and its contact with the undersurface of younger beds. The term is used by the petroleum industry in the reconstruction of the paleogeography of an area.

subcrustal convection Gross movement of semifluid magma in the Earth's mantle resulting from convection, which in turn causes the movement of the crustal plates. *See* plate tectonics.

subduction zone (destructive plate boundary) An area on the Earth's surface where ocean floor is destroyed by one lithospheric plate overriding another. The overriding plate may be of oceanic or continental material. The oceanic plate that is overridden is pushed down into the mantle, its position being marked by a series of earthquake foci. Such zones are marked by oceanic trenches.

subglacial Describing the region underneath a glacier or ice sheet.

subhedral Describing crystals with partially developed crystal forms. *Compare* anhedral; euhedral.

sublimation The process by which ice is converted to vapor and water vapor into ice with no intervening liquid state. Chemically, the term is used only for the first process.

submarine bar An offshore accumulation of sand, more or less parallel with the coastline, that never becomes exposed by the tide. Submarine bars occur in areas in which the tidal range is extremely small and usually form in groups of two or three. They are formed at the point at which steep storm waves break. Seaward of this point material is moved toward the land, then after the wave has broken movement is seaward. Thus sand accumulates from two directions, although an optimum size is reached eventually, beyond which no further increase in height is possible, owing to the effect of the bar on waves passing it. The outermost bar, in deepest water, is infrequently fully developed and is formed by the rare strong storm waves; it is unaffected by the smaller storm waves responsible for the bars nearer the shore. The innermost bar tends to move its position as wave heights vary; movement is seaward with increasing wave height and landward with decreasing height.

submarine canyon A deep steep-sided valley or trench cut into the continental shelf or continental or insular slope. Many are deeply incised into the solid rock of the sea floor, with V-shaped cross profiles, and some form deep winding gorges. Well-known submarine canyons include those off the mouth of the Congo River, the Hudson River, and Cape Breton in S France. Canyons may have tributary incisions and and these sometimes display a dendritic pattern; others occur singly or as part of a system of subparallel incisions. These features are thought to have originated through erosion by submarine TURBIDITY CURRENTS. *See also* sea valley.

submarine eruption A volcanic eruption from a vent in the oceanic crust on the sea floor. Most of these eruptions take place on SEAMOUNTS or along MID-OCEAN RIDGES.

submarine fan A fan-shaped mass of sediment normally found at the lower end of a SUBMARINE CANYON. Much of this material may have been deposited under the action of submarine slides and turbidity flows. For example, a very large fan exists at the lower end of the Congo Canyon, where slides or turbidity flows occur about 50 times annually. The Mississippi Delta is flanked on its western side by a trough valley that leads, at a depth of some 1800 m, into a large submarine fan. In both cases, very substantial amounts of muddy sediments are carried to the coast by the rivers.

submarine valley *See* sea valley.

submergence The covering of land by the sea, either because the land has sunk or sea level has risen. *See also* cycle of erosion; fiord; ria.

subsequent stream A stream that has developed on a weak substructure, such as a clay vale, or in regional joint or fault patterns, at right angles to streams consequent to the dip of the slope. Subsequent streams owe their development to the accelerated rate of headward erosion they are able to perform in these weaker areas, rapidly extending themselves and often coming to integrate and dominate the drainage pattern. *See also* consequent stream; obsequent stream.

subsidence The widespread downward movement of air associated with surface DIVERGENCE. The rate of descent is usually fairly slow, of the order of a few meters per hour, but this results in adiabatic warming and a low relative humidity. Near the ground, mixing takes place so that extreme values of relative humidity are unusual. The weather associated with large-scale subsidence is always dry although cloud layers may result from mixing with moister air near the ground.

subsoil *See* profile.

subsolidus A mixture of compounds, such as minerals, below their melting points in which chemical reactions may still take place (in the solid state).

subsurface eluviation The process by which water falling on the soil and sinking in eventually moves downslope as a subsurface flow; as it does so it dissolves or transports soil and rock material with it. This process has tended in the past to be overlooked in studies of slope evolution. However, evidence has shown that in places it can be very effective.

subtropical high One of the anticyclones in the belt of semipermanent high pressure in the subtropics.

succession The vertical sequence of rock units in a particular place. A number of local successions may be combined to produce an idealized succession demonstrating the sequential order of strata over a larger region.

suevite A type of BRECCIA consisting of angular fragments of rock in a glass matrix.

sugar loaf An INSELBERG, especially in the S coastal area of Brazil.

sulfur A native element (S) of characteristic yellow color found around volcanic vents and hot springs.

sulfur bacteria Bacteria that release elemental sulfur as a result of their metabolism. They may, for example, oxidize hydrogen sulfide (to obtain energy) or release sulfur through a type of PHOTOSYNTHESIS. *See also* iron bacteria.

sulfur oxidation A process in soil science analogous to nitrification, in this case the bacterium *Thiobacillus* converting inorganic sulfur compounds into the sulfate form. This may occur solely by chemical means but most appears to be biochemical in nature.

summer solstice *See* solstice.

sun crack *See* desiccation crack.

sunshine Direct radiation received from the Sun. It is one of the climatological elements recorded at most observing stations.

sunshine recorder Sunshine is normally measured by the Campbell-Stokes recorder, which consists of a glass sphere that focuses the Sun's rays onto a card, burning a trace on the card whenever bright sunshine occurs. The length of the trace indicates sunshine duration. Allowance is made for the seasonal variation in declination of the Sun. As it records only bright sunshine many hours are not recorded, especially in early morning or late evening, or in conditions of severe haze. No distinction is made between the different heating capacities of winter and summer sunshine.

sunstone A translucent type of oligoclase (*see* feldspar) that appears to glow with a reddish light, caused by tiny parallel inclusions of HEMATITE. *See also* moonstone.

supercooling The cooling of a liquid to a temperature below its normal freezing point. This is a very common phenomenon in the atmosphere because ice does not form in cloud droplets at 0°C because of the sparsity of suitable CONDENSATION nuclei. Supercooling is essential to the BERGERON-FINDEISEN THEORY of precipitation formation.

superficial deposits Material of Recent and Pleistocene age that lies on top of the solid geology, being extraneous material transported and deposited by various geomorphological processes. They include glacial drifts, terrace gravels, alluvium of rivers, raised beach deposits, windblown sand and loess, solifluction deposits, etc. Often various relict weathering horizons are included in this term, but this is not strictly correct because these materials are related to the rocks beneath.

supergroup The largest division in the hierarchy of the lithostratigraphical classification of bodies of rock (*see* lithostratigraphy; stratigraphy). A supergroup is formed of two or more associated and adjacent GROUPS.

superimposition The most common process by which a drainage pattern becomes discordant. Drainage originates on a land surface and develops a conformable pattern with the structures in that surface; as dissection proceeds this surface is eroded away, and the drainage is let down onto the newly exposed rocks below. These have different structures and lithologies from the original surface, and so the drainage pattern no longer conforms to the structure over which it flows. Hence it has been superimposed from a former cover onto the one it currently occupies. Once superimposition has occurred, the drainage may adjust to the structures of its new surface, and mask its true origin. On Dartmoor, England, the E–W and N–S rectangular pattern superimposed from the former Eocene and Cretaceous cover has been supplemented by a NW–SE and NE–SW pattern following the joint pattern of the granite. Current drainage is therefore a mixture of accordant and discordant streams.

superposition The principle, propounded by William Smith, that if one set of strata occurs on top of another in a succession the upper unit was formed later (except in the case of strata that have been overturned tectonically). This principle provides the basis of understanding the sequence of events in time as shown by a particular succession of rocks.

supersaturation The state of air that contains more water vapor than the amount required to saturate it. However, the abundance of condensation nuclei usually prevent this situation arising in the atmosphere.

superstructure The higher levels within an orogenic belt, which have deformed in a brittle manner as a result of their near-surface position. These levels have suffered only minor metamorphism. *Compare* infrastructure.

supervolcano An exceptionally large volcano that begins as a boiling reservoir of magma risen from the mantle to within the

Earth's crust, building in pressure until it finally erupts in a massive and devastating explosion. Unlike the majority of volcanoes, which are cone-shaped, supervolcanoes may be immense CALDERAS and can be hard to detect. For example, the magma-filled caldera (about 70 km by 30 km) of Yellowstone National Park was only detected in the 1960s through infrared SATELLITE imagery. Yellowstone has been on a regular eruption cycle of approximately 600 000 years, the last eruption being some 640 000 years ago. Calculations have been made that indicate that during the 20th century parts of the caldera rose by over 70 cm, raising concerns that the overdue eruption may be imminent.

During the largest volcanic eruptions forces are great enough to eject volcanic dust, debris, and gases, including sulfur dioxide, into the upper atmosphere. The sulfur dioxide combines with water to form droplets of sulfuric acid, which form a reflective barrier to incoming solar radiation with the effect of reducing temperatures. The last supervolcano to erupt was Toba, in Sumatra, 74 000 years ago. The volcanic winter that ensued created a global catastrophe and is believed by some geneticists to have pushed human life to the brink of extinction.

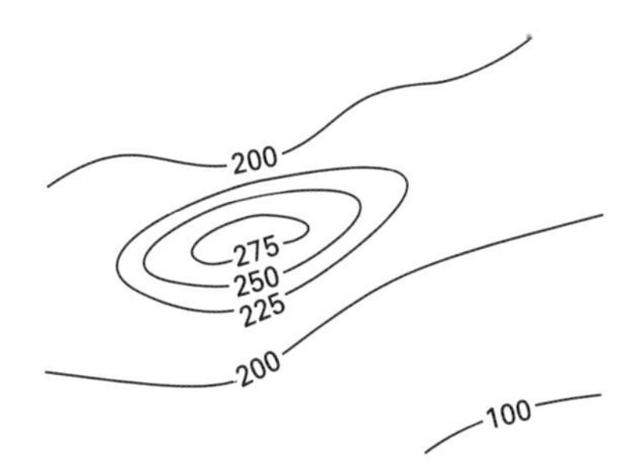

100 ft contour interval with supplementary contours to show a small hill

supplementary contour In areas of detailed or irregular relief patterns it is sometimes necessary to show supplementary contours as well as those at the standard interval for the area. For example, in an area where the land is generally level but there are several isolated small hills, a contour interval of 200 ft (100 m) may not show up the hills. (See diagram.)

supratenuous folding A form of folding that results from differential compaction. Sediments are bent around more competent materials such as concretions or coral reefs.

surf Broken water resulting from turbulent wave activity between the outer limit of the SURF ZONE and the swash-backwash zone on the beach.

surface runoff The part of rainfall that runs off the surface of the land and does not filter into the ground or evaporate into the atmosphere.

surface water Water that stays at or near the surface of the land, as opposed to GROUNDWATER.

surf beat A type of long wave activity that is evident at the coast by irregular oscillations of water level in the surf zone. These oscillations have periods of several minutes. A possible cause is the arrival of groups of higher-than-average waves, which have the effect of piling up water in the nearshore zone and causing a reflection of long-wave energy. The presence of surf beat has been demonstrated by precision wave recorders set up in the offshore zone.

surf zone (breaker zone) The strip of water along the shore in which breaking waves are actively dissipating their energy. They do this by a process that involves surface turbulence at a time when the waves or swell have become unstable and break. Usually this is because of their encountering very shallow water, but it may also be due to their meeting with opposing currents or wind. This complete breakdown of waves or swell must not be confused with breaking seas or so-called white horses, in which there is only partial collapse, at sea, of wave crests (*see* breaker). The surf zone is characteristically wide off gently sloping coasts and relatively narrow off steeply shelving coasts. The width of the surf zone fluctuates with constantly changing wave conditions; also, it is usually relatively wide in the case of sand beaches and narrower in the case of shingle or boulder beaches.

surge *See* sea surge.

surging glacier A glacier that has a brief period of comparatively rapid flow. It occurs when accumulated ice in an ice reservoir reaches a critical amount and suddenly surges downward at up to a hundred times the normal speed.

surveying The construction of maps and plans by accurately recording the relative positions and heights of features on the Earth's surface, and plotting them to some suitable scale. This is achieved by measuring distances, directions, and heights, the major surveying methods being: TRIANGULATION, TRAVERSING, CHAINING, PLANE TABLING, TACHEOMETRY, and LEVELING. *See also* geodetic surveying.

suspended load **1.** The sediment carried in SUSPENSION within a body of water, beneath waves or in current flow, as opposed to the sediment that moves wholly or intermittently in contact with the bottom (bed load and saltating load). Shoreward of wave break-point, a great deal of fine material may be carried in suspension beneath sea waves. (This also occurs in deeper water offshore provided that the material is sufficiently fine-grained.) Experiment has shown that at the break-point of waves and in the swash zone, sandy material held in suspension is fairly evenly distributed from the sea surface down to its bed. Suspended load transport is of great importance in the dispersal of pollutants in the sea.
2. The fine sediment carried in suspension by a body of flowing air.

suspension A type of transport in which fine particles of sediment are carried in water. The particles remain in suspension, held up by eddies caused by turbulence. The material being carried in this way is called the SUSPENDED LOAD. It is the suspended particles that make some rivers look milky or muddy.

swale *See* full.

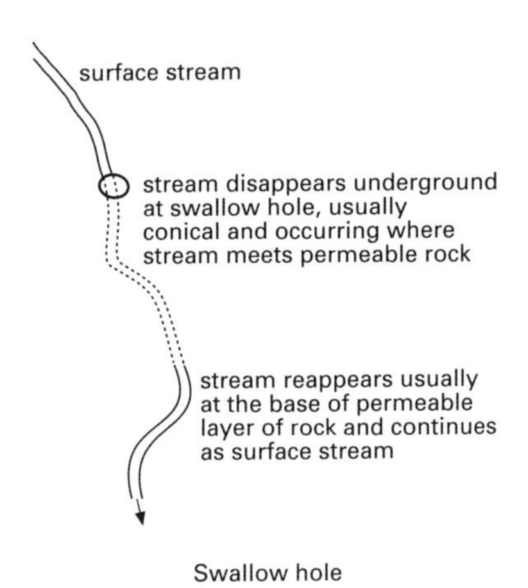

Swallow hole

swallow hole A SINKHOLE that is usually the site at which a stream disappears underground. See diagram.

swamp An area of soft wet land that has poor drainage and is generally waterlogged. The dominant vegetation consists of trees, such as eucalyptus, mangroves, maples, palms, and willows, depending on the climate. *See also* bog; marsh.

swash The movement of a fairly thin layer of turbulent water up a beach, following the breaking of a wave. This water can move material, the extent of the movement depending on the nature of the waves and of the beach. Most landward movement is achieved by flat waves, within which the oscillation of water particles tends to be elliptical. The oscillation in steeper waves is more circular, and the breaking water strikes the beach material from directly above, giving equal opportunity for up and down beach movements.

Some of the swash water infiltrates into the beach material and the rest returns to the surf zone as BACKWASH. The edge of swash action is frequently obvious from such signs as the wetted area of beach and the line of seaweed, driftwood, mica flakes, etc., left at the upper limit of swash flow. Swash action on a falling tide may lead to the formation of swash channels, as water tends to become channeled on its return flow down the beach.

S wave *See* secondary wave.

swell **1.** Waves that, having developed under wind stress in the generating area, have sufficient energy to travel beyond the wind field or to survive after the wind has dropped. They may travel into an area in which the winds are far weaker or absent or they may ultimately travel into another wind field. Such swell waves are usually fairly regular in form, with somewhat rounded crests. Their energy may enable them to travel considerable distances. In general, swell decreases in height and increases in length as it travels away from a wind field. Swells from several directions may be superimposed and swell may also carry relatively short and sharp-crested waves, which are generated by local winds.

2. A dome or anticline, esp. one that rises from the sea floor without reaching the surface.

syenite A coarse-grained intermediate igneous rock in which the proportions of alkali feldspar or feldspathoids are dominant over sodic plagioclase; in fact many syenites contain no plagioclase. The intermediate plutonic igneous rocks and their volcanic equivalents are divided as shown in the table.

Syenites range from oversaturated quartz-bearing types through saturated types containing neither quartz nor feldspathoid to undersaturated strongly alkaline varieties containing little or no feldspar. Division may also be made according to the composition of the feldspar into potassic (orthoclase, microcline), sodipotassic (perthites), and sodic (albite, oligoclase) types. The mafic minerals are usually hornblende and biotite but the more alkaline varieties contain sodic pyriboles. Common accessory minerals include magnetite, apatite, and sphene.

Nordmarkite is an oversaturated sodipotassic syenite containing microperthite, quartz, aegirine, and riebeckite. With an increase in the amount of quartz to over 10%, nordmarkite passes into alkali granite. *Perthosite* and *pulaskite* are sodipotassic leucocratic syenites. In the former, the feldspar is perthite and in the latter, antiperthite. *Larvikite* is a distinctive variety containing dark blue feldspars displaying strong SCHILLER. The feldspars are antiperthitic oligoclase and orthoclase and, in addition, the rock contains aggregates of mafic minerals, titanaugite, olivine, and biotite. *Shonkinite* is a saturated potassic melasyenite containing orthoclase feldspar and mafic minerals. Nepheline is the commonest feldspathoid present in undersaturated syenites but other varieties contain sodalite, analcime, and leucite. Some of the many different kinds of alkali syenite are:

Borolanite, containing orthoclase and nepheline-orthoclase intergrowths termed pseudo-leucite, thought to have been formed from the breakdown of leucite; *malignite*, containing potassic feldspar and

INTERMEDIATE PLUTONIC ROCKS WITH VOLCANIC EQUIVALENTS

Plutonic (*sodic plagioclase*)	*Volcanic* (*andesine/oligoclase*)
syenite	trachyte, phonolite
(dominant alkali feldspar/feldspathoid)	
syenodiorite (monzonite)	trachyandesite (latite)
(alkali feldspar = plagioclase feldspar)	
diorite	andesite
(dominant plagioclase feldspar)	

nepheline; *foyaite*, containing a perthitic feldspar and nepheline; *litchfieldite*, containing potassic feldspar, albite, and nepheline. Syenites containing nepheline and albite are called *monmouthite* or *mariupolite*; in the former nepheline predominates, in the latter, albite. The feldspar-free types constitute the IJOLITE series.

Microsyenites are the medium-grained equivalents and the plutonic rock names apply with the prefix micro-. Most are porphyritic and contain phenocrysts of orthoclase. The volcanic equivalents of syenites and alkali syenites are trachytes and phonolites.

Syenitic rocks are found in intrusive complexes often associated with continental rifting, in eroded volcanic islands of alkaline affinity, and within differentiated dikes and sills. *See also* diorite; syenodiorite.

syenodiorite (monzonite) A coarse-grained acid and intermediate rock that lies chemically and mineralogically between syenites and diorites. Syenodiorites are characterized by approximately equal amounts of alkali feldspar and plagioclase of oligoclase-andesine composition. The mafic minerals are usually either hornblende or biotite, or both, but in some varieties augite occurs, often rimmed by hornblende. Syenodorites containing accessory quartz are called quartz-monzonites and with an increase in the amount of quartz pass into adamellites. With a change to plagioclase of more calcic composition and an increase in the amount of mafic minerals, syenodiorites pass into syenogabbros.

Syenodiorites usually occur in small volumes associated with granite masses. The medium-grained varieties are termed *micromonzonites* or *microsyenodiorites* and the volcanic equivalents are the *trachyandesites* or *latites*. The alkali feldspar in trachyandesites is usually sanidine, occurring as microlites in a pilotaxitic or trachytic groundmass set with plagioclase phenocrysts. Trachyandesites occur in association with andesites in calc-alkaline orogenic volcanic suites. *Compare* diorite; syenite.

syenogabbro *See* alkali gabbro.

symbiosis (mutualism) The close association that exists between two organisms of different species in which each partner benefits from the relationship. A common example of symbiosis is the relationship between flowering plants and some insects, in which the insects feed on the nectar of the plant, pick up grains of pollen, and later transfer them to another flower (thus pollinating it). *See also* parasite.

symmetry *See* crystal symmetry.

symplectite An intergrowth of two minerals, one mineral being riddled with complex wormlike inclusions of the other, as in MYRMEKITE.

syncline A basin-shaped fold in which the beds dip toward each other. See diagram at FOLD.

synclinorium A syncline of regional extent, which is composed of a series of minor folds.

synecology The scientific study of interactions within ecological COMMUNITIES, and between communities and their environment.

synform A synclinal structure composed of sediments whose precise stratigraphic relationships are not known. See diagram at FOLD.

syngenetic Describing mineral deposits that formed at the same time and by the same process as the rock in which they occur.

synkinematic Describing minerals that are formed during a period of deformation.

synoptic chart A map that summarizes the weather conditions in a given area during the previous day. It includes isobars to show atmospheric pressures, temperature measurements, wind speeds and directions, cloud cover, etc.

synoptic index A method of assessing the sequences of pressure systems in a quantitative manner. Instead of isolated wind readings the overall direction of system movement is taken into account, then aggregated to give 10-day or monthly totals.

synoptic meteorology The section of meteorology concerned with the description and analysis of the *synoptic chart.* This is built up from surface weather observations of such elements as wind velocity, air temperature, cloud amount, dew-point temperature, and pressure tendency. Isobars are constructed from pressure observations and fronts inserted where appropriate. It is most directly concerned with day-to-day weather forecasting.

syntexis *See* assimilation.

system A division of rock in the Standard Stratomeric scale of stratigraphic classification (*see* chronostratigraphy). It indicates the body of rock that has formed during one PERIOD. System and period generally have the same name; for example, the Triassic System is the body of rock formed during the Triassic Period. A system, which does not have a standard uniform thickness, is formed of a number of SERIES grouped together.

T

tabular Describing a crystal or rock formation that is thin, wide, and long – shaped like a table.

tacheometry (tachymetry) A surveying method for finding the location and height of points using a theodolite (set up at a point of known height and position) and a leveling staff (placed at the unknown points). The directions of the unknown points are obtained by measuring horizontal angles from a known direction, while the distance from the theodolite position is calculated using a formula involving the difference in staff readings for two cross-hairs on the diaphragm of the theodolite, and the angle of elevation or depression, derived from the vertical circle reading. Another formula, using the staff reading of the central hair and the angle of elevation, permits the height of the point to be calculated. The height of the theodolite must always be taken into account whenever this method is being used.

tachylite A dark basic glass formed by the rapid chilling of BASALT and found particularly at the margins of dikes.

taconite A fine-grained iron-rich laminar sedimentary rock, commonly having layers of chert. The iron may be in the form of carbonate, oxide, silicate, or sulfide. It is used as a low-grade ore of iron.

taenite An alloy of iron and nickel (with 27–65% nickel) that occurs in iron METEORITES.

taiga (Boreal forest) A large region of coniferous forests that occupies N latitudes from Canada, across Scandinavia, to Siberia. It has cool summers and very cold winters, with frozen subsoil for much of the year. Fallen needles from the trees create PODZOL soil.

talc A white or pale green mineral with a mica-like layered structure and composition $Mg_3(Si_4O_{10})(OH)_2$. Talc is greasy to the touch and very soft, having a hardness of only 1 on the Mohs' scale. It is formed during the hydrothermal alteration of ultrabasic and basic rocks and during low-grade thermal metamorphism of siliceous dolomites. Rocks known as soapstone or steatite consist almost wholly of talc.

talus *See* colluvium.

tangential folding A type of folding produced in homogeneous rocks, analogous to the bending of a block of rubber with circles drawn on it; the circles become distorted but there is no sliding movement of any kind present.

tangential stress *See* shear.

tantalite A black mineral form of mixed tantalates and niobates of iron and manganese, $(Fe,Mn)(Ta,Nb)_2O_6$. It crystallizes in the orthorhombic system, and occurs in alluvial deposits, pegmatites, and granites. It is the chief source of tantalum.

taphonomy The study of the processes affecting an organism from its death to its possible fossilization. Taphonomy elucidates the many differences that exist between a fossil assemblage and the community (or communities) of living plants or animals from which it came (*see* thanatocoenosis). It is a necessary preliminary to investigations in PALEOECOLOGY. *See also* actuopaleontology.

taphrogeosyncline *See* rift valley.

tarn A small lake, usually occupying a CIRQUE.

tar pit A deposit of natural BITUMEN in a depression at the surface of the land. Animals sometimes become trapped in such pits, which are a good source of recent fossils.

tar sand A type of sedimentary rock that contains commercially useful amounts of ASPHALT.

taxonomy The study and practice of naming organisms and classifying them into a hierarchy of groups based on the similarities between them. It is implied that this also represents their evolutionary relationships. Thus the members of a lower group show more similarities and are therefore more closely related than those of a higher group. The group at the base of the hierarchy is the SPECIES. The groups above the level of species, in ascending order, are GENUS, FAMILY, ORDER, CLASS, PHYLUM, and KINGDOM. Any of these groups may be subdivided into smaller ones; for example a class may contain several subclasses.

tear fault (wrench fault, lateral fault, transcurrent fault) A fault in which the movement is horizontal, i.e. along the strike of the fault. This results from a stress configuration in which the maximum and minimum principal stresses are horizontal, whereas the intermediate principal stress is vertical. See diagram at FAULT.

tectogene A buckling of the Earth's crust that results in the down-buckling of granitic rocks deep into the Earth's crust to form the roots of mountains. It also results in the upward-buckling of the shallower levels in the crust to form orogenic mountain belts.

tectonic breccia A type of BRECCIA that has been formed through movements of the Earth's crustal plates (*see* plate tectonics).

tectonics The study of the Earth's deformational movements and their effect on sedimentation and geomorphology.

tektite *See* meteorite.

Teleostei The group of bony fish dominant since the Cretaceous, occurring in both freshwater and marine environments. The body covering is usually reduced to thin scales of bone. Teleosts show great morphological variation and many fossil species are known.

telluric current A naturally occurring electric current at or near the Earth's surface.

tellurometer A surveying instrument for precise distance measurement. Like the GEODIMETER it is indirect in its measurements. The time required for radio microwaves to travel from one point to another is determined and corrections applied for the meteorological conditions to convert this time to that in a vacuum. The distance is then calculated from the known velocity of the waves, and further corrections applied for slope and altitude. The best results are obtainable along a line with few obstacles, in moderate sunshine, with a light breeze and low relative humidity.

temperate Describing a moderate climate, such as that of the mid-latitudes north and south of the Equator, characterized by warm summers and cool winters. It extends from the Tropics of Cancer and Capricorn to the Arctic and Antarctic Circles, and thus lies between the tropics and the frigid zone.

temperate glacier A glacier containing considerable amounts of water above, within, and beneath the glacier ice. Water at the ice-rock boundary promotes easy sliding and consequently such glaciers flow faster than other types. This movement produces erosion of the abrasive type, while the presence of water allows freezing and thawing at the ice margins, an important factor in GLACIAL PLUCKING.

temperate grassland A type of GRASSLAND that occurs in the world's TEMPERATE regions. It includes the PRAIRIES of North America and the STEPPES of Russia. They have warm wet summers and cold dry winters, making them ideal pasture and, when plowed, good land for growing cereal crops.

temperature An index of the heat content of a substance, which determines the flow of heat between one substance and another. Because the specific heat capacity of different substances varies appreciably, temperature cannot be equated directly with heat. A variety of scales are in use. In most parts of the world the Celsius (centigrade) scale is used. For scientific work, the ABSOLUTE TEMPERATURE was formerly used but temperature is now measured in KELVINS. *See also* maximum temperature; minimum temperature; soil temperature.

temperature inversion *See* inversion layer.

tension crack (gash, fissure, *or* fracture) A crack that develops in competent beds as they are deformed. The rock is put under tension and fractures. This usually happens at several sites and sets of tension cracks develop en echelon. These fractures are usually infilled by secondary minerals such as calcite.

tension joint A joint produced as a result of tension, usually developed parallel to the fold axis of an anticline.

tephra (ejectamenta) Fragmental materials thrown out by a volcano, including bombs, ash, cinders, pumice, and lapilli. *See* pyroclastic rock.

tephrite An olivine-free type of ALKALI BASALT.

terminal moraine An accumulation of TILL material that develops against the front of a glacier. Such moraines are best formed when the front is slowly advancing or is stationary, because in these circumstances the front will be high and fairly steep. Such forms will finally be deposited on the retreat of the glacier, thereby marking its farthest limit from the source.

termite Ants and termites are abundant in tropical soils, their activity being conspicuous in the form of termite mounds, which generally are about one meter high. Their main benefit appears to be in improving the structure of the soil.

terra fusca A soil that, like TERRA ROSSA, develops on limestone in a warm subcontinental climate. Terra fusca are brown clay loam soils, neutral or slightly alkaline in reaction, occurring under cork oak or maquis vegetation. Beneath the brown A horizon is a relict red B horizon, suggesting that they may be degraded terra rossas.

terra rossa Red soil that develops on hard limestone in warm subcontinental climates. These soils typify the karst areas adjacent to the Mediterranean, and also occur in S Australia, Texas, central Spain, and Israel. They typically have a clay-loam texture, with some free sesquioxides, a variable depth, and low base status; they pass directly and abruptly to the parent material below, with perhaps a humus surface horizon. Their origin is uncertain: they were formerly thought to represent a weathering residue. They may develop from TERRA FUSCA on deforestation, because they occur only under garrigue vegetation. They are related in some way to RENDZINAS, which seem to occur more typically on soft limestone (e.g. chalk); it may also be that the terra rossa soils are more mature, because their lower base status and content of free sesquioxides shows more intense weathering.

terra roxa A Brazilian red earth (*see* ferrallitic soil) with a high percentage of titanium and manganese oxides and a low aluminum oxide:iron oxide ratio, developed on a parent material of basic igneous rocks. *Compare* terra rossa.

terrigenous deposit A sediment from land washed into the sea as mud, where it slumps down the continental shelf under gravity or the affect of currents. The deposits are named after their colors, and include blue mud, gray mud, green mud, and red mud. The color, in turn, depends on the composition.

Tertiary A period of the CENOZOIC Era that followed the CRETACEOUS nearly 65 million years ago, and lasted some 63 million years until the beginning of the QUATERNARY. The Tertiary was formerly regarded as an era, but because of its short length it is now classified as a period. The divisions of the Tertiary, which previously formed periods, are now regarded as epochs. They are the PALEOCENE, EOCENE, OLIGOCENE, MIOCENE, and PLIOCENE. Some authorities have suggested that the Tertiary should be regarded as a subera, which would consist of two periods: the PALEOGENE and the NEOGENE.

During the Tertiary the spatial distribution of the continents was similar to that of today. There was intense volcanic activity in NW Scotland, N Ireland, the Faeroe Islands, and Greenland associated with the rifting and separation of Eurasia and North America between Scandinavia and Greenland. Massive eruptions of basaltic lava occurred on the Deccan trap. Australia and South America separated from Antarctica. Episodes of orogenic movements extended through the period with the formation of major mountain belts, including the formation of the Alps and Himalayas (attributed to the movement north of India and its collision with Asia), the Rockies, and the Andes of South America. Extensive volcanic activity took place along the west coast of North and South America.

The Tertiary Period saw the evolution and gradual increase in abundance of modern invertebrates and mammals, with a corresponding reduction of primitive groups. The modern angiosperms became the dominant plants.

teschenite A nepheline-free type of ALKALI GABBRO.

Tethys Sea (Tethyan Ocean) Before the northward movement of Africa relative to Europe, a large ocean existed, of which the Mediterranean Sea is the surviving remnant. Within this ocean, which existed for a long period of time, the sediments of the Alps and Himalayan mountains were deposited. As the ocean closed as a result of SEA-FLOOR SPREADING this wedge of sediments was compressed into the present-day mountain chains and associated orogenic features.

tetragonal *See* crystal system.

tetrahedrite A gray sulfide mineral that contains copper and antimony, sometimes also with iron and arsenic, typical composition $(Cu,Fe)_{12}Sb_4S_{13}$. It crystallizes in the terahedral system, and occurs in veins associated with copper, silver, and zinc. It is used as a source of copper.

texture The small-scale structures recognized in hand specimens and THIN SECTIONS of rocks, originating from the geometrical relationships of the minerals that constitute the rocks. Some of the textural terms in common use are:

1. Terms describing grain size: coarse-, medium-, and fine-grained, APHANITIC, HYALINE.
2. Terms describing the degree of crystallinity: HOLOCRYSTALLINE, HYALOCRYSTALLINE, HYALINE.
3. Terms describing grain shape: EUHEDRAL, SUBHEDRAL, ANHEDRAL, IDIOMORPHIC, ALLOTRIOMORPHIC, HYPIDIOMORPHIC.
4. Terms describing the relationships between grains: EQUIGRANULAR, INTERGRANULAR, OPHITIC, POIKILITIC, PORPHYRITIC, APHYRIC, poikiloblastic, PILOTAXITIC, SPHERULITIC, interstitial.

thanatocoenosis (death assemblage) An assemblage of fossils composed of the remains of animals that have been accumulated by various agencies after their death and thus may not have lived together in life. They may have become affected by scavengers and by mechanical breakage in the process. Most fossil occurrences are thanatocoenoses. *Compare* biocoenosis.

theodolite An accurate surveying instrument used for the measurement of horizontal and vertical angles, consisting of a telescope that can be rotated in both the horizontal and vertical planes.

theralite An analcime-free type of ALKALI GABBRO.

thermal A volume of air that possesses buoyancy relative to its cooler surroundings. Thermals tend to arise as a result of strong insolational heating at the ground or a steep environmental lapse rate. Some will reach the condensation level producing cumulus clouds, others may lose their buoyancy earlier. The thermal rises by slowly mixing with its cooler surroundings at its upward margin and by entrainment in its wake. The mixing with environmental air means that strictly the thermal is not cooling adiabatically, but in practical terms the difference is rarely significant.

thermal depression An area of low pressure produced by intense solar heating of the ground surface. The heating reduces the density of the air, which often results in rising air and a fall in surface pressure. The best examples of thermal depression are the monsoon lows, such as those over the Thar Desert. These do not give rise to rain because they are capped by strong anticyclones, which prevent all thermals reaching condensation level.

thermal efficiency A concept devised by Thornthwaite (*see* Thornthwaite classification) for use in his climate classification scheme. It is a measure of the amount of heat given to a specific area expressed in terms of the potential evapotranspiration that would result.

thermal equator The line of highest mean surface air temperature. Because land areas tend to absorb heat more than oceanic areas, the thermal equator has a mean position in the N hemisphere. In July, it reaches about 20°N over the continents and in December is located at about the actual Equator except in E South America where it does extend farther south over Amazonia.

thermal high (low) A closed center of high (low) values of THICKNESS on a thickness chart. It represents a high (low) temperature averaged through the isobaric layer covered by the chart.

thermal metamorphism *See* contact metamorphism.

thermal spring (hot spring) A spring that produces hot water from underground, heated by geothermal energy. If it ejects steam as well as boiling water, it is termed a GEYSER. *See also* smoker.

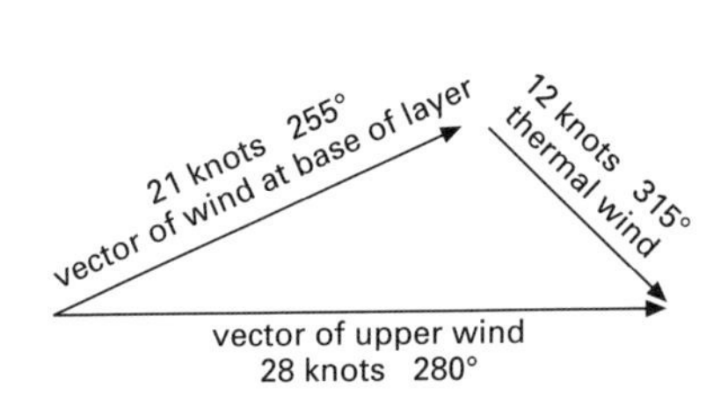

Thermal wind

thermal wind A theoretical wind used to indicate the horizontal temperature field in a layer of air. Actual winds change in the vertical as a result of variations in the thermal field at higher levels. The thermal wind is a measure of the difference between the wind at the top and at the base of the layer. It is obtained by subtracting the geostrophic wind vector at the base of the layer from that at the top. See diagram.

thermocline A subsurface layer of water within the ocean in which there is a marked decrease in water temperature as the depth increases. The water layer above the thermocline is essentially homogeneous or isothermal: the differences in temperature and salinity in the vertical column are minimal. This uppermost layer seldom exceeds 100 m in thickness. Hence, the thermocline is located at the bottom of the surface water masses. The maximum vertical temperature gradient is generally found at depths between 100 and 200 m. Some distance beneath the thermocline, in depths greater than 200 to 300 m, the water masses are surprisingly constant in their characteristics, the temperature gradient rapidly decreasing toward the ocean floor.

thermodynamic diagram A diagram on which the properties of a parcel of air may be indicated in terms of its pressure, temperature, and humidity. Adiabatic lapse rates are also frequently included on the diagram, from which it is possible to make assessments of atmospheric stability and cloud base.

thermograph A self-recording thermometer, which consists of a temperature-sensing mechanism and a system of time portrayal. A continuous record of temperature can be obtained from it.

thermohaline circulation The circulation of sea water within the oceans that arises from differences of salinity and temperature, which cause differences in density. Ocean water circulation tends to be controlled by two principal factors: the wind stresses that affect the sea's surface and the distribution of density within the water masses. These two factors are closely related and influence one another. Thermohaline convection, largely due to the cooling of surface waters at high latitudes, gives rise to pronounced currents operating in the oceans.

thermometer An instrument for measuring temperature by recording a change in a heat-dependent property of a substance. The property chosen will depend on such factors as the accuracy required and the range of temperatures to be measured. The most common method relies on noting the length of a column of mercury (or alcohol where low temperatures are involved) enclosed in a sealed glass capillary rising from a small bulb. The length of the column varies as the mercury expands and contracts with changing temperature. Other properties used include electrical resistance (*resistance thermometer*), the variation in the pressure of a gas kept at constant volume (*gas thermometer*), and the magnitude of the EMF produced by a bimetallic junction (thermocouple).

A *dry- and wet-bulb thermometer* consists of two thermometers side by side, one of which has its bulb enclosed in a wet muslin bag. The difference in readings between the two thermometers enables the relative humidity of the atmosphere to be calculated from standard tables (the wet-bulb thermometer will have a lower temperature due to the cooling effect as the liquid evaporates).

thermophile A plant that can survive in very hot climates.

thermoremanent magnetization (TRM) The permanent magnetism possessed by igneous rocks after they have solidified from the molten state. It is produced in the rocks by the Earth's magnetic field, and the direction of the magnetism in ancient rocks provides evidence of the history of the Earth's geology.

thermosphere The layer of the Earth's atmosphere above the mesosphere, in which the temperature increases with height. The pressure ranges from about 0.01 mb to 10^{-8} mb. The thermosphere falls within the IONOSPHERE.

thickness (in meteorology) The vertical thickness of air between any two specified pressure levels. The most common layer is the 1000 mb–500 mb, but others could be used. The thickness of the layer is directly proportional to its mean temperature, warm temperatures being associated with a thick atmosphere.

thin section A thin slice of rock, cut and ground to a uniform standard thickness (usually 0.33 mm) and mounted on a glass slide for petrographical study under a microscope.

tholeiite One of the three main types of BASALT.

thomsonite A white mineral form of hydrated sodium calcium aluminum silicate, $NaCa_2(Al_5Si_5O_{20}).6H_2O$. It crystallizes in the orthorhombic system, and occurs in crevices in basic igneous rocks and in amygdales within lava. It is a member of the ZEOLITE group of minerals.

thorn forest An area of thorny SCRUB that occurs in regions that have prolonged dry weather. The plants generally have long roots (to reach underground water), thick bark (to prevent water loss), and sharp thorns (to discourage browsing animals).

Thornthwaite classification A system of classification based on climatic efficiency, or the capacity of a climate to support the growth of plant communities. It is based on available moisture, the annual variation of temperature, and the degree of association between the temperature and precipitation regimes. The climate of an area is described by a four-digit label from which it can be identified and its characteristics in terms of rainfall, temperature, moisture index, and concentration of thermal efficiency understood. Because of its complexity and difficulties, it has not become widely used since its final version appeared.

threshold The lip of a CIRQUE, which may act as a dam, holding back water after the ice has melted and creating a tarn.

threshold wind speed The lowest wind speed that permits air to pick up grains of rock dust, sand, or soil. *See* aeolian transport.

throughflow The rainwater that flows down a hillside through the soil. It occurs when more rain falls on the surface of the ground than can be absorbed quickly downward by the soil.

throw The vertical change in level of a previously continuous bed of rock as a result of faulting.

thrust A low-angle reverse fault (see diagram at FAULT) that extends over a large distance. In thrusting large rock bodies can be displaced over each other, with the overriding block moving upward as it passes over the lower block. It results from a principal stress configuration in which the maximum and intermediate stresses are horizontal, while the minimum stress is vertical. The Moine Thrust of NW Scotland, which is 150 km long, carries Precambrian rocks far over Cambrian and other Precambrian rocks. These large displacements are often facilitated by gravity and the presence of fluids in the pores of the rocks undergoing deformation.

thulite A pink manganese-containing type of zoisite. *See* epidote.

thunder The noise heard accompanying a lightning flash resulting from the sudden heating and expansion of air as the flash passes through the ionized atmosphere. Because the speed of sound is far less than that of light, thunder is always heard after the flash. An approximate measure of distance from the storm is 1 mile for every 5 seconds between flash and thunder.

thunderstorm A storm that is accompanied by thunder and LIGHTNING. It involves large convection currents, which can result in very heavy rain, often causing flooding. Convection rain accompanied by thunder occurs nearly every day in tropical regions. If the convection currents carry drops of water high enough, they freeze and as a result hail may accompany the rain.

tidal current **1.** The periodic horizontal flow that occurs in response to the rise and fall of the tide. Such a current, like the vertical oscillation of the tide, arises from the gravitational attraction between the Earth, the Sun, and the Moon. Near the coast, tidal currents tend to be rectilinear and reversing. Farther offshore, they tend to flow in a more rotary manner. Tidal currents are responsible for the transport across shelf areas of large quantities of sediment, partly along the sea floor, partly in suspension above it. Because they are reversing cur-

rents near the coast, the residual flow in one or other direction often determines the net direction of sediment transport. Tidal currents, which in limited situations attain speeds even in excess of 9 or 10 knots, tend to flow most rapidly at the sea surface, but rather more slowly near the seabed. Bottom friction accounts for this difference. Tidal currents create several distinctive morphological features, including tidal banks and channels, and depressions eroded out of soft rocks.
2. The movement of water between two places differentially affected by the rise and fall of the tides. Two tidal regions may operate in close proximity, such that the water level at one point is, at any particular time, higher than at a neighbouring point. A tidal current will be produced causing water to move from the high to the low area. Although moving with some velocity, tidal currents are not noted as erosive agents, but may be important in the transport of material thrown into suspension by waves. The most notable tidal currents operate within straits where high tide at each end occurs at different times.

tidal flat A wide flat area of barren or marshy land that is covered and uncovered at high and low tides. It is made up of unconsolidated sediments. *See* marsh; mudflat.

tidal hypothesis A theory of the origin of the Solar System, which suggested that a star approached close to the Sun, and that the tidal attraction coupled with the Sun's unstable nature caused some of the Sun's mass to be torn off, later to cool as the planets. The theory has fallen into disfavor. *See* nebular hypothesis.

tidal limit The highest point in an estuary or river inlet that is reached by sea water at high tide.

tidal prism The quantity of sea water that floods into or ebbs out of a harbor, estuary, or other sea inlet as a result of tidal flow (not taking into account the freshwater discharge from streams, rivers, or canals). It is usually measured in cubic meters. The stability of tidal inlets is largely determined by the relationships that exist between the tidal prism and littoral drift: the ebb flow, reinforced by freshwater discharge, tends to evacuate sediments from inlets whereas the flood flow, reinforced by the littoral drift, tends to infill inlets with sediments. Dredging is often necessary to maintain the desired balance because most inlets tend toward progressive siltation.

tidal range The difference in height between the water level at high and low tide. This range is never a fixed value because wherever tides occur there is always variation between great ranges, associated with the especially high and particularly low tides of spring type, and small ranges associated with the neap tides. Whenever a single value for tidal range is quoted for a place, it is a mean value. A major geomorphological importance of the tidal range is that it controls how much of a beach profile will come under the influence of wave action, and therefore the nature of the coastal features.

tidal stream A movement of water into and out of bays, estuaries, and other restricted coastal openings, associated with the rise and fall of the tides. These streams may move comparatively rapidly and can cause considerable erosion both in flood and ebb. In the fine sediments of estuaries different streams are utilized by the rising and falling tide, and the results of their erosion are complex patterns of interdigitating channels, termed flood and ebb channels according to their mode of formation. *See also* tidal current.

tidal wave *See* tsunami.

tide The regular rising and falling of water level resulting from the gravitational attraction that exists between the Earth, the Sun, and the Moon. Depending on their relative positions, either SPRING TIDES or NEAP TIDES result. Tidal predictions can be made, although other influences such as local meteorological conditions make very accurate predictions difficult. With the rise and fall of the tide, the horizontal movement of water gives rise to TIDAL CURRENTS. The response of the oceans to tide-producing forces is based on a series of amphidromic points (*see* amphidromic system), around which the tide oscillates in the manner of a PROGRESSIVE WAVE. Diurnal tides are those with one low water and one high water during the tidal day; semidiurnal tides have two low waters and two high waters during the tidal day.

tie *See* chaining.

tierra caliente The hottest regions of Central and South America. They occur on the coastal plains, on the lower slopes of the Andes near the Equator, and on the mountainsides at altitudes of up to about 1000 m. The temperature is constant and rainfall heavy throughout the year – conditions that favor the growth of tropical rainforest.

tierra fria The cold regions of Central and South America. They occur higher in the Andes, at up to 3000 m near the Equator. Rainfall is spread throughout the year, and the natural vegetation is coniferous forest on the lower slopes with scrub at higher altitudes.

tierra templada The TEMPERATE regions of Central and South America. They occur at altitudes of between 1000 and 2000 m in the equatorial Andes. Natural forest covers much of the land, although some areas have been cleared (and the mountain slopes terraced) for agriculture.

tiger's-eye A yellow-brown type of QUARTZ, colored by iron oxide impurities. Fibrous inclusions of crocidolite give the mineral its shiny banded appearance, making it popular as a semiprecious gemstone.

tight fold A type of fold with parallel or slightly diverging limbs, with an interlimb angle of less than 30°. See diagram at FOLD.

till (boulder clay) The unstratified material deposited by glaciers and ice sheets. Characteristically unsorted, it contains angular material ranging from clay-sized particles to huge boulders. These may exhibit STRIATIONS and can be derived from a wide variety of rock types. *See also* ablation till; lodgment till.

till fabric The general pattern of orientation of elongate stones within TILL material. Analysis of the positioning of pebbles has shown that in many cases a large proportion are located with their long axes parallel with the direction of former ice flow, and with a low angle of dip downstream. Frequently another minor concentration exists, consisting of stones whose long axes are perpendicular to the direction of movement. It is believed that the former were moved by sliding and the latter by rotation about their long axes. Tills of different ages have been distinguished on the evidence of their different stone orientations.

timberline (treeline) The highest altitude on a mountainside or the highest latitude at which trees will grow. Trees do not grow beyond this line because it is too cold, there is insufficient soil, or there is not enough rainfall. *See also* snow line.

time-distance curve (T.D. curve) A graph that shows the relation between the arrival times of various seismic waves and their distances to the epicenter of an earthquake. Such curves provide seismologists with various kinds of data, such as information about discontinuities in the rocks through which the waves pass.

time series A set of data arranged in order of occurrence with equal time intervals between each value. Such series are common in climatology because standard records are taken at equal intervals and extend over many years. A time series can be for individual values or for averaged values such as annual mean temperature. Time series can be analyzed by a variety of statistical techniques such as spectrum or harmonic analysis.

tinguaite An old term for PHONOLITE.

titanaugite A titanium-rich variety of AUGITE. *See* pyroxene.

titanite *See* sphene.

tombolo A form of coastal SPIT, composed of sand or shingle, that links an offshore island to the mainland. Two tombolos (*double tombolos*) may link one island to different parts of the coast, enclosing a lagoon between them. The usual existence of comparatively shallow water in the straits between island and coast assists in the reduction of wave energy in this area, thereby promoting deposition of material. Tideless areas seem especially suited to tombolo formation, the Mediterranean containing many examples. Where the coastal outline is suitable, normal longshore movements of beach material may produce a spit, which eventually extends to reach an island.

tonalite A quartz-diorite. *See* diorite.

topaz A variably colored orthorhombic mineral of composition $Al_2(SiO_4)(OH,F)_2$ found in acid igneous rocks, such as granites and pegmatites. Some types are used as semiprecious gemstones.

topoclimatology The study of the interaction between topography and climate at the local scale. Stress is placed on the varying slope conditions (as they affect radiation receipt), the local water balance, raindrop trajectories, the geostrophic wind flow, and gravitational winds. These modifications of the heat and water balances produce a distinctive topoclimate, which may be widely different from the macroclimate as recorded in a Stevenson screen on a horizontal surface.

topography The relief, drainage, roads, vegetation, and cultural features of the Earth's surface.

topography correction (terrain correction) The correction of gravity measurements for the effects of local topography, because hills and valleys cause variations in the strength of the Earth's gravitational field.

topset bed *See* delta deposit.

topsoil The dark fertile soil at the surface, the A HORIZON.

tor A small hill projecting abruptly from a gently undulating land surface. Tors usually consist of stacks of joint blocks, which remain in situ and attached to bedrock at depth, together with many collapsed blocks. The most favored explanation of tor formation postulates two cycles: the first involves initial differential DEEP WEATHERING, the least jointed rocks being the least weathered; the subsequent period of EXHUMATION exposes the former undulating WEATHERING FRONT and leaves the unweathered blocks as tors. Many tors display UNLOADING sheets, which must have developed as exhumation proceeded.

torbanite A type of dark brown OIL SHALE that contains 70–80% of carbonaceous material.

tornado (twister) A violently rotating column of air characterized by a funnel-shaped cloud, which may reach to the ground surface, accompanied by a roaring noise. On a local scale the tornado represents the greatest wind intensity of all surface weather conditions. Most anemometers and even barographs are damaged if a tornado passes overhead, but falls of up to 200 mb and wind speeds of up to 500 km per hour are believed to be experienced. The tornado moves across country with the wind flow and direction of the wind at a higher level, leaving a swath of destruction wherever the funnel cloud has reached the ground. The precise origins are not fully understood, but in their severest form the following environmental conditions are required: warm moist air at low levels, an INVERSION at about 2000 m, a tongue of dry air between 850 and 700 mb, and a triggering mechanism to initiate the individual storm. These factors occur most frequently in the mid-West of the USA, where tornadoes are a great problem.

However, tornadoes of lower intensity occur in many countries. Tornadoes and other severe weather are tracked in the USA by a system of radar stations that use Doppler radar.

torrid zone The tropical areas between 23.5°N and 23.5°S.

tourmaline The tourmaline minerals all have the general formula NaR_3^{2+}-$Al_6B_3Si_6O_{27}(OH,F)_4$ where R = Fe^{2+}, Mg, or (Al + Li). They have trigonal symmetry and are variably colored although black varieties are common. Tourmaline is found in veins and pegmatites associated with granites.

The pneumatolytic alterations of granitic minerals caused by the introduction of boron is called *tourmalinization.* The rock luxullianite is a tourmalinized granite in which the biotite and much of the alkali feldspar has been replaced by radiating aggregates of acicular tourmaline. Quartz is the only mineral to survive tourmalinization and the end product is a tourmaline-quartz assemblage known as *shorl* rock. *See* pneumatolysis.

tourmalinization *See* pneumatolysis; tourmaline.

trace element (in geology) An element that occurs in a rock or mineral in very small quantities (much less than 1%), and is not regarded as an essential part of its composition.

trace fossil A fossilized remnant of the effects of an organism in the past, rather than the remains of the organism itself. Trace fossils may be the fossilized feces of an organism or structures produced by it that have been preserved in the sediment, such as tracks and burrows. ACTUOPALEONTOLOGY is used in relating these fossils to the species that produced them. Trace fossils are classified in a taxonomic system analogous to that applied to the organisms themselves, characteristic trace fossils being described as *ichnospecies.* They are often used in elucidating the paleoecology of a species.

trachyandesite A type of SYENODIORITE.

trachybasalt The fine-grained equivalent of syenogabbro (*see* alkali gabbro), displaying mineralogical and chemical features intermediate between TRACHYTES and ALKALI BASALTS. *Hawaiite* and *mugearite* are members of the alkali basalt volcanic suite containing andesine and oligoclase feldspar, respectively, together with basaltic mafic minerals. These lavas are distinguished from andesites largely on chemical grounds and field associations. *Benmoreite* is chemically intermediate between mugearite and soda trachyte.

trachyte The volcanic equivalent of syenite, ranging from oversaturated to undersaturated in composition. Some varieties contain sanidine and oligoclase feldspars, others contain a single anorthoclase feldspar. Feldspars occur both as phenocrysts and as close-packed microlites in the groundmass. The microlites usually have a subparallel flow orientation and swirl around the phenocrysts, imparting a characteristic trachytic texture. Mafic minerals include biotite, clinopyroxene, and hornblende. In the sodic varieties, soda pyriboles and fayalite are common. Trachytes may contain up to 10% accessory quartz. Oversaturated sodic varieties are termed pantelleritic trachytes which, with an increase in the amount of quartz, pass into pantellerites (alkali rhyolites). Varieties containing accessory nepheline are termed phonolitic trachytes. When the nepheline content reaches 10%, phonolitic trachytes pass into phonolites. Kenytes are a variety of sodic phonolitic trachyte containing distinctive rhombic phenocrysts of anorthoclase.

Trachytes are the intermediate members of the alkaline basalt volcanic suite and are found mainly in ocean islands and areas of continental rifting and vulcanism.

traction load *See* bed load.

trade winds The predominantly easterly winds that blow steadily over the ocean areas and less steadily over continental interiors in the tropics, converging toward the Equator. They play a very important role in the GENERAL CIRCULATION OF THE ATMOSPHERE by evaporating moisture in large quantities from the tropical seas, which helps maintain the global heat balance by removing surplus heat from these areas. On the eastern side of the tropical oceans, the trade winds are very steady in direction and speed and have a very strong low-level inversion, preventing the vertical development of cloud. These areas, such as the Canary Islands or the Galapagos Islands, tend to have low rainfall. Crossing the oceans increases the depth of the moist layer near the surface, the inversion becomes less intense, and disturbances more frequent, so that on the western sides of the oceans the trade winds provide an adequate rainfall in most areas.

trajectory (in meteorology) A line drawn to indicate the actual movement of a particle of air over a certain time interval. Ideally the path should be three-dimensional, but as little is normally known in detail about the vertical component, the horizontal projection of the path is portrayed. The term is also used for the downward motion of a precipitation particle as far as it can be assessed from upper air information.

transcurrent fault *See* tear fault.

transform fault (conservative plate boundary) A fault present within the Earth's oceanic crust (see diagram at FAULT). Such faults have strike-slip displacement and are orientated at right angles to the mid-ocean ridges (constructive plate boundaries), which they offset for several tens or even hundreds of kilometers. These faults lie parallel to small circles the axis of which is the axis of rotation for the relative motion of the plates on each side.

transgression Flooding of land area by a positive movement of BASE LEVEL, resulting in onshore migration of the high-water mark. A transgression results in the shortening of rivers by the drowning of former valleys, leading to the creation of fiords, rias, and estuaries: these features are relatively shortlived because rapid aggradation soon fills them up with silt. Past transgressions at a particular point can be identified by studying the deposits left. Fossil remains may also show changes from freshwater to marine types. In the postglacial period there have been two major transgressions, the first one in the Boreal period (7600–5500 BC) being the most important, which have elevated sea level to its current extent after the low stand of the last glacial period. Marine transgression can be local or worldwide.

transport (in geology) The link in the geomorphological system between sites of erosion and sites of deposition. The media of running water, wind, and currents transport material in varying proportions by bed creep, saltation, and suspension. In rivers bed creep is dominant; in air saltation is most important. Solution can also be important in rivers. Waves transport by beach drift of pebbles and sand in the surf zone, and the setting up of LONGSHORE CURRENTS. Ice transports as a solid medium, by shunting material in front, incorporating material in its base, and carrying slope-eroded material on its top. Gravity is inherent in most transport; on slopes it is directly responsible for downslope movements and it provides energy for flowing water.

transverse Mercator projection A map projection constructed in the same way as the Mercator projection but having the cylinder tangential to a meridian rather than to the Equator. It is used mainly for small areas with a north–south orientation, and for all British Ordnance Survey maps. Rhumb lines on this projection are curved.

transverse valley A valley that cuts across the prevailing geologic structure of the land, or at right angles to the general alignment of the underlying rock strata.

trap A geologic structure in which hydrocarbons accumulate. Hydrocarbons migrate away from their source rock under pressure and accumulate where an effective

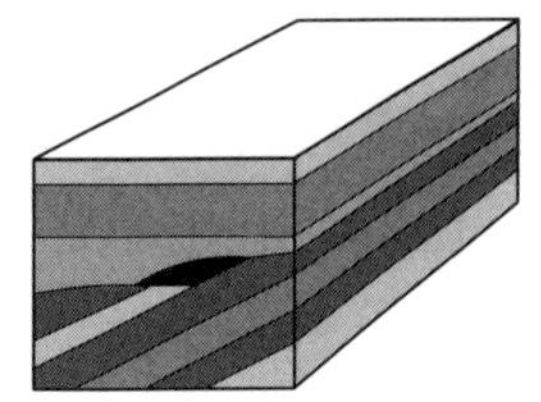

Fig. 1: Unconformity trap

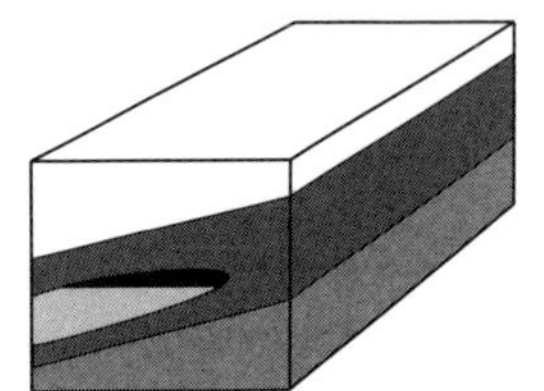

Fig. 2: Facies change

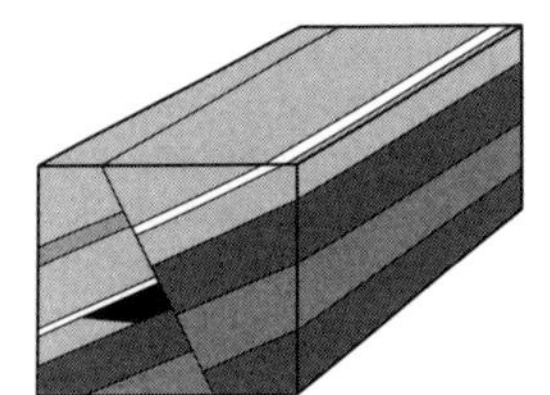

Fig. 3: Fault trap

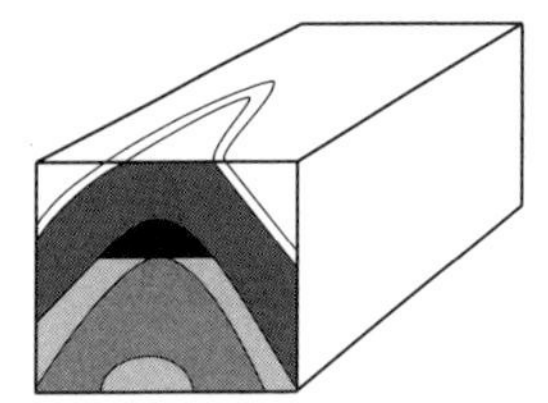

Fig. 4: Anticlinal trap

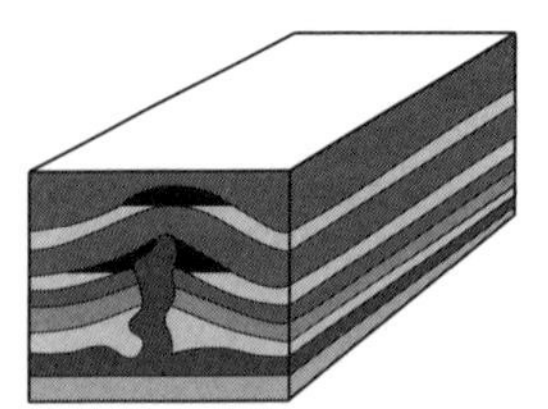

Fig. 5: Salt dome trap

Types of trap

barrier to their further migration exists. Traps can occur in a variety of ways. Since hydrocarbons, both oil and gas, are less dense than water, they tend to accumulate in the highest points of the trap, displacing most of the water previously present in the rock pores. The barriers to further migration can result from tectonic activity, as in the case of fault, unconformity, and anticlinal traps, and from lateral changes in the properties of the rock body, as in the case of facies change traps and reef traps. When more than one type of trap exists in the same place it is termed a *composite trap*. *See also* salt dome.

traversing A surveying method for finding position by measurement of direction and distance, using either a prismatic compass and tape (*compass traverse*), or a theodolite and leveling staff (*tacheometric traverse*). If the starting and finishing points are of known position, whether a single or two distinct stations, then the traverse is *closed*. If only one station, at one end of a traverse line, is of known position the traverse is *open* or *unclosed*. The former type is better because it allows easier correction if errors are made. Unlike a compass traverse, a tacheometric traverse will not only fix position but also height. To obtain heights for points fixed by compass and tape, a leveling traverse must be run, starting and finishing at a point of known height. *See* tacheometry.

travertine Deposits of calcium carbonate formed by precipitation from hot springs.

treeline *See* timberline.

trellis drainage A drainage pattern that develops where two sets of structural controls of a different type occur at right angles. If a steep regional slope is crossed by varying hard and soft lithologies, short stream segments will follow the slopes, while the soft lithologies will rapidly develop long subsequent streams that will come to dominate the pattern. One example is in the Jura Mountains of France; another is the Ridge and Valley province in

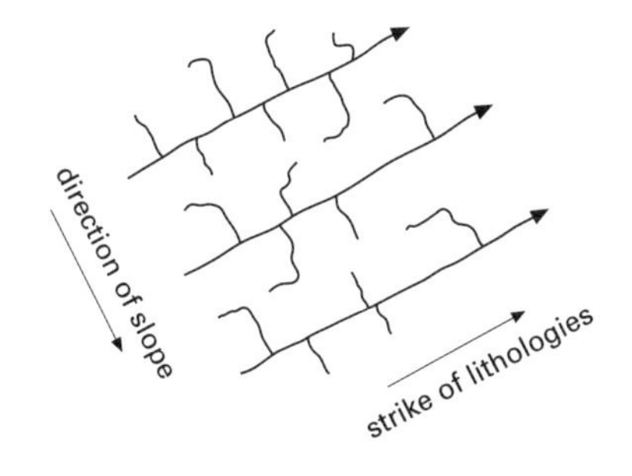

Trellis drainage

the Appalachian mountains of the eastern USA.

A similar pattern can also arise if glaciation of an area of former consequent streams proceeds at right angles to the initial drainage. Subsequents will develop following the grain of the glacial advance, as expressed in fluted ground moraine or drumlins, and come to dominate the system.

tremolite A monoclinic mineral of the AMPHIBOLE group.

trench (deep-sea trench, foredeep) A long narrow deep furrow in the Earth's crust, generally developed at the margin of a continent where an oceanic plate is being subducted (*see* subduction zone), and often bordered by an ISLAND ARC. Trenches can also develop where one oceanic plate overrides another. They mark the site of a destructive PLATE BOUNDARY and the surface expression of a BENIOFF ZONE. Most trenches have steep V-shaped walls and are often terraced, with a flat floor that is formed by a sediment infill. They are the deepest parts of the oceans: the greatest known depth (10 911 m) was recorded in the Mariana Trench, and similar depths occur in the Tonga and Kuril Trenches.

trend General alignment, direction, or bearing. See diagram at FOLD.

triangulation The accurate location of a number of points by dividing the area containing them into a series of triangles for which the values of internal angles and the lengths of sides are ascertained. Only one side of one triangle need be measured physically, because once all the angles are known, the side lengths can be computed using trigonometry. After a BASELINE has been measured very accurately, the angles are measured using a THEODOLITE, usually a one-second instrument, which is set up in turn at all the stations to be fixed. Triangulation schemes are the basis of all surveying because they provide the locations of the few initial control points, from which subsequent detail mapping may take place using other methods.

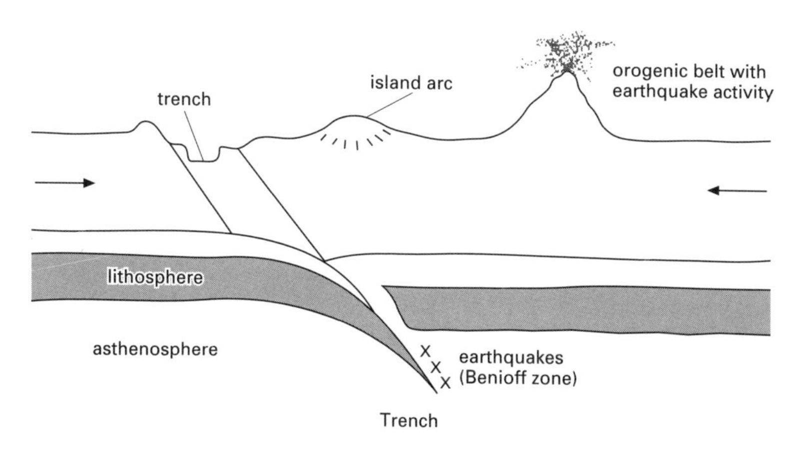

Trench

Triassic (Trias) The first period of the MESOZOIC Era, beginning about 246 million years ago, at the end of the PERMIAN, and lasting for some 40 million years until the beginning of the JURASSIC. The Triassic System has three main divisions. The Lower Triassic consists of the Griesbachian, Nammalian, and Spathian Stages; the Middle Triassic comprises the Anisian and Ladinian Stages; and the Karnian, Norian, and Rhaetian Stages form the Upper Triassic. The name reflects the threefold division of these rocks in Germany: Bunter, a lower unit of continental red-colored sediments; Muschelkalk, a middle unit of marine limestone, sandstone, and shale; and Keuper, an upper unit of continental rocks. A wide variety of sedimentary rocks is evident from the Triassic, with some igneous. Toward the end of the Triassic the supercontinent of Pangaea, which extended from pole to pole, started to show signs of breaking up with rifting along the Tethys seaway.

Triassic faunas, following the Late Paleozoic extinctions, show a typically Mesozoic character. Modern corals appeared, ammonites developed, and bivalves replaced the declining brachiopods. The evolution of the Reptilia produced such diverse types as the dinosaurs and the marine ichthyosaurs and plesiosaurs.

tributary A secondary stream or river that flows into a larger river. There may be many tributaries, all of which collect rainwater and groundwater from the main river's catchment area. Rivers that flow through semiarid or arid areas tend to have few tributaries. *See also* distributary.

triclinic *See* crystal system.

tridymite A white crystalline high-temperature derivative of QUARTZ, SiO_2. It crystallizes in the orthorhombic system, and occurs in cavities in acid volcanic rocks. *See* silica minerals.

trigonal *See* crystal system.

Trilobita An extinct class of the phylum ARTHROPODA whose members are common and widely distributed as fossils. Trilobites had an oval flattened segmented body divided longitudinally into three lobes and transversely into three regions: an anterior *cephalon*, a *thorax*, and a posterior *pygidium*. They possessed a large number of appendages, including one pair of antennae. Most trilobites were about 50 mm in length although a few reached over 500 mm. The first fossils are found in lower Cambrian rocks; the group declined from the Ordovician Period onward and by the end of the Paleozoic it was extinct. The trilobites formed a widespread, diverse, and rapidly evolving group and their fossils are of great value in the calibration and correlation of Lower Paleozoic rocks.

triple junction A point at which three lithospheric plate boundaries meet.

tripoli A pale-colored porous sedimentary rock containing silica. It is powdered and used as an abrasive for polishing.

trochoidal wave *See* cycloidal wave.

troctolite A GABBRO consisting of olivine and plagioclase feldspar of labradorite composition.

trona A white, gray, or yellowish mineral mixture of hydrated sodium carbonate and sodium hydrogen carbonate, $Na_2CO_3.NaHCO_3.2H_2O$. It crystallizes in the monoclinic system, and occurs as tabular crystals or evaporite deposits. It is used as a source of sodium compounds.

trondhjemite A coarse-grained igneous rock consisting mainly of quartz and plagioclase, with some biotite; it lacks feldspar. *See* granite.

trophic level The position an organism occupies in a FOOD CHAIN. The main levels are PRIMARY PRODUCERS (such as plants, at the lowest level), PRIMARY CONSUMERS (such as herbivores, at the next level), and secondary consumers (such as carnivores, usually at the highest level).

tropical black soil A black heavy-textured soil characteristic of tropical lowland areas, either plainlands or the downslope ends of CATENAS, which are subject to seasonal drying and cracking due to the shrinking and swelling of the dominant montmorillonitic clay during dry and wet periods respectively, caused by the ability of the clay to take up water in its lattice. During the wet swelling periods, soil is pushed up to form swells (*gilgai*) and in the dry phases cracks develop, which infill themselves from the surface, thereby maintaining a kind of cycle, which makes the soils self-mulching or inverting. There are possibly 40 or so different names for these soils, including vertisol, mbuga, tir, black cotton soil, grumusol, morgatitic, black earth, black turf, and regur, each of which denotes a regional variant differing in some small way. As a group, they are similar to CHERNOZEMS, but have less organic matter and differences in the surface horizons, due mainly to the higher temperature and greater evaporation of their environment. Tropical black soils have indistinct horizons due to their self-mulching nature, often with calcareous or gypseous concretions; silt and clay content is high (possibly 85%+), pH is neutral or alkaline, and parent materials typically basic, e.g. marine alluvium, limestone, or basalt.

tropical cyclone A low pressure system of tropical latitudes accompanied by strong winds. It is a general term used to describe a cyclonic storm of any intensity, distinctions of wind strength being made by other terms such as tropical depression, tropical storm, and hurricane. The diagram shows the distribution of tropical cyclones, which are clearly restricted to certain areas. They develop only in oceanic areas where the sea temperature is above 27°C and at least 5° of latitude away from the Equator. It appears that the warm sea surface must provide energy through evaporation, 27°C representing its lower limit, and as the CORIOLIS EFFECT is zero at the Equator, the storm must be at least 5° away before rotational components of airflow can develop. The precise origins of tropical cyclones are debatable, but some form of initial disturbance in the airflow seems to be a prerequisite.

tropical easterlies *See* trade winds.

tropical grassland A type of GRASSLAND that occurs in the world's tropical regions.

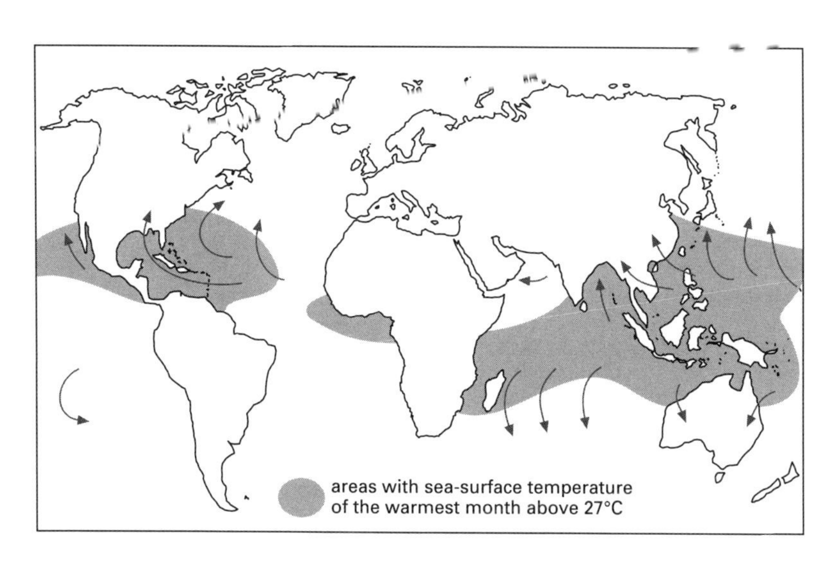

Distribution of tropical cyclones

These areas have seasonal rainfall followed by prolonged droughts preventing the growth of trees. *See also* savanna.

tropical podzol (groundwater podzol, geant podzol) A soil developed in low-lying tropical areas, often adjacent to riverine, marine, or deltaic areas, subject to a fluctuating water table. They are characterized by their striking colors and overdeveloped A_2 horizon, which can be two meters deep. They develop by the leaching of sesquioxides and organic material to the B horizon, with precipitation taking place at the level of the water table.

tropical rainforest (equatorial rainforest) Lush evergreen forest that grows in the tropics, generally in regions between 5° north and south of the Equator. There are constant high temperatures and it rains throughout the year. Such forests support a very large number of different species of animals and plants. The soil, however, is shallow and poor, and easily eroded away if trees are felled. Even if the soil is used for agriculture after the trees have gone, the soil nutrients are consumed after only a few seasons and the ground becomes no longer suitable for crops.

Tropic of Cancer The line of latitude 23° 30′ N. At the summer SOLSTICE in the N hemisphere, the Sun is directly overhead at the Tropic of Cancer, the most northerly latitude at which this occurs.

Tropic of Capricorn The line of latitude 23° 30′ S. At the summer SOLSTICE in the S hemisphere, the Sun is directly overhead at the Tropic of Capricorn, the most southerly latitude at which this occurs.

tropopause The boundary between the troposphere and the stratosphere. It represents the point at which temperatures stop falling before the isothermal conditions of the lower stratosphere. On occasion, there may be more than one tropopause. The height of the tropopause varies, being higher above warm air than cold. Consequently its mean height is greatest over tropical latitudes at above 16 km but only about 8 km over polar latitudes. It also changes because of different weather conditions. The tropopause is not a continuous surface between the tropics and poles, but is broken at the latitude of the subtropical jet stream and is suddenly lower on the poleward side. These breaks enable mixing to take place between tropospheric and stratospheric air, which would otherwise not occur because the isothermal layer of the stratosphere acts as a stable inversion.

tropophyte A plant that can survive in both dry and wet climates, such as deciduous trees.

troposphere The lowest layer of the atmosphere, where almost all weather phenomena develop. It takes its name from the Greek word tropos meaning a turn: it is the atmospheric layer where turning and convective mixing is dominant.

trough A pressure system on a weather chart distinguished by an isobaric pattern of low pressure having much greater length than width and a concave curvature of the isobars toward the main depression center. All fronts occupy troughs of low pressure, but not all troughs are frontal. They are normally associated with precipitation and cloud.

trowal A trough of warm air aloft. This indicates the presence of a layer of warmer air in the upper atmosphere, which is not in contact with the ground surface. This type of situation occurs in an occlusion.

truncated spur A SPUR that formerly protruded into a valley but has been partly or completely separated from the high ground by the action of a moving glacier.

tsunami A seismic sea wave generated in the ocean by submarine earthquakes, explosive volcanic eruptions, or mass slides underwater. Although fairly rare, such waves can have catastrophic effects when they do occur. Perhaps of limited height (about a meter) in the open ocean, they may travel hundreds or even thousands of kilometers, and they attain considerable heights (up to tens of meters) in shallow coastal water and at breakpoint. Their length at sea may be 150 km or more, and they usually travel at high speeds, probably between 300 and 500 knots. They have caused immense havoc and destruction and heavy loss of life, especially along low-lying coasts. Tsunamis are sometimes incorrectly called *tidal waves.*

tufa Deposits of calcium carbonate formed by precipitation from water and including stalagmites, stalactites, flowstone, and travertine.

tuff A type of PYROCLASTIC ROCK made up principally of small fragments of consolidated volcanic ash.

tundra A flat or undulating region located north of the TAIGA in the subarctic lowlands of North America, Europe, and Asia. Summers are cool although they may be sunny; for more than six months of the year in winter temperatures never rise above freezing and there is a thick layer of PERMAFROST. No trees grow on the tundra and the main vegetation is grasses, lichens, and mosses.

turbidite A sediment deposited in water by TURBIDITY CURRENTS. Because the speed of a turbidity current is proportional to the square root of effective density, its lower part will be more heavily charged with sediment and will move faster than the less heavily charged upper part. For this reason the particle size of the sediment transported decreases from the base of the flow upward, a consequence being that a form of sediment grading, or graded bedding, occurs in the deposit ultimately laid down. A typical sequence in such a deposit is for coarse material at the base to give way to finer material above, and possibly be covered in turn by lutite. If a series of turbidity flows occur in the same area, a rhythmic pattern of sedimentation arises, with superimposed layer upon layer of graded turbidite material. Many sandy turbidites contain a significant amount of silt and clay.

turbidity **1.** (in meteorology) The property of the clear atmosphere that leads to a reduction of solar radiation, usually due to particles of dust and smoke, which lead to attenuation rather than molecular scattering. It is therefore a measure of atmospheric pollution either natural or resulting from human activities.
2. (in oceanography) The stirring up of sediment by water.

turbidity current A flow of dense sediment and water. Such currents develop with the stirring of sediment, resulting in the formation of a layer that is denser than the surrounding water. Such a situation occurs if strong wave action, a submarine slide, or an earthquake disturbance affects the sea floor in an area in which sediments, especially the finer-grained ones, are reasonably abundant. In the case of a flat seabed and in the absence of currents, the suspended material would tend to settle out again, but where the seabed slopes (e.g. on CONTINENTAL SLOPES) the suspended material usually begins to flow, most rapidly when the slope is steep, and thereby initiates a turbidity current. Turbidity currents can be very dense and may possibly attain speeds of 80 km per hour or more. They have been powerful enough to break submarine cables and to erode channels in the deep-sea floor. They lead to sediment deposits in characteristic morphological forms.

turbulence Irregular movements of a fluid, such as air in the lower atmosphere or water in a rapidly flowing river. *See* turbulent flow.

turbulent flow The usual type of movement characteristic of air and water in nature, in which the net forward movement of the fluid has superimposed upon it a chaotic pattern of secondary eddies, carrying molecules from one layer in the flow to either higher or lower layers. This constant exchange of molecules speeds up slower layers and slows down faster ones; the eddies are principally responsible for creating hydrodynamic lift, so important in sediment transport. LAMINAR FLOW becomes turbulent when the viscosity of the water falls as temperature increases, releasing the liquid from the confines of flowing purely as laminar layers. Alternatively, increased velocity, increased bed roughness, or a decrease in water depth can all promote turbulent flow. In waterfalls and rapids the greatly increased velocity and lowering of the water surface promotes a special type of turbulence, "shooting" flow, which accelerates erosion and leads to rapid lowering of the falls or regrading of the rapids.

The scale of turbulence in the atmosphere is wide, ranging from small eddies rising above a strongly heated ground surface to mid-latitude depressions, which represent large-scale turbulence in the westerlies. It is important in the atmosphere as an effective mechanism of dispersal.

turquoise A complex mineral form of hydrated, copper aluminum phosphate, $CuAl_6(PO_4)_4(OH)_8 5H_2O$, valued for its light blue or green color. It crystallizes in the triclinic system, and occurs in nodules, small masses, or thin veins in various rocks. It is used as a semiprecious gemstone.

twilight The period of weak daylight after the Sun has set below the horizon in the evening or before it rises in the morning. It can last for up to an hour in Arctic and Antarctic regions, but is over in a few minutes near the Equator.

twinning The growth of two or more crystals so that they join together, either intergrown or in contact, with different orientations of the crystal axes. It can occur with several minerals, such as calcite, fluorite, gypsum, feldspar, staurolite, and rutile.

twister *See* tornado.

type locality **1.** (in paleontology) The place from which a fossil that is the TYPE SPECIMEN of a species comes.
2. (in stratigraphy) The site of the outcrop of rock used to define a particular stratigraphical division. *See* formation; lithostratigraphy; stratigraphy; type section.

type section The particular outcrop of rock that has been selected as the standard section for the definition of the limits of a lithostratigraphical unit because it clearly demonstrates the characteristic lithological features of the unit. The place in which it occurs is known as the TYPE LOCALITY. *See also* formation; lithostratigraphy; stratigraphy.

type species The species regarded as typifying the GENUS to which it belongs. The type species is often the species whose characteristics were used in defining the genus.

type specimen A specimen of an organism that was orginally selected and preserved to define the morphological characteristics of a SPECIES. It may not necessarily be typical of the species as a whole.

typhoon A tropical cyclone or hurricane with winds above 117 km per hour (64 knots or Force 12 on the BEAUFORT SCALE). The term is used for such storms in the W and N Pacific Ocean and comes from a Chinese word meaning great wind.

U

ugrandite A chemical series of GARNET minerals.

ulexite A white mineral form of a hydrated sodium calcium borate, $NaCaB_5O_9 \cdot 8H_2O$. It crystallizes in the triclinic system as rounded masses of silky hairlike fibers, and occurs as evaporite deposits in arid regions and lake basins. It is used as a source of boron.

ultisol One of the ten soil orders from the SEVENTH APPROXIMATION classification, denoting highly weathered soil formed in subtropical to tropical climates, which has a surface horizon containing residual iron oxides and an illuvial horizon beneath rich in clay. Base status and fertility is low. It is so named because it contains soils at the ultimate stage of weathering and includes the red-yellow podzolics, red-brown lateritics, and associated hydromorphic variants.

ultrabasic rock An igneous rock containing less than 45% silica (by weight). Most ultramafic rocks (consisting largely of ferromagnesian minerals) are ultrabasic but some pyroxenites may have more than 45% silica and are basic rather than ultrabasic in composition. Current usage has tended to equate the two, but strictly, ultrabasic and ultramafic are chemical and mineralogical terms respectively. *See* acid rock; ultramafic rock.

ultramafic rock An igneous rock consisting largely of FERROMAGNESIAN MINERALS. Such rocks may be divided into three groups:

1. Peridotites, in which olivine is dominant and feldspar absent (dunite, harzburgite, lherzolite, wehrlite).
2. Pyroxenites, consisting wholly of orthopyroxenes and clinopyroxenes (bronzitite, websterite).
3. Picrites, which contain accessory feldspar (oceanite, ankaramite).

With an increase in the amount of plagioclase, picrites grade into gabbros. Most ultramafic rocks are plutonic and occur in layered intrusions, ophiolite complexes, zoned ultrabasic bodies, and as nodules in basaltic lavas.

ulvöspinel *See* spinel.

umber A greenish-brown earthy pigment that contains oxides of iron and manganese, together with silica and lime. Known also as *raw umber*, it can be roasted to give dark brown *burnt umber*. *See also* sienna.

unaka An area of high ground that rises above a PENEPLAIN (which has been eroded away around it), larger than a MONADNOCK.

uncompahgrite *See* ijolite.

unconformable *See* discordant (def. 2).

unconformity A surface representing a period of nondeposition or erosion separating rocks of different ages. Some unconformites show a marked angularity, the beds above and below the unconformity surface having different dips and strikes, whereas other unconformities can be detected only by paleontological means. *Compare* disconformity.

underground drainage In limestone areas solution can take place to such an extent that joints, bedding planes, and other fissures become progressively enlarged, forming hollows (SINKHOLES), UVALAS, or POLJES) on the surface, and cavern systems underground. Linked cave systems can lead the water of a stream from a sinkhole at one point to a spring some considerable distance away, thereby eliminating surface drainage for that stretch.

underground stream A stream that flows underground for all or part of its course, most commonly in limestone regions. The erosive action of such streams can carve various formations underground. *See* underground drainage.

undersaturated rock A type of igneous rock that is deficient in silica and therefore contains FELDSPATHOIDS. Undersaturated rocks include ijolites, nephelinites, and alkali basalts. *See* silica saturation.

underthrust A low-angled fault resulting in the movement of the foot wall relative to the hanging wall.

undertow A seaward flow of water beneath, or in the vicinity of, breaking waves. Some authorities claim that the backwash of waves constitutes a periodic sheet-flow seaward, describing this as an undertow. The term has also been used (erroneously) to mean the RIP CURRENT flows that occur within fairly confined zones.

uniformitarianism (actualism) The theory, now generally accepted, that all geologic changes have occurred by the gradual effect of processes that have been operating over a long period of time and are still going on today. *Compare* catastrophism.

units For general scientific purposes, SI UNITS are now widely used throughout the world, However, in some disciplines older systems of units still persist. For example, in meteorology and climatology many elements are still measured in Imperial units and knots continue to be used for wind velocity and ocean currents in most countries. Although the SI unit of pressure is the pascal, it has been agreed that the millibar will continue to be used in meteorology.

unloading (sheeting, pseudo-bedding, pressure release) A weathering process resulting in the division of rock masses into sheets separated by fractures. These sheets tend to increase in frequency and decrease in thickness toward the top of an exposure; they are concentric, and essentially follow the topographic outline of the ground surface. The usual explanation is that the fractures are due to the expansion of the rock on reduction of the confining pressure, achieved by the removal of overlying materials by erosion. These features occur most frequently in granites.

unmixing *See* exsolution.

upslope fog A form of advection fog resulting from the cooling of moist air to saturation point as it is forced to rise up the windward slope of uplands. It has the appearance of stratus cloud from below.

upthrown Describing the side of a FAULT that appears to have moved upward relative to the other side (by an amount called the upthrow). *See also* heave.

upwarp The uplifting or uparching of a large area of the Earth's crust, usually in response to isostacy. The area of Scandinavia has gradually been rising after having been depressed by the weight of the ice during the last ice age.

upwelling An ascending water current, not necessarily from the bottom of the sea, which transports colder water up into the surface layers of the sea. Such currents occur particularly in tropical and subtropical seas, and in the water flanking the Antarctic, especially in those regions where surface currents diverge (DIVERGENCE zones). Upwellings also occur in some inland or marginal sea areas, including the Caspian Sea and Black Sea. They may stem from strong offshore winds that tend to drag the surface water layer seaward so to be replaced by colder water from beneath. Upwellings occur off the coasts of California, Peru, and Ghana. They tend to operate seasonally and to form cold zones within otherwise warm sea areas. They are of primary importance in bringing to the surface waters that are rich in nutrients; these in turn may support thriving fisheries. Shifts in the zones of upwellings have proved disastrous to the fishing industry.

uralitization The late-stage or metasomatic alterations (*see* metasomatism) of primary igneous pyroxenes to fibrous pale green amphiboles.

uraninite A black radioactive uranium oxide mineral, UO_2, found as an accessory in acid igneous rocks and in hydrothermal veins.

urban climate The distinctive modification of the macroclimate produced by extensive urban areas. The marked difference in the nature of the surface in rural and urban areas alters the local heat balance to such an extent that, even in an area of uniform topography, the city evolves its own climate. This has become the subject of much research in recent years and to distinguish it from the macroclimate determined from Stevenson screens at standard exposures, it is called the urban climate. *See also* heat island.

Urstromtäler In N Europe, wide trenches or valleys eroded by meltwater from the ice sheet as it retreated northward at the end of the last ICE AGE.

urtite *See* ijolite.

U-shaped valley A well-developed glacial valley with a characteristic U-shaped cross profile, as opposed to the V-shape of fluvially developed valleys. Such development can take place by simple widening of preglacial valleys, or by a combination of widening and deepening, the latter process depending upon the glacier's ability to cut into the bedrock. The typical steep sides of glacial valleys reflect the fact that whereas rivers occupy only the bottoms of their valleys, a glacier tends to fill much of it, erosion taking place wherever there is an ice-rock contact. The extent to which a valley can be deepened depends greatly upon the amount of weathering active in the valley before the appearance of the glacier, since GLACIAL PLUCKING is the major process of rock removal.

uvala A large depression found in limestone areas, resulting from solution (*see* limestone solution). They may be several kilometers in diameter, and frequently result from the coalescence of a number of adjacent DOLINES. In such a case the uvala may have a scalloped margin. *See also* polje.

uvarovite A rare bright green member of the GARNET group of minerals, $Ca_3Cr_2Si_3O_{12}$. It occurs in chromite-rich serpentinites. It is used as a semiprecious gemstone.

V

vadose water Groundwater above the water table in permeable rock, e.g. limestone. Underground streams in this zone flow with free air surfaces. *Compare* phreatic water.

valley A long depression, lower than the surrounding terrain, generally formed by the erosive action of a glacier or river. Its width and depth may be as little as a few meters or many kilometers. The cross-sectional shape depends on the prevailing rock type and its origin (*see* U-shaped valley; V-shaped valley). Young valleys, in mountainous areas, tend to be steep and narrow, whereas in lower regions the valleys are generally broader and less steep. *See also* river.

valley breeze The daytime equivalent of the MOUNTAIN WIND. It is formed by greater solar heating on the inclined mountain slopes relative to the air at the same level over the valley. Because wind speeds tend to be higher during the day as a result of general turbulence and convection, the valley breeze is more easily masked than the mountain wind.

valley glacier An accumulation of ice moving down a preexisting valley and restricted in width by the valley walls. Such glaciers may develop as a result of the merging of a number of CIRQUE GLACIERS in valley-head locations at times of deteriorating climate, in which case they are known as the *Alpine* type, or they may be developed at the very edge of an ICE CAP or ICE SHEET, when they are known as the *outlet* type. If the supply of ice to the source areas is sufficient, valley glaciers can extend down-valley to comparatively low altitudes.

vanadinite A brilliant orange, red, or brown mineral, a chloride and vanadate of lead, $Pb_5(VO_4)_3Cl$. It forms hexagonal crystals or fibrous masses in deposits of lead ores, and is used as a source of vanadium.

vapor pressure The pressure exerted by the molecules of a liquid or solid that escape from the surface. In meteorology it is that part of the total atmospheric pressure exerted by any water vapor that is present. Vapor pressure is measured indirectly from dry- and wet-bulb temperatures using tables or a humidity slide rule to obtain the precise value.

variolitic Denoting a texture occurring in basaltic glasses similar to SPHERULITIC texture in acid glassy rocks. Variolites usually consist of radial or sheaflike aggregates of plagioclase crystals.

Variscan The period of mountain-building in late Paleozoic time that affected Europe and includes both the Armorican and Hercynian phases.

variscite A greenish mineral form of hydrated aluminum phosphate, $AlPO_4.2H_2O$, which occurs as nodular masses.

varve A thin bed of sediment in glacial lakes that represents a seasonal increment. Spring and summer glacial melting produces a sudden influx of coarse sediment upon which finer materials settle during the remainder of the year. These annual rhythms can be used in dating events within the PLEISTOCENE.

vauclusian A spring that flows from underground in a limestone region.

veering In meteorology, a clockwise change in wind direction, such as from southwesterly to westerly. *See also* backing.

vegas In Spain, the local name for irrigated lowland areas used for agriculture.

vein A thin sheetlike deposit of minerals in a crack, fault, or joint in rock. Most minerals in veins were deposited after volcanic activity by hot fluids (gases or liquids) that solidify when they cool.

veld (veldt) A large area of grassland that occurs in southern Africa, where it is too dry for trees to grow. Some has been plowed to grow cereal crops, such as corn and wheat.

vent (conduit) The subterranean passage from the underlying magma chamber through which volcanic products, i.e. lava, ashes, and vapor, are discharged at the Earth's surface. Where a volcano has only one such hole, debris accumulates around it as a roughly symmetrical cone: this is termed a central-vent volcano. Where there is more than one eruptive site, usually along a large fracture in the crust, it is termed a fissure volcano.

ventifact A pebble or stone worn and polished by the action of wind and sand, and found in the gravel-strewn plains of the deserts. If a ventifact becomes worn in such a way that a roughly triangular cross section evolves, it is known as a dreikanter (German = three sides).

vermiculite *See* clay minerals.

vernal equinox *See* equinox.

Vertebrata The subphylum of the Chordata to which FISH, AMPHIBIA, REPTILIA, AVES (birds), and MAMMALIA belong. They possess an internal skeleton of cartilage or bone, including a jointed backbone formed of vertebrae. Vertebrates are active animals and consequently are usually bilaterally symmetrical. The sense organs and feeding apparatus are grouped anteriorly, where they and the brain are protected by a skull. Vertebrates have a dorsal hollow nerve cord and, in most higher groups, two pairs of limbs. Fossil evidence indicates that vertebrates appeared in the Ordovician (*see* Ostracodermi), since when they have evolved to occupy a great variety of environmental niches.

vertical corrasion Abrasive erosion (corrasion) on a rocky river bed that acts downward and forms POTHOLES, producing a gradual lowering of the bed.

vertisol One of the ten soil orders of the SEVENTH APPROXIMATION classification, denoting mixed and inverting soils of the TROPICAL BLACK SOIL group. Vertisols are split into uderts, which crack open for less than three months in a year, and usterts, which open and close more than once a year.

vesicle A small rounded cavity within lavas formed by bubbles of gas coming out of solution during the solidification of the lava. The concentration of vesicles at the top surfaces of lava flows serve as a way-up criterion. Elongate hollow tubes formed by the escape of steam through lavas flowing over wet ground are called pipe vesicles. Highly vesicular lava fragments are termed pumice. *See also* amygdale; pyroclastic rock.

vesiculation The formation of small cavities in molten igneous rock as it cools. The gases, previously in solution, form bubbles as the pressure decreases when the magma rises nearer the surface. If a great deal of gas is released, an explosive volcanic eruption may occur.

vesuvianite *See* idocrase.

viscous flow Flowage that occurred in once molten rocks, for example lavas, or imperfect fluids, such as asphalt. This results in highly complex fold patterns, each fold being distinct from and related to its neighbors.

viscous lava Slow-flowing acidic LAVA, which tends to build up steep volcanic cones. *See also* lava flow.

visibility (in meteorology) The greatest distance that the eye can see. If there is a marked variation with direction then the lowest visibility is recorded, because visibility is mostly used for aviation purposes and therefore the minimum value is important. Reports at night are based on distances to unfocused lights of moderate or known intensity.

vitrophyre A PORPHYRITIC lava in which the groundmass is wholly glassy.

vogesite A type of LAMPROPHYRE that contains biotite and hornblende, and in which ORTHOCLASE predominates.

volatile (fugitive) A substance, normally gaseous, that is dissolved under pressure in a magma. The main volatile constituents of magmas are water and carbon dioxide, together with smaller quantities of chlorine, hydrochloric acid, fluorine, hydrofluoric acid, hydrogen sulfide, sulfur, and boron compounds.

The solubility of volatile substances increases with pressure so that during the ascent of magma and the attendant lowering of pressure, the magma is unable to retain these constituents in solution. Volatiles concentrate at the top of the magma chamber in a residual fluid from which pegmatites, hydrothermal veins, and mineral deposits may originate. During escape, volatiles effect mineralogical changes on the early-formed crystals (PNEUMATOLYSIS).

Volcanic eruptions are accompanied by the release of large quantities of steam and carbon dioxide. The explosive release of these gases produces pyroclastic rocks. Water and gases are also released at minor vents, geysers, hot springs, fumaroles, and solfataras.

volcanic ash Fragments of PYROCLASTIC ROCK less than 2 mm cross.

volcanic bomb *See* pyroclastic rock.

volcanic breccia A type of BRECCIA consisting of angular fragments of volcanic rock more than 65 mm across.

volcanicity (vulcanicity) The various processes involved with the formation of magma and how it moves beneath and through the Earth's crust.

volcanic neck (volcanic plug) A cylindrical column of solidified magma that occupied the core of an inactive volcano. *See* neck.

volcanic rock Any igneous rock that is formed on the surface of the Earth as a result of volcanic action. The chief basic volcanic rock is BASALT; its acidic equivalent is RHYOLITE. Because of their rapid cooling, such rocks consist of very small microscopic crystals; even more rapid cooling (as with a submarine volcano) results in the formation of glassy OBSIDIAN.

volcanism (vulcanism) The various processes involved in the ejection of molten rock (magma, which becomes lava) from a volcano or hot water and steam from a FUMAROLE or GEYSER.

volcano A fissure or vent on the Earth's surface connected by a conduit to the Earth's interior, from which lava, gas, and pyroclastic material are erupted. Each volcano has its own characteristics, but they can be classified generally as:

1. *Hawaiian* Fairly quiet eruptions in which fluid lava is erupted from fissures or pits, with gas being liberated freely. When the eruption is accompanied by spurting gases, incandescent spray is thrown into the air.
2. *Strombolian* More violent than the Hawaiian type, the eruption taking place more spasmodically as trapped gas escapes from a more viscous lava confined in a crater. Eruptions may be every few minutes. During violent activity bombs are ejected, which may be accompanied by lava flows.
3. *Vulcanian* Characterized by viscous lava whose surface rapidly solidifies. Beneath this crust gas accumulates and builds up

pressure until the crust shatters. This results in large quantities of pyroclastic deposits ranging from large bombs to fine ash.
4. *Vesuvian* Similar to 2 and 3 but with gas-charged lava being shot violently up into the air, emptying the lava column to a considerable depth.
5. *Plinian* A culmination of type 4 in which a violent blast of gas rises to a height of several kilometers. The gas and vapor on reaching this height spreads into a large cloud. The fallout of ash is low, being confined to material removed from the conduit.
6. *Pelean* Very violent eruptions accompanied by nuées ardentes and hot avalanches of incoherent constantly expanding self-explosive lava, lubricated by hot gases and vapors. Such eruptions are characteristic of highly viscous lava, whose gas content cannot readily escape.
See also supervolcano.

volcanology (vulcanology) The scientific study of volcanoes, including their origins, action, structure, and classification. Volcanologists are also conducting research into the prediction of volcanic eruptions.

vorticity The three-dimensional rotation of a fluid about an axis, measured as a vector of twice the local rate of rotation of an individual fluid element. As rotation occurs in most aspects of atmospheric motion, vorticity is very important. Even if air is still relative to the Earth, it will still be rotating in space because of the Earth's rotation. To distinguish these two types, *relative vorticity* is taken as rotation relative to the Earth's surface; it is positive if cyclonic, i.e. rotating in the same direction as the Earth, and negative if anticyclonic. *Absolute vorticity* is the relative vorticity plus the component of the Earth's rotation about its axis; this is zero at the Equator and a maximum at the poles.

V-shaped valley Traditionally, fluvial valleys of youthful and mature rivers are said to have a V cross profile, with steep walls and narrow bottoms, while by senility, the broadening floodplain produces more of a U-shape. Irrespective of age, most valleys display a V-shape near their source, with some trend toward a broadening bottom near their mouth. Streams with sandy and gravelly banks tend to have a V-shape as the material slumps on erosion of the foot slope to its angle of rest, and rainfall landing on the valley sinks in and does not run off and sculpture the gentle side walls; conversely, valleys in cohesive silty or claylike material tend to maintain their walls as the river cuts sideways, leaving a flat bottom, with steep side walls and hence a U-shape. Climate can also be important: shady (north- and west-facing in the N hemisphere) valley sides have a different microclimate from sunny (south- and east-facing in the N hemisphere) walls and this can influence the rate and type of processes acting, hence lowering the two side walls at different rates to produce an irregular valley shape. If precipitation is high, the valley walls become clothed in dense vegetation, which slows down erosion, maintaining a V-shape, whereas in arid zones where there is no vegetation U-shapes often result from increased erosion of valley walls.

vugh (vug) A cavity within a volcanic rock lined with deuteric or SECONDARY MINERALS. *Compare* amygdale; miarolitic cavity.

W

wacke A type of young SANDSTONE with poorly sorted grains in a matrix of clay and fine silt.

wackestone A type of limestone consisting of grains of calcium carbonate in a matrix of fine-grained lime mud.

wadi A normally dry valley in a desert or semidesert environment. In such locations rain falls very infrequently but when it does so it is often in the form of violent downpours of limited duration. As normal fluvial activity does not exist, large amounts of weathering debris accumulate in situ. Channel flow following a desert cloudburst can have a considerable erosional effect, cutting these narrow valleys with nearly vertical walls into the weathered rocks. Wadis may be relicts from times of wetter climate.

wake The trail of eddies or vorticities that develop on the leeward side of an obstruction to wind flow. Large vertical velocities may be induced in essentially horizontal flow by this method, as on the lee side of hills, buildings, or wind breaks.

waning slope The gentle concave footslope lying beneath the FREE FACE. The waning slope is equivalent to the pediment, by which name it is more commonly referred to. It comprises the gentle slope of the valley floor, terminating in a drainage line or floodplain. Waning slopes are said to become more widespread at the latter part of the cycle of erosion, as the dominant process becomes retreat of hillsides (backcutting), rather than downcutting of valleys. *See also* waxing slope.

warm front A strong thermal gradient in the atmosphere in which, from its direction of movement, warm air is replacing cold air. The slope of a warm frontal surface is about 1 in 150, and so weather phenomena associated with it precede the surface front by many kilometers. The first indication is the spreading of cirrus clouds across the sky, followed by cirrostratus, altostratus, then nimbostratus as the rain starts falling. Temperatures rise, relative humidity increases, and the wind veers as the front passes. Warm fronts are a major contributor to annual rainfall totals in coastal locations of temperate latitudes. *Compare* cold front.

warm rain Rain falling from clouds that are entirely warmer than freezing and which therefore must have developed by the COALESCENCE process alone.

warm sector The region of warmer air existing in a DEPRESSION between warm and cold fronts. It eventually disappears from the surface during the evolution of the depression as the cold front catches up the warm front (*see* occlusion). Weather in the warm sector is usually noticeably milder than the preceding and following air streams. Precipitation is very variable, varying from none if there is a strong ridge of high pressure to heavy, especially over mountain areas, if the warm sector is potentially or conditionally unstable.

wash *See* overland flow.

washout A filled-in river channel that cuts into preexisting sediments. It is usually developed on deltas and other areas of slow-flowing rivers. The channels are usually infilled with sands, which often contain fragments of the eroded bed. They are very common in the Coal Measures and deltaic Jurassic beds. The sand body infilling the channel is sometimes known as a *horse*.

wash slope (alluvial toe-slope) That part of a hill slope lying at the foot of the CONSTANT SLOPE, formed by the washing out of fines from the scree in the constant slope. This fine material, often lying in the river valley, has a lesser angle of rest than the coarser constant slope debris, and so there is a gentle break of slope between the two components.

waste mantle *See* regolith.

water balance (in meteorology) The movement of water between and within the atmosphere and ground surface on a global scale. It involves the balance between precipitation, evaporation, advection of moisture in the air, ocean-current circulations, and river runoff on land. There is a long-term balance, so that no area of the Earth is continuously losing or gaining moisture. The values of the components of the water balance are often known only very approximately, especially over the ocean areas.

water cycle *See* hydrologic cycle.

waterfall A vertical fall of water at a steep break in the long profile of a stream. This may be the product of base-level fall producing a large KNICKPOINT, local displacement due to an earthquake, or irregularities in the long profile produced by hard bands of rock through which the river has been unable to grade a smooth profile owing to lack of erosive power or insufficient time. In areas of horizontal strata, breaks through individual rock beds will often result in a steep fall. Once created the falls immediately begin to retreat upstream and become less pronounced owing to lowering of their height through the concentration of erosion on the lip. Deposition below the falls also obscures the break and tends to restore a smooth long profile.

water gap A pass through a ridge of mountains through which a stream or river flows.

watershed *See* divide.

waterspout The oceanic equivalent of a TORNADO. It occurs over seas or large lakes when a funnel-shaped cloud descends from a cumulonimbus cloud to produce violent agitation of the sea surface and a very local intense rotation of wind. Waterspouts are never as severe or long-lasting as the variety of tornado found in the mid-West of the USA.

water table *See* groundwater.

water vapor The distribution of water vapor in the atmosphere varies greatly, with largest amounts in the humid tropics and least in polar regions and to a lesser extent the desert areas. As well as being necessary for precipitation, water vapor has another important meteorological role in absorbing long-wave or terrestrial radiation. With carbon dioxide, it acts as a very effective mechanism for preventing the loss of radiation that would otherwise occur. The mean temperature of the Earth is therefore partly a response to the presence of water vapor; without it, it would be appreciably lower. Water vapor also represents the highest energy state of water and so it can act as an energy store. When condensation to liquid or sublimation to solid takes place, energy is released, which may give rise to an increase of temperature.

wave A water wave is an undulation or deformation of a water surface. The period of water waves ranges from a second or so to several hours. The size of waves ranges from tiny capillary waves (only a few centimeters long) to the large storm waves, tidal waves, and long waves over 150 km in length. Many types of waves are wind-induced, directly or indirectly; others arise because of submarine earthquake disturbances or submarine sediment slides. *See also* constructive wave; destructive wave; ocean wave; sea wave.

wave base The lowest limit of orbital motion beneath waves. Below the sea surface in deep water, the orbits of water particles remain, as with surface particles, almost circular, but the diameter of the orbits decreases with depth. For every 1/9th of the wavelength in depth, the size of the water particle orbits are roughly halved. At a depth that is equivalent to the wavelength, the orbit size is 1/535th of the orbit size at the surface. Where orbital motion becomes almost negligible, wave base is reached. The concept of wave base assumes importance in problems of sediment transport beneath waves, because bed movements due to waves are influenced by the wave-induced oscillatory currents experienced close to the seabed.

wave-cut platform An irregular gently sloping bare rock platform extending out to sea and usually backed by cliffs. The mechanism of platform erosion is still not totally understood: it is not clear whether the waves are themselves responsible for the rock breakdown or merely for the removal of debris weathered by subaerial processes. Whichever processes are active, however, it is the wearing back of the cliffs that causes the enlargement of the platform. Because these forms will not develop where there is a covering of beach material, the most favorable sites are headlands.

wave cyclone A low-pressure system developing as a wave along a front. Most mid-latitude depressions form in this way.

wave diffraction (in oceanography) The transfer of the energy of wave crests as they negotiate a structure, such as a breakwater or sharp promontory. Diffraction usually leads to a marked reduction in wave height. The various diffraction patterns that result from, for example, the passage of sea waves through a restricted harbor entrance, can be calculated mathematically provided the characteristics of the waves and the receiving basin are known. Diffraction may allow some of the energy of incident waves to affect the water area in the lee of a breakwater.

wave forecasting The prediction of future wave conditions using empirical observations of wave conditions that have already occurred or theoretical calculations. Such calculations take into account predicted meteorological conditions and such factors as FETCH distance. Of the various methods of wave forecasting that have been devised, two main types have emerged: those that take account of the significant height and period of the predicted waves (*see* significant wave), and those that take account of the total WAVE SPECTRUM. Wave forecasting is of great importance to navigation and to studies of expected coastal changes.

wave hindcasting A consideration of wave conditions that may have applied at an earlier period and in a particular area. The method employed usually involves synoptic wind charts and other wind data for the period in question, and (if available) hydrographic charts covering that same period. The method is useful if coastal changes have been carefully documented and have to be accounted for.

wave ray *See* orthogonal.

wave refraction (in oceanography) The turning of wave crests so that they become more parallel to the coastline as they approach it. The process is due to the effects of the shallowing of water: as the depth of water decreases so the wave velocity decreases; consequently, a wave crest approaching the land obliquely will be slowed initially at the end nearest the shore, while the other end will continue moving faster until the crest has become almost parallel to the coastline. Wave energy can be represented diagrammatically by lines drawn perpendicular to the crests. As refraction takes place on an indented coast more energy becomes concentrated on the headlands than the bays, thereby favoring coastal straightening.

wave spectrum **1.** The complete array of waves present on a water surface in a particular place at a particular time. Careful measurements at sea or in a scale model

enable graphic wave records to be produced, which show such an array of waves. A typical wave spectrum in the open sea will nearly always consist of waves of many different heights and periods superimposed one upon the other. The measured wave records enable their size and frequency distributions to be determined.
2. A graphic classification of waves, varying from capillary waves with a period of less than 0.1 second to the transtidal waves with periods of 24 hours or more. Most of the energy of such a spectrum would lie in two of the period ranges: the ordinary tides and the ordinary gravity waves.

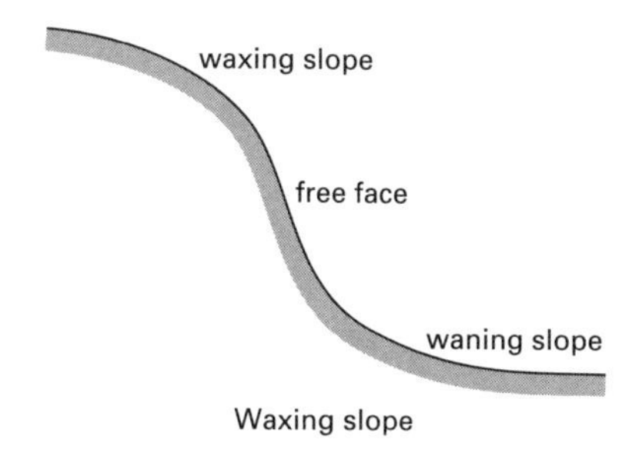

Waxing slope

waxing slope The convex slope at the upper part of a hill-slope profile. According to King, the waxing slopes are most prominent at the early stages of the cycle of erosion, when the main processes are the incision, rather than widening, of valleys. These slopes are said to be dominated by CREEP. *See also* slope convexity; waning slope.

way-up structures Sedimentary structures that can be used to determine the way up of a series of unfossiliferous beds. Typical way-up structures are cross bedding, mud cracks, ripple marks, graded bedding, included fragments of older beds, convolute bedding, burrows, roots, and geopetal cavities.

weather The state of atmospheric conditions at any one place and time. The total synthesis of these individual weather states is CLIMATE. The state of the weather has been classified into a code form for use in synoptic meteorology. In this way, current and past weather can quickly be identified, although greatest stress is placed on precipitation or other weather factors likely to be considered important in forecasting or in their impact on humans.

Efforts have been made to modify the weather, but without much success. The most publicized work has been on CLOUD SEEDING, but hurricane steering and fog clearance have been attempted. Because of the vast amounts of energy involved in the weather and the large area affected, the prospects for large-scale modifications are not hopeful. Unintentional modifications are far more numerous, such as the effects of towns on weather (*see* urban climate), or the effect of irrigating arid lands.

weathering The process of breakdown and alteration of rock on the Earth's surface in response to the changes in environmental conditions since the time of their formation. During formation many rocks were subjected to great pressures or high temperatures, away from the effects of atmospheric air and water. Under present-day conditions these rocks are at comparatively low temperatures and pressures and will naturally be influenced by the presence of air and water. The resulting products of weathering are materials more nearly in equilibrium with their environment than those from which they were derived. *See also* chemical weathering; deep weathering; differential weathering; mechanical weathering; organic weathering.

weathering front The junction between sound and weathered rock within a weathering profile. Certain rocks, especially dense types with well-developed jointing, such as granite and basalt, display extremely sharp transitions, whereas in porous or fissile rocks the junction is frequently so unclear that there is no real weathering front.

websterite An ultramafic rock consisting of hypersthene and diopside.

wedge (in meteorology) A region of high pressure that extends from an ANTICYCLONE. It is narrower than a RIDGE OF HIGH PRESSURE.

wehrlite An ultramafic rock consisting largely of olivine with accessory augite.

welded tuff *See* pyroclastic rock.

wernerite Another name for SCAPOLITE.

westerlies The main winds blowing between 40° and 70° latitude. It is in the westerly circulation that depressions form and maintain the vital meridional heat exchange, and in this zone the strongest undisturbed wind flows are found.

West Wind Drift A circumpolar current that constitutes one of the Earth's large and significant permanent ocean currents. In the S part of the Pacific Ocean, it occupies a wide tract of water and generally flows in an easterly direction to the south of the subtropical convergence zone. At about latitude 55°S, the Antarctic Convergence cuts across the West Wind Drift. In the S Atlantic Ocean, the West Wind Drift occupies the wide tract of water that lies approximately between latitudes 35°S and 63°S. It moves with surprising constancy, and is often characterized by very large sea waves and swell.

wet-bulb depression The temperature difference between the dry- and wet-bulb THERMOMETERS in a Stevenson screen. It is a measure of the amount of cooling resulting from evaporation of water on the wet-bulb thermometer. Relative humidity of the air can be calculated from this value using tables.

wet day A period of 24 hours commencing 9 a.m. GMT in which 1.0 mm or more of rainfall is recorded. The concept has its limitations because 1 mm could fall from a single heavy shower or be the result of prolonged drizzle throughout the day. *See also* rain day.

wetland *See* marsh; swamp.

wet spell A period of 15 or more consecutive days with at least 1 mm of precipitation.

wetting-and-drying weathering The mechanical breakdown of fine-grained rocks through alternate wetting and drying, causing surface FLAKING or major splitting of blocks into two or more large pieces. A possible explanation involves the fact that water is a polar liquid, with positive charge on the hydrogen atoms and negative charge on the oxygen atom. Lines of mutually attracted water molecules may develop within the rocks on wetting and drying and these may subsequently exert the expansive forces that cause fracture.

whirlpool A small rotating area of water in a lake, river, or the sea; larger whirlpools may become funnel-shaped. It may be caused by the shape of rocks on the bed or result when two opposed currents meet and circulate around each other.

whirlwind A small revolving vertical eddy of air, which whirls around a local low-pressure area. Stronger whirlwinds may be heard, but they are more frequently noticed because they pick up small pieces of litter or debris. They are formed during periods of local heating and instability, appearing as small-scale TORNADOES.

white alkali soil *See* solonchak.

white ice Ice that contains trapped air, located at or near the surface of the ground, often on a glacier. As it descends under further layers of ice, the pressure forces out the air and the ice appears blue. This change is often accompanied by melting (*see* regelation).

white-out Drifting snow or blizzard conditions on a preexisting snow surface in polar regions. The horizon disappears and everything then becomes white with no shadows or identifiable landmarks.

wind The horizontal movement of air relative to the Earth's surface. Air movement results from thermal differences, which produce pressure variations, or by dynamic factors such as divergence of the air flow itself. Wind is one of the basic elements of weather. It is measured in knots and its direction is that from which it blows. Wind speeds at the ground surface can be very variable from absolute calm during anticyclones up to 200 knots in a tornado.

windbreak Any barrier having the deliberate effect of reducing surface wind speeds. It can be artificial or natural, although the latter is more frequently called a *shelter-belt.* The degree of shelter that a barrier offers depends upon its height, its degree of permeability, and distance. For a dense barrier, wind speeds fall dramatically immediately to the leeward, but soon recover to the prevailing wind speed. With a barrier that offers about 50% permeability, the effect reaches a maximum in terms of overall reduction of wind speed and distance affected downwind. Even at a distance of forty times the height of the barrier there is still some reduction.

wind-chill index An index of the effects upon living creatures of cold winds. The stronger a cold wind is blowing the more rapid will be the rate of heat removal from a mammal. Consequently, cold but calm conditions are more physiologically acceptable than warmer temperatures in a strong wind. The wind-chill index was devised for the US army to determine clothing requirements under different climatic conditions; it is measured as a cooling effect in kilocalories per square meter of exposed flesh.

wind drift current A current that results from wind stresses on the sea's surface, part of the wind energy producing surface waves and part generating these currents. Indirectly, the wind is responsible for most of the currents in the surface and near-surface waters. Because of the Earth's rotation and the resulting Coriolis effect, the wind-induced currents in the open ocean have a mean direction approximately 45° to the wind direction (somewhat less than this near the surface, and greater than this at depth). The currents flow to the right of the wind in the N hemisphere and to the left of the wind in the S hemisphere. A wind blowing fairly hard for a long period may produce a thick homogeneous layer of isothermal water, for example, under the influence of the steady trade winds. In the shallower water near the coast, onshore winds tend to drag the surface water shoreward, while offshore winds cause the reverse to happen; in both cases, the related currents near to the seabed may be flowing in the reverse direction to the surface wind-induced current.

wind erosion *See* aeolian erosion.

wind gap A gap (col) or notch in a ridge or hilltop, through which the wind can blow and may whistle. It may have been formed by RIVER CAPTURE or glacial action.

window (fenster) An outcrop of rock lying beneath a recumbent fold or thrust. Such rocks are exposed as an enclosed outcrop, as a result of erosional processes cutting down through the overlying rock mass to the younger rock beneath.

wind rose A diagram for illustrating the frequency of wind directions over a specified time period. It is normally split into eight or sixteen points of the compass and can also give an indication of the frequency of wind speeds within specified ranges by direction, as shown in the diagram overleaf.

wind set-down The lowering of the still-water level on the down-wind flank of a body of water because of wind stresses acting on the water surface, or the difference in levels between the up-wind and down-wind flanks resulting from wind stress. Like WIND SET-UP the term is usually applied to the phenomenon as it affects reservoirs, lakes, estuaries, and other bodies of water that are relatively limited in size. One example is the Plate River estuary of Argentina: here winds blowing up (or

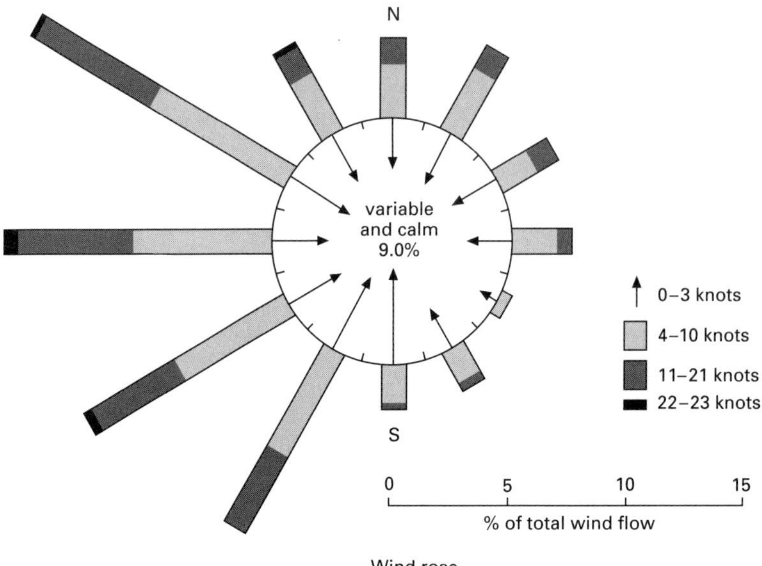

Wind rose

down) the estuary cause set-up (or set-down) and also change the current patterns. Statistical analyses of wind data show that for some 220 days per year, tidal levels differ by up to a half a meter or more from the predicted values.

wind set-up The raising of the still-water level on the down-wind flank of a body of water because of wind stresses acting on the water surface, or the difference in levels between the up-wind and down-wind flanks resulting from wind stress.

wind shear The local rate of change of wind velocity, usually at right angles to the horizontal air flow. Vertical shear may also be considered.

windward Describing the side of a hill or other feature that faces into the prevailing wind.

winter solstice *See* solstice.

witherite A white, yellowish, or gray mineral form of barium carbonate, $BaCO_3$. It crystallizes in the orthorhombic system, and generally occurs in veins or in association with the lead ore GALENA. It is used as a source of barium. *See also* carbonate minerals.

wold In E or NE England, a chalk hill.

wolframite A red-brown to black mineral form of iron manganese tungstate, $(Fe,Mn)WO_4$. It crystallizes in the monoclinic system, and occurs mainly in pegmatites and hydrothermal veins in quartz. It is the chief source of tungsten. *See also* scheelite.

wollastonite A white triclinic calcium silicate mineral of composition $CaSiO_3$. It is found in metamorphosed limestones and in some alkaline igneous rocks, such as nepheline SYENITES.

woodland An area where the dominant vegetation is trees that form a canopy, but not as dense as a forest. In temperate climates, the mild moist climate favors woodland of mixed deciduous trees (replaced by evergreens in the Mediterranean area), which grow between the tropical forests farther south and the coniferous forests farther north.

worm cast *See* earthworm.

wrench fault *See* tear fault.

wulfenite A yellow or orange mineral form of lead molybdate, $PbMoO_4$. It crystallizes in the tetragonal system, and occurs as earthy or granular aggregates in lead deposits. It is used as a source of molybdenum.

WWW World Weather Watch, a program planned by the World Meteorological Organization to provide a worldwide coverage of surface and upper atmospheric data. This is then used to provide an initial detailed data set for numerical models of the atmosphere.

xanthophyllite A brittle yellowish form of hydrated calcium magnesium aluminosilicate. It crystallizes in the monoclinic system, and is a member of the MICA group of minerals.

xenoblastic Describing crystals in metamorphic rocks that are anhedral, exhibiting no crystal faces.

xenocryst A crystal superficially resembling a PHENOCRYST, which is not in equilibrium with the other minerals in an igneous rock. It is commonly anhedral, having suffered resorption due to reaction with the magma.

xenolith An inclusion within an igneous rock body. The inclusion may be a block of country rock that has been caught up in the intrusion, but has not been completely assimilated. It may alternatively be a block of the igneous body itself that solidified at an earlier period and therefore has a slightly different composition.

xenomorphic Describing an igneous or metamorphic rock whose mineral crystals do not have characteristic crystal faces, caused by disturbance during crystallization.

xerophyte A plant that can tolerate drought or extremely dry climatic conditions. Their adaptations include long roots (to reach underground water), small or thick leaves (to reduce water loss during transpiration), water storage cells in fleshy stems, and dense hairs (to trap moist air). Cacti are typical xerophytic plants.

Y

yardang A narrow steep-sided ridge that occurs in arid regions. Yardangs are caused by erosion by windborne sand (corrasion), and often lie parallel to the direction of the prevailing wind.

yazoo stream A tributary stream that flows for a considerable distance parallel to the main stream, from which it is separated by a natural levée, before joining the main stream. Yazoo streams are named for the Yazoo River, which flows alongside the Mississippi River for 320 km before joining it near Vicksburg, Mississippi.

yellow ocher *See* ocher.

yield point The point at which a rock ceases to deform elastically to an applied force, and beyond which further force will cause it to fracture.

Young's modulus A stretch modulus (Y), equal to the ratio of the stress on a cross-sectional area of a rod of material to the longitudinal strain.

Z

zenith The point in the sky that is vertically above an observer, It is frequently used in radiation geometry for calculating solar radiation input.

zenithal projection *See* azimuthal projection.

zeolites A large group of minerals with a general formula $(Na_2,K_2,Ca,Ba,Sr)((Al,Si)O_2)_n.xH_2O$. Zeolites have a structure similar to feldspars and feldspathoids and consist of a very open framework of linked $(Si,Al)O_4$ tetrahedra, with the metal cations filling the large cavities in the structure, which also contains loosely held water molecules. The compositions of some of the commonly occurring zeolites are as follows:
natrolite $Na_2(Al_2Si_3O_{10}).2H_2O$
mesolite $Na_2Ca_2(Al_2Si_3O_{10})_3.8H_2O$
scolecite $Ca(Al_2Si_3O_{10}).3H_2O$
thomsonite $NaCa_2(Al_5Si_5O_{20}).6H_2O$
heulandite $(CaNa_2)(Al_2Si_7O_{18}).6H_2O$
phillipsite $(\frac{1}{2}Ca,Na,K)_3(Al_3Si_5O_{16}).6H_2O$
harmatome $Ba(Al_2Si_6O_{16}).6H_2O$
stilbite $(Ca,Na_2,K_2)(Al_2Si_7O_{18}).7H_2O$
chabazite $Ca(Al_2Si_4O_{12}).6H_2O$
laumontite $Ca(Al_2Si_4O_{12}).4H_2O$

Most zeolites are colorless or white and are relatively soft, varying in hardness from 3.5–5 on Mohs' scale. Natrolite, mesolite, scolecite, and thomsonite have fibrous habits. Zeolites are commonly found as late-stage minerals in AMYGDALES in basic lavas. They also occur as alteration products after feldspars and feldspathoids. The occurrence of zeolites on a regional scale in volcanic rocks and sediments is considered to indicate a very low-grade metamorphism termed the zeolite facies.

zeugen A type of PEDESTAL ROCK whose column comprises tabular slabs. The variation in profile is the result of differential erosion of the rock by windborne particles.

zinc blende *See* sphalerite.

zincite An orange to deep-red mineral oxide of zinc and manganese, (Zn,Mn)O. It crystallizes in the hexagonal system, and occurs as masses in metamorphosed limestone. It is used as a source of zinc.

zinnwaldite *See* mica.

zircon A zirconium silicate mineral of composition $ZrSiO_4$, found as an accessory mineral in intermediate and acid igneous rocks. Being very hard and resistant to weathering, it is a common detrital mineral in sediments.

zoisite A member of the EPIDOTE group of minerals that crystallizes in the orthorhombic system.

zonal circulation Wind circulation with a dominant west-east directional component. The main zonal circulations are the WESTERLIES of the mid-latitudes and the EASTERLIES of the tropical oceans. Easterly winds are normally defined as being negative. *Compare* meridional circulation.

zonal index A measure of the strength of the zonal circulation in a specified area and time period. The index most commonly used is that for the N Atlantic Ocean between latitudes 35°N and 55°N based on the differences in mean pressure along these lines across the ocean.

zonal soil A soil occurring over a wide area because of the dominance of the bioclimatic factor in soil formation, which determines soil-forming processes. Where the bioclimatic influence is locally replaced by relief and drainage, INTRAZONAL SOILS occur. A century ago the Russian Sibertzev conceived soil as a climatic succession running from the tundra soils to laterite. This simplification is increasingly regarded as outmoded because of the location of many soils outside their zone through the local influence of some other factor (e.g. podzols on sandy parent materials in the brown earth zone) and the climatic change of the recent past, which has varied zonal boundaries.

zone 1. (in metamorphism) A spatial division in an area that has undergone METAMORPHISM, based on the first appearance of successive index minerals within the metamorphosed rocks in the progression toward rocks of the highest metamorphic GRADE. The progressive nature of metamorphism was developed by Barrow working in the Scottish Highlands, where he identified a number of zones in pelitic rocks. The index minerals occurring in PELITES in order of increasing grade are: chlorite, biotite, almandine garnet, staurolite, kyanite, and sillimanite. A line drawn on a map representing the first appearance of a particular index mineral is termed an *isograd.* These are lines of equal grade and link rocks originating under similar physical conditions of metamorphism. The almandine zone, for instance, is the zone occurring between the almandine and staurolite isograds. However, in rocks of other compositions, an index mineral may appear at a higher or lower grade than that corresponding to its first appearance in pelitic rocks. Different mineral zones have been correlated with different rock types and this correlation is embodied in the concept of metamorphic FACIES.

Sequences of metamorphic zones other than those proposed by Barrow have been identified in other areas. In NE Scotland, the Buchan zones are defined by the index minerals staurolite, cordierite, andalusite, and sillimanite. The Buchan zones represent a different metamorphic gradient involving relatively lower pressures than those represented by the Barrow zones.

2. (in stratigraphy) The fundamental division used in biostratigraphical methods of calibrating and correlating rock successions. A zone is demarcated by the fossils it contains and is usually named after a ZONE FOSSIL; when applied to a zone the name of the fossil should be italicized. Examples of various types of zone are as follows. An *assemblage zone* is characterized by the occurrence of a particular assemblage of fossils; an *acme zone* is characterized by the particular abundance of a single species or fossil group; a *total-range zone* comprises the entire thickness of rocks in which a fossil occurs; a *concurrent-range zone* consists of the strata in which the ranges of particular fossil species or groups overlap. In some situations a zone in biostratigraphy may correspond to a CHRONOZONE in the chronostratigraphic scale.

3. *See* seismic zone.

4. *See* crystal zoning.

5. *See* morphogenetic zone.

zone fossil (index fossil) A fossil species that is used in BIOSTRATIGRAPHY to delimit a ZONE. Ideally a zone fossil should have a limited vertical range in the succession and should in life have been rapidly and widely distributed to permit correlation over a large area. Several species, forming a whole zonal assemblage, may be used in the definition of a biostratigraphical division.

zone of aeration The zone below the ground surface and above the water table in which the pore spaces and openings within the soil, sediments, and rock contain mainly air.

zone of saturation A zone below the ground surface where all cracks and pore spaces in sediments or rock are filled with water. The zone's upper boundary forms the water table (*see* groundwater).

zooplankton Small, often microscopic, animals that float passively in the sea or other bodies of water (*see* plankton).

zweikanter A pebble with two facets, formed by the erosive action of windblown sand in desert regions. *See also* dreikanter.

APPENDIX

The Chemical Elements

(* *indicates the nucleon number of the most stable isotope*)

Element	*Symbol*	*p.n.*	*r.a.m*	*Element*	*Symbol*	*p.n.*	*r.a.m*
actinium	Ac	89	227*	fluorine	F	9	18.9984
aluminum	Al	13	26.982	francium	Fr	87	223*
americium	Am	95	243*	gadolinium	Gd	64	157.25
antimony	Sb	51	112.76	gallium	Ga	31	69.723
argon	Ar	18	39.948	germanium	Ge	32	72.61
arsenic	As	33	74.92	gold	Au	79	196.967
astatine	At	85	210	hafnium	Hf	72	178.49
barium	Ba	56	137.327	hassium	Hs	108	265*
berkelium	Bk	97	247*	helium	He	2	4.0026
beryllium	Be	4	9.012	holmium	Ho	67	164.93
bismuth	Bi	83	208.98	hydrogen	H	1	1.008
bohrium	Bh	107	262*	indium	In	49	114.82
boron	B	5	10.811	iodine	I	53	126.904
bromine	Br	35	79.904	iridium	Ir	77	192.217
cadmium	Cd	48	112.411	iron	Fe	26	55.845
calcium	Ca	20	40.078	krypton	Kr	36	83.80
californium	Cf	98	251*	lanthanum	La	57	138.91
carbon	C	6	12.011	lawrencium	Lr	103	262*
cerium	Ce	58	140.115	lead	Pb	82	207.19
cesium	Cs	55	132.905	lithium	Li	3	6.941
chlorine	Cl	17	35.453	lutetium	Lu	71	174.967
chromium	Cr	24	51.996	magnesium	Mg	12	24.305
cobalt	Co	27	58.933	manganese	Mn	25	54.938
copper	Cu	29	63.546	meitnerium	Mt	109	266*
curium	Cm	96	247*	mendelevium	Md	101	258*
dubnium	Db	105	262*	mercury	Hg	80	200.59
dysprosium	Dy	66	162.50	molybdenum	Mo	42	95.94
einsteinium	Es	99	252*	neodymium	Nd	60	144.24
erbium	Er	68	167.26	neon	Ne	10	20.179
europium	Eu	63	151.965	neptunium	Np	93	237.048
fermium	Fm	100	257*	nickel	Ni	28	58.69

Element	*Symbol*	*p.n.*	*r.a.m*	*Element*	*Symbol*	*p.n.*	*r.a.m*
niobium	Nb	41	92.91	selenium	Se	34	78.96
nitrogen	N	7	14.0067	silicon	Si	14	28.086
nobelium	No	102	259*	silver	Ag	47	107.868
osmium	Os	76	190.23	sodium	Na	11	22.9898
oxygen	O	8	15.9994	strontium	Sr	38	87.62
palladium	Pd	46	106.42	sulfur	S	16	32.066
phosphorus	P	15	30.9738	tantalum	Ta	73	180.948
platinum	Pt	78	195.08	technetium	Tc	43	99*
plutonium	Pu	94	244*	tellurium	Te	52	127.60
polonium	Po	84	209*	terbium	Tb	65	158.925
potassium	K	19	39.098	thallium	Tl	81	204.38
praseodymium	Pr	59	140.91	thorium	Th	90	232.038
promethium	Pm	61	145*	thulium	Tm	69	168.934
protactinium	Pa	91	231.036	tin	Sn	50	118.71
radium	Ra	88	226.025	titanium	Ti	22	47.867
radon	Rn	86	222*	tungsten	W	74	183.84
rhenium	Re	75	186.21	uranium	U	92	238.03
rhodium	Rh	45	102.91	vanadium	V	23	50.94
rubidium	Rb	37	85.47	xenon	Xe	54	131.29
ruthenium	Ru	44	101.07	ytterbium	Yb	70	173.04
rutherfordium	Rf	104	261*	yttrium	Y	39	88.906
samarium	Sm	62	150.36	zinc	Zn	30	65.39
scandium	Sc	21	44.956	zirconium	Zr	40	91.22
seaborgium	Sg	106	263*